BEHOLD YOUR BODY

Anatomy and Physiology Anyone Can Enjoy

Volume One

DRY BONES OR WET NOODLES ?

Volume Two

BELLY UP OR BUST ?

Volume Three

BONE HEADS OR JELLY BRAINS ?

This book is for you!

Our body can make us, or break us, but how much time do we give it. Do we understand it, appreciate it, or care for it? If you have always been curious to know what is cooking inside, but endless terminology smothers your fire, then this book is for you.

If your present anatomy skills go no further than knowing 'your ass from your elbow', then this book is for you.

If you want to have fun and discover your own body, then this book is for you.

If you ponder how the macrocosm of the universe can duplicate itself in your own cellular mechanisms, then this book is for you.

If you would like to greet the skeleton that holds you up, then this book is for you.

If you wonder if exercise is really all a spoof, then this book is for you.

This book is for you and about you! This book is a self study guide of yourself.

If you are a curious explorer, then you can use this book to behold your own body and its intriguing world.

Many blessings along your journey, my honored guest!

Charlene Penner MD

BEHOLD YOUR BODY

Anatomy and Physiology Anyone Can Enjoy

Volume One

DRY BONES OR WET NOODLES ?

Rose Bud Publishing Ramona, California

Illustrator: Charlene Penner MD
Editor: Gloria Woodley
Proofreader: Ken Houghton
Printer and Binder: Patterson Printing
Cover photos: Ken Howard
Cover design: Blaize Mekinna

Behold Your Body, Anatomy and Physiology anyone can enjoy
Volume One: Dry Bones or Wet Noodles?

Printed in the United States of America.

Rose Bud Publishing
17126 Mt. Woodson Rd.
Ramona, California, 92065, USA.

Library of Congress Catalog Card Number 96 - 68996

Penner, Charlene
Behold Your Body
Volume One: Dry Bones or Wet Noodles?

Includes bibliography and index
1. Anatomy, Human 2. Holistic Health 3. Physical examination - Methods 4. Physiology, Human

ISBN 1-888515-53-8

Front Cover Illustration: Charlene Penner with a skeleton owned by the International Professional School of Bodywork

BRIEF CONTENTS

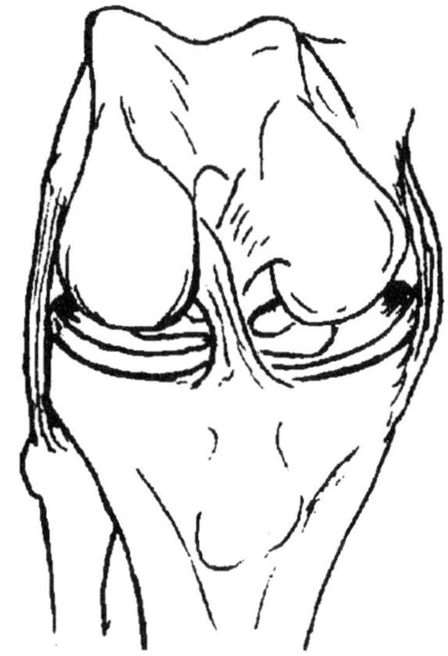

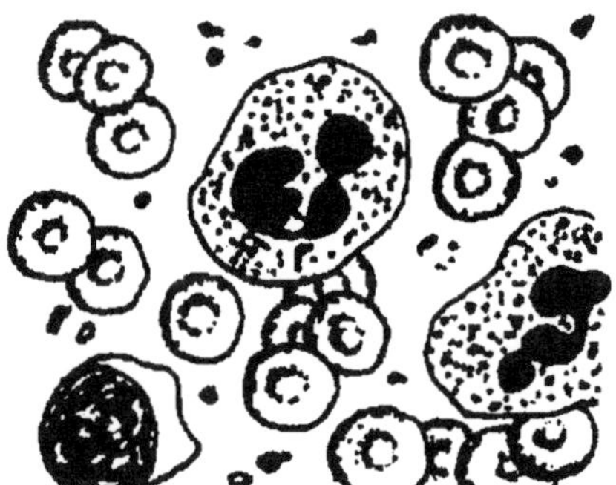

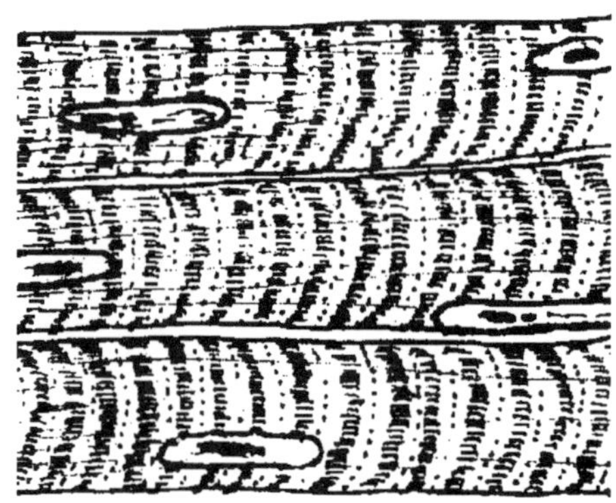

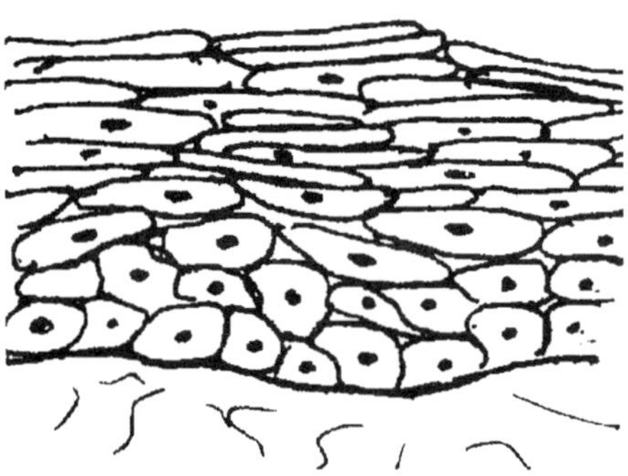

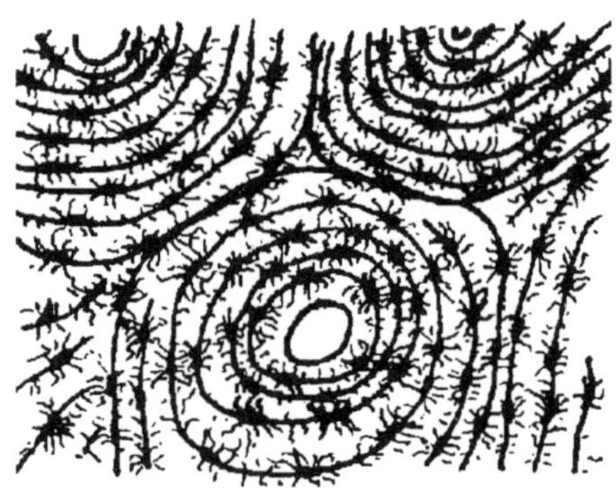

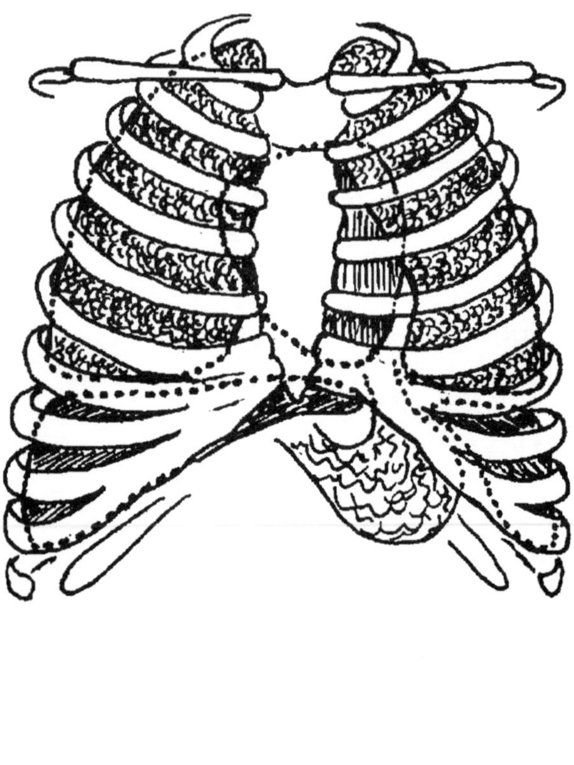

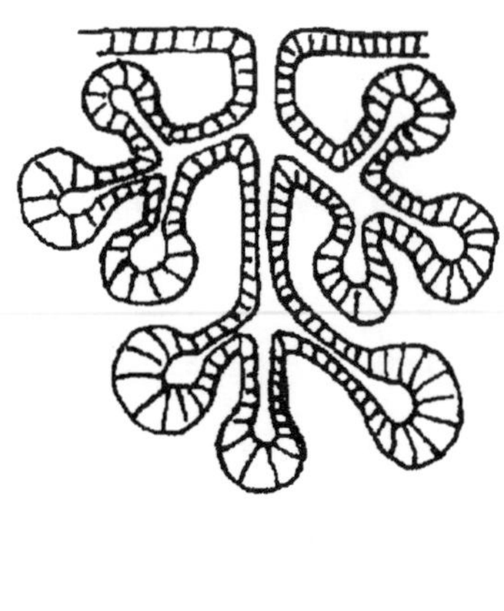

EXPANDED CONTENTS

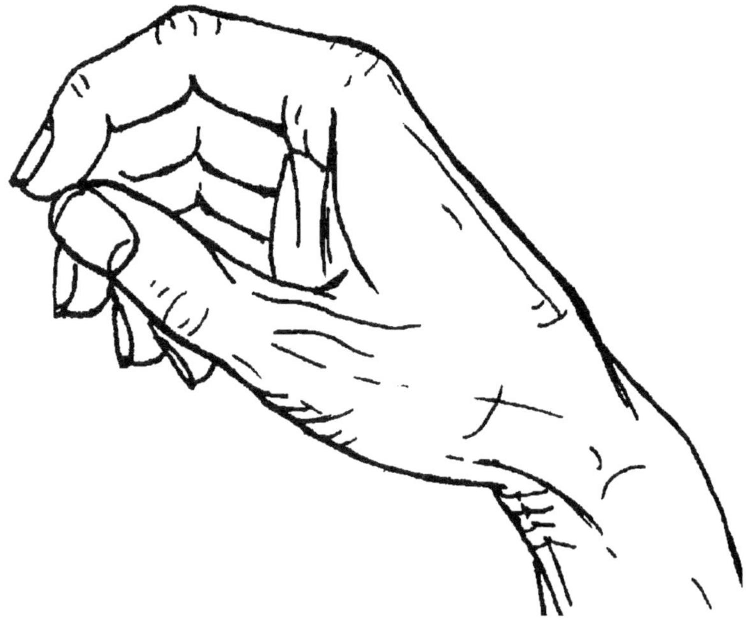

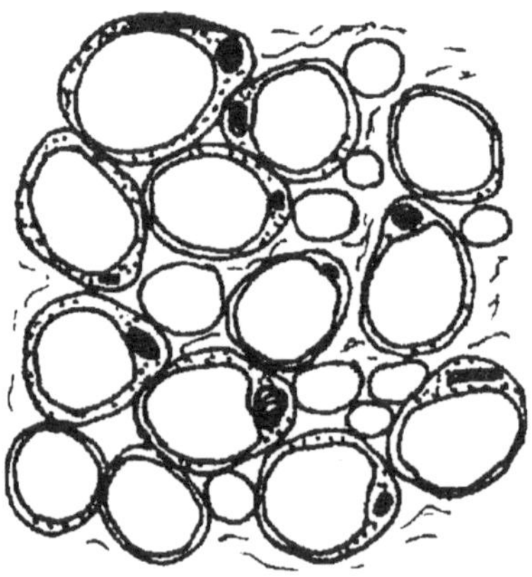

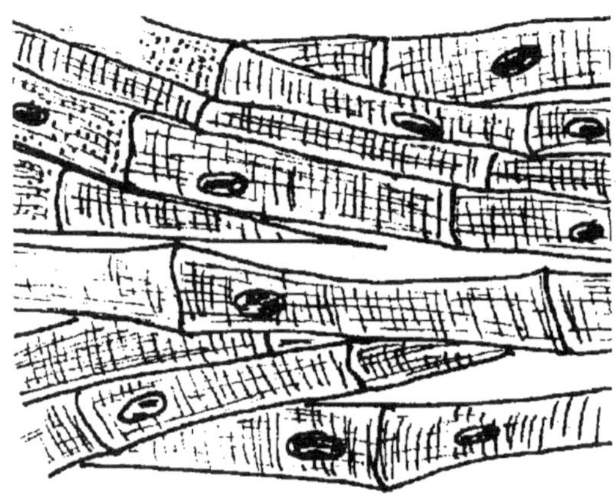

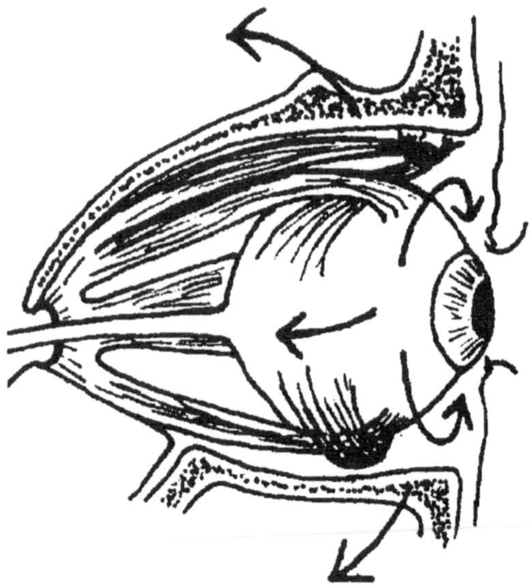

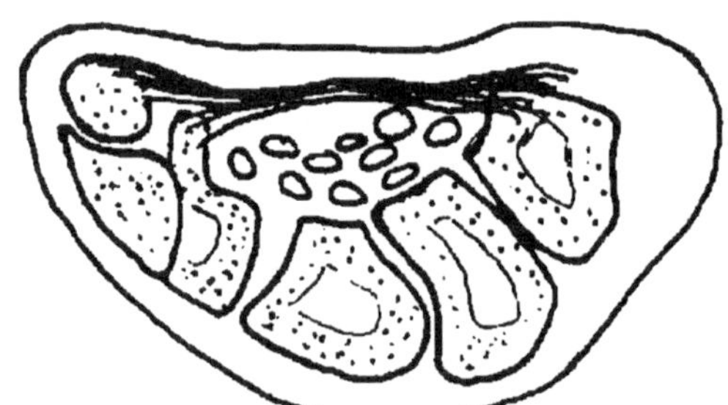

BOOK OVERVIEW

PURPOSE OF THIS BOOK
i.e., its physiology

In this book we explore basic concepts of the anatomy and physiology of the human body. There are hundreds of current, colorful books available with this same purpose, so why develop another one? I had to answer this question thoroughly before it made sense to embark on the writing of Behold Your Body.

• First and foremost, in this book we explore basic concepts of the anatomy and physiology of the human body, exemplified specifically in our own body. Most anatomy books are written about the body, the head, the arms, the muscles. This book is written about our own body, our own arms that we can lift up in front of our own face. This book is written about our own muscles that we can watch and feel rippling under our own skin. Our body tags along almost every where we go, but what do we know about it?

It is amazing how difficult and elusive it can be to apply the printed page to the contours of our own body. Behold Your Body presents accurate information like a story in which you can participate. A unique feature of this book is its instructions, guiding you to locate and feel things inside your own body. This is called **palpation**. I hope you will take the time to follow these instructions, to increase your kinesthetic sense of your own body, to confirm the information as you find it alive inside your own body, to experience the wonders of the body as part of yourself, just as writing this book has deepened my discovery of my own body.

• Secondly, this book makes learning anatomy and physiology easy and fun, available to any curious explorer.

It is true that our body is an encyclopedia of complex intricacies. It also is true that our body follows very simple, basic principles common to the whole animal kingdom. To learn this simple foundation is our first task, then the details can fall into place as needed.

Learning about our body can be an amazing adventure. Many of you embarking on this adventure have little or no scientific background, so dipping into a hard science like anatomy can be intimidating, overwhelming, and confusing. For many folks, studying anatomy means spending senseless hours memorizing mile-long, meaningless words.

For me, more important than naming a structure is to understand why it is there and how it works. Behold Your Body is loaded with illustrations and analogies that keep concepts memorable and light. Terminology is minimized and explained; concepts are kept simple and basic. At the same time, the accuracy of the information always is maintained, never sacrificed for fun or ease.

I never imagined that I would spend years of my life harmonizing the humming, clickety bleep of my hard, cold, computer with the cackling laughter of my own voice. Our bodies are more than something to get around in or swear at when they break down. Our bodies are meant to be enjoyed. I do not expect anyone else to find this book as precious and humorous as I do. I only hope you will catch a small portion of the immense delight that creating Behold Your Body has been for me.

• Thirdly, in response to my disappointing search for a textbook for my classes, I have selected and applied information to meet the specific needs of hands-on body therapists and other blossoming therapeutic modalities which are not as strongly based on hard science. Some anatomy books are much fatter than this one, loaded with many more exciting facts and details. Some anatomy books are thinner, quicker to read, and easier to digest; however, after you know them cover to cover, you will have many unanswered questions. I have used 15 years experience in teaching massage students to filter and discriminate useful information.

In some areas, especially in the naming and locating of specific bones and muscles, the amount of information included may be more than the interest of most curious laymen or even some therapists. I chose to include this information so that more specialized therapies would be informed of these details in all parts of our body.

- A fourth purpose in writing this book is my belief that we all benefit from knowing the basic structure and function of our body. The more we touch and are in touch with our body, the more we are aware of and understand our whole being, the more we learn to know and feel the effects our life has upon us.

Our body will tell us what it likes and needs. What we have to do is pay attention. Then we can learn to know what we need and how to satisfy our needs. Our knowledge helps us develop guidelines for good maintenance for our whole being. It motivates us to follow the guidelines, to exercise, eat healthy, reduce stress, and so forth. Then we feel alive and flourish.

Our knowledge also can help us to know when we need help from others and how to wisely select that help.

Therefore, in presenting the concepts and guided self-study, it is my endeavor to inspire a healthier life style in us all.

In this light, I encourage you to participate in this book. Follow the instructions, finding and feeling your bones and muscles, watching your changing shapes as you move. Use this book to become familiar with your own body, and to increase your range of communication and learning about yourself.

BASIC DEFINITIONS

The content of Behold Your Body can be explored in three basic definitions.

1. Anatomy is the study of the form and structure of any living thing. Anatomy examines what it is made of, the content of the construction materials and how they are arranged to make up an organism. Anatomy locates where the different parts of an organism are relative to adjacent parts and the whole.

2. Physiology is the study of the normal or appropriate functions of the various organs or parts of any living thing. Physiology explains the job or purpose performed by the organs.

3. Pathology is the study of the disease states in an organism, their causes and effects. The root of pathology may be in defective structure or function. Pathology examines what happens in an organism when a structure is broken, ruptured, or built or located incorrectly. Pathology examines what happens in an organism when a job either is not performed or is performed abnormally.

For example, think of a square wooden dinner table. The anatomy or structure of the table describes that it has four vertical wooden supports, one beneath each corner of a square, horizontal, wooden platform. The physiology or function of the table is to hold up elbows and plates of food during the process of eating. One possible pathology is that a leg breaks. The table tips and everything falls to the floor. Because of a structural problem, the table can no longer perform its function.

In our body, we go through the same process, taking in turn each part of our body, examining where it is located, what its structure consists of, what function it serves for our whole body, and what happens in different disease states.

Behold Your Body is primarily an anatomy/physiology book. We only glimpse sporadically and briefly at pathologies when they: **1.** are very common, **2.** illustrate a specific point, or **3.** are important to a hands-on body therapist in the sense of taking special care or being especially effective. These discussions are brief. They skim only the surface of a disease and should not be considered as comprehensive information. Likewise, therapeutic tips are just bits and pieces included because they are commonly overlooked or misunderstood or illustrate a point. There are no comprehensive guidelines for any situation.

Peeks at pathology and therapeutic tips are often put in the highlighted tidbit boxes.

ORGANIZATION OF THIS BOOK
i.e., its anatomy

The organization of Behold Your Body follows the basic organization of our body, beginning at the cellular level.

Unit I, chapter 1 is about the **Cell**, describing the internal structures, general functions, and basic life support systems which are common to all cells in our body. In subsequent chapters we will detail how a specific cell differentiates, specializes, or elaborates in some way. For many, chapter 1 is the most difficult because we cannot see or touch an individual cell. The cell is the smallest level covered in this book, from this point on subject matter becomes larger and more tangible. The cell is, however, the structural foundation of our body, so it is important to understand it at a basic level.

Unit II is about our **Tissues**, describing their structural characteristics, general functions, locations, sources of nourishment, cellular reproduction potentials, and repair mechanisms. **Chapters 2 through 5** discuss each of the four kinds of tissue. Understanding the tissues simplifies the remaining process of learning about our body. When we understand the tissues, we only need to learn their arrangment in an organ to have a preliminary understanding of that organ's potential. We do not have to start from scratch with each organ.

The rest of the book is dedicated to the systems, one unit per system.

Unit III is about our **Skeletal System. Chapter 6** describes general concepts of bones and joints, including their structure, function, development, and repair. **Chapters 7 through 9** describe the details of the individual bones of our body, their connections to each other, and their palpation.

Unit IV is about our **Muscular System. Chapter 10** describes general concepts of skeletal muscles, including their structure, function, development, nerve innervation, and principles of action. **Chapters 11 through 13** describe the details of the individual muscles of our body, including their bony attachments, specific actions, and palpation.

Unit IV is followed by an **Appendix** filled with my biased opinions on ways to be with our illness process and the roles of our healer.

A **Glossary** provides definitions of many anatomical words and word parts used throughout the text. The parenthesized number indicates the text page where that word appears in bold and is most fully discussed.

Last but not least is an extensive **Index**. I believe that a book of this nature is only as good as its index so that we can easily find what it contains. Words followed by a 'g' are defined in the Glossary. The bold number indicates the text page where that word appears in bold and is most fully discussed.

This concludes **Volume One**.

The plan for **Volume Two** is to have one unit for each of our systems whose major organs are in the trunk of our body, namely our **Cardiovascular System, Lymphatic System, Respiratory System, Digestive System, Urinary System, Reproductive System,** and **Skin**. We will look at the general functions of each system and the location, structure, and specific functions of its individual organs.

The plan for **Volume Three** is to have two units, one for our **Nervous System** and one for our **Endocrine System**. Again, we will look at the general functions of the systems and the location, structure, and specific function of the organs which comprise them.

TIDBIT BOXES

Throughout the text, some information is presented in boxes that are highlighted like this one, with diagonal lines on their right and bottom edges. I call these **Tidbit boxes**. This information is a tangent. It adds some detail or concept related to the main topic, but is not essential to understand subsequent information. Therapeutic tips, peeks at pathology, chemistry clarifications, and my own pet topics are some of the tidbits tucked away in these boxes.

BOLD FACE

Whenever an anatomical term is used for the first time, it is printed in **bold face** and defined or described at that point. This page number appears in the Glossary and in bold in the Index.

UNDERLINING

Key concepts or summary statements are underlined to help your eye pick them out more easily.

In Unit IV, underlining is used to mark the description of the origin, insertion, action, and palpation of individual muscles.

SCALE DRAWINGS

In Unit III on our skeleton, the basic illustrations of our bones are drawn to scale, giving us a sense of their relative sizes. The scale is one-half average size. Double the size of the picture to get a sense of the actual size of your own bones. Enlarge or shrink them to extrapolate from the average to your own size.

TO MEMORIZE OR UNDERSTAND

Vocabulary and terminology are important. Any specialized skill has its own vocabulary and definitions. To participate in that skill and converse with others, requires a certain familiarity and ability with the vocabulary, whether it is horseback riding, gardening, cooking, practicing some form of hands-on body therapy, or living wisely in our body. A certain amount of vocabulary has to be memorized.

For me, however, when practicing any skill, more important than memorizing vocabulary is understanding its basic concepts and operating principles. Before it is valuable to command an extensive vocabulary, the conceptual framework for its use must be established. In Behold Your Body the emphasis is always to understand the process rather than to memorize a bunch of words. Our first goal is to present simple, basic concepts of the structure and function of our body. In this process we will introduce elementary anatomical vocabulary, dividing words into their parts and defining meanings.

THE VOCABULARY FEAR

Many people, students and the curious public alike, cringe at the thought of exploring anatomy and physiology. One fearful aspect is the vocabulary. The words are long and unfamiliar.

The sheer length of many words can be overwhelming. A really long string of letters may form a single word. But many of these endless words are built of several smaller **word parts**. If we can hold our composure long enough to seek out these word parts, the vocabulary becomes a lot easier. Throughout the text, I break composite words apart and describe the meaning of each part.

For better or worse, the study of anatomy was in hot pursuit in the days when Latin and Greek dominated the scientific world. The words used to name different body parts are often simple descriptions of their shape, color, size, or function, but it all looks like Greek to us. And in fact, it is, either Greek or Latin. So in a sense, studying anatomy involves learning two different languages.

When we do learn the words, however, they tell us about the thing they are naming. They paint a picture that can help us learn and remember. For example, the endoplasmic reticulum is a certain anatomical structure. 'Endo' means inside. 'Plasma' refers to fluid. A 'reticulum' is like a net or weaving. So the name already gives us a picture of a network, woven through the fluid, inside the cell.

More good news is that Latin and Greek are very consistent languages. Words or word parts always mean the same thing wherever they appear. And believe me, the same word parts show up over and over again. When we learn their meaning, we can put them together in any specific word and already have information about the thing being named. For example, relating to the definitions given above, reticular proteins are proteins that form a net or weaving and endocardium is inside the cardium (whatever that is).

Lucky for us, Latin and Greek are part of the foundation of the English language, so some words and word parts are already familiar. For example, the 'sub' from submarine, means under and the 'bi' from bicycle means two. Learning anatomy can even expand and sophisticate your ability with English, so warn your friends.

When long, unfamiliar words do appear, say them out loud. It drives their evil spirit away. When you can say them and hear them, they are much less formidable. None of them bite and some of them have beautiful rythmic sounds that will help you learn and remember them.

THE CHEMISTRY FEAR

Chemistry is another scary aspect of trying to understand the amazing world in our body. We dread chemistry with its infinite, invisible details.

Understanding a little chemistry forms a foundation which enables a lot of other information to fall into place. But a little chemistry is quite simple. Water is a chemical. We all have some familiarity with water. We bathe in water. We drink tea. We water a plant. Our body weight consists of 60 percent water. The characteristics of water determine a lot about our body's processes.

In Behold Your Body we mention a little chemistry, only as it is useful to clarify other information. There is no special chapter on chemistry. It is simply included wherever it enhances a specific topic. But don't worry, it is so ordinary, you will have learned it before you even stop to label it as chemistry.

IMAGES AS TOOLS

In Behold Your Body, images are used as tools. They can help us remember. They can increase our understanding. They can help us leap from familiar to unfamiliar or from known to unknown.

Some images in this book also are used to stimulate our imaginations, to begin playing with metaphysical extrapolations from our body. 'Meta' means beyond, afterward, following something in a series, or searching for. 'Physical' pertains to the material aspect or substance of our person, that which can be known through our five senses. We all ask questions, searching for existence or meaning beyond our physical constitution. What are we more than our flesh, blood, and bones?

In this book, all the images and their extrapolations are my personal experience or hypothesis. I have included them as examples or guides, to generate ideas, to expand possibilities, to demonstrate one possible path from physical to more than physical. If a particular example fits you, push it as far as you like. If a particular example does not fit for you, do not impose it on yourself. Bend and twist it as you like. Use it to spark your own ideas. Ask your own questions. Discover your own correlates or references.

SCIENTIFIC THEORY - The Facts Of Life?

Many of us look at scientific information as the one sure thing in life, based on hard facts, what we see, hear, touch, smell, and taste. Science is the truth we hang our hat on and we limit what can live under our hat to scientific facts. Beware. The hat is floating on thin air, about to fall.

All scientific information is based on assumed, unseen theories. These theories are based on, or arise from, accumulated evidence, but no one really knows what is underneath. All facts are open to question and change and that includes all the facts presented in Behold Your Body. In this book, facts are based on my understanding, selection, and interpretation of current scientific assumptions.

For eight continuous months, with the same partner, I played three games of ping-pong every day. We arduously collected a wide variety of evidence in hopes of predicting who would win each match. We watched a long list of things: the kind of toothpaste we used, the color of our socks, the time of day we played, what we had for breakfast, who we said "Good morning" to, which end of the table we stood at, the color of the paddles we used, who had played just before us, etc. etc. etc. In spite of our long list of observed factors, there was a much longer list of unobserved factors, many of which we could not even know, let alone name. Still, it was amazing how long some seemingly silly thing would be consistent. Mathematical calculations would verify it to be statistically significant. At that point we could have developed a scientific theory about how to win ping-pong.

This is how scientific experiments happen. A situation is set up to collect evidence about something. Certain conditions are controlled and observed. After a certain time the evidence is considered to be sufficient to be proof. No one knows what would have happened the next day. No one knows all the uncontrolled, unobserved things that were influencing the situation. At some point it is important to make a statement about what seems to be happening, but the basis of it should never be forgotten.

In my ping-pong experiment it may be weeks after the formation of a hypothesis that it would no longer hold true. If we had quit the experiment after the hypothesis, we never would have known its fallacy.

THANKS and ACKNOWLEDGEMENTS

As a teacher, I am created in part by my own fascination and delight in my own body and my desire to share this with other health professionals and people in a down to earth, pragmatic, experiential way, so that we might understand and commune with our body rather than be encaged by it.

Another essential ingredient in my creation is my students. Their questions stimulate my exploration and bring me into their world. Their blank faces challenge me to find metaphores, examples, and new ways of presenting things. Their sighs and groans help me discriminate what is and is not useful or meaningful. Their excitement and self-discovery encourages me, validates me, and feeds me. Thank you students.

Thank you to a host of friends for your support and honor. Your continuous question "How is your book coming?" kept me working.

Thank you to Joe Stewart, who was my guinea pig, model, and consultant countless times. He would answer anything. "What does this statement mean to you? What do you see in this illustration? Which is clearer, 1 or 2? How do cartoon eyebrows show excitement? Put your arm out like this and let me draw your back."

Thank you to Ken Houghton, a teaching colleague who said "Don't be afraid to ask for help". Ken's immense enthusiasm for this book and his invincible assurance that the purpose of this book is accomplished spurred me along. He proofread the entire manuscript, sparing you thousands of misspelled words.

Thank you to Gloria Woodely, a former student and professional editor. Her gracious perserverence in reviewing every miniscule detail of this manuscript ensures its integrity and professional quality. Whenever these are broken, it is because I overrode her suggestion.

Thank you to Tyme, a former student, whose computer knowledge and equipment accomplished a formidable task, scanning all my drawings and interfacing my antiquated system with a modern printer.

For the cover, thank you to: the International Professional School of Bodywork for space and the use of their skeleton, Lucy, who endured a bath for the occassion; Ken Howard for his expertize as a photographer; Jane Blount for her ability with makeup, and for being eyes and muscles for Lucy's most vulnerable condition; and Blaize Mekinna for her receptivity and skill as a graphic artist, designing the cover.

Thank you to Barbara Clark, Milton Trager, Ed Maupin, and David Werner who graciously reviewed this manuscript and are willing to put their words and name in support of this book.

Thank you to the voice of Willie Nelson and the words of Kris Kristofferson in the song "Songwriter". For two years this song was my daily mantra, until one day I realized that I did believe in myself.

BODY OVERVIEW

OUR VITAL INTEGRITY

The study of anatomy, physiology, and pathology is specifically focused to examine the physical **aspect** of our being, our body. This does not mean that the existence of other aspects of our being, such as emotional, mental, moral, spiritual, energetic, or imaginary, is questionable or less significant. It simply means that, right now, we are selecting the physical manifestation of our being.

Our physical aspect is our material substance. The part which we can discover through our five senses. The part that science as a whole validates. The part that, at first glance, follows the laws of physics and chemistry.

At second glance, our physical aspect is much more complex than the laws of physics and chemistry. In fact, a second glance at these laws reveals more complexity than we often dare to admit.

The scientific preponderance of our culture supports our tendency to believe that our physical aspect is our most or only real self. Increasingly, scientific evidence demonstrates the existence of our other aspects. Our physical aspect simply tells physically what is in our thoughts, emotions, judgments, spirit, energy pattern, and imagination. All our aspects are equally important. All of our aspects constitute who we are, they form our vital integrity.

Each of our aspects affects all the others. Any physical condition always affects our energetic, imaginary, moral, spiritual, mental, or emotional aspects. We often hear "The pain is so bad, I can't think." or "She is pretty irritable these days. She's been sick for a long time."

Likewise, any emotional, energetic, moral, spiritual, imaginary, or mental pattern affects or shapes our physical body. Peptic ulcers of the digestive canal have long been called stress ulcers (although stress management is not so often part of the therapy). We only have to abate our tremendous fears and we already feel much better and our pain is much less. When it is still only in our imagination that something bad will happen, we can already start feeling sick about it and literally get a headache. When we are elated, our body moves differently than when we are depressed. When we are confident, we move and position our body differently that when we are insecure. This is the basis of body language. We are the interweaving of all our aspects.

ORGANIZATION OF OUR BODY

In the study of anatomy and physiology, we already have selected the physical aspect of our being as our object of study. Even so, it is difficult to explain everything in our body at once.

We facilitate the study of our body by dividing it into smaller segments (Fig. 1) and exploring each segment in sequence. The first step is to divide our body into its eleven major **systems**. Each system is a collection of mutually dependent parts that work together to perform some basic function for our body.

Each of the parts of any system is called an **organ**. An organ has a more localized, contained structure with distinct boundaries. An organ is an object that you could pick up and hold. The function of an organ is more specific and limited. For example, our stomach is an organ within our digestive system. The general function of our digestive system is to bring nutrients into our body. One specific job of our stomach is to digest protein and fat. Our small intestine is another organ within our digestive system. It functions to absorb the digested food.

Figure 1

We can divide our body into systems, which are made of organs, which are made of tissues, which are made of cells and intercellular substances, which are made of molecules, which are made of atoms, which are made of vibrating particles, and so forth. The parts in each level of division are smaller, and at the same time, are a whole organization of their own.

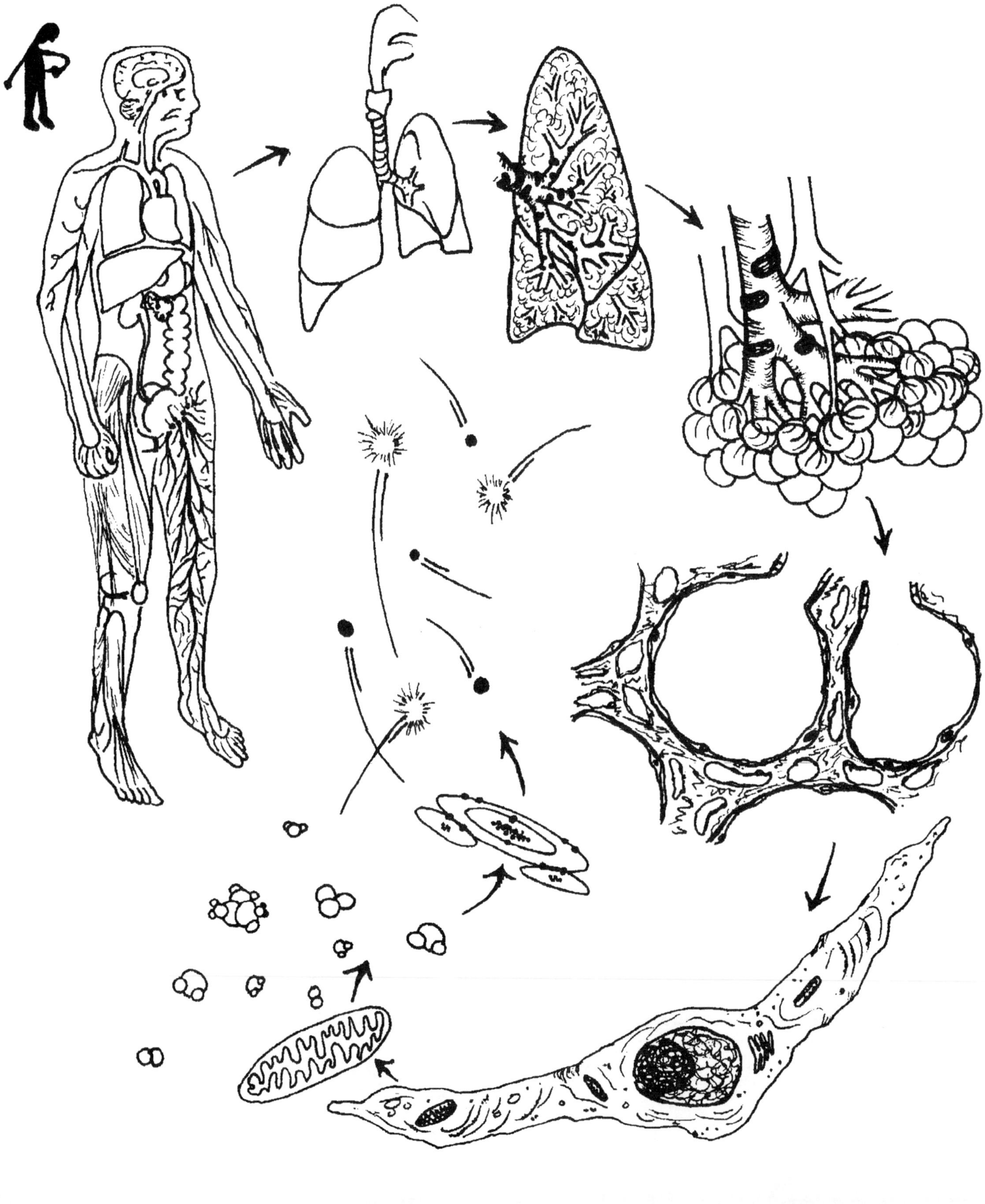

Any organ is, itself, made of a collection of different **tissues**. A tissue is a collection of many similarly specialized **cells** and **intercellular substances** ('inter' means between, so intercellular substances are the substances or chemicals filling the space between the cells).

The cell is the smallest living unit. It contains many tiny structures working to keep the cell alive. These tiny structures are made of **chemicals** or molecules such as water, fats, proteins, minerals, and vitamins.

As we pass through these levels of organization, an interesting thing happens. We start with one whole being which can be divided into many aspects. We pick the physical aspect and divide it into eleven systems. Each system has many organs. As we go through the levels of divisions, we identify more and more parts. Each part is smaller and more specialized. It is like a funnel, all the organs pouring into systems, pouring into our body, pouring into our whole person.

In the division from organs to tissues, an amazing thing happens. Only four different tissues form all the hundreds of organs in our body. The expanding diversity reverses. The number of choices is markedly reduced. To make any organ there are only four tissues from which to choose.

There is a similar narrowing of choices in stepping from tissues to cells. All the cells in our body have the same basic life support systems and general characteristics. We are back to one. In fact, all the cells in the whole animal kingdom follow the same basic principles. Different cells may emphasize or refine a certain process, but they are more alike than they are different.

Our body is amazing in its intricate complexities. What can be even more amazing, however, is the simplicity of its basic rules and patterns.

In a sense we are back at the beginning. Each cell in our body is like an independent individual doing its particular job within the community of cells that make up our body, just as we each are an independent individual doing our particular job within the community where we live. We like to think of ourself as independent, but we are very dependent on the community as a whole, as well as each individual within the community doing his/her job, and each person is dependent on us. Likewise in our body, each cell has certain dependencies on other kinds of cells and the integrity of our whole structure, and our whole structure depends on the integrity of each cell. The more differentiated and specialized our cells are, the more they depend on each other.

CHEMICAL BASIS OF OUR BODY

Looking at the chemical structure of our body by weight, we are about **60% water**, **15% fat** or **lipid**, and **20% protein**. Knowing the characteristics of these three familiar substances enables us to know the major potentials in our body. The remaining 5% includes all the other stuff - iron, calcium, iodine, sodium, vitamins, potassium, trace minerals, and so forth. As essential as it may be, all this stuff is a relatively small quantity.

When we say that 15% of our body weight is **fat** or **lipid**, we are talking about healthy, average-weight people where fat molecules build an important part of their body structure. These fats are in our food, entering our body through our digestive canal, flowing through our blood to be built into our tissues and cells.

In this context, fat and lipid are interchangeable and belong to a large family of chemicals, defined by certain characteristics.

Fat in the sense of needing to lose weight is when we have taken in more nutrients than we need for fuel or structure. Be it protein, carbohydrate or fat, our body converts all our excess nutrients into fat molecules and stores them in special cells called **adipose cells** which are grouped together into **adipose tissues,** for a fuel reserve and protection. Everyone has adipose tissue in their skin and around many internal organs, an overweight person has excess.

Looking more closely, we find that water contains only two elements - **hydrogen** and **oxygen**. The basic building blocks of fats are called **fatty acids** and **glycerol** and contain only three elements - hydrogen, oxygen, and **carbon**. The basic building blocks of proteins are called **amino acids** and contain four elements - hydrogen, oxygen, carbon, and **nitrogen**. So when we add it all up we do not go very far across the chemical periodic table. Except in very small, but essential quantities, our body consists of hydrogen, oxygen, carbon, and nitrogen.

Basic nutritional requirements include all the substances we have mentioned above - water, fats, proteins, minerals, and vitamins - plus one more - **carbohydrates**. Carbohydrates form the greatest portion of our food intake, but very little of the actual structure of our body. Carbohydrates are our main source of fuel, providing most of the energy for all our body functions. Starches, sugars, and cellulose are all carbohydrates. Their basic building blocks are simple sugars. The most common sugar in our body is **glucose**.

HOMEOSTASIS

All this elaborate structure in our body has one purpose, to keep our body's internal environment as consistent as possible. Sounds boring. What a rut. That is life for our body. Our cells thrive best in a specific environment. They do their life preserving functions best at a specific temperature, a specific acid/base balance, a specific oxygen concentration, a specific blood pressure, and a specific electrolyte balance. The list could fill pages. Normal lab values reflect this best environment.

Normal lab values have a range which is considered normal. This range shows our body's ability to tolerate a certain deviation from the best. Sometimes lab results show that we are markedly deviated from the optimum, and we still do not feel that bad. Our body has amazing reserve and tolerance.

We have said that the greatest part of the structure of our body is made of water, protein, and fat. How do these three substances get along? Water and protein are great companions. Proteins are generally water soluble, they dissolve and flow in the water of our body. Water and fats do not get along. Fats do not dissolve in water, the two repel each other. When placed in water, fats clump together, separating from water as much as possible.

In a sense, all the chemicals of our body can be divided into two teams - water and water soluble substances against fat and fat soluble substance. Both kinds of substances are essential in our body.

If our body were made exclusively of water and water soluble substances, we would have no ability to create a barrier to the flow of water through us, into us, or out of us. During a walk in the rain, we would absorb any drop that hit our skin. We would need to be dry-cleaned. Nothing could stop us from just dissolving in our bath water and flowing down the drain. It is the fats of our body that provide us with this barrier. Fats form walls that limit the flow of water. We will see that the basic structure of our cells is fat, so within all the water of our body, our cells can hold their integrity. Our cells will not dissolve.

The fact is that cells and their fatty structure are located from head to toe, all over our body. We need to carry fatty building blocks to all these cells. This is where the problems start. The solvent of our body's transportation systems is water. The only way to get from place to place within our body is by dissolving in the water of our blood or lymph. As our fats step into the flow of water, they freak out, clump together, lodge in the walls of our blood vessels, and cause hardening of our arteries - the basic pathology in heart attacks and strokes.

What a dismal prospect. We either swirl down the tub drain or get a heart attack. Our body has a way through this impasse. Since proteins are generally water soluble, special proteins are made to bind to and carry fats within our blood. These proteins with their loads of fat are often called **lipoproteins**.

Many of our lipoproteins fit into one of two categaories, they are either **high-density lipoproteins (HDL)** or **low-density lipoproteins (LDL)**. I have an analogy which helps me remember how this works. I think of cargo boats on a river, different sizes of boats, with different sizes of loads. The river is our blood stream or lymph. The cargo boats are our carrying proteins. Their load is our fat. While proteins are the buoyancy factor in this case, they are more dense than fat.

High-density lipoproteins then, are large boats with relatively small loads (Fig. 2). The relatively large amount of protein leads to the high density as well as increased buoyancy. High-density lipoproteins are beneficial since they give us the fats we need to build our body in a way that is easily carried in our blood and so reduces arterial disease.

Low-density lipoproteins are small boats with big loads (Fig. 2b). The relatively small amount of protein leads to the low density as well as reduced buoyancy. Low-density lipoproteins are more difficult for our body to transport. They tend to sink and run aground, causing more arterial disease.

There are certain low-density lipoproteins which are essential to build our structure, so they need to be eaten. The trick is to eat enough fat but not too much.

Figure 2

(a) High-density lipoproteins are like large boats with small loads, they easily stay soluble in the water of the blood. **(b)** Low-density lipoproteins are like little boats with large loads, struggling to stay afloat, to not run aground.

(a)

(b)

The environment around us is always changing and affecting us. Inside our body, we release heat, we lose water, we ingest salty popcorn, and so forth. Chemicals are constantly being consumed and created. Glucose and oxygen are constantly being burned for energy, leaving water and carbondioxide as wastes. Our internal environment is always being thrown off normal.

The job of every system and organ has something to do with bringing the environment back toward optimum. Our sweat glands sweat and the blood vessels in our skin dilate to get rid of the heat. Our kidneys get rid of excess salt. Our respiratory system brings in more oxygen and gets rid of carbondioxide, and so on. To finish this list, we would have to say every function of every system and organ in our body. **Homeostasis** is the name for this on-going process of our body as it moves toward its optimum environment. 'Homeo' means same or like and 'stasis' is standing, so together it says holding the same stance or like standing. Any deviation can be either toward the excess (too much, too high), or toward the deficient (too little, too low). Too much and too little are both stressful and pathological. No substance escapes this principle.

It is easy to see how running out of a substance can cause problems, but too much of that same substance can cause equally bad problems. Potassium is good for our body, but that does not mean that more and more potassium is better and better. Too much potassium can cause destructive processes in our body, such as rupturing our red blood cells. Oxygen is another vital chemical that our body consumes in large quantities, but too much oxygen can cause blindness.

To call something **toxic** or to label a state as toxic is very common these days, but what does it really mean to be toxic? My definition is that toxicity develops whenever any chemical or state is greater than the normal lab value, when it has deviated to the excess. It is not a substance itself which is toxic so much as the quantity of it. So we can have potassium toxicity, oxygen toxicity, and water toxicity. Each substance has its own threshold where toxicity would be reached. Some thresholds are high, some low. Even arsenic has a tolerable quantity in our body, small as that may be.

Our body does an amazing feat just to stay alive with all we throw at it and into it.

UNIVERSAL ORGANIZATION

In the structuring of our universe, there are many levels of organization, just as we have referred to levels of organization within our body. It appears that certain requirements and principles are repeated in each level, from the tiniest to the most expansive. Each level has certain requirements of ingestion and elimination to maintain its life force and flow. This is sometimes referred to as the microcosm within the macrocosm and the macrocosm

Figure 3

We are a little speck in the expanding levels of organization beyond ourself.

within the microcosm. We have seen an example of this when comparing a cell in our body to one of us in our community. Similarly, while each of our cells plays one little role within our whole drama, likewise, inside itself each cell is directing its own drama. Each cell has a life of its own. Its tiny structures are like its systems and organs. It is true that each one of us is just a drop in the ocean, however, inside of us there is a whole planet full of oceans, each one full of drops.

Where do we sit in the whole scheme of things? We are each a member of a community, which is part of a county, which is part of a state, which is part of a country, which is part of a planet, which is part of a solar system, which is part of a galaxy, which is part of a universe (Fig. 3).

In the other direction, our body is composed of systems, which are composed of organs, which are composed of tissues, which are composed of cells, which are composed of molecules, which are composed of **atoms**, which are composed of vibrating particles, and so on (Fig. 1). In each direction, our ability to go further is limited by our power to observe or find. A cell is too small to see with our naked eye, but from the perspective of an atom, a cell is like a huge city.

- A hierarchy of dependence and independence exists. Individual members of a certain level are not so dependent on each other, but each member is dependent on the levels below and above it. For example, two adjacent counties are not so dependent on each other, but each county is very dependent on the communities of which it is made and the state of which it is a part.

In our body, two adjacent liver cells are not so dependent on each other, but they are very dependent on the molecules they are made of and our body as a whole. The greater the diversity and specialization of any member, the more it depends on its greater collective to meet all its needs.

Figure 4
As our body changes position, we change the words describing the orientation of its parts.

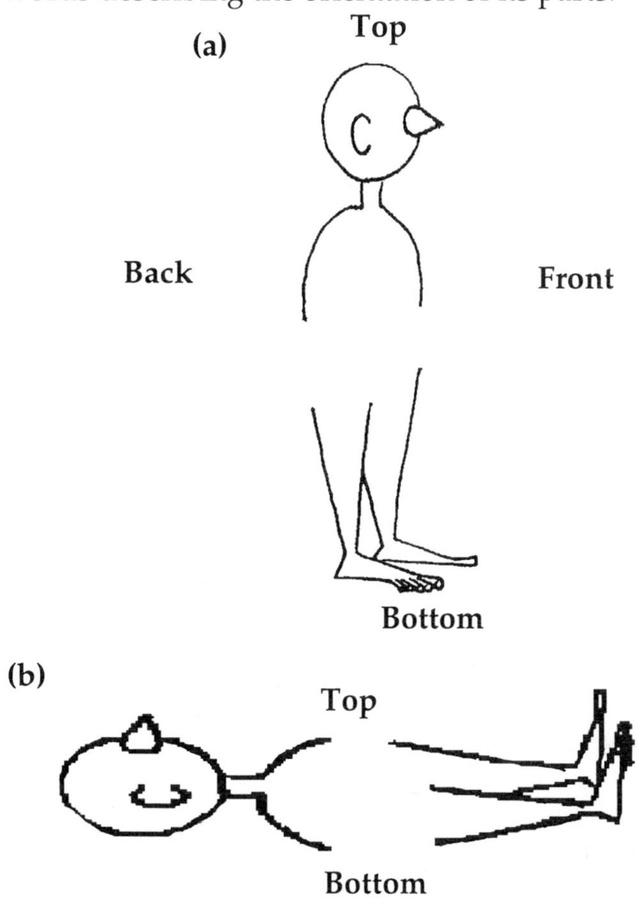

ANATOMICAL ORIENTATION

The study of anatomy includes describing the location of body structures relative to each other.

If the man from Mars calls us on the phone and asks "Exactly what do humans look like?", what would we say? Using words like top, bottom, front, back, and side, we would describe the location of one body part relative to the others. It might sound like this, "A human is topped by a firm spherical structure, supported from below on a narrow stalk. This sphere has a cone-shape protruding from its center front, while a C-shaped flap decorates each side. etc. . . . two tapering pedestals project downward, side by side, from the bottom of the body. Underneath each pedestal is a horizontal wedge, tapering forward, splitting into five separate rolls. These rolls are sequenced by size, the biggest being central, the smallest to the side." (Fig. 4a). The more detail we added, the more clearly the Martian could sketch out the human form, standing in an upright posture.

If we had been thinking of a person sitting in a chair, some of our descriptions would have been different, for example, the pedestals would project forward from the bottom of the body. If we had been thinking of a person lying on his/her back, the sphere would be at one end, supported from one side, with the cone protruding center top. The wedge would be tapering upward and so forth (Fig. 4b). A lot of things would be described differently even though it was the same body.

• Any time we change our body position, the relative orientation of body structures also changes. When we stand upright, our feet are underneath our hips, but when we sit on the floor, our feet might be in front of or to the side of our hips. Our thumb may be described as attaching to the outside, front, or inside of our hand, depending on the rotation of our arm. Consequently, before we can locate body structures relative to each other, we need to describe the basic position of our body.

Figure 5

Anatonical position with arrows showing the terms used to describe the different directions.

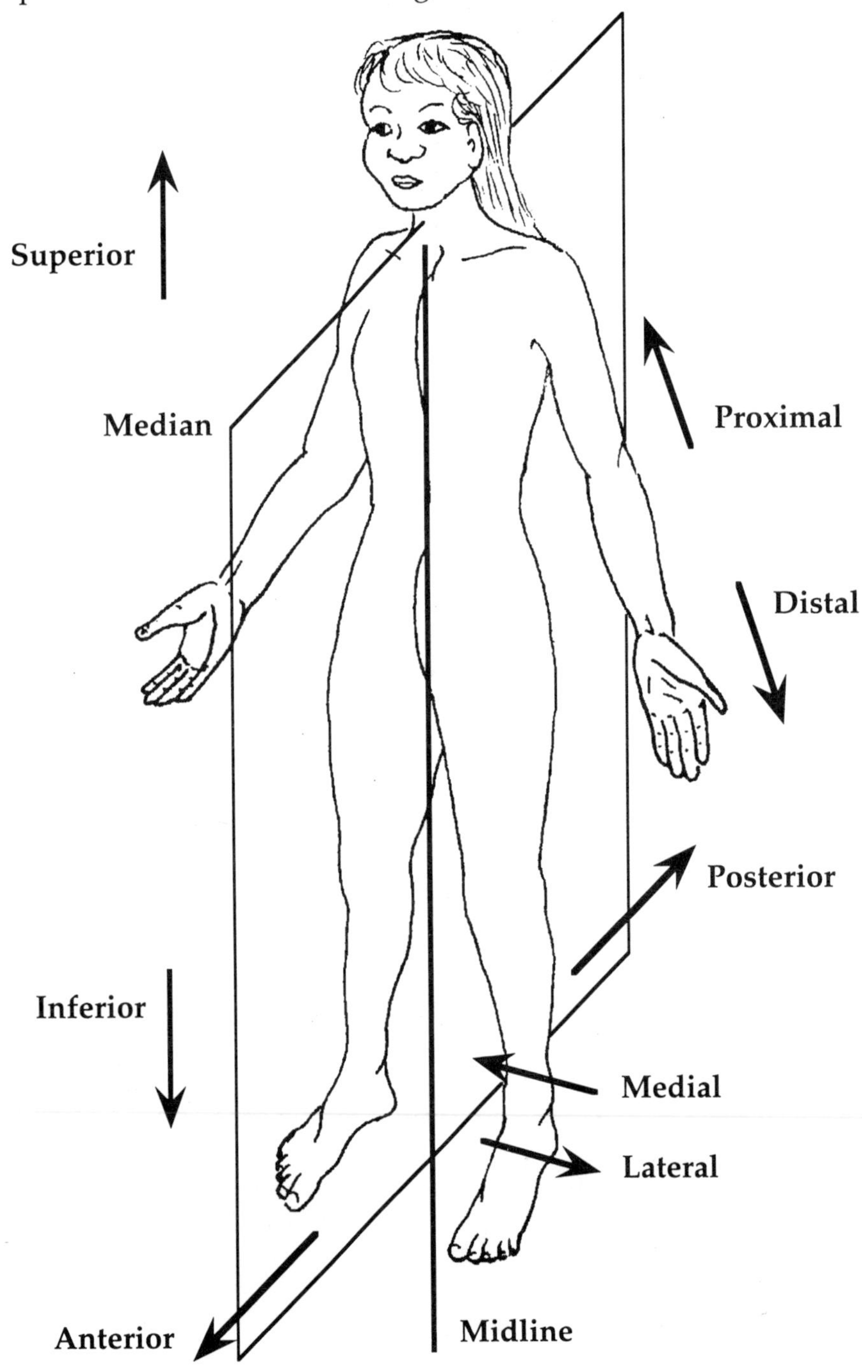

In western anatomy, a specific body position is chosen and defined as the standard position. This position is call **anatomical position**. Whenever locations are described, we can assume that they are based on this standard, even if it is not said every time.

Western anatomical position is standing upright with our nose and toes pointing forward and our arms down by our sides, rotated so our palms face forward (Fig. 5). (Western is used in contrast to oriental, where anatomical position varies.)

The descriptive words we have used, top, side, underneath, and so forth, are perfectly clear and correct, but ancient anatomists chose specific words for the different directions, just like north, south, east and west are used in geography. We will add these words to our repertoire since they are an important foundation of anatomical vocabulary, frequently being used as parts of longer words. Their meaning in each place is the same, so adding them to our vocabulary will increase our understanding of many terms, and impress the Martians even more.

One pair of words is used for each of the three directions in space.

- In the vertical direction through our body, anything that is above something else is said to be **superior** to it, while anything that is below, is said to be **inferior** (Fig. 5). For example, our nose is superior to our mouth, or our mouth is inferior to our nose.

We would never just say "Our mouth is inferior". If we did, the question would arise, "Inferior to what?" To our nose? Yes. To our belly button? No.

It does not matter where the structures are in terms of left to right or front to back, these words always can be used to describe their vertical placement. For example, we can say that our mouth is inferior to our eyes or that our mouth is superior to our shoulders.

An example of these words being used as word parts is two muscle names infraspinatus and supraspinatus. Whatever the 'spinatus' is, these muscles are respectively inferior and superior to it.

- Imagine a vertical plane that divides your body into equal right and left halves. It would pass through your nose, mouth, navel, backbone, and so forth. This plane might be called a **median**, and any vertical line in it, a **midline** (Fig. 5). When midline is used in an anatomical context, it refers to the center of our body from right to left as contrasted with top to bottom or front to back.

In a horizontal direction from side to side, the words used are **medial** and **lateral**. When something is medial to something else, it is closer to our median (Fig. 5). When something is lateral, it is further away from our median. For example, our eyes are medial to our ears or our ears are lateral to our eyes.

- In a horizontal direction from front to back, **anterior** means in front of and **posterior** means behind (Fig. 5). For example, our belly button is anterior to our back bone or our back bone is posterior to our belly button. Also, our nose is anterior to our ears and our ears are posterior to our nose. As we mentioned earlier, how things are positioned in the other directions does not change the use of these words.

With these three pairs of words we can describe the relative position of any structure in all possible directions. There are, however, a few other words that come in handy. We always like choices.

- One pair of words is used primarily for our arms and legs. Visualize where each extremity attaches to your body. When describing the relative positions of any two structures in an extremity, the one that is closest to this attachment is said to be **proximal,** while the one that is further away is **distal** (Fig. 5). Proximal has the same root as proximity, meaning nearest to. Distal and distant have the same root meaning far away. We can say that our elbow is proximal to our wrist or our wrist is distal to our elbow.

Proximal and distal also may be used when referring to an end of an internal organ. These are usually tubular organs where something flows through. In this case, the proximal end of the organ is where the flow enters, while the distal end is where the flow leaves the organ.

Figure 6

A baby curled up in fetal position is protecting its ventral surface, shown in white, and exposing its dorsal surface, shown as dotted.

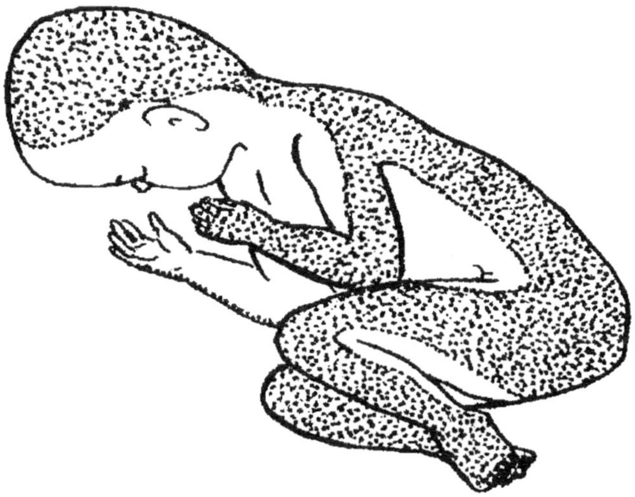

• A last pair of words is **ventral** and **dorsal**. These words relate to our body surfaces. Our ventral surface is our most vulnerable and sensitive surface, the one we would instinctively protect. Our dorsal surface is our less sensitive surface, the one we would use to be the protector. Our face, abdomen, genitalia, the palms of our hands, backs of our knees, and soles of our feet are our most sensitive surfaces, while we would turn our backs, posterior forearms, or shins to ward off a blow. A little baby, curled up in fetal position, is folded into its ventral surface, leaving its dorsal surface more exposed (Fig. 6).

The protective instinct and sensitivity of our ventral surface is already developed in our early embryonic state as our nervous system develops. At this time our arms and legs both protrude laterally from the trunk of our body (Fig. 7). There is even a little tail extending between our legs. We look like a little wee lizard with an oversized head. At this point our whole anterior surface is our ventral surface, they are essentially the same thing.

As we develop, our arms do a slight medial rotation and swing down by our sides. Anatomical position, however, reverses this rotation in our hand. Our legs do a greater medial rotation and swing down under our body so we can walk on them. If we think of standing on our hands, our arms rotate in the same way.

Our sensory nerves are already in place before this development, so they move with it. In our arms and legs, then, our sensitive surface is no longer anterior. Ventral no longer equals anterior.

From anatomical position then, our ventral surface is the anterior surface of our head, chest, abdomen, and hands, but the medial surface of our arms and thighs, the posterior surface of our knees and lower legs, and the soles of our feet (Fig. 8). Sweep your ventral surface by passing your hand over these structures, from your face to the soles of your feet.

Our dorsal surface is always opposite to our ventral, including the posterior aspect of our head, backs, buttocks, and hands, our lateral arms and thighs, anterior knees and shins, and the top of our feet (Fig. 8). Sweep these surfaces. Bend over and place your hands on the floor beside your feet as if you were walking on all fours (Fig. 9). See how your arms rotate and the ventral and dorsal surfaces in your arms and legs are similarly placed. This position helps to remember dorsal and ventral.

• The muscle name latissimus dorsi is another example of anatomical orientation words being used as word parts. Do you see two of them? Lateral and dorsal. This name in itself tells us that this muscle is moving in a lateral direction somewhere on the dorsal surface of our body.

Figure 7

(a) An anterior view and **(b)** a lateral view of early embryonic development. The white portion indicates the ventral surface while the dotted portion indicates the dorsal surface.

(a) **(b)**

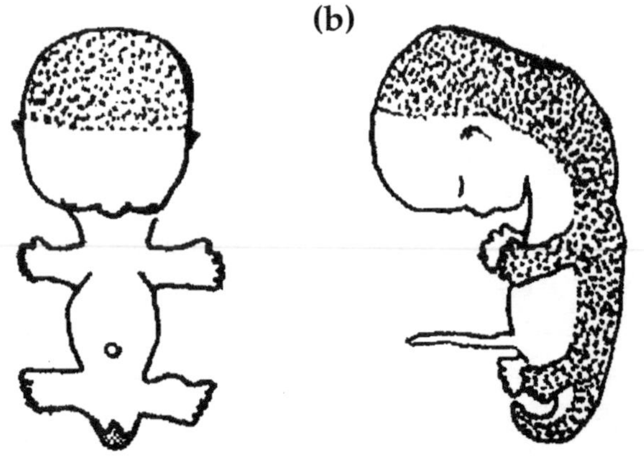

Figure 8

An **(a)** anterior, **(b)** lateral, and **(c)** posterior view of the body showing the ventral surface as white and the dorsal surface as dotted.

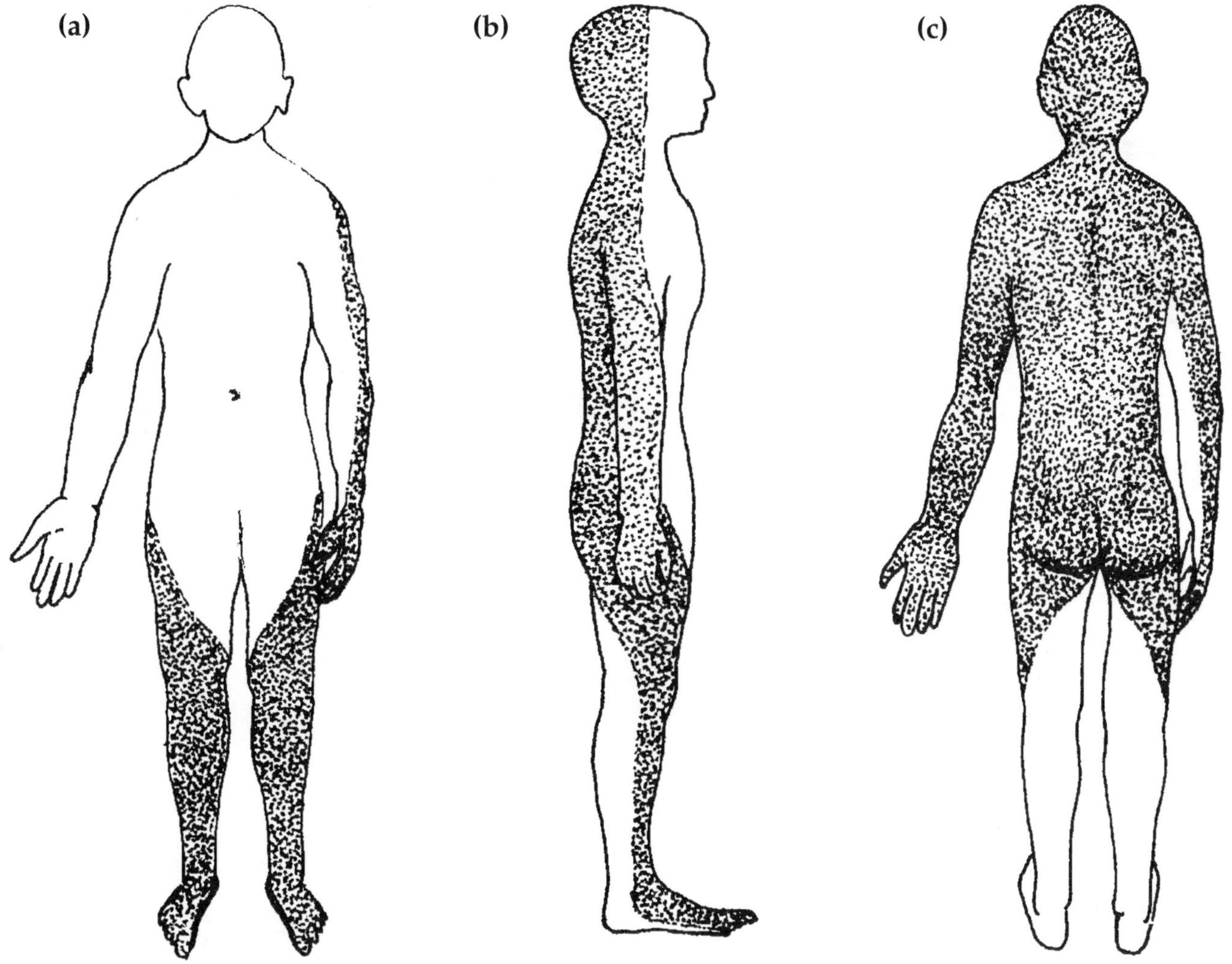

Figure 9

A lateral view of a person standing on all fours, showing the similarity in the rotation in both extremities. The ventral surface is white, while the dorsal surface is dotted.

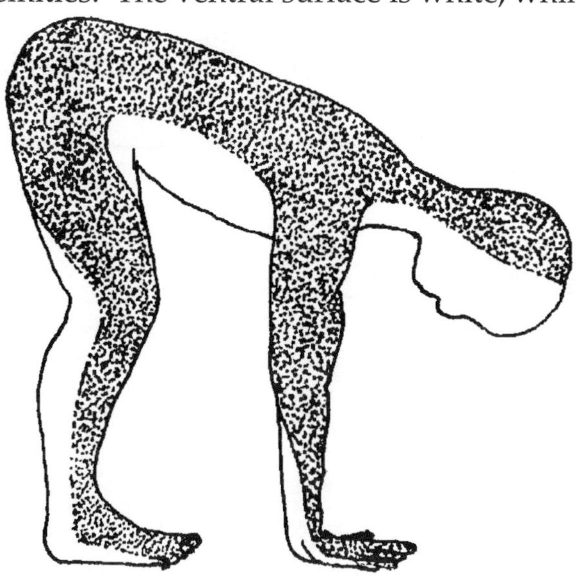

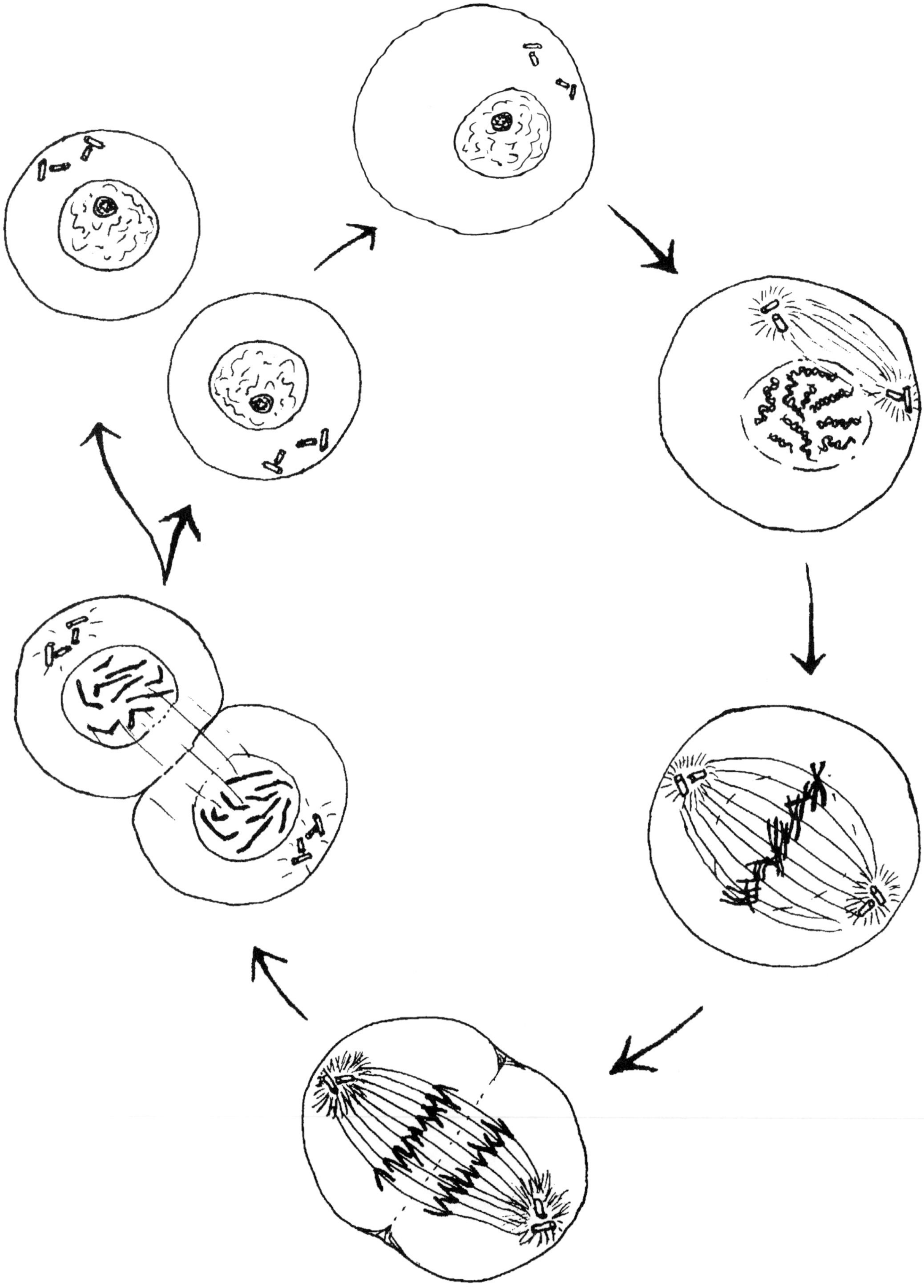

Unit I: **THE CELL**

On this planet, the cell is the smallest structure capable of maintaining life and reproducing. The cell then, is the basic unit of composition of all living forms. For some microscopic animals, such as amoebae or bacteria, their whole body is just one cell. For humans and many larger animals, trillions of cells are included in each body. It is the potential structure, function, and coordination of all its cells, that prescribes the potential of any living being. Activities essential to life originate at the cellular level of an organism, as do many disease processes. Understanding the cell, then, is basic to understanding all anatomy, physiology, and pathology.

As humans, we each begin as a single cell, a fertilized egg. This cell and all its descendents divide over and over again to establish and maintain our adult cell population. At specific points in this reproductive process, the two daughter cells of one cell division differentiate, manifesting different characteristics. So it is that we have many different kinds of cells in our body, each doing different, specialized jobs. Every cell, however, comes from the same source and has the same basic life support mechanisms.

In this unit we will present the basic characteristics and life support systems common to all cells in our own body, the whole animal kingdom, and much of the plant kingdom. In subsequent units we will go into the specializations of specific cells.

Direct

chapter 1: **The Cell**

The **cell** is the smallest, independent living structure in our body. Each type of cell has special characteristics which differentiate it from other cells, but there are basic structures and functions which are the same for all cells, whether they are part of the human body, other animals, fish, plants, or bacteria. In this chapter we will discuss a generalized cell, the structures which all cells have in common and the functions which all cells perform. As we discuss the specific tissues and systems of our body, we will identify special differentiating aspects of their cells when they are important.

Cellular organization has all the same needs as an organization at any level, whether it is a family, a community club, or a nation with land, natural resources, and people. Any organization needs the following six aspects. **1.** A code or guidelines that define itself, that say who belongs, that allow the group to distinguish who is in and who is out. This may include a physical meeting place or a structural barrier to the surrounding world, allowing control of who enters and leaves. These guidelines and barriers are the basis of the organization's integrity. **2.** A system of

Figure 1 - 1

The cell and its organelles.

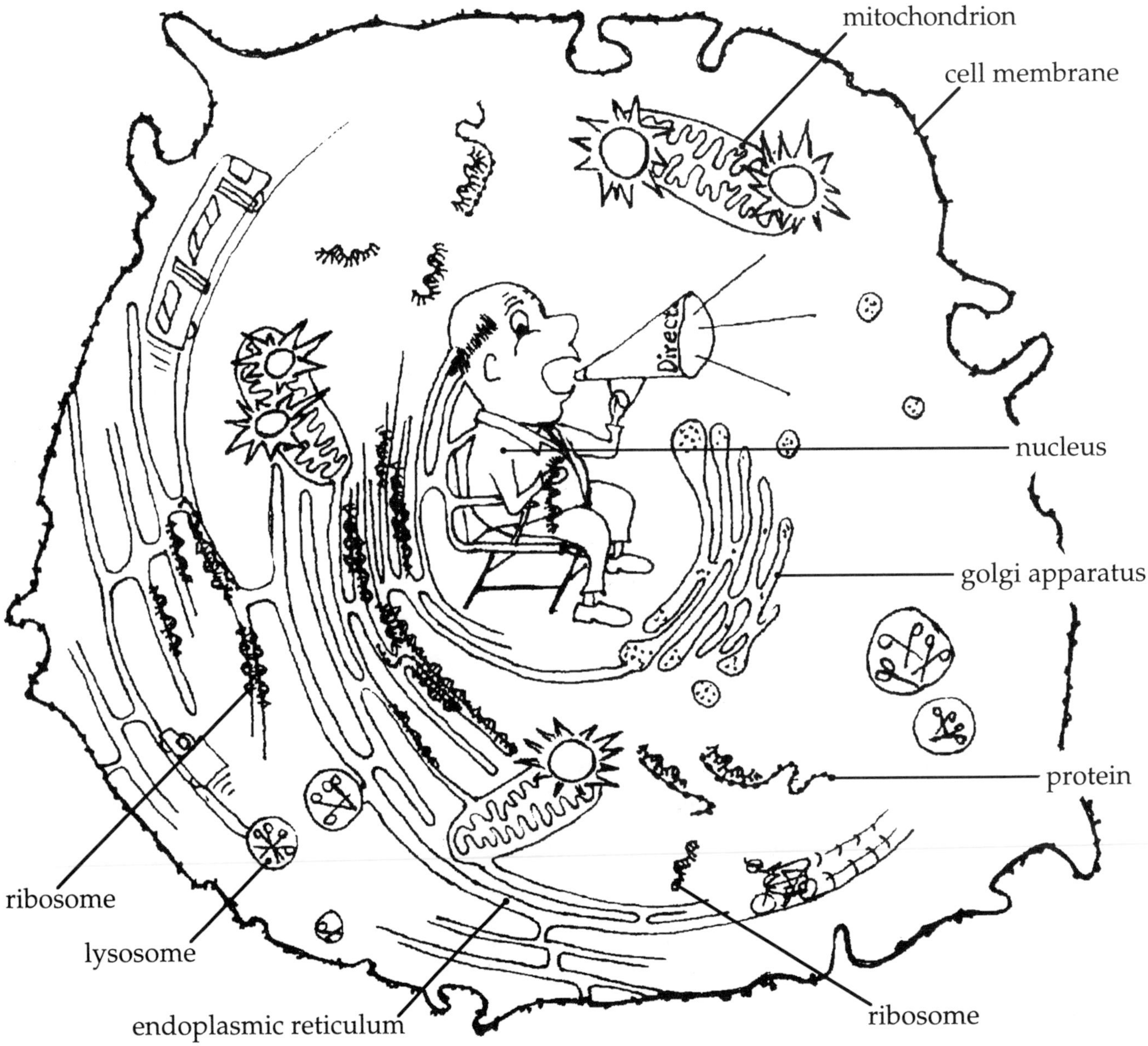

communication with other similar levels of organization, diplomats, representatives, regional meetings. This may include exchange of ideas or products. **3.** A system of self-defense to protect its boundaries, its properties, and all its members. A community club may not experience that much aggression or invasion. On a national level, self-defense is a major issue. **4.** A method of making decisions, such as a person who is in charge, a board of directors, or government which organizes and controls the organization, influencing what direction it takes and how fast it will proceed. **5.** A system of communication and transportation within itself, such as a phone chain, a newsletter, a bulletin board, or a network of roads and highways. **6.** Energy to operate and a source or method to generate that energy, whether it is the participation of its members or money to buy services, build and maintain equipment, communicate among its members, and reward special efforts of its members.

These are the basic needs of any organization or person, and these are the needs of each cell in our body. Each cell has surface and internal structures, called **organelles** which perform these basic needs.

Cells come in all different shapes and sizes. We will use roundish as a general shape.

ORGANELLES

CELL MEMBRANE

If a figure is drawn as a line that completely encloses a space and is labeled as The Cell (Fig. 1-2a), most people would just assume that the line and the enclosed space are the cell and everything outside the line is not the cell. This is correct. For the cell, this line is called the **cell membrane** (Fig. 1-1). The cell membrane is made of two materials: **1.** a **phospholipid barrier** and **2.** randomly placed **protein.**

Phospholipid Barrier

The **phospholipid barrier** meets the first basic need of the cell. The structural integrity of phospholipid defines, limits, and maintains the cell's integrity. The phospholipid forms a barrier to the surrounding world, segregating inside from outside (Fig. 1-2b).

A cell can be killed simply by poking a hole through its phospholipid barrier. Then the inside and outside of the cell get all mixed together (Fig. 1-2c). The cell loses control of its internal environment and its integrity. The cell dies. This is how many antibiotics and your own antibodies kill invading bacteria.

The position of the phospholipid barrier defines the size and shape of the cell. While the line in the figure is stationary, the membranes of many cells are in constant motion, writhing, spreading, withdrawing, accommodating to its work at that moment.

Figure 1 - 2
The cell membrane defines and maintains the integrity of the cell. A hole in the cell membrane may kill the cell.

(a) THE CELL

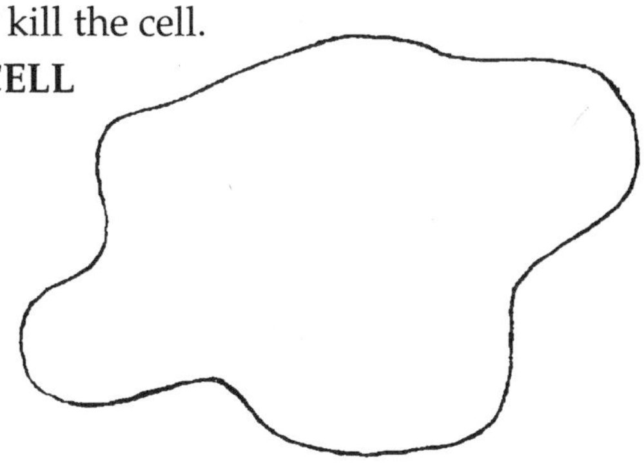

(b)

(c)

• Why phospholipid? What is the most abundant substance in our body? Water. Water is the most popular chemical inside and outside our cells. The water inside our cell is called the **intracellular fluid.** 'Intra' means within. The water outside our cell is called the **interstitial fluid.** 'Inter' means between and 'stitium' refers to things that are standing close together like people in a crowded stadium. Usually, where there is one cell, there are many cells together. Our interstitial fluid fills the crevices between adjacent cells.

Our cell membranes separate our intracellular and interstitial waters. To keep these two puddles separate, the structure of our cell membrane cannot dissolve in water. A barrier made of salt or sugar would be useless. One day, while taking a bath, our cells would just drain away with the bath water.

Which of the three basic nutrient groups repels water and will not mix with it? The fats or lipids. Wash greasy dishes in clear water without soap. The grease clumps together and does not mix with the water. This is exactly what we need to form a barrier between the intracellular and interstitial water in the body.

The structural integrity and barrier of our cell membrane is built of phospholipid, a molecule of fat with phosphorous attached to one end like a popsicle on a stick. The cell membrane is a double layer with millions of these molecules, precisely organized so that the fats are sandwiched in the middle, while the phosphorus covers the internal and external surfaces (Fig. 1-3).

While our phospholipid forms a barrier, it is not like a plastic bag, blocking all movement. It allows certain things to come and go, according to our cell's needs. Our cell controls most movement across its membrane, taking in what it wants and putting out or keeping out what it does not want. This is called **selective permeability.** Specific methods of transport across the cell membrane are discussed at the end of this chapter.

The walls of all organelles are made of the same phospholipid sandwich described as the cell membrane structure. The cell cannot afford to have any of its organelles dissolve.

Figure 1 - 3
The cell membrane.

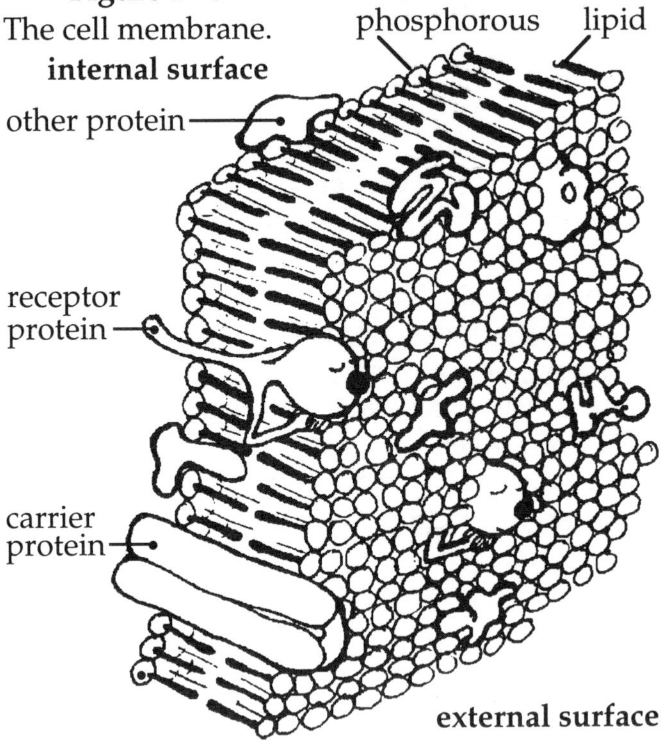

Proteins

Once our cell has a barrier to define and contain itself, its second need is to exchange and communicate with what is outside around it. To this end, our cell membrane has **proteins** which imbed into its internal or external surface, or dangle through it (Fig. 1-3).

• One of the protein's jobs is to assist with transportation in and out of our cell. These are called **carrier proteins.** They can open passageways through the cell membrane without threatening the integrity of the cell (Fig. 1-5). Molecules can be transported in either direction through these passages.

• Other cell membrane proteins are called **receptor proteins.** Sometimes a very large molecule from another place in our body carries a message or instructions for a cell. To let this molecule through the cell membrane would disrupt the cell's integrity and kill it. Receptor proteins pass all the way through the cell membrane (Fig. 1-3). On the outside, they chemically bond to the large messenger molecules. This bonding is like a door bell being pressed by someone who has a message for the cell. This bonding in itself transfers the message, rather than the molecule, into the cell (Fig. 1-4). Each specific messenger molecule has its own specific receptor protein. This is how nerves and many hormones influence cell function.

Figure 1 - 4
A receptor protein **(a)** bonding and **(b)** transferring a message across the cell membrane.

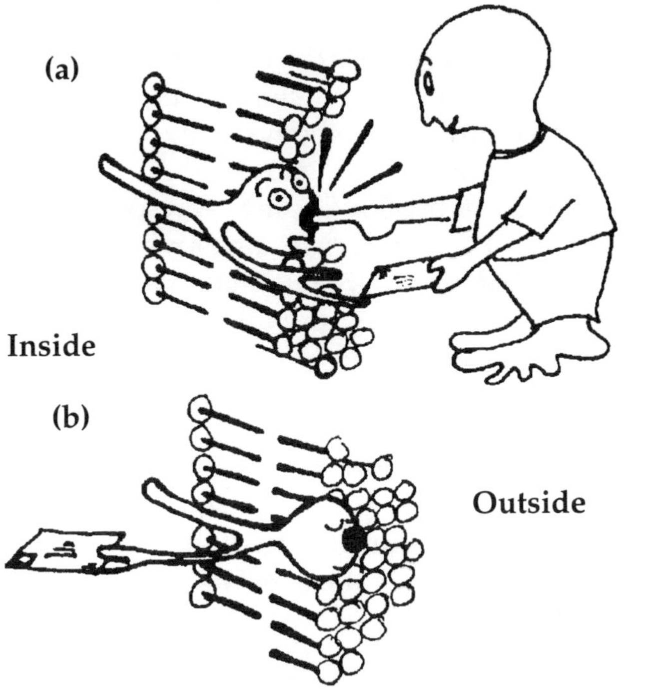

Figure 1 - 5
A carrier protein creates a passage through the cell membrane.

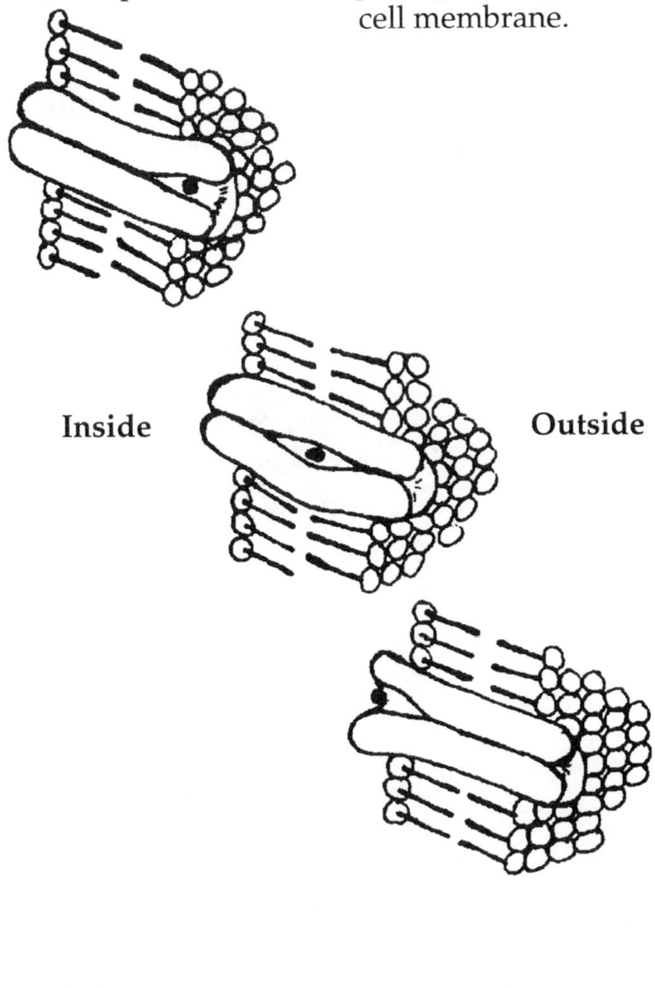

• While the basic phospholipid structure of every cell membrane is the same for all organisms, the cell membrane protein constitution in every individual is different from all others. The external surface of the cell membrane protein serves as the name tag or bar code of that cell and its organism. It is the uniform which identifies who the cell belongs to, like the particular wrapping of a package which makes it quickly recognizable from the outside (Fig. 1-6). This is the basis of our cell's self-defense and our whole body's immune response. This is how cells recognize if a cell is foreign, an invader which does not belong (Fig. 1-7). The protein constitution of the cell membrane meets the third need of our cells.

A cell's surface proteins are called its **antigens.** Within our self, our own antigens are recognized as part of self and tolerated. Anybody else's antigens found in us are recognized as foreign and that cell is destroyed (Fig. 1-8). In anybody else, our antigens and cells should receive the same treatment.

Figure 1 - 6
The particular wrapping of a product makes it quickly recognizable from the outside.

Blood is typed according to its antigens. A, B, and Rh are the names of common antigens.

When blood is cross-matched, or donors are scrutinized for organ transplants, they are searching for antigens that are close enough to those of the recipient to slip in unnoticed. If they are recognized as foreign, the donated blood or organs will be destroyed.

A cell's protein is designed by its genetic material. Every cell in one person's body has the same genetic material and therefore the same protein potential. In contrast, the genetic material of every organism is somewhat different from the next, yielding different protein and therefore different antigens. In identical twins it is the most similar because they have the same source. In siblings it is quite similar. In different races it is more different. The more divergent the species, the more different their genetic material and, consequently, their cell membrane proteins All human beings have some differences in their cell membrane protein, but that of a bacteria is very different from any human being.

Figure 1 - 7
Foreign invaders are recognized on the basis of their surface protein.

In **autoimmune diseases** there is a loss of self-recognition and self-tolerance, with the result that the body looks at some of its own tissue as foreign and destroys it.

The species to which human antigens are most similar is the pig. Lay that on your pretentious, snub-nosed acquaintance.

LYSOSOME

The **lysosome** is the weapon of our cell's self-defense. When something is recognized as foreign or unwanted, the lysosome applies its tools of destruction (Fig. 1-8). 'Lysis' means dissolution or breaking apart. A lysosome is a package of **catabolic** or digestive enzymes which break apart organic material. If a foreign organism or any organic material enters our cell, our cell can expose it to the lysosomal enzymes and destroy it, like the disinfectant Lysol destroys bacteria.

When a substance is not organic or cannot be broken down, like mercury or asbestos, our cell is in trouble. Our cell may not be able to get rid of the substance and may be damaged or even killed by it.

The wrapping of the lysosome enzymes package is a phospholipid sandwich similar to the structure of the cell membrane.

Figure 1 - 8
When something is recognized as foreign or unwanted, the lysosome destroys it. This might appear brutal, but our own life depends on it. Ultimately it is them or us.

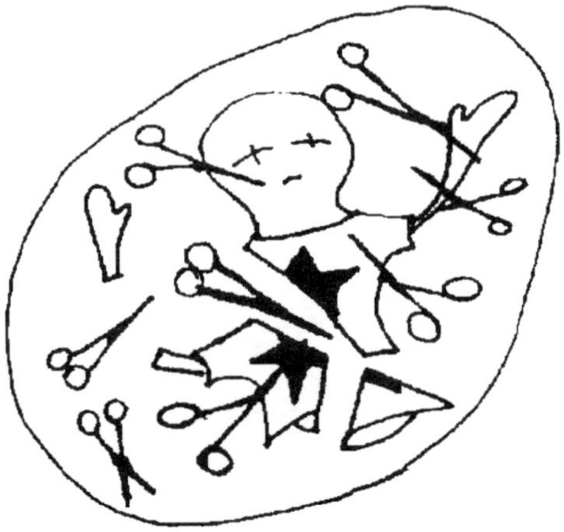

The basic building blocks for most organisms are the same as we use for ourself, so the broken down bits and pieces of organic invaders may be recycled as our own cell nutrients.

NUCLEUS (nuclei in plural)

The **nucleus** is the decision making center of our cell, the person in charge, the board of directors, or the brain of the cell. It contains all the instructions for all the cell's activities (Fig. 1-9). This information is in the form of the genetic material which we get from our parents, half from our father and half from our mother. (In this sense, our parents always control us, as their parents did them.) In a fertilized egg, the genetic materials of the sperm and egg fuse to form the nucleus of the cell.

'Nucleus' means kernel. The nucleus of a cell is like a single kernel near the center of the cell. Its genetic material is wrapped in a phospholipid sac.

• Genetic material is stored in extremely large molecules called **chromosomes.** Each human nucleus contains 46 chromosomes. The chemical format for chromosomes is called **DNA,** the initials of DeoxyriboNucleic Acid.

DNA is shaped like a twisted ladder. The side-pieces of the ladder are a chain of phosphorus and sugars. The cross-pieces are bases in a very specific sequence (Fig. 1-10). Every three bases is the pattern for one amino acid (Fig. 1-12), so the exact sequence of amino acids to build a certain protein is specified by the DNA base sequence. Each protein has a differect and very specific base sequence.

Thus, in each cell, genetic material is translated in specific proteins. Proteins form an important part of the structure of many body tissues. Proteins are also enzymes which catalyze or stimulate all the chemical reactions or processes in our cell and therefore control our cell's activity.

In this way the nucleus controls the cell. If the nucleus wants more of a certain activity, it causes the cell to make more of the protein enzyme which will enhance that activity.

Figure 1 - 9
The nucleus contains the instructions for the cell's activity.

We all begin as a single cell, a fertilized egg. The first step of every cell division, making our billions of cells, is to duplicate the nucleus so that each cell gets all the genetic material. Each cell uses only what is appropriate for its own function, but contains the instructions for all the different kinds of cells in our body. This is what makes cloning possible. This is how the cells in the amniotic fluid can tell us about a baby's genetic makeup.

Figure 1 - 10
The structure of DNA.

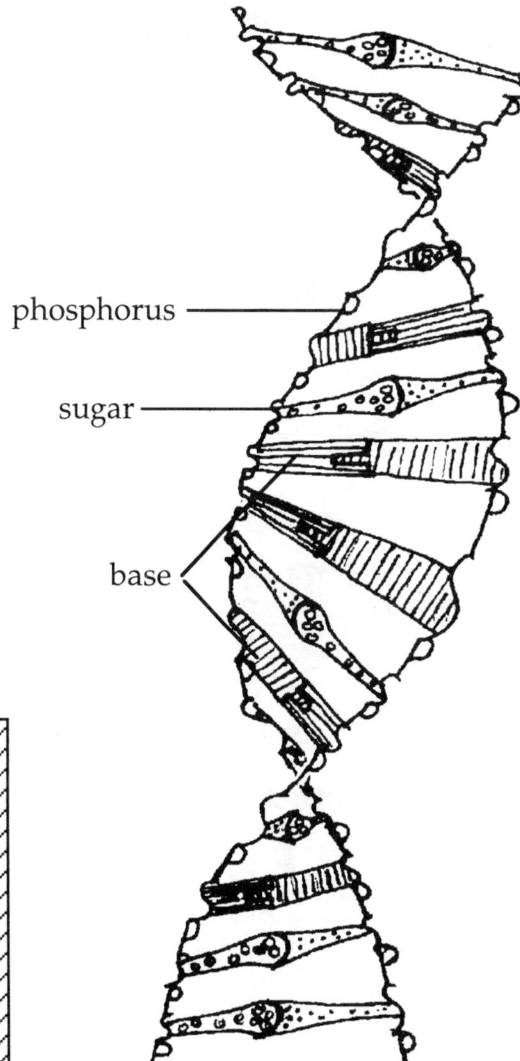

The pattern for any one protein is a tiny speck, but there are millions of proteins. The patterns for as many as 10 million proteins are included in one chromosome.

> In cancer, the nuclear instructions of the cell are broken or changed and therefore the cell's structure and behavior is changed. In reference to the nucleus as the brain or mind of the cell, one might say that the cancer cell is deranged or has gone insane.

RNA AND RIBOSOMES

While chromosomes never leave their central headquarters, proteins are made elsewhere in the cell. The nucleus has a busy copy shop. It makes copies of its protein patterns.

Figure 1 - 11

A part of the chromosome is being copied.

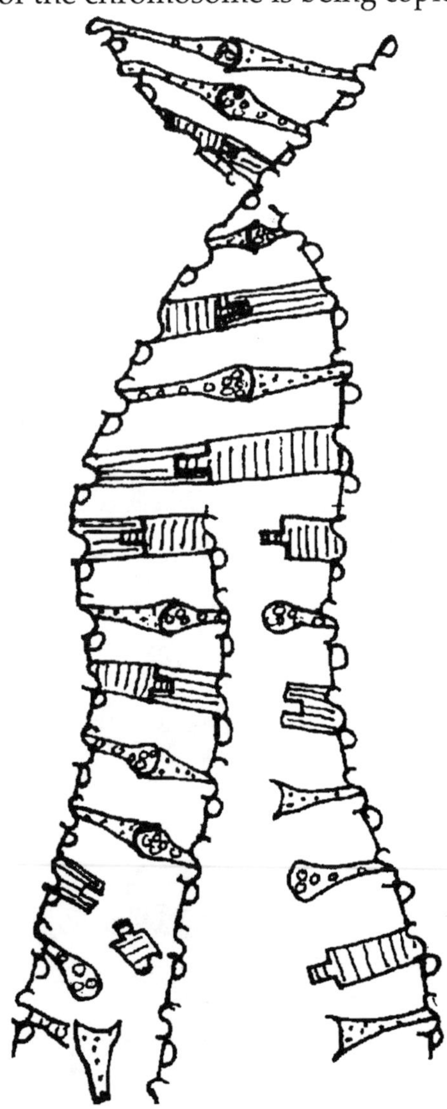

(Fig. 1-11). Any part of the whole chromosome can be copied over and over again, like making a bunch of copies of one page of this book.

It's like the dentist making an impression of a tooth so an exact copy of that tooth can be made. The impression is sent off to the laboratory where the tooth is actually made. This can be done over and over again.

The copies of the protein patterns have a different sugar than chromosomes so they are called **RNA,** the initials for RiboNucleic Acid. One RNA molecule usually would contain the pattern for only one protein. The RNA leaves the nucleus to go to the lab of the cell where the proteins are made.

- **Ribosomes** are special RNAs which leave the nucleus with the protein patterns to help the cell make the protein from the pattern.

Figure 1 - 12

The amino acid sequence for a certain protien is specified by the base sequence of the RNA.

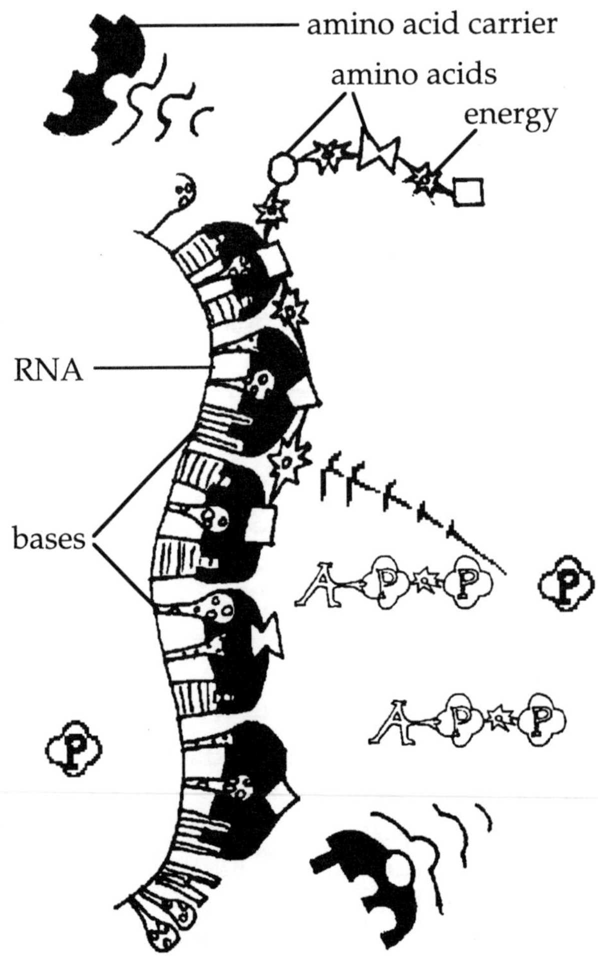

ENDOPLASMIC RETICULUM

A 'reticula' is a weaving or net. 'Endo' means within. 'Plasma' is fluid, the implication in this case is our cell's fluid. So our **endoplasmic reticulum** is a network within our cell's fluid. It connects with the cell membrane, the nucleus, and many other organelles (Fig. 1-13). The cord of this net is a tubule made of the phospholipid sandwich.

• The endoplasmic reticulum is the communication and transportation system of this organization. The telephone wires and streets of our cell are combined. Messages from the cell membrane can travel along the tubules to all parts of the cell, spreading the word about what is happening or what to do next. Messages travel in the form of changing electrical potentials across the phospholipid wall of the tubule. Molecules can travel within the tubules to and from all parts of the cell.

• The endoplasmic reticulum also forms the lab or a workbench for building many proteins and fats. Proteins are made of long, tangled strings of amino acids, specified by the RNA patterns. The outer surface of the endoplasmic reticulum forms a huge surface area on which ribosomes can spread out their patterns to facilitate the arranging and bonding of amino acids (Fig. 1-12). These ribosomes prefer to sit at a table to string their beads rather than float in the middle of a teeming lake.

> You may wonder why a microscopic structure like a cell needs its own intercom or shuttle bus. Why can't the cell membrane just whisper the message and everyone hear? Why can't an amino acid take a few steps across the room to the protein work bench? When we put things into size perspective, the answer is obvious. An amino acid in a small cell is like a person in a city 4 miles long and 4 miles wide, where all its buildings are super skyscrapers 4 miles high. To do business on the top floor across town, a phone or shuttle service sounds like a good idea to me. A sodium ion is like a fly finding its way in the same city.

Figure 1 - 13

The endoplasmic reticulum is a tubular network within the cell.

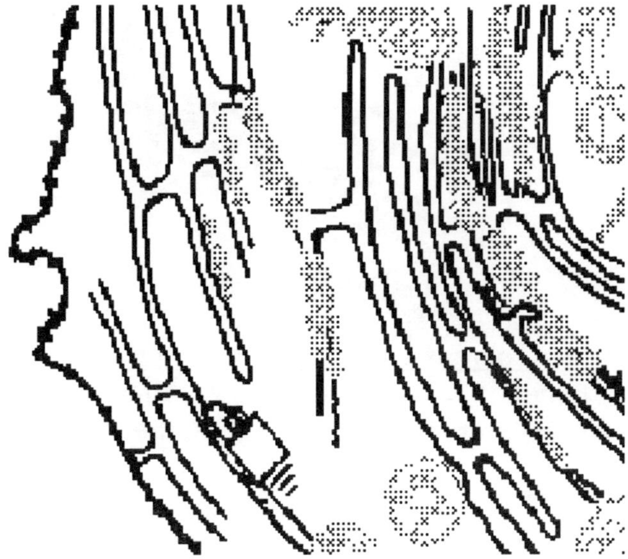

MITOCHONDRIA (mitochondrion in singular)

'Mitos' means thread and 'chondros' refers to cartilage. For someone in authority, the **mitochondria** in our cell looked like a thread of cartilage. Mitochondria are actually short and stubby for a thread, more like a fat sausage (Fig. 9-14). Their length is only two to three times their width. In width, 200 mitochondria could stand shoulder to shoulder across a small cell. In our 4-mile-wide cell, a mitochondrion would be the width of two city lots (100 feet).

The following is a non-literary approach to link the name and function of mitochondria. 'Mit' stands for might, power and energy. Making energy available for all cell activity is the job of our mitochondria.

Mitochondria have a double wall, each one being made of the phospholipid sandwich. The outer wall is the smooth sausage shape. The inner wall is extensively folded (Fig. 1-14), creating a large surface area for the work of our mitochondria.

• All cells of our body consume energy in all their active functions, whether it is bonding amino acids together to build proteins (Fig. 1-12), transporting a substance across the cell membrane (Fig. 1-5), or communicating a message throughout the cell. The more active the cell, the more energy it uses.

Figure 1 - 14
A mitochondrion.

• There is energy in any kind of bonding, whether it is a mother to her infant or two papers stapled together. In the chemical world, energy is stored in the form of chemical bonds. Energy bonds atoms together to make molecules. If the atoms are pulled apart, that energy is released.

Most organizations do not have a mint, they cannot print their own money. They only can cause existing money to be transferred into their organization. As they buy goods or services, they continue to circulate the money.

In the whole universe there is no mint. Energy and matter can never be made or destroyed. Energy and matter only can be transformed into each other or transferred from place to place. We have to snitch energy from someone who has it, just as everyone at a party is energized by the life of the party.

In animal cells (including humans), **glucose** is the life of the party, the stalwart of energy transference. Glucose is a simple sugar, shaped like a Christmas wreath or tambourine. Its core is five **carbons (C)** and one **oxygen (O)** bonded together in the shape of a circle. One more carbon and bunches of **hydrogen (H)** and oxygen sprig outwards, ornamenting the circle (Fig. 1-15). The chemical formula for glucose is $C_6H_{12}O_6$. Energy holds all the atoms together to make the glucose molecule. When any of the atoms are pulled apart, energy is released.

Our cell makes energy available for its own purpose by pulling apart the carbon ring of glucose. Our cell snatches the energy released from the carbon to carbon bonds and uses it for its own functions.

• Mitochondria are the structures in the cell which have the enzymes to open the carbon ring. Oxygen is a necessary assistant and is consumed in this process (Fig. 1-17). This is the major use of oxygen in our body.

While there are many other simple sugars, glucose is the most common simple sugar in our blood. Blood glucose levels are often referred to as blood sugar levels.

When glucose is not available, many cells will pull apart proteins and/or fats for the energy to keep going. Generally the waste products of this process are harder for our body to get rid of, but that is better than no energy at all.

Our nervous system is the most dependent on glucose for energy. It cannot use any alternative. This is why low blood glucose causes drowsiness and very low blood glucose can lead to a coma. There is not enough energy for the brain to function.

Figure 1 - 15
A glucose molecule with energy ( ) bonding together its six carbons (C), six oxygens (O), and twelve hydrogens (H).

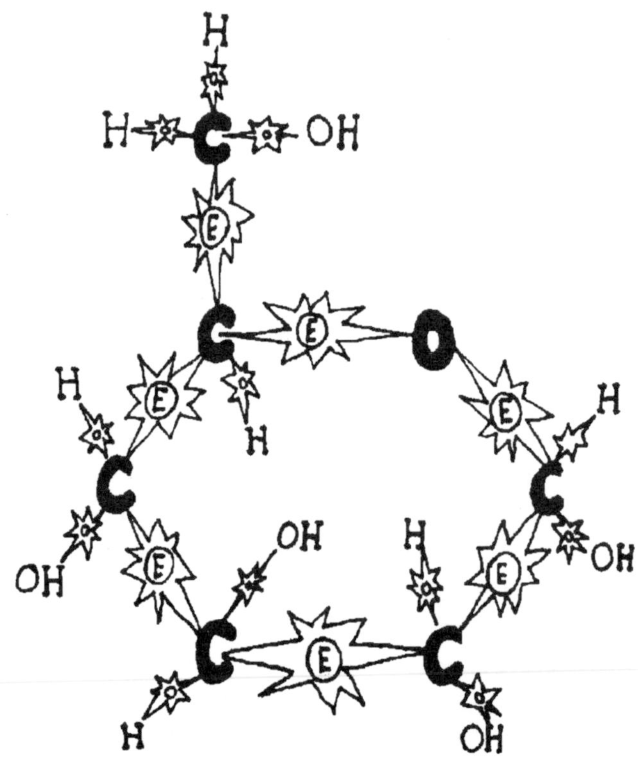

• Approximately 75 percent of the energy from opening the carbon bonds escapes in the form of heat (Fig. 1-16). This heat is our internal fire which keeps us warm. The other 25 percent is captured in a chemical bond that any part of our cell can open and use. This bond attaches a third **phosphate (PO_4)** onto **ADP** (Adenosine Diphosphate) to make **ATP** (Adenosine Triphosphate) (Fig. 1-16). When the cell needs energy, it removes the third PO_4 and uses that energy, leaving ADP and a separate PO_4 (Fig. 1-12), which can be reattached with more energy from glucose. The PO_4 just keeps hopping on and off according to the energy availability and needs in our cell.

• In this way, every cell is responsible to acquire the energy it needs for its own maintenance and function. ATP cannot be transferred from cell to cell.

• After our mitochondria have done their number on the glucose, separating all the carbon bonds, the leftover products are **carbon dioxide (CO_2)** and **water (H_2O).** Our body is able to deal easily with both these molecules. The CO_2 is a waste that has no other use and is breathed out from our lungs. The water is added to the whole pool of water in our body. If too much water collects, our kidneys dump the excess into our urine (Fig. 1-17).

• Mitochondria can replicate themselves. If cells get busier, more mitochondria are made to keep an adequate flow of energy in the cell.

Figure 1 - 16

Some of the energy ( ) from glucose can be used to attach another phosphate ( ) onto ADP. Some energy escapes in the form of heat ( ).

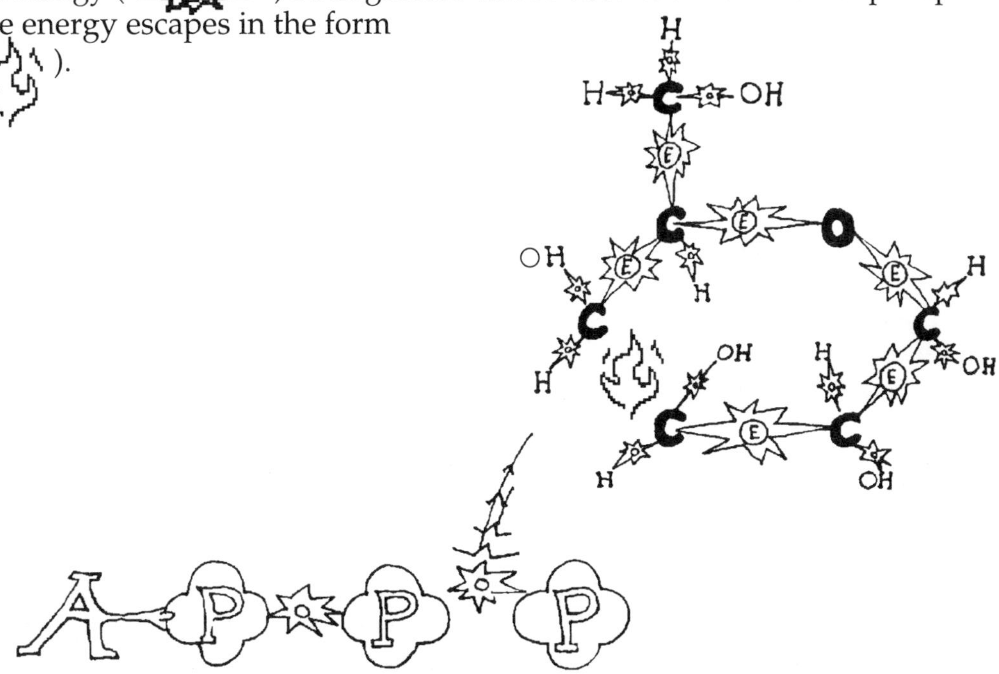

> ATP is relatively unstable. If the energy is not quickly put to use, the third PO_4 just falls off and its energy becomes heat. Before this happens the third PO_4 and its energy may be bonded to a **creatine** molecule. This bond is more stable and stores longer in our cells.
>
> If a cell has lots of ATP, it does not need to break apart all the incoming glucose. Glucose may be stored in the form of **glycogen**, a stash in case of hard times.

> It sounds like our cells need lessons in fuel efficiency, putting only 25 percent of their energy to good use. This is only 5 to 10 percent less than our modern, high performance engines, which lose so much energy to heat that they would burn human flesh. Of course, if we ran 5 miles up hill in 5 minutes, we also might be too hot to touch.

When oxygen is not available, our cells can begin to pull apart glucose, but it cannot complete the process. The resultant waste product is **lactic acid.** If too much lactic acid builds up in a cell, it interferes with the cell's normal chemical activity. This disruption injures the cell, leading to inflammation and pain.

Lactic acid sometines is called an **oxygen debt.** It allows our cells to get some energy out of glucose, but the only way to get rid of the lactic acid and its toxicity is to add oxygen back to it.

If oxygen becomes available to an indebted cell, it may pay its own debt, continuing to pull apart the lactic acid into less harmful wastes.

Lactic acid also may move out of the cell where it was made into the blood. It circulates throughtout our body to become some other cell's problem. While the cells of our body cannot release energy for each other, they do share each other's debts. Hopefully any cell which takes in the lactic acid will have the oxygen to break it down. In this way the whole body can pay the debt for one overworked muscle or tissue.

Performing gentle movements after heavy work helps to distribute the lactic acid of one tissue throughout our whole body. This is why atheletes do cool down exercises after rigorous events.

From whom does glucose snitch its energy? Who gives up energy to bond carbons together to make glucose? On this planet, the **sun** is the life of the party radiating its energy out to us.

Glucose and all carbohydrates have their beginning in the plant kingdom. Plants have it easy. They do not have to hunt down animals or dig and cut at other plants for food. They never feel guilty about killing a life form for their own survival. Plants are vulnerable in another way. They are immobile and depend on the weather to bring them what they need.

Plants absorb CO_2 from the air and H_2O from the ground. Then they hang out in the sun for what looks, on the outside, like a lazy day. Inside, it is actually very busy. Plants have the tools to snip off bits of the sun's energy and use it to bond CO_2 and H_2O together. It takes six of each chemical to make one glucose molecule or other similar sugars. Six O_2 are left over after the process and breathed out by the plant (Fig. 1-17). Plants use these sugars and more sun energy to build their carbohydrate structure.

Then we, or some other animal, come along, tear up the plants and munch them down. We digest the carbohydrates in our digestive canal, convert all the simple sugars into glucose, and absorb it into our blood. The glucose moves from our blood into our cells. There, with the help of O_2, our mitochondria take apart the basic building blocks of glucose, making the energy of those bonds available for our cell's activity, leaving behind a waste pile of CO_2 and H_2O (Fig. 1-17).

So in the end, we operate off the sun's energy, cycled through the plant kingdom. When we try to use the sun's energy directly, the best we can do is get a nice tan, but too often it is a lobster-red burn.

Animals are dependent on plants to capture the sun's energy and release O_2 for their function. Plants are dependent on animals to renew the supply of CO_2 and H_2O to build new plant structure. This is one of the perfect circles or complete cycles in nature. Plants and animals both survive on each other's products and leftovers. We all depend on the sun for its constant input.

Figure 1 - 17

Animals depend on plants to capture the sun's energy and release O_2, while plants depend on animals to renew the supply of CO_2 and H_2O. This is one of the perfect circles in nature.

GOLGI APPARATUS

The **golgi apparatus** is our cell's center of benevolence, making substances which are given away for the benefit of outsiders. Many organizations have no equivalent, but in our bodies, all our cells work for each other and know that their own well being depends on it. While most cells have a golgi apparatus, it is most developed in cells which secrete products into other tissues.

Our golgi apparatus is named after the Italian histologist who first spotted the little contraption squatting near the nucleus. Our golgi apparatus is like a stack of flattened sacs (Fig. 1-18). The walls of the sacs are made of the phospholipid sandwich and contain the enzymes to build complex carbohydrates and combine them with protein. These molecules are called **glycoproteins** and usually are in secretory products such as mucus. Our golgi apparatus also may combine lipids and proteins, called **lipoproteins,** in preparation for secretion from our cell.

When materials are ready for secretion, they push into a corner of a golgi sac. This corner with its contents pinches off and floats away to the cell membrane where the contents can be extruded from the cell (Fig. 1-21). This is how our pancreas makes and secretes its digestive enzymes, and our salivary glands their saliva.

Figure 1 - 18

The golgi apparatus sitting near the nucleus.

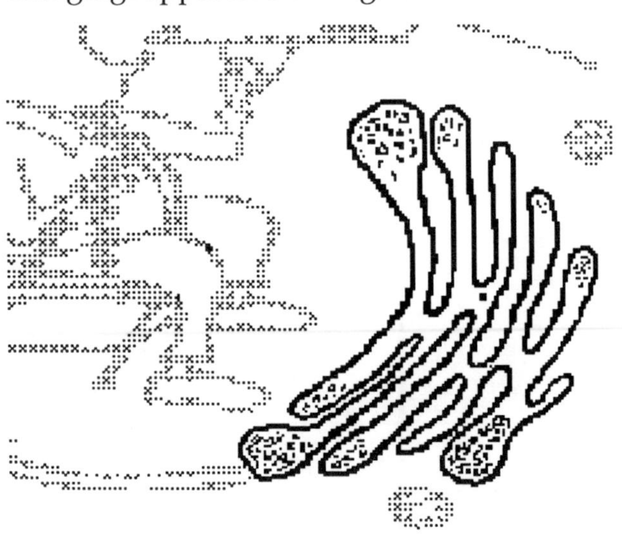

CELL MEMBRANE TRANSPORT

We have said that our cell membrane forms a semipermeable barrier between our intracellular and intersititial fluids, allowing certain movements in and out of our cell. These movements are called **transport** across the cell membrane. There are two kinds of transport: **active** and **passive** (Fig. 1-19). These words mean the same in their physiological usage as in ordinary conversation.

If we are actively involved with something, we work to make it happen, we put time and energy into it. Our suggestions and ideas influence the direction of the project and give us some control or say over what happens.

If we are passive in a situation, we put in no time, no energy, no ideas, and have no control. We just go along with whatever happens or let it happen to us without lifting a finger. When cells transport actively or passively, the same principles apply.

ACTIVE TRANSPORT

Active transport, in either direction across our cell membrane, consumes cell energy. Our cell must have glucose, oxygen, and functioning mitochondria.

In exchange, our cell can control the transport. Our cell governs what is transported and in what direction, regardless of other conditions such as molecular size, concentration gradient, electrical potential, and so forth. There already could be one thousand potassium ions inside the cell and only five outside, but the cell still could actively bring those five into itself (Fig. 1-19).

There are four mechanisms of active transport: the **carrier proteins, phagocytosis, pinocytosis,** and **exocytosis.**

Carrier Proteins

We already have discussed the **carrier proteins** present in our cell membrane (Fig. 1-5). As long as they are paid for their work, (i.e., are given the energy they need) they will attach to molecules and bring them across our cell membrane, in or out, according to our cell's demand.

Figure 1 - 19

(a) Active and **(b)** passive transport acoss a cell membrane.

(a) **(b)**

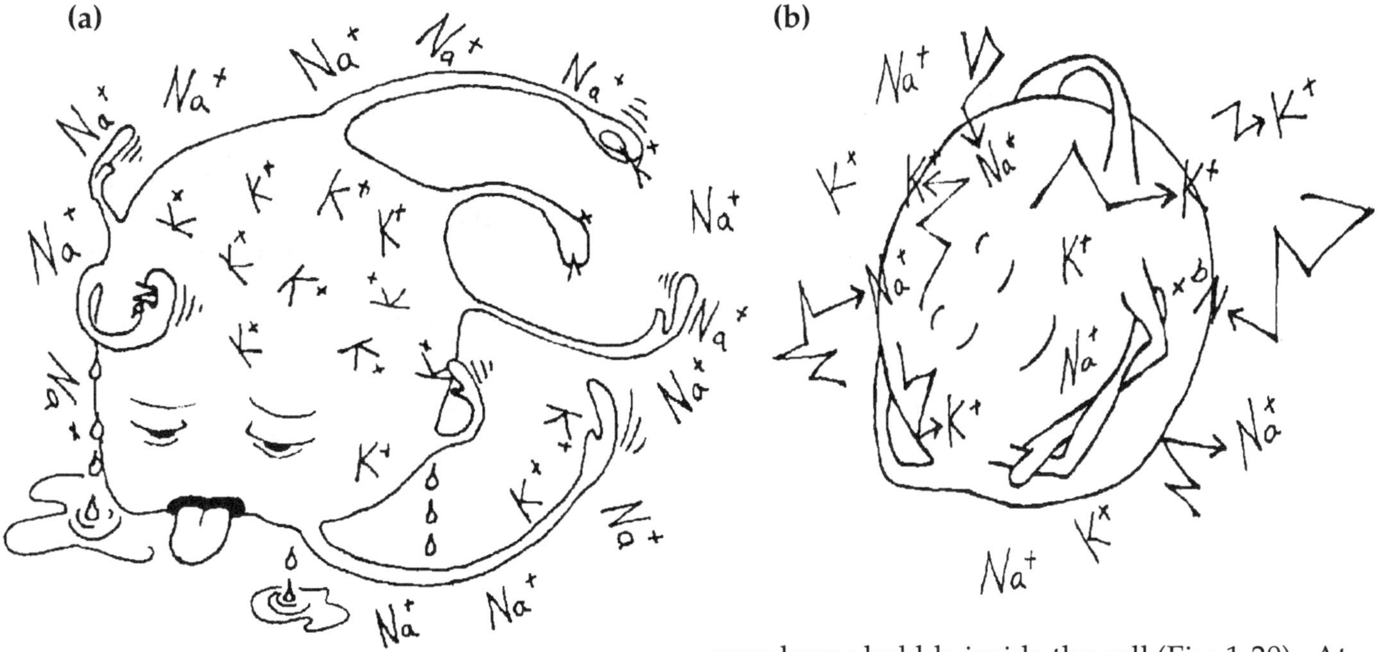

Phagocytosis

The word **phagocytosis** is made of three word parts, 'phagein' means to eat, 'cyt' refers to cell, and 'osis' means the condition or process of. So literally translated, phagocytosis means the process of a cell eating. A cell eats much like we do. When it spies a tasty morsel close by, it reaches out, projecting its cell membrane in the shape of a cup, like a big open mouth which surrounds the morsel (Fig. 1-20). The handy thing for the cell is that it can open a mouth anywhere on its surface. The lips of the mouth close over the morsel, which then is completely contained within a cell membrane bubble inside the cell (Fig. 1-20). At that point phagocytosis is complete and the morsel is at the mercy of our cell. What the cell does depends on what kind of cell it is and what the morsel is.

Pinocytosis

'Pinein' means to drink, so **pinocytosis** is the process of a cell drinking. The engulfing process is similar to phagocytosis, only fluids or small particles are taken in, the mouth is smaller, and the lips do not project as far. It is more like the cell membrane folding over what is against its outer surface. Our cell takes a dainty sip instead of a big chomp.

Figure 1 - 20

The process of phagocytosis.

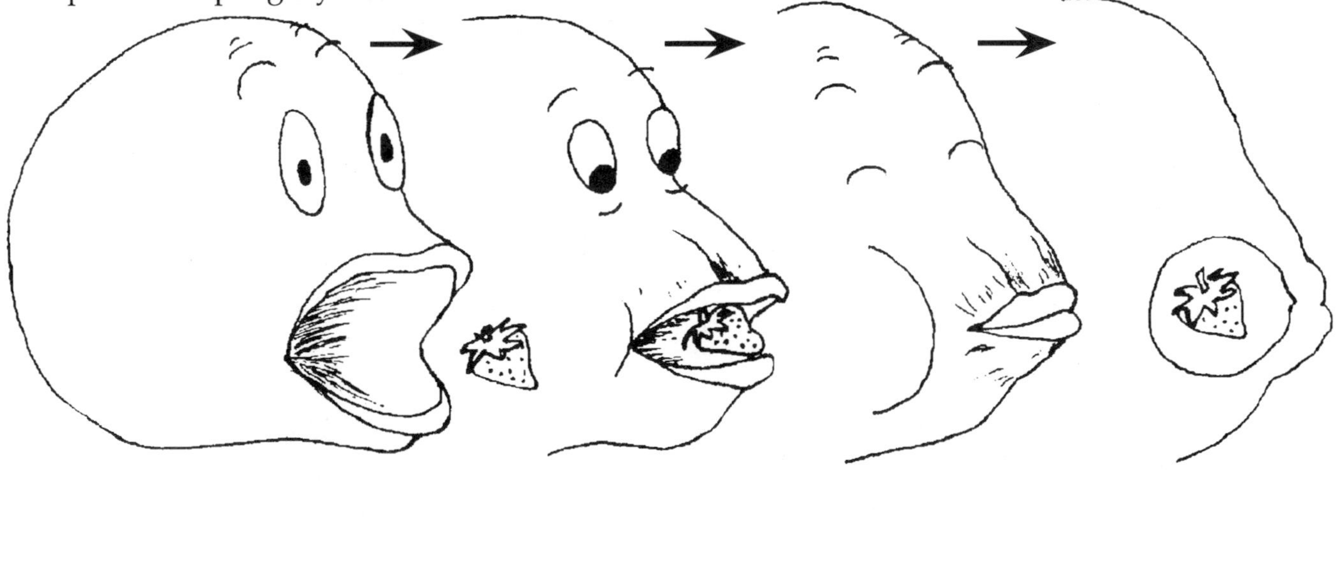

Exocytosis

Phagocytosis and pinocytosis transport into our cell. Both these processes can be reversed to spit things out of our cell. This is called **exocytosis,** 'exo' meaning outside. Products made by our cell for export can be wrapped in a phospholipid bubble and floated off to touch our cell membrane (Fig. 1-21). At the point of contact, the bubble and cell membrane fuse. Fusion is easy since they have the same phospholipid structure. The bubble opens flat to become part of the cell membrane. The bubble contents then are outside our cell. This is how our golgi apparatus secretes its products.

The bubble may also serve as a garbage bag when the cell has wastes to dispose of.

All these manipulations of the cell membrane consume cell energy and are therefore active transport by the cell.

Figure 1 - 21
The process of exocytosis.

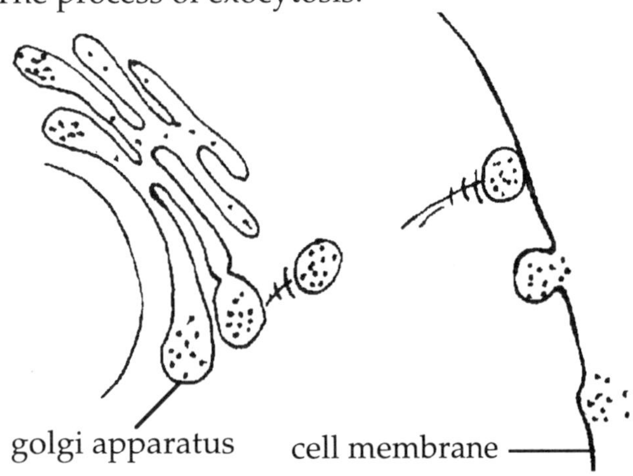

> When I draw cell organelles as lines on a page, it gives the sense of a static or rigid cell. This is far from true. I have already hinted at the flexible nature of cell organelles. Since they all are made of the phospholipid sandwich, a single segment of phospholipid can become the structure of any organelle in the cell. At one moment it may be a golgi sac, then fuse into the cell membrane, then engulf a pinocytosed bubble, then join a lysosome as its contents are digested, and so forth. Within an active cell, there is a constant writhing exchange.

PASSIVE TRANSPORT

Passive transport, in either direction across a cell membrane, requires nothing from the cell (Fig. 1-19). A cell could be dead and the movement still would happen. The energy and control for passive transport come from something other than the cell. The cell cannot help it or stop it. The cell cannot speed it up or slow it down. Whatever happens happens. In passive transport the cell's hands are tied. If the cell does not like what happens, its only option is to use its own energy to undo it.

There are four kinds of passive transport across the cell membrane: **diffusion, osmosis, facilitated diffusion,** and **filtration.**

Diffusion

In **diffusion,** the energy for transport comes from the atom or molecule that is moving.

Built into any atom or molecule is a vibrational energy called **kinetic energy.** Kinetic energy gives the atom the jitters, the heebie-jeebies. The atom never can quite sit still. Even in solid material, atoms and molecules vibrate. This vibration is the molecule's source of energy for diffusion across our cell membranes.

If we consider a molecule jittering about in the vicinity of a cell, there is a random chance that it bumps into the cell membrane. It then may bounce away from the cell, or it may hit at just the right angle to skip between two phospholipids, through the membrane, and into the cell (Fig. 1-22). The molecule then is transported into the cell by its own kinetic

Figure 1 - 22
When a molecule bumps into the cell membrane, it may bounce away or skip through.

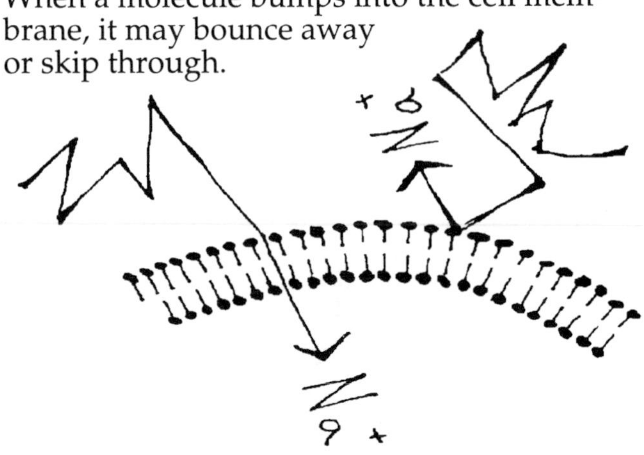

energy with no cellular input. This is called diffusion.

Our cells may benefit from diffusion. A nutrient, like oxygen, may just happen to skip in. On the other hand, diffusion can be a bother for our cells. Things the cell has no use for may move in and clutter the space, or even worse, interfere with efficient cell function. Pleasure or pain, the cell can like it or lump it, but the cell cannot change the process of diffusion. The molecule is putting out the effort, so the molecule makes the decisions.

> The kinetic energy of substances in a **solid state** is low, so the movement is quite small. The molecules do not push each other very far apart. That is why the substance appears solid. If we try to pass our hand through a solid object, we come up against the compact collection of vibrating molecules. There is not enough space for us to pass between them, except of course, if we are Casper the Ghost.
>
> It is possible to pass solid objects through each other when there is clear understanding of the object's internal molecular patterns of movement. This may explain phenomena like psychic surgery.
>
> In a **liquid state** the energy is increased. We add heat energy to melt a solid into a liquid. The molecules are more rambunctious and need more elbow room. We can sweep our hand through water. As we move, we feel the many water molecules bumping against us, but there is enough space for them to move over and let us through.
>
> In a **gaseous state** the energy is the greatest. We add more heat to evaporate a liquid into a gas. The vibration of a gas is the largest, requiring the greatest space between molecules. As we sweep our hand through the air, it passes through the space between the molecules. Our hand is bombarded by so few molecules, we can hardly feel them.
>
> Our body is mostly in a semi-solid or liquid state with significant vibration of all its molecules.

• Kinetic energy is the impetus of diffusion, but there are certain things which limit the potential for diffusion. Certain conditions must be present for diffusion to occur.

The structure of the cell membrane is a major limiting factor. Its structure determines to some extent who goes. This is the only control that cells have over diffusion.

Lipid soluble molecules have the most freedom across the phospholipid cell membrane. A lipid soluble molecule of any size can dissolve in phospholipid and pass through it.

Water soluble substances will not dissolve in the cell membrane. In order to diffuse, they need to be small enough to fit through the tiny spaces between the phospholipid molecules. Examples of water soluble molecules which diffuse are water, oxygen, carbon-dioxide, sodium, and potassium. Glucose and amino acids are too big under ordinary circumstances (Fig. 1-26a).

If the structure of the membrane changes, then who can diffuse through it may also change.

• A specific kind of molecule may be simultaneously diffusing in both directions across a cell membrane. The most popular direction of movement is given a positive number. The opposite direction is given a negative number. The sum of the two numbers is the **net direction of diffusion** of that molecule. It is comparable to total income minus expenses equals the net income.

The net direction of diffusion is a matter of random chance, determined by the concentration gradient across the membrane. Net movement always goes from an area of high concentration or pressure to an area of low, until both areas are equal (Fig. 1-23).

For example, if there are one thousand sodiums (Na^+) immediately outside a cell, but only two inside, chances are that outside Na^+ will skip in more frequently than inside Na^+ will skip out. From time to time an inside Na^+ may skip out, but the net direction of movement will be inwards, from the area of high concentration towards the area of low concentration.

Figure 1 - 23

The net direction of diffusion is from an area of high concentration toward an area of low concentration.

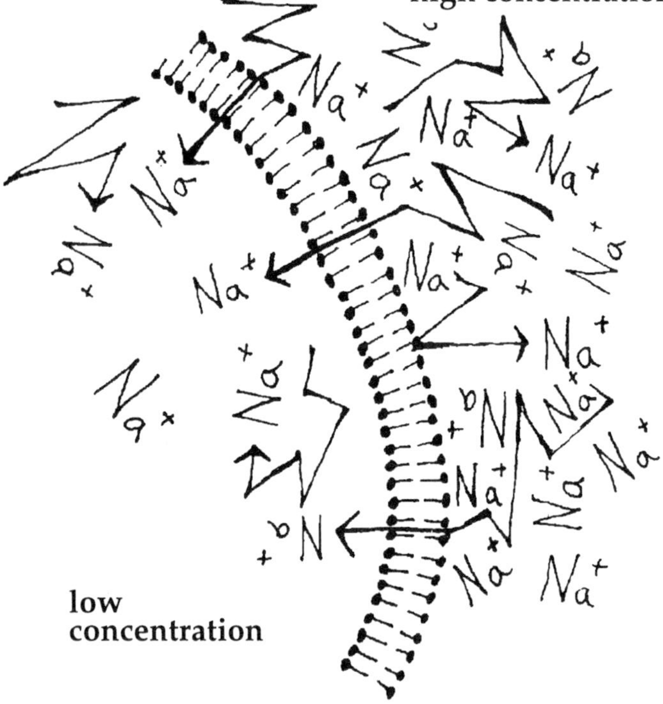

• The **rate of diffusion** is determined by the quantity of the difference in concentrations across the cell membrane. The greater the difference, the faster the diffusion. The smaller the difference, the slower the diffusion. In our example above, when the difference is 1000 - 2 = 998, inward diffusion is rapid.

As diffusion occurs, however, Na^+ is leaving the outside of the cell and collecting inside the cell. After a net of 200 Na^+ have moved in, the difference will be 800 - 202 = 598. Now the difference across the cell membrane is less, so diffusion is slower. The chance that inside Na^+ skips out is increasing while the chance that outside Na^+ skips in is decreasing.

When the concentrations on both sides of the membrane are equal, inward and outward movement is equal. The net direction of diffusion is zero and the rate of diffusion is zero, although movement is still ongoing.

The rate of diffusion is also influenced by temperature. Increased temperature will speed up the vibration and the rate of diffusion, so equilibrium will happen faster. Colder temperatures will slow the whole process down.

Electrical potentials across a cell membrane (i.e., when the electrical charge on the inside and outside of a membrane is not the same) and the electrical charge of a molecule also will influence diffusion. Generally, opposite charges attract each other, while like charges repel each other.

Osmosis

When only water can diffuse across a membrane, this diffusion is called **osmosis.**

All the principles of diffusion apply to osmosis. Kinetic energy is the impetus for the water's movement. The structure of the membrane determines who goes, in this case only water is allowed. The differences in concentration, temperature, and electrical potential determine the net direction and rate of the movement.

In osmosis, water will move toward the area where there is less water. The concentration of other substances may be greater, but wherever water is less, that is where water will go (Fig. 1-24).

Figure 1 - 24

In osmosis, water moves toward the area where there is less water.

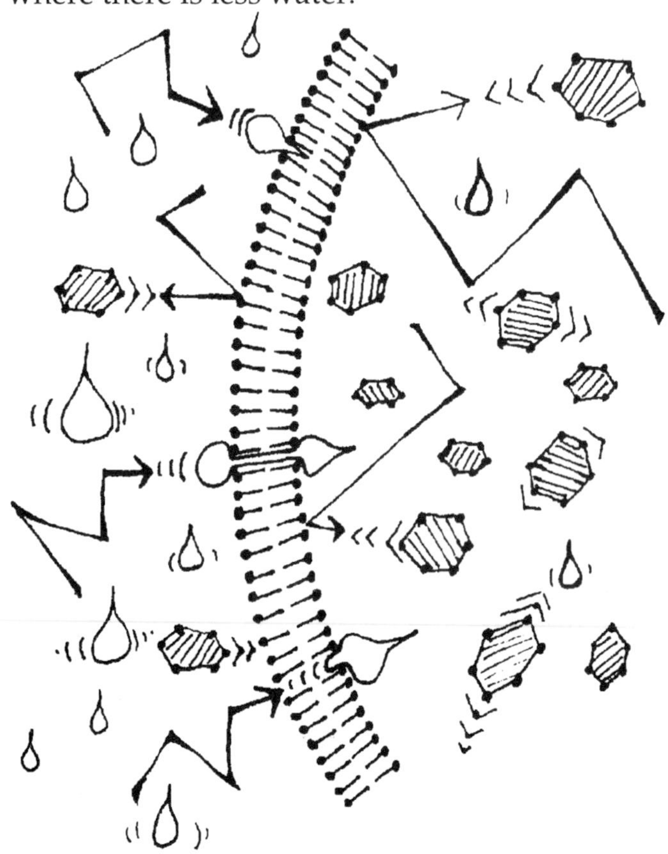

Some molecules have more space or need for water than others. They like to have water around them and will pull it in their direction (Fig. 1-25a). They are terrified of droughts. The force that moves water in this way is called **osmotic pressure.** Na^+, glucose, and plasma proteins are a few molecules which pull water toward themselves, creating an osmotic pressure.

Whenever Na^+, glucose, or plasma proteins travel, they like to keep a gourd of water tucked under their arm. Of course, cell membranes do not always allow this and the molecule may have to go thirsty (Fig. 1-25b).

Figure 1 - 25

(a) Sodium pulls water towards itself, and **(b)** likes to take it along wherever it goes. Sometimes this is not allowed and the sodium may have to go thirsty.

(a)

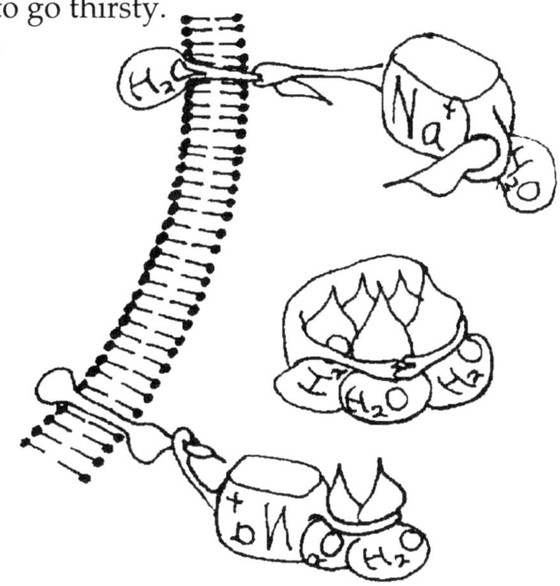

(b)

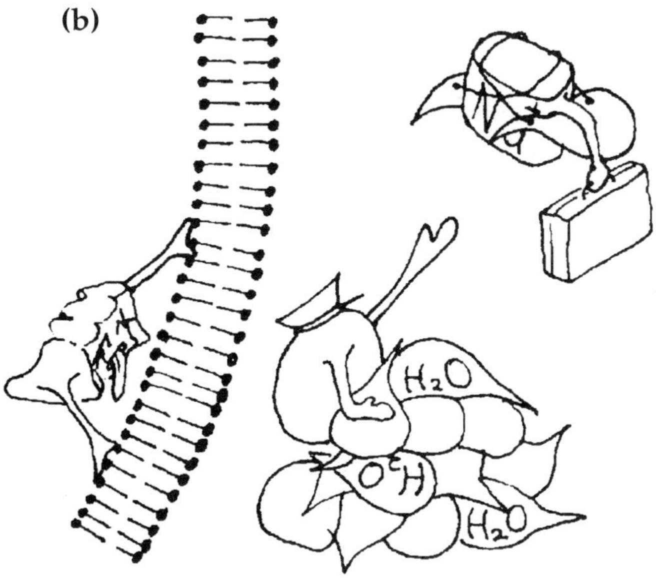

Figure 1 - 26

(a) Glucose is too big to diffuse through cell membranes, but **(b)** insulin facilitates, opening a space that glucose can skip through.

(a)

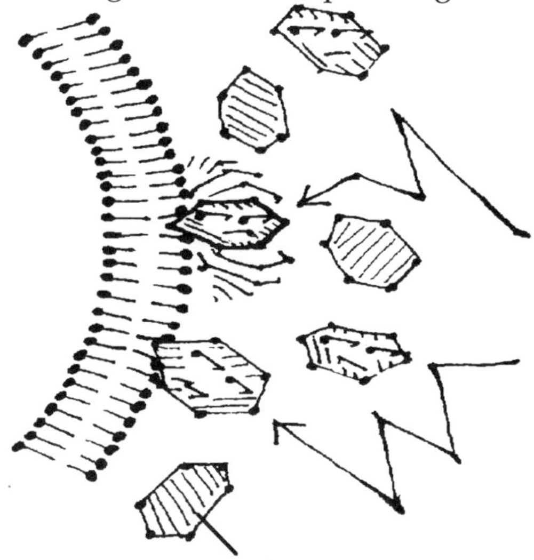

(b)

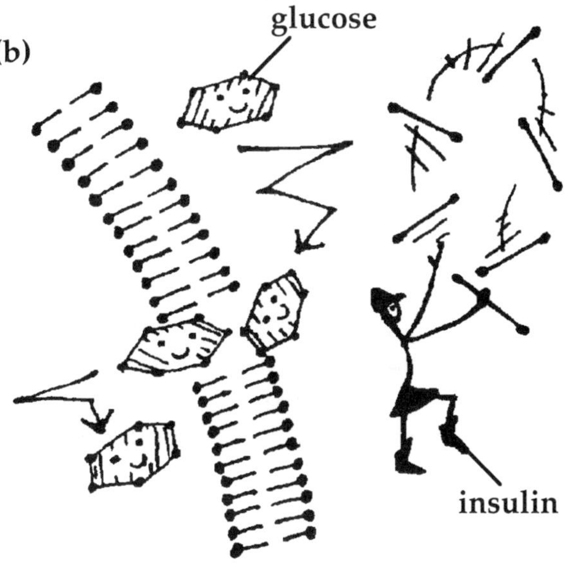

Faclitated Diffusion

Some molecules move across cell membranes by **facilitated diffusion.** All the basic principles of diffusion apply to this transport, but something extra is needed. A particular molecule wants to diffuse but for some reason the cell membrane structure says "Not you!" (Fig. 1-26a). The molecule goes for help. It finds a facilitator who convinces the cell membrane to change its structure to let the molecule skip through (Fig. 1-26b). Each specific molecule has its own specific facilitator, and the cell membrane will only maintain that change as long as the facilitator is present.

Glucose moves into most cells by facilitated diffusion, insulin being the facilitator.

Glucose is a willing martyr for our cell's energy. It tries, of its own accord, to diffuse through our cell's membrane, but is a speck too big. Insulin comes to the rescue, and facilitates by opening the membrane spaces to let glucose skip in (Fig. 1-26).

This is the only way to get glucose into the cells of our body. Cells have no active process for transporting glucose. As our blood level of glucose rises, more insulin is secreted by our pancreas to help the extra glucose into our cells. Without insulin, our blood can be bursting with glucose and our cells can be starving, but there will be no transport of glucose (Fig. 1-27a). This is why **sugar diabetes,** a disease of deficient insulin, can be so catastrophic.

On the other hand, it does not matter how much insulin there is, if our blood level of glucose is low, the diffusion of glucose into our cell will be slow (Fig. 1-27b). Insulin only opens the space. The direction and rate of movement still depends on the concentration gradient of glucose across the cell membrane. The cells cannot help the process. This is why **hypoglycemia,** a state of low blood sugar, leads to marked fatigue.

When insulin is present, glucose moves fastest into cells which have the least glucose since that produces the greatest concentration gradient.

On their own, most atoms or molecules are neutral in their electrical charges. They have equal numbers of positive and negative charges. When dissolved in water, many atoms and molecules are more stable if they either give off negative charge, leaving them with excess positive charge, or take on extra negative charge, leaving them with excess negative charge. Atoms or molecules which are more stable when they are positively or negatively charged in water are called **ions** or **electrolytes.** This is indicated by superscripting + or - behind their names.

Figure 1 - 27
(a) No matter how much glucose there is, it cannot diffuse into the cell without insulin. **(b)** Even when insulin is present, the rate of diffusion of glucose depends of the concentration gradient, low concentration yeilding slow movement. **(c)** It takes insulin and normal blood glucose levels to give the cell the energy it needs.

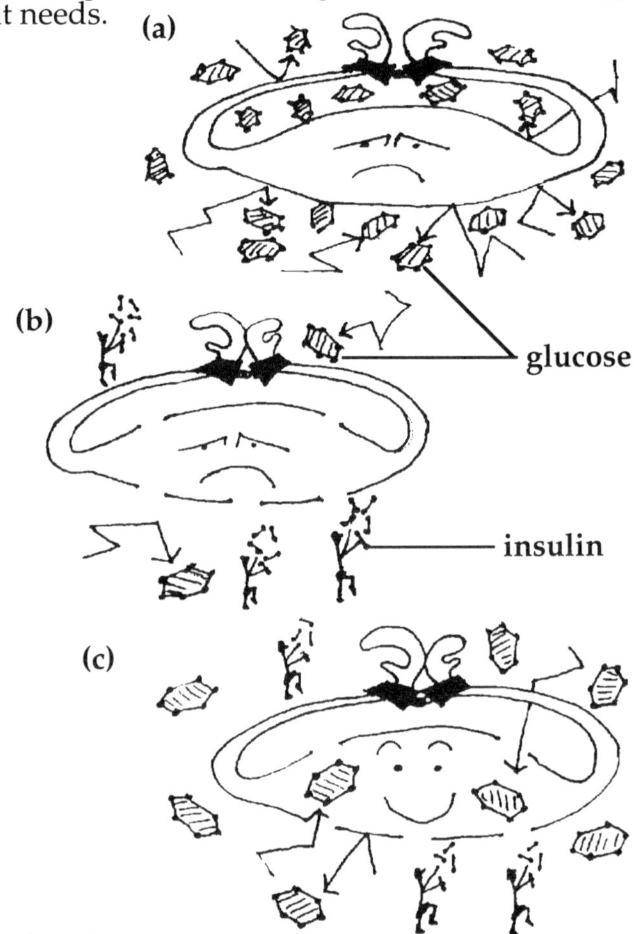

Filtration

When molecules move across a cell membrane by **filtration,** the cell still is passive and the molecules moving still have the jitters, but the main energy for movement comes from some other mechanical pressure such as gravity, osmotic pressure, blood pressure, or the fluid pressure of some other body space. When filtering our morning coffee, the mechanical pressure which causes the movement is gravity (Fig. 1-28a). When making espresso, it is steam. In our body, many nutrients move from our blood into our tissues by filtration, motivated by our blood pressure. The beginning of making urine in our kidneys is filtration, motivated by blood pressure and the osmotic pressure in our kidneys and blood.

Figure 1 - 28

Gravity is frequently the force of filtration. What is filtered changes as the filtering membrane changes, **(a)** being a fine paper and **(b)** being a thin cheesecloth.

(a) **(b)**

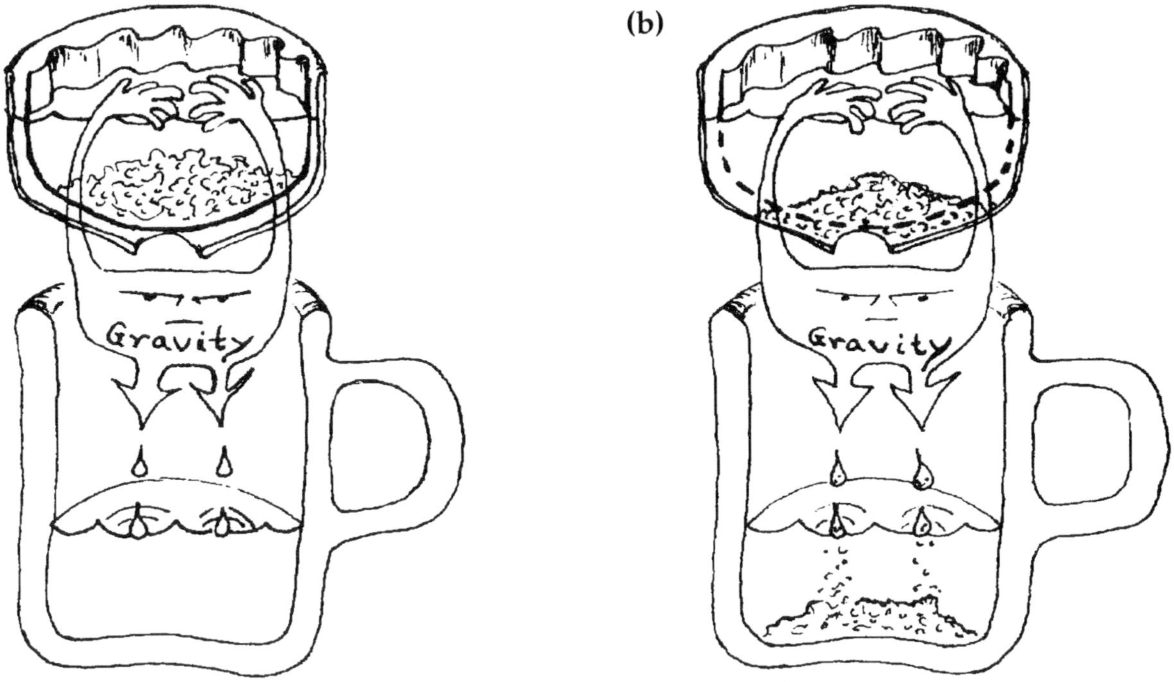

Usually water and many small molecules will move together during filtration. As in diffusion, exactly what is filtered depends on the structure of the membrane through which the filtering takes place. The results will be quite different if we filter our morning coffee through a plastic bag, a fine paper (Fig. 1-28a) or a thin cheesecloth with larger holes (Fig. 1-28b).

SODIUM - POTASSIUM PUMP

Sodium (Na^+) and potassium (K^+) are two ions whose concentrations strongly influence cell function. Our cells work best when the majority of Na^+ is outside the cell and the majority of K^+ is inside the cell. Even though the number of atoms on each side of the cell membrane is not equal, this is the correct balance for our cells. If the number does become equal, the electrolytes are imbalanced.

The problem is that both Na^+ and K^+ diffuse through the cell membrane (Fig. 1-19b). They go where they want, dependent on concentration gradients rather than our cell's likes or dislikes. When the majority of Na^+ is outside our cell, this already tells us that Na^+ will be rapidly diffusing into the cell and similarly, K^+ will be diffusing out. This diffusion disrupts the preferred working environment of our cell.

Since our cells cannot stop the diffusion, they have created an active process of kicking out all the Na^+ that skips in, and dragging back in all the K^+ that sneaks out (Fig. 1-19a), leaving the net movement of Na^+ and K^+ across the cell membrane at zero. This active process is called the **Na^+ - K^+ pump.** Pump is a good word for this endless enterprise since it consumes cell energy just as any pump consumes fuel.

Our Na^+ - K^+ pump is part of the survival maintenance of our cells. If a cell is so low in energy that its Na^+ - K^+ pump falls behind, the subsequent electrolyte imbalance in the cell may cause the cell to burst and die.

Our Na^+ - K^+ pump may briefly speed up or slow down to cause a fleeting electrolyte imbalance. This is the basis of the changing electrical potentials which transmit messages through cells or along nerves.

Providing the energy for the Na^+ - K^+ pump is part of the basic metabolism of our cells and the escaping heat maintains the **basal metabolic temperature** of our body.

Unit II: **THE TISSUES**

A tissue is a collection of similar cells and the substances between the cells. These substances are called intercellular substances, 'inter' meaning between. This collection works to accomplish some special task for an organ in our body.

It is amazing that, while our body has great variety in organs and tasks, it is all made and done by only four different kinds of tissues: **epithelial tissue**, **connective tissue**, **muscle tissue,** and **nerve tissue**. Whether it is crystalline bone, soft, squirmy intestines, or oily skin, it is made of these four tissues. Whether it kills bacteria, bends a joint, or oxygenates our blood, it is made of these four tissues. Explaining the structural and functional characteristics of these tissues, then, summarizes the basic potential of our whole body.

The structure of most organs incorporates a bit of each of the first three tissues listed, epithelial, connective, and muscle. Nerve tissue, on the other hand, touches many organs but does not become a part of the structure of any organ except those of our nervous system, including our nerves, spinal cord, and brain.

In this unit we present the unique characteristics which differentiate the four tissues. In ongoing units we will see how these tissues organize to form the organs of our body.

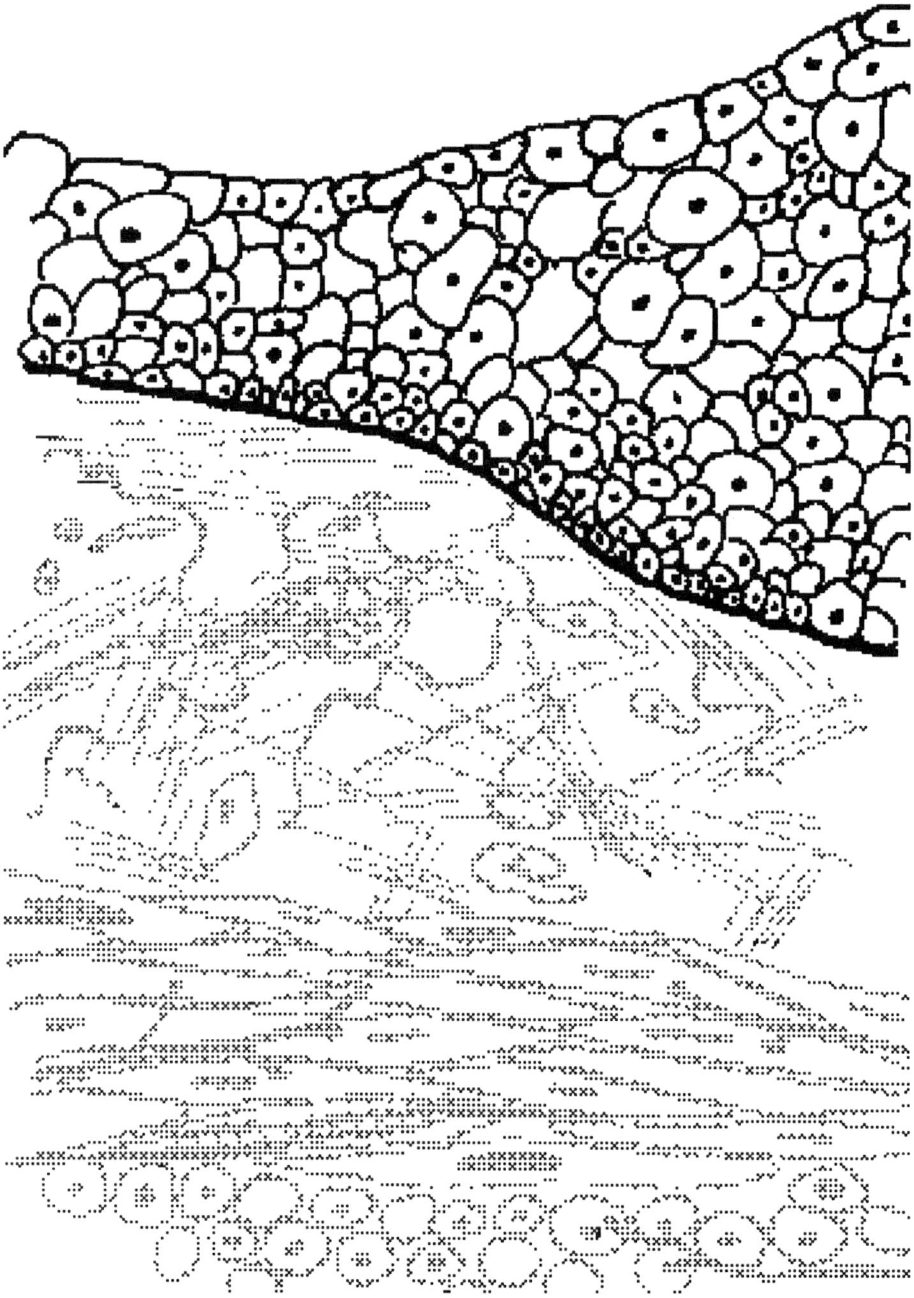

chapter 2: **Epithelial Tissue**

In the chapter title, 'epithelial' is used as an adjective, describing tissue. We also can use the noun form of this word and simply say the **epithelium.** 'Epi' means upon or over and 'thele' means nipple. We will see that epithelium covers our nipples as well as many other surfaces inside and outside our body.

STRUCTURAL CHARACTERISTICS

- The cells of **epithelial tissue** snuggle neatly side by side, touching shoulders with all their neighbors. The space between cells and the amount of interstitual fluid is miniscule. In fact, for a short distance on the touching surfaces of adjacent cells, there is no space between the cells (Fig. 2-1). A short segment of adjacent cell membranes is modified and fused as if the cells shared the same membrane, like Siamese twins sharing a patch of skin. At this point of sharing, something must be either part of the cell's membrane or within one of the cells. There is no space in between.
- Epithelial tissues generally sit on top of some other tissue, usually connective tissue. They are separated from the underlying tissue by a loose mesh, like a cheese cloth or gauze. This mesh is called the **basement membrane** (Fig. 2-1) and is made by the epithelial and underlying cells. It is made out of sugars and proteins.

The epithelial cells in contact with the basement membrane are called the **basal layer** or **basal cells** (Fig. 2-1).

- Blood vessels do not pass from underlying tissues through the basement membrane into epithelial tissue. With no space between cells, how could they? Epithelial tissue does not have direct blood supply and is described as **avascular** (Fig. 2-1). (To be vascular means to have blood vessels. In Latin, putting an 'a' in front of a word negates it, like 'un' in English.)

The avascular condition is perfect for epithelium on a surface exposed to abrasion, like our skin or inside our digestive canal. Injury must be severe enough to go through the epithelium before blood is lost. If we slide along the sidewalk and get a scrape where we can see that a layer of skin is missing but there is no bleeding, the injury was confined to epithelial tissue not penetrating the basement membrane.

Figure 2 - 1
Structural characteristics of epithelial tissue.

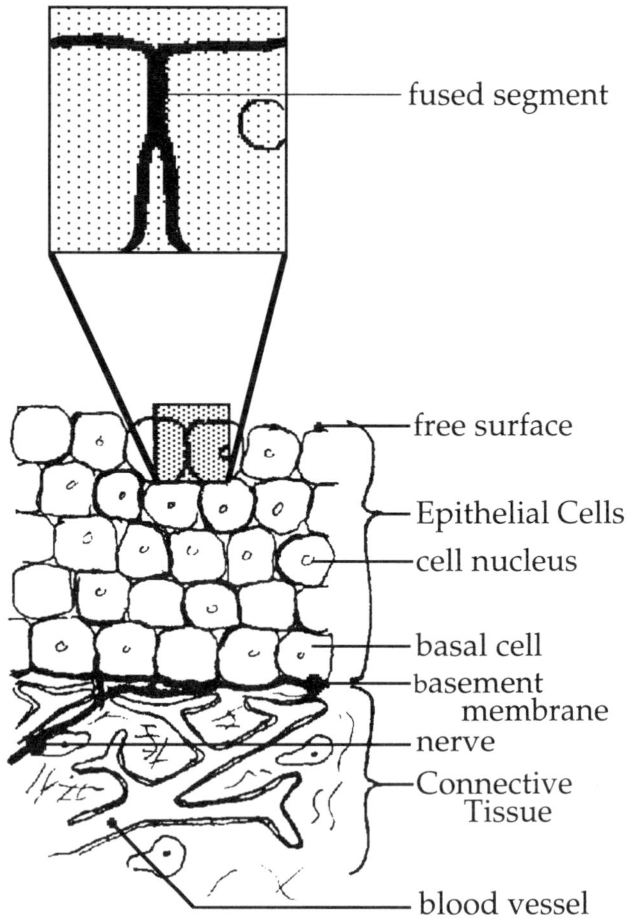

Later we will see the disadvantage of the avascular condition when nourishing epithelium which is many cells thick. In thicker epithelium, the outer cells are a long ways from their blood supply.

> The epithelium lining the inside of the **uterus** takes exception to avascularity. It is vascular, forming the maternal side of the placenta when pregnant.

> In a **skin graft,** it is the epithelial layer which is transplanted. Since it never has a direct blood supply, there is no worry of connecting blood vessels. The epithelium is simply laid over the underlying tissue. It is familiar with a distant blood supply.

• A few, but only very few, nerves from underlying tissues (mainly carrying the sense of pain) may penetrate the basement membrane into epithelial tissue. They go only as far as the bottom of the basal cells (Fig. 2-1). This is how we can be without pain as we thread a needle through superficial skin or pick open the surface of our skin to catch the end of a sliver.

CLASSIFICATION

• Functionally, epithelial tissues can be divided into two kinds: **1.** flat sheets of cells that line the outer or inner surfaces of organs or body cavities, and **2.** globular, pouched, or branching collections of cells that form secreting glands. Glands vary in size from a single cell to large collections of cells (Fig. 2-2). Small glands may be located within a sheet of epithelium.

• Structurally, there are different kinds of epithelium depending on their **1.** thickness, and **2.** cell shape. Any epithelium is described in both these ways. Table 2-1 illustrates the different kinds of epithelium and indicates where they are located in our body.

Figure 2 - 2

Examples of epithelial glands.

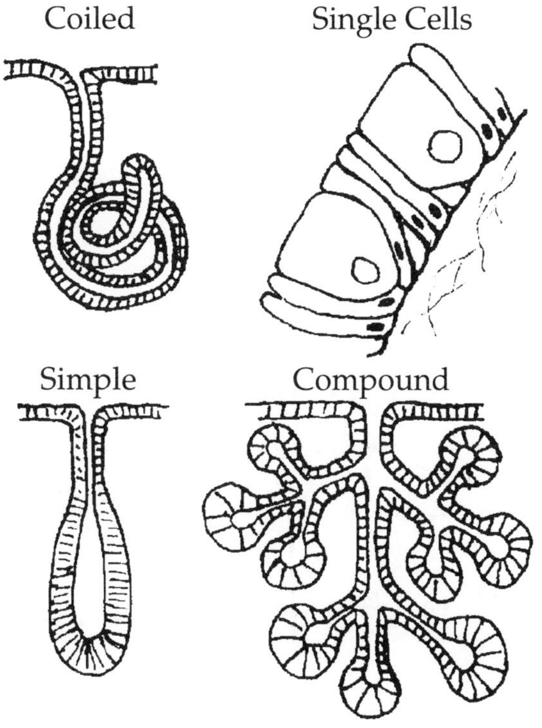

• When classifying epithelium, the first word designates thickness, how many layers of cells are stacked on top of the basement membrane. **It** is done in one of three ways.

In many organs, the epithelium is a single layer, with each cell touching the basement membrane. This is called **simple** epithelium.

In other organs, there is more than one layer, with only the bottom layer touching the basement membrane. This is called **stratified** epithelium. Stratified tells you there is more than one layer. It does not say how many layers, and this varies from place to place.

Some epithelial tissues give the false impression of being stratified and are called **pseudostratified** epithelium. This happens when cells of varying height bend and lean over each other. A straight cut through the tissue may go sequentially through three or four cells. This gives the impression that the cells are stacked on top of each other, but they all actually begin at the basement membrane.

• The second word, when classifying epithelium designates the cell shape**.** Epithelial cells come in one of three shapes. Cells that are flat and thin, like pancakes or scales, are called **squamous.** Cube-shaped cells are called **cuboidal.** Cells that are tall and slender, like columns, are called **columnar.**

In stratified epithelium, the shape of the cell may be different in the different layers. In that case, the epithelium is named by the shape of the cell in the outermost layer or the **free surface** (Fig. 2-1).

Transitional epithelium is specialized to change its shape and thickness to accommodate for distention of an organ.

GENERAL FUNCTIONS

Epithelial tissues perform five functions in the body.

• We have said that, at a certain spot, there is no space between epithelial cells. How or where might this be useful in our body? With no space between cells, it does not matter how clever, crafty, tricky, or tiny something is, it cannot sneak between the cells to get to the other side of the tissue.

Table 2-1 Classification of Epithelial Tissue

Simple Squamous

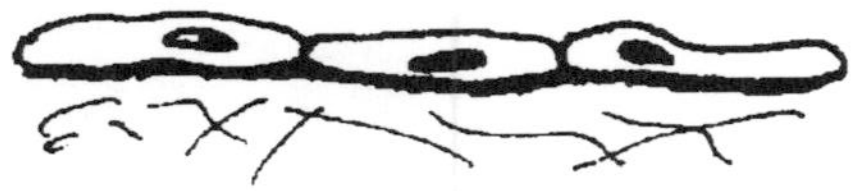

Function: lubrication and establish barriers allowing passive transport.

Location: 1. inside blood & lymph vessels (not shown); 2. ear drum; 3. pleura; 4. air sacs in lungs; 5. pericardium; 6. inner surface of abdominal cavity; 7. part of kidney tubule; 8. outer covering of a. liver, b. stomach, c. spleen, d. small intestine, & e. transverse colon.

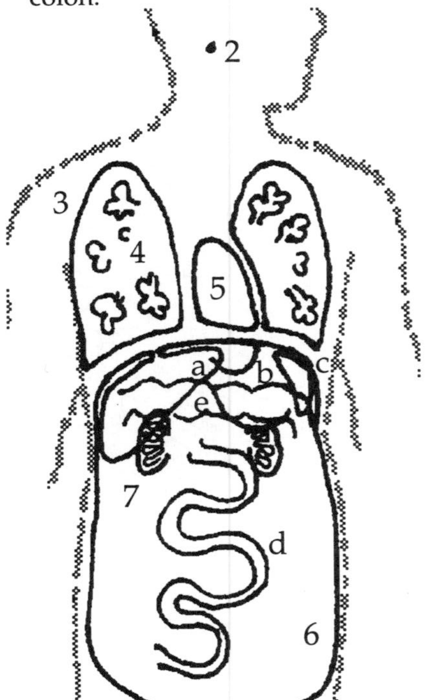

Simple Cuboidal

Function: secretion and establish barriers with controlled active transport.

Location: 1. choroid plexus; 2. retina; 3. auditory tube; 4. bronchioles; 5. part of kidney tubule; 6. outer covering of uterus, uterine tube, & ovary; 7. testes; 8. many glands a. tear, b. thyroid, c. liver, d. pancreas, e. adrenal, f. prostate, g. seminal vesicle, h. sweat, & sebaceous (not shown).

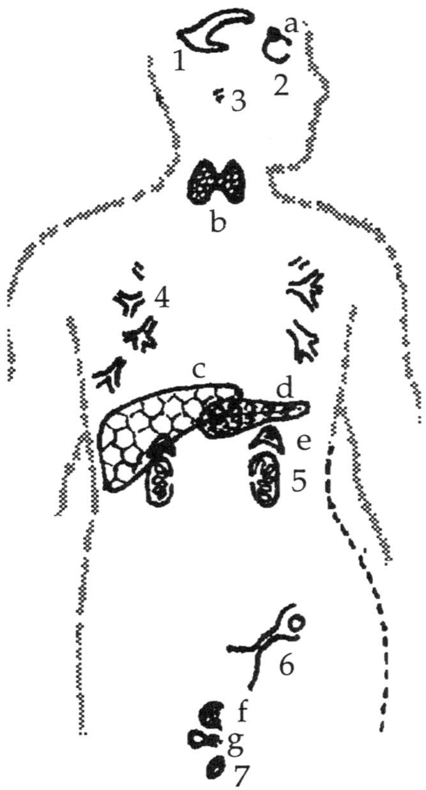

Simple Columnar

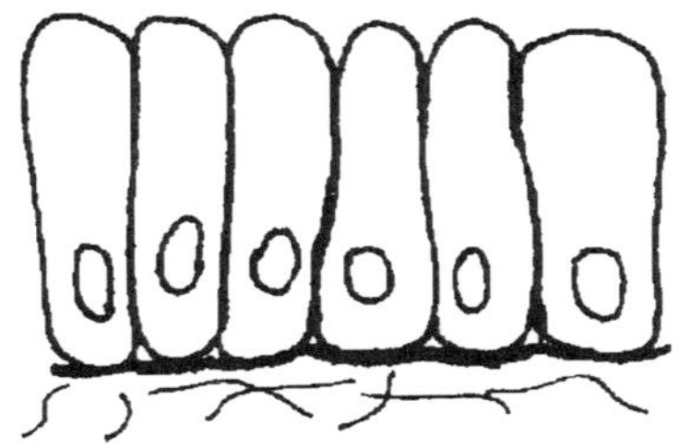

Function: secretion and establish barriers with controlled active transport (absorption).

Location: inside lining of a. ventricles of brain, b. stomach, c. small intestine, d. colon, e. gall bladder, f. uterus & uterine tube.

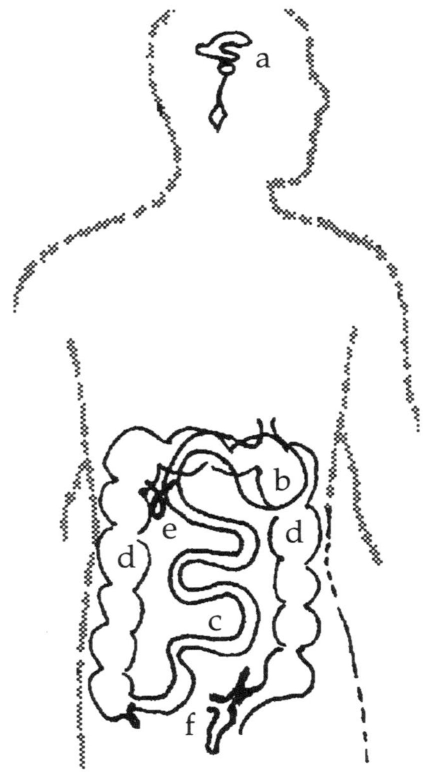

Psuedostratified

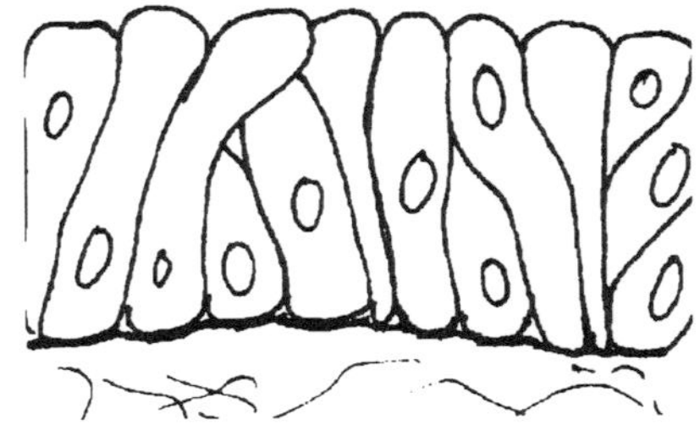

Function: mucus secretion and establish barriers with no transport.

Location: inside lining of a. air sinuses, b. nose, c. throat, d. trachea, & e. large bronchi.

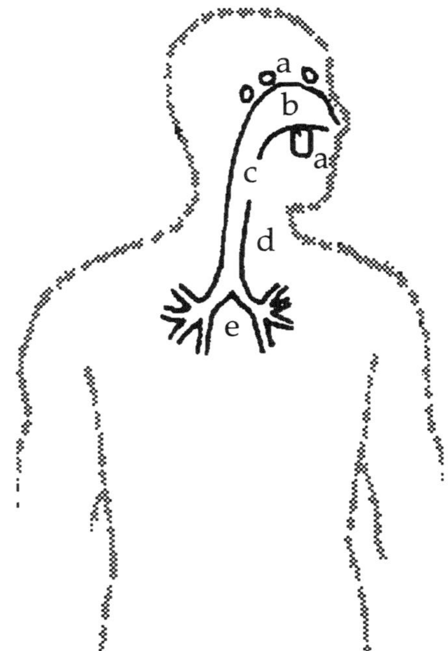

Stratified Squamous

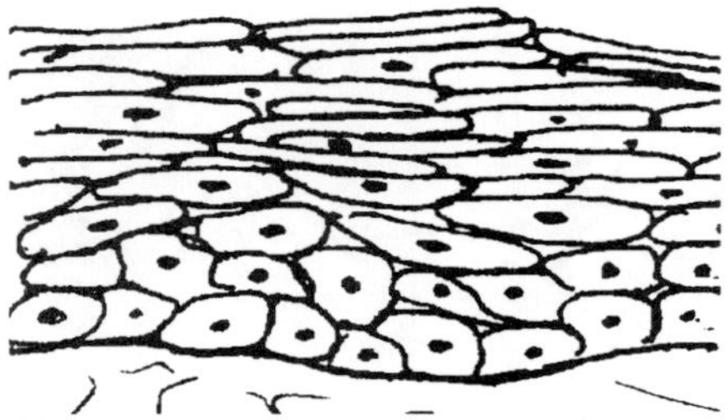

Function: protection and establish barriers with no transport.

Location: 1. outer layer of skin; 2. conjunctiva; 3. inside lining of a. mouth, b. upper esophagus, c. anus, d. vagina, e. lower urethra, & ducts of sebaceous glands (not shown).

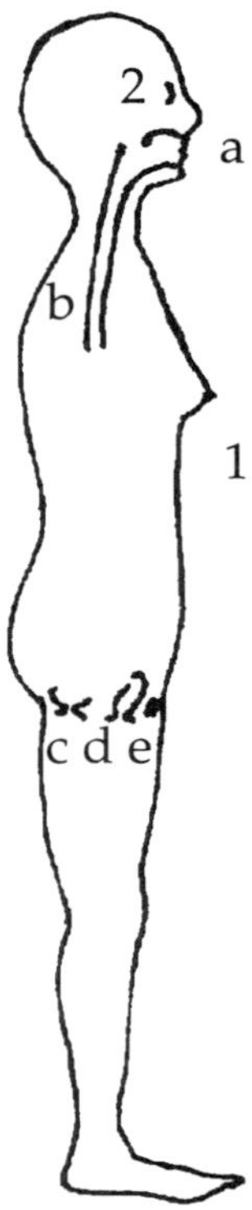

Stratified Cuboidal

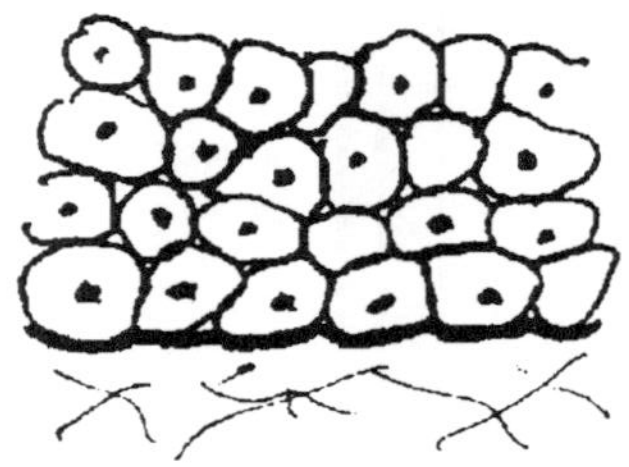

Function: protection and establish barriers with no transport.

Location: inside lining of a. throat & b. ducts of sweat glands (not shown).

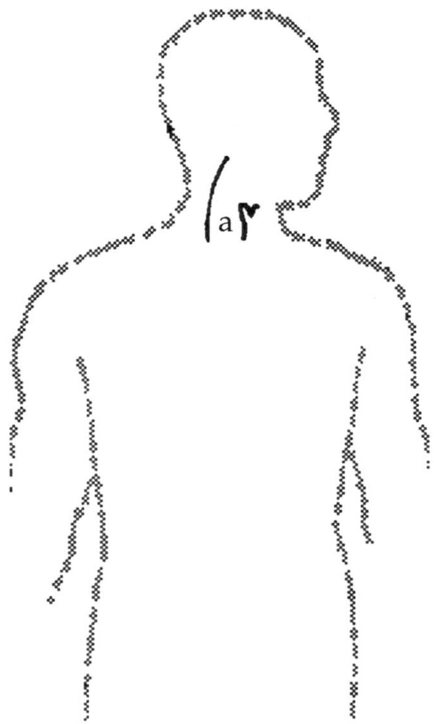

Stratified Columnar

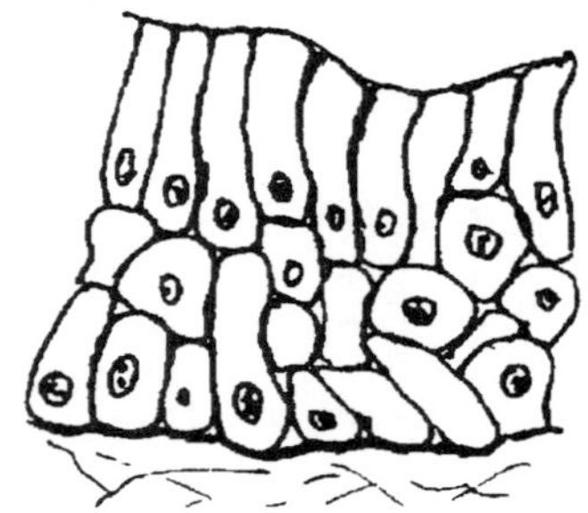

Function: protection, secretion, and establish barriers with no transport.

Location: 1. conjunctiva; 2. inside lining of a. larynx, b. ducts of mammary glands, c. vas deferens, & d. epididymus.

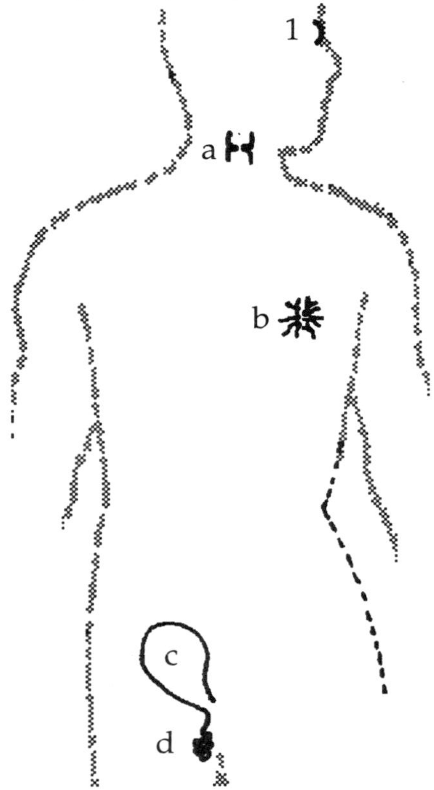

Transitional

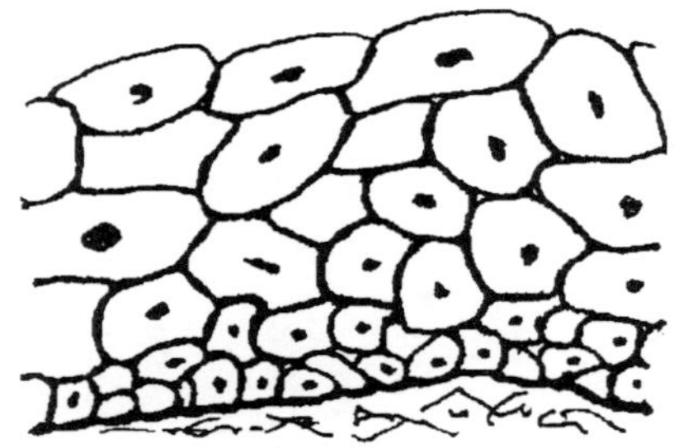

Function: protection and establish barriers that permit distention and have no transport.

Location: inside lining of a. ureters, b. urinary bladder, & c. upper urethra.

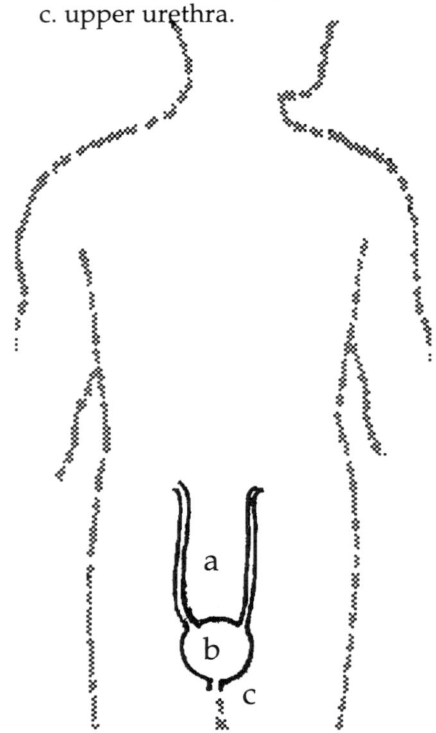

This allows sheets of epithelium to establish effective barriers (Fig. 2-3). Wherever our body needs to enclose a space, separate two spaces, or contain a substance, a sheet of epithelial tissue is perfect for the job. A sheet one cell thick can be adequate. "No Trespassing" is rigidly enforced. The basement membrane does not contribute to the barrier function of epithelial tissue, since most molecules can easily diffuse through it.

The following are three examples of epithelial barriers in our body. **1.** Food, as we swallow it, is not ready for our blood. It needs to be digested first. The inner surface of our digestive canal is lined with epithelial tissue to contain our food until it is adequately digested. **2.** Blood belongs within our blood vessels. The inner layer of blood vessel walls is epithelial tissue, containing our blood inside our vessel. **3.** Urine is full of body wastes. We do not want these wastes to seep back into our body as the urine waits in our urinary bladder, so our bladder is lined with epithelial tissue.

- Epithelial barriers are not like plastic bags, blocking all movement. In our digestive canal, once our food is adequately digested, it needs to be transported through the epithelium into our blood. Blood itself would be quite useless if nutrients and wastes could not move appropriately in or out through the epithelium of the vessel walls.

Generally there are two paths across a tissue, between cells or through cells. Slipping between cells is not available in epithelium, but the cells may pass a substance through themselves. When epithelial cells have a good reason, they transport a substance in one side and out the other, passing it through the barrier (Fig. 2-3). Cells can control active transport through their own cell membranes. They choose what they will and will not transport, absorbing certain chemicals and blocking others. Their sense of discrimination is good.

Thus, epithelial tissues produce controlled movement across the barrier. Digested food is absorbed. Our epithelial kidneys throw out just what our body does not need. Not everything in the air goes through the epithelium of our lungs into our blood. (Thank God!)

> Only molecules that can diffuse passively through cell membranes go where they please, overriding the limit set by epithelial tissue. Molecules that diffuse are small and simple such as water, oxygen, or sodium and potassium ions. As we have seen in the cell, passive diffusion is often useful and sometimes problematic.

Figure 2 - 3

Epithelium establishes barriers with controlled movement.

Figure 2 - 4

An epithelial sac between the lung and ribs lubricates the movement of breathing.

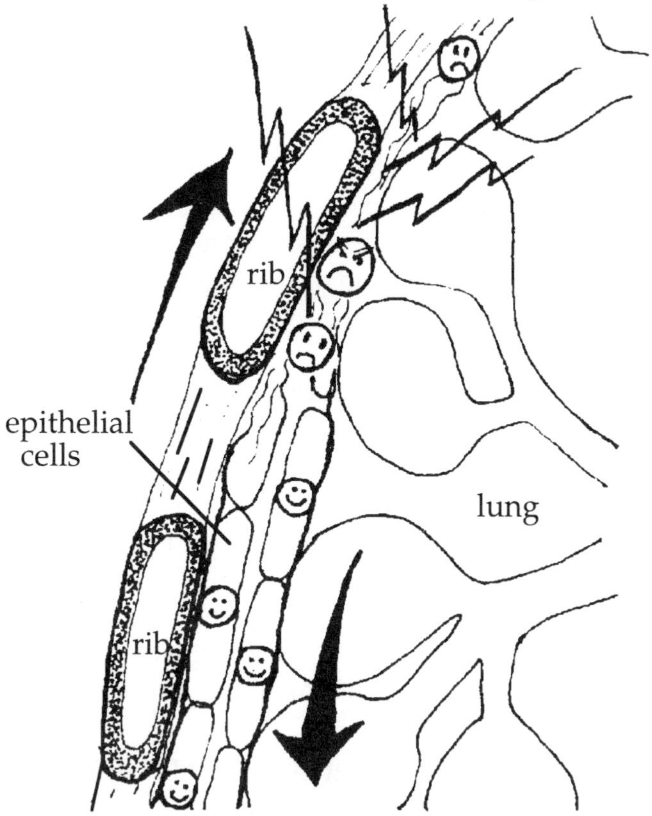

Figure 2 - 5

Mammary glands use nutrients from the blood to make milk which is secreted into a duct.

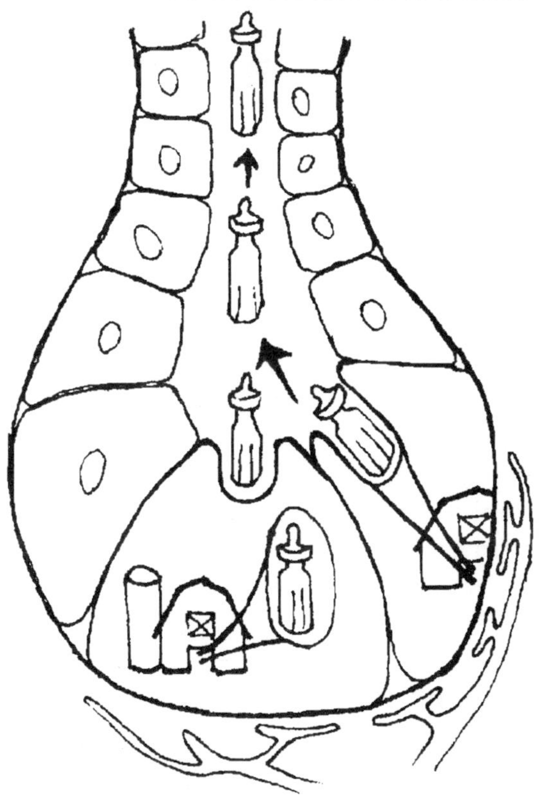

• Epithelial sheets form protective coverings. The outer layer of skin is epithelial tissue, the most extensive protective covering of our body. It will not stand up to sidewalks, but it is amazingly tough considering its softness and flexibility. Motorcyclists know this and classically use leather jackets to save their own hide in case of a spill. The inside of many organs such as our throat, rectum, and vagina are protected by epithelium.

• Epithelial sheets form smooth, lubricated surfaces which reduce friction during physical movement. The epithelium lining the inside of blood vessels reduces the resistance and turbulence of blood flow. The outside of our small intestine is covered by epithelium to reduce friction as it squirms over itself. An epithelial sac surrounds our lungs to lubricate their movement against our ribs as we breathe (Fig. 2-4).

• Many epithelial cells make substances which are used elsewhere in the body. The cells spit these substances out into ducts or adjacent tissues to be carried away (Fig. 2-5). This is called secretion, an important function of glandular epithelium.

The following organs, whose main tissue type is epithelium, are noted for their secretions: sweat glands, sebaceous glands, mammary glands, our liver secreting cholesterol and bile, our kidney secreting urine, and our endocrine glands including our pituitary, thyroid, adrenals, pancreas, and ovary/testis.

To say that a structure is a **gland** means that it secretes something. All glands are made of epithelial tissue and most secretions are made by epithelial cells.

Specific shapes of epithelium specialize in specific functions. Squamous cells form barriers, cuboidal cells secrete, and columnar cells absorb. These are only preferences, not hard and fast job assignments. Their labor unions are not rigid. Many epithelial tissues are workaholics, doing several or all of the possible jobs. Table 2-1 indicates the job preferences of specific epithelial tissues.

LOCATION

• Our body places epithelial tissue wherever we need an epithelial function. We have already seen many examples. Table 2-1 locates more examples and indicates what kind of epithelium it is. As we discover our body, we will complete the epithelial picture.

SOURCE OF NOURISHMENT

• Existing without direct blood supply, epithelial tissues receive their nutrients from the blood of the underlying tissue (Fig. 2-6). Nutrients easily diffuse through the basement membrane to the basal cells. In simple epithelium, it is that simple. The basal layer is as high as it goes. In stratified epithelium, the basal cells still have it easy. The feast is spread out below them. It is the cell layers above that grumble. "This is it? Nothing more for lunch?" Remember, "no sneaking between cells" also applies to nutrients. The second cell layer has to survive on nutrients tossed out by the first layer. The third layer only gets the leftovers from the second. The fourth layer, leftovers from the third, and so on. You can imagine, after a few layers, nutrients are picked down to nothing. No matter how excessive Thanksgiving dinner is, leftovers eventually come to an end.

• The more active a cell is, the greater is its nutrient requirement. In epithelial tissue, then, the lower cell layers are the only ones fed well enough to work hard. The further away from the blood, the more inactive and debilitated the cells become. In our skin, there are so many layers, the outer epithelial cells literally die of starvation.

Since epithelial tissue is healthy and happy on a distant blood supply, it does not notice immediately when nutrients are deficient, but the quantity and quality of blood eventually influences epithelial activity.

> Our body uses its dead outer layer of skin to our advantage. If we slide gently along the sidewalk, besides not bleeding, we may only lose dead tissue.

Diseases that thicken the basement membrane, narrow blood vessels, or reduce nutrients will reduce epithelial cell function and reproduction.

CELL REPRODUCTION

• Throughout one's life, epithelial cells have the potential to reproduce or divide to make new cells. This is so important for tissues which cover outer and inner surfaces of our body and its organs, since they suffer all kinds of abuse and damage.

Some epithelial tissues only reproduce to replace cells lost through injury or disease.

Many epithelial tissues continually divide, rejuvenating themselves, pushing older cells up to the surface to be discarded or sloughed off. (Quite the social behavior!) This happens in our skin where towels or clothing rub off the outer cells. In the digestive canal, sloughed epithelium forms about one third of the volume of feces.

Cell reproduction is a strenuous process, requiring lots of nourishment, so it is mainly basal cells that reproduce. Reduced nutrients, for any reason, soon will reduce their reproductive potential. So while we say that epithelial tissue can reproduce, how well it is done depends on the nutrient supply.

TISSUE REPAIR

• Since the cells of epithelial tissue are so closely arranged, injury or disease to epithelial tissue always damages cells. Whenever cells are lost, the remaining cells, at the edge of the wound, increase their rate of reproduction. They push each other across the stripped or denuded area until they contact epithelial tissue cells from the other side (Fig. 2-7). Then the epithelial covering is reestablished. The wound is repaired. Cell reproduction slows down to the maintenance rate.

The control of the rate of cell reproduction during tissue repair is called **contact inhibition.** When an epithelial cell is in contact on all sides with other epithelial cells, cell division is inhibited. When contact is lost by injury or disease, cell division is stimulated.

Figure 2 - 6

Epithelial cells receive nutrients from the blood of underlying tissues.

Figure 2 - 7

Epithelial tissues repair by reproducing new cells and pushing them across the wound.

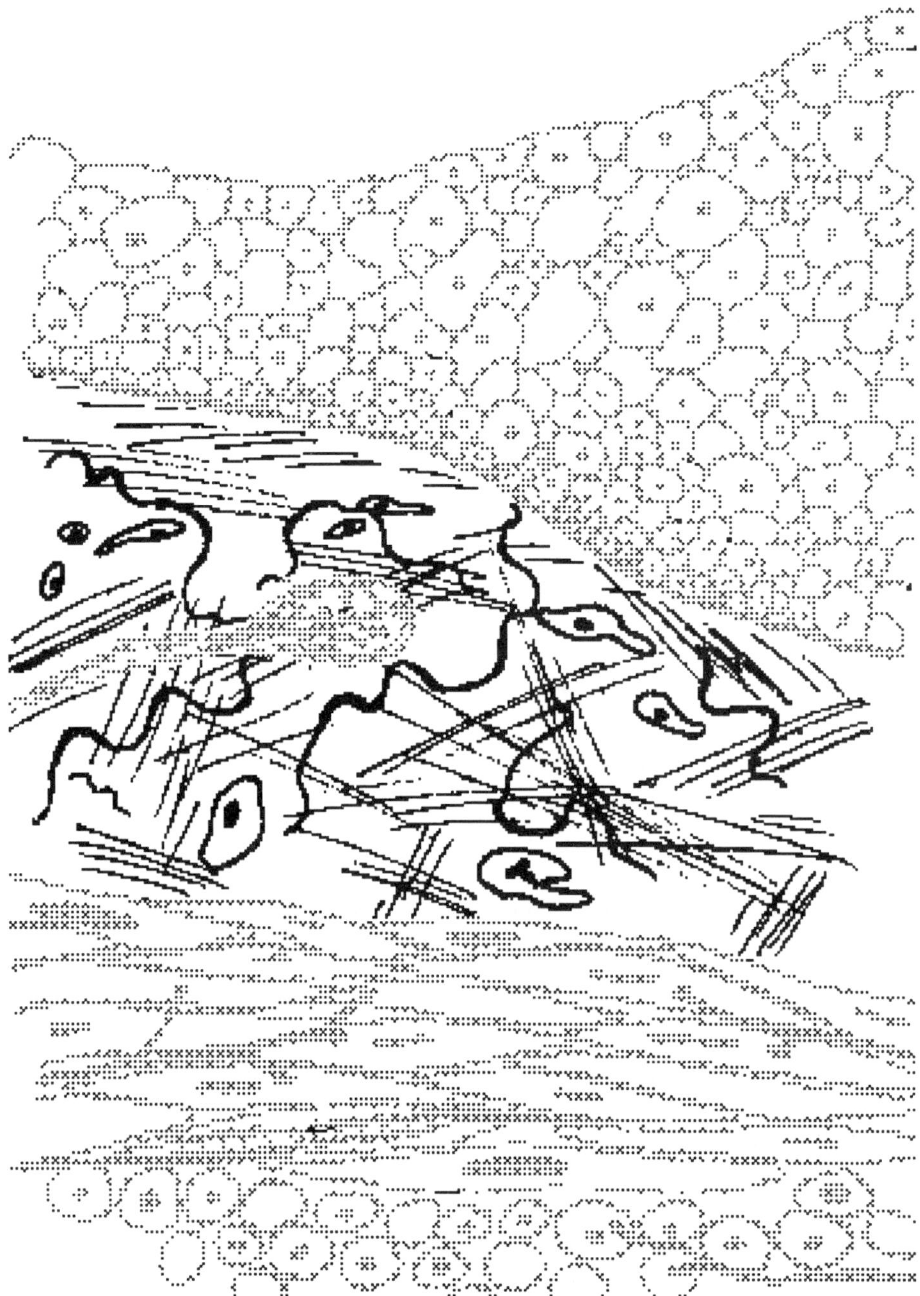

chapter 3: **Connective Tissue**

GENERAL FUNCTIONS

- The name **connective tissue** tells us the main function of this tissue. It connects, holds together, and supports things. Bones, ligaments, tendons, and fascia are all connective tissue (Fig. 3-1). Our skin is supported by connective tissue. The connective tissue of a blood vessel wall contains our blood under high pressure (Fig. 3-2). Internal organs are held in place by connective tissue.
- Connective tissue stores certain substances as reserves for a sparse day. Adipose is connective tissue which stores extra calories in the form of fat. Bones store minerals. Any connective tissue can be consumed for energy in a time of crisis or shortage.
- Connective tissue transports many products throughout our body, blood being an important connective tissue.
- Connective tissue protects. Even in skinny people, little, soft pillows of adipose tissue surround and cushion vital organs such as the kidney and eye (Fig. 3-3a). Our white blood cells and lymphatic system are both connective tissues which form our body's major defense (Fig. 3-3b).

Figure 3 - 1
(a) A tendon is connective tissue which attaches muscle to bone. **(b)** A ligament is connective tissue which attaches bone to bone.

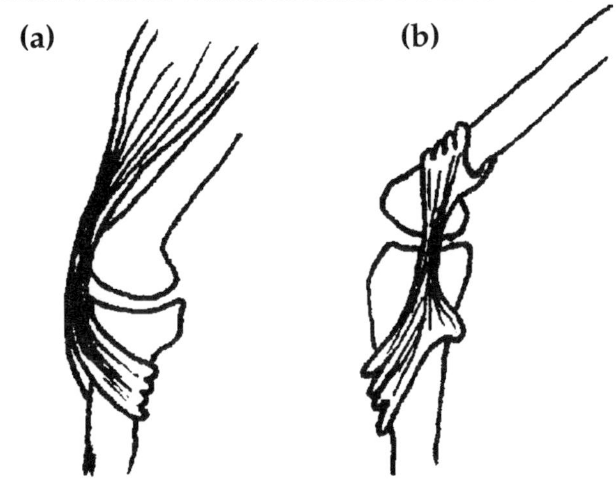

Figure 3 - 2
The connective tissue of the blood vessel wall contains the blood.

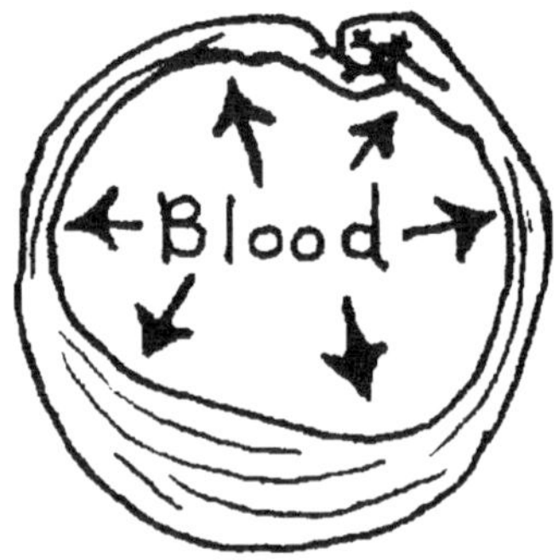

Figure 3 - 3
Connective tissue protects, **(a)** adipose tissue cushions the eye, and **(b)** white blood cells engulf and destroy harmful agents.

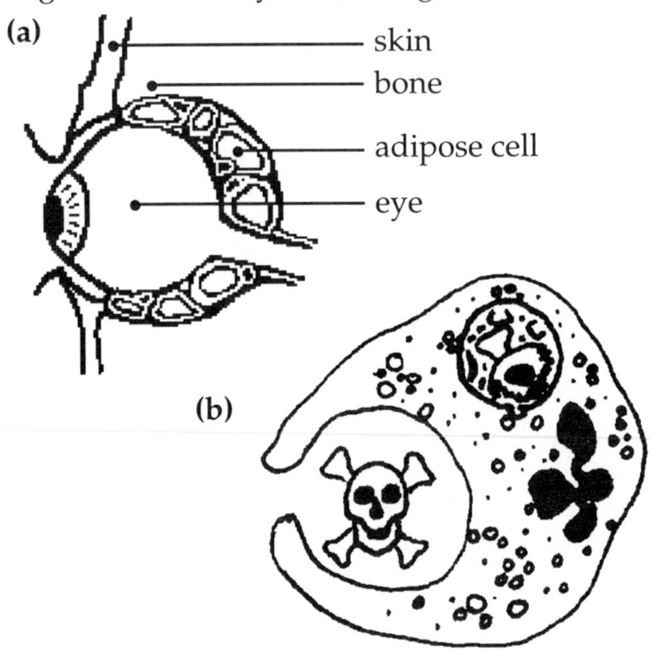

STRUCTURAL CHARACTERISTICS

Of our four tissues, connective tissue is the most diverse. The above examples of connective tissue included bone, tendon, adipose, and blood. What could these things possibly have in common to all fit into the same category of tissue? Believe it or not, they all have the same three structural components: **ground substance, extracellular protein,** and **cells.** Any tissue that has these three things may be called a connective tissue. Let us take a closer look at the three common structural components of any connective tissue.

Ground Substance

Ground substance is whatever stuffs the spaces left between our proteins and cells. The substance used for ground substance adds great variety to connective tissue. It can be anything from water to hard crystals (Fig. 3-4).

Figure 3 - 4

Ground substance stuffs the space between the cells and proteins. It can be anything from **(a)** water to **(b)** hard crystals.

(a)

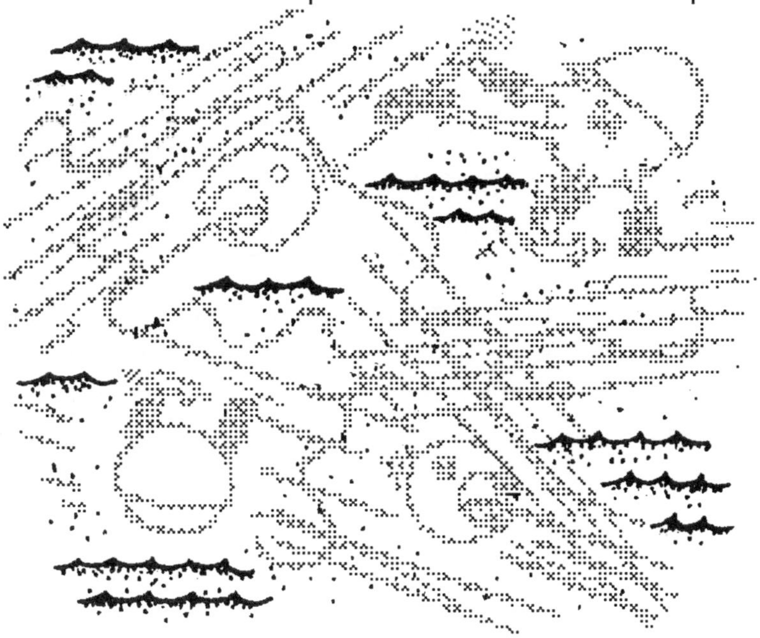

(b)

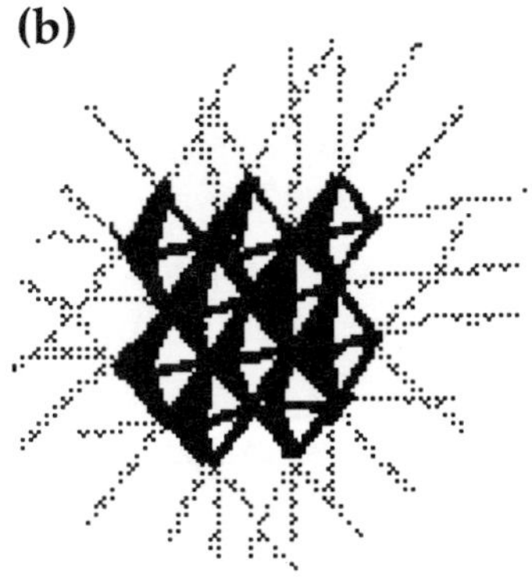

In blood, it is water. In cartilage, it is a firm, springy gel. In bone, it is calcium-phosphate crystals. In many soft connective tissues, it is a slippery, sticky compound which contains **hyaluronic acid.**

- **Hyaluronic acid** functions to lubricate our tissue, to hold water in our tissues and to assist the movement of nutrients and wastes between our capillaries and cells.

Hyaluronic acid is a distinctive molecule made of amino acids and sugars. Its consistency is affected by temperature and activity. When cold and immobile, it is firm, fixed, and coagulated (Fig. 3-5a). With warmth and movement, it becomes soft, pliable, and slippery (Fig. 3-5b). The leftovers of a well-cooked soup demonstrate this quality. When it comes out of the fridge, it is cold and coagulated like a jelly. As we heat or stir the contents, it again becomes runny like a soup.

When connective tissues are made soft and pliable before strenuous physical work or exercise, the risk of injury is reduced.

Warmth and movement can be added to connective tissue in two ways: **1.** by using our own muscles to move ourself about, or **2.** by someone else rubbing, chafing, shaking, or in any way manipulating our tissue. Most hands-on therapies benefit our tissues in this way.

Figure 3 - 5

Hyaluronic acid is **(a)** fixed when cold, and **(b)** pliable when warm.

(a)

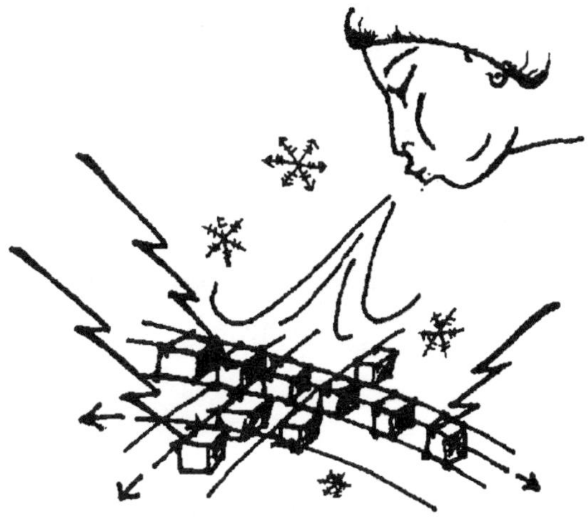

(b)

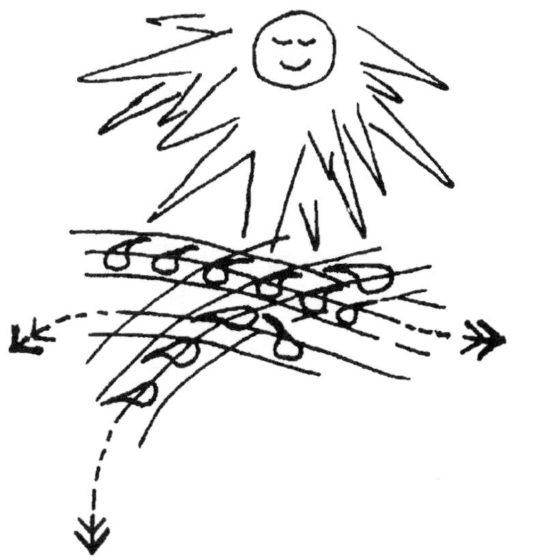

This is one of the things accomplished by athletic warm up exercises or pre-event massage.

Extracellular Protein

• The **extracellular proteins** of connective tissue occupy our extracellular space, outside our cells, between our cells and blood vessels, amongst the ground substance. These proteins are long thin molecules, like a thread or string or the fibers of a cloth and are frequently referred to as the **fiber** of connective tissue (Fig. 3-6). The protein does the first function of connective tissue, it is the strength of connective tissue, providing support and connections, holding things together.

• Three different types of protein are available, depending on the requirements of a specific connective tissue.

1. Collagen is a strong, dense, non-stretchy fiber. It is used in a tissue that puts up with a lot of physical stress, where strong support is required. Bones, ligaments and tendons have lots of collagen. Most connective tissues have at least some collagen. It has been estimated that about one quarter of all the protein in our body is collagen.

2. Elastic fibers are not so tough as collagen, but they do as their name implies, they allow stretching and then snap back to their original length. Elastic fibers support where stretch and flexibility are important. Our skin is supported with elastic fiber so it can accommodate movement.

> Ultraviolet light from the sun or any artificial source destroys elastic proteins leaving the skin more wrinkled.

3. Reticular fibers form a light net or spider web structure which becomes the skeleton for organs like our liver or spleen. 'Reticular' has the same root as 'reticulum' in endoplasmic reticulum and implies a loose weaving or net.

• As all proteins of connective tissue mature, they slowly shrink and stiffen, reducing our body's flexibility and movement. This is one of the inevitable rules of connective tissue. Young children can bend their bodies into circles. At age 55, low and behold, after bending as far as possible, our toes may still be a long ways away (Fig. 3-7). Either we are working to maintain the length of our connective tissues and our current range of motion or it becomes a thing of the past.

Figure 3 - 6

The proteins are long thin molecules outside the cells, amongst the ground substance.

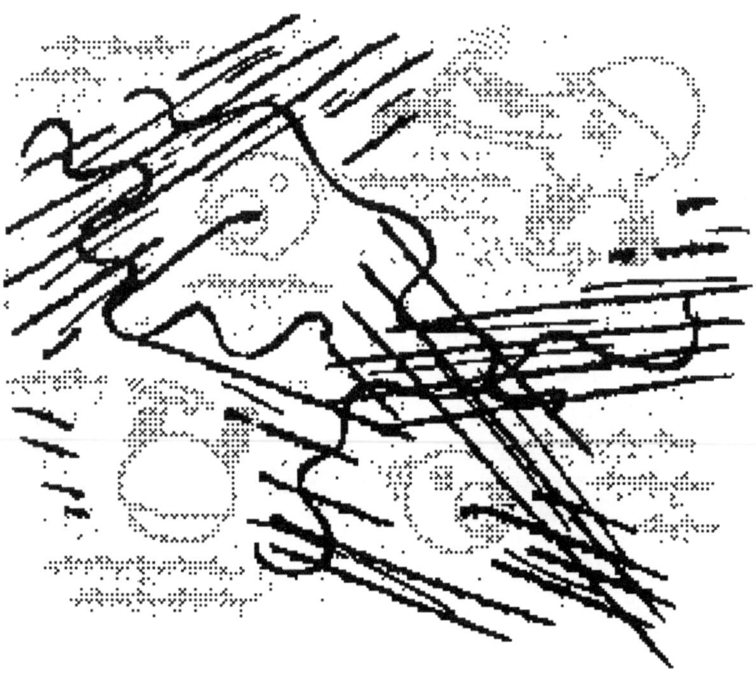

Figure 3 - 7
As time passes, connective tissue proteins shrink and stiffen, reducing flexibility and movement.

How does shrink happen to connective tissue? Generally, connective tissue proteins can, to some degree, slide over each other. As time passes, our proteins become more and more pinned to each other by chemical bonding. This causes the shortening and stiffening of our connective tissue.

The rate of shrink can be diminished by constantly reminding connective tissues how long they can be. If we bend and stretch out our connective tissues every day, we can inhibit the shrinking process and maintain greater mobility.

Increased sugar in our blood and tissues is thought to stimulate the protein bonding and increase the rate of shrink and stiffening.

> Length and softness in our connective tissues enhances the range of motion of our joints as well as the function of internal organs. For example, the flexibility of arterial walls accommodates the pulse of blood as our heart pumps.

Cells

- The cells of connective tissue stand relatively randomly placed at arm's length from each other (Fig. 3-8). They can move around independently in their immediate neighborhood, having a sphere of influence, keeping track of a small range of connective tissue.
- The job of the cell is to build and maintain the extracellular protein and in many cases, the ground substance. This involves two processes and often is done by two different cells. One cell is a **nibbler.** It keeps busy nibbling, digesting the broken, frazzled proteins, worn by use and time. The other cell is a **builder,** constantly constructing new proteins and stringing them around in the extracellular space, replacing the ones which have just been nibbled away (Fig. 3-8). Even in adults, after growth is complete, connective tissue cells never say that a job is done. They never build a bone or tendon and then take a vacation till something breaks or tears. The protein fiber which first built our bones is not there today. Our cells are always working, continually rejuvenating our protein to reduce the chance of breaks and tears.
- It follows then, that the quantity of fiber in any connective tissue can change with time, depending on the rate of building relative to the rate of nibbling. Three conditions influence these rates:

1. Our nutritional status. A healthy diet provides the nutrients required by our cells to build protein. This includes adequate but not excessive protein, vitamins, minerals, and fuel. Excess protein may make our body acidic and remove protein out of our tissues.

Figure 3 - 8

The cells are randomly organized and spaced. They function to build and rejuvenate the protein.

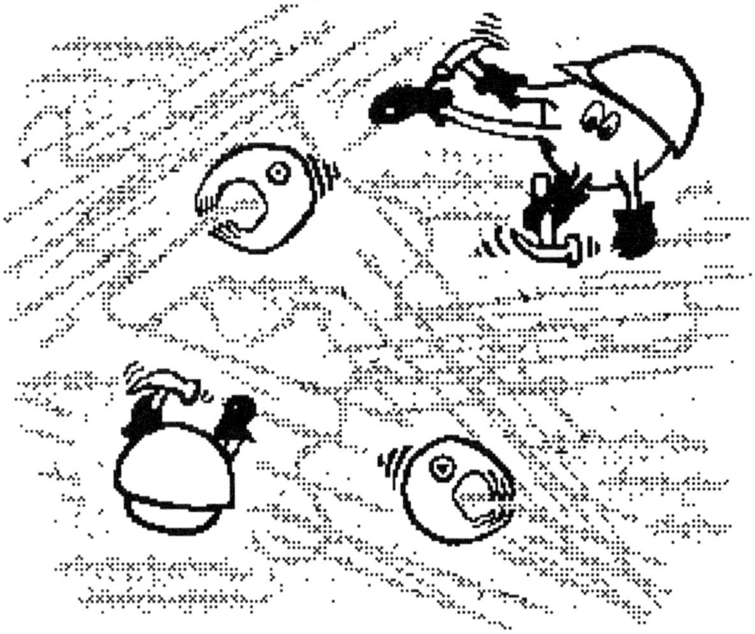

2. Our hormone balance. Three hormones have a powerful influence on connective tissue cell activity.

Growth hormone and **the sex hormones** stimulate our builders, increasing protein density and therefore, the strength of our connective tissue.

Cortisone inhibits our builders and stimulates our nibblers, reducing the amount of protein and the strength of connective tissue.

When we discuss these hormones in the Endocrine chapter we will describe their effect more fully.

3. Our level of physical activity. Connective tissues respond to a test of their strength. Muscles tugging, body weights compressing, skin stretching, any physical activity challenges our connective tissue and stimulates our cells to build protein.

The more heavily connective tissue is used, the more protein it makes. Under physical stress, our cells gulp, "Wow, if we have to carry this load, we better get more support behind us", so they lay down more fiber.

The less connective tissue is used, the less protein it makes (Fig. 3-9). Never asked to perform, our cells grunt "who needs all this stuff" and digest fiber without replacing it. Connective tissue atrophies to the level of use. This is another inevitable rule of connective tissue. Either we are working to maintain the strength of our connective tissues or we are losing it. There is no other choice. We cannot just sit like a bump on a log and maintain strong connective tissue. We do not save our body by resting it, we maintain it by using it.

- As cells make protein, they must decide how to organize the protein within the connective tissue. Again, physical activity is the

> **Postmenopausal women** have reduced sex hormones and so may develop weakness in all their connective tissues. This is one argument supporting sex hormone supplementation postmenopausally.

> The **anabolic steroids** taken by athletes to increase their size and strength are often derivatives of the male sex hormone, testosterone.

> **Cortisone** is used therapeutically in many inflammatory diseases, but people taking it are likely to develop weakness in their connective tissues.

Figure 3 - 9
When connective tissues are not used, they digest their protein but do not replace it.

guide. Our builder cells constantly monitor the direction of the forces applied through our connective tissue. Then they use a little common sense. Protein only can provide strength or prevent tearing when it lies in the direction of the stress. A fiber perpendicular to stress supports nothing and adds unnecessary bulk. If the stress across a tissue is always in one and the same direction, all the proteins in that tissue are built in that direction. Tendons experience one direction of pull from the muscle to the bone, so all their proteins lay in that direction (Fig. 3-10). If the stress comes from many directions, the proteins are woven in all these directions to support all the forces experienced by the cells. Our skin experiences multi-directional stress and its protein goes in every direction to give diffuse support (Fig. 3-6). If the stress through a connective tissue changes, the whole shape of that connective tissue may change. This is even true of bone.

A limb with a broken bone may be casted to initiate the bone's mending. This immobilization leaves the healing area with no stress, so the bone can begin to knit together. It does, however, allow the whole bone and all the surrounding tissues to weaken. Also, there is no sense of the normal stress pattern to guide the cells in laying down new protein. After taking the cast off, one has to slowly work up to heavy activity, rebuilding all the structures. One cannot just bounce out as if it is all better.

Figure 3 - 10
Tendons experience pulling in one direction, and so build all their proteins in that direction.

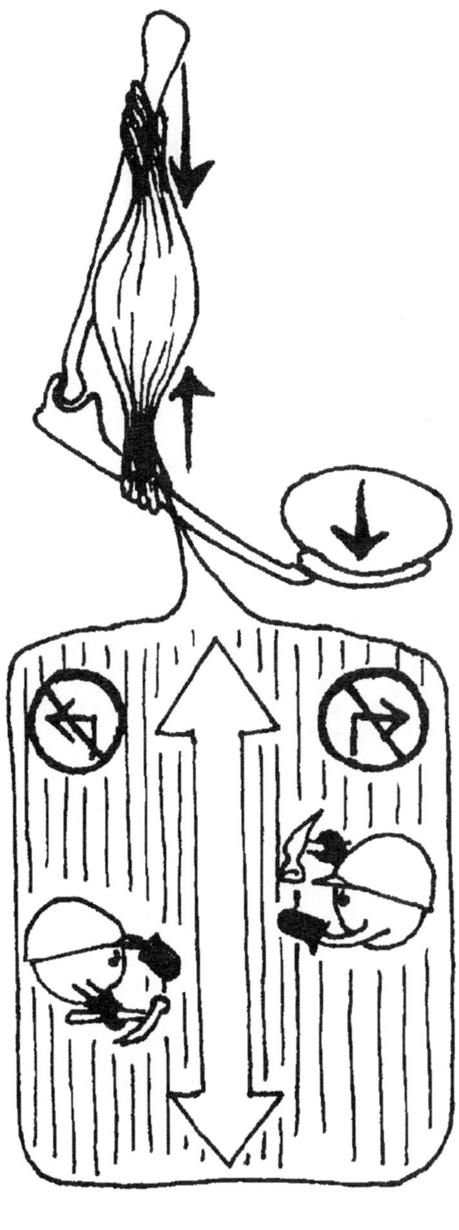

• So we see, physical activity benefits the length, strength, and organization of connective tissue.

There are many reasons for reduced physical activity. Illness can immobilize anyone. Jobs and transportation often are sedentary. Older folks may be less active. The weightlessness of outer space travel reduces the physical stress on connective tissue. In all these situations connective tissues become short and weak. Short connective tissues only increase the immobilization and weak connective tissues provide poor body support. Bones break, ligaments and tendons tear, blood vessels rupture and bleed, the skin becomes thin, and so forth. Within reason, activity should be maintained as much as possible.

The older we are, the more quickly our connective tissues deteriorate and the slower they rebuild. One more inevitable rule is that we are getting older, so every day physical activity becomes more important. Don't panic, that increases your cortisone. Just do something!

• We have discussed the three basic components of connective tissue: ground substance, extracellular protein, and cells. The relative quantity of each component is different in different connective tissues. Adipose tissue is mostly cells storing fat. Ligaments are mostly strong proteins holding bones together, with one molecule of water here and there between the proteins. Blood is mostly the ground substance water, to make a runny transport medium.

• In all this we see how connective tissue can have so much variety. To summarize, connective tissues vary with: **1.** the type of material used for ground substance **2.** the type, quantity, and organization of the protein, and **3.** the relative quantity of ground substance, protein, and cells.

CLASSIFICATION

• Connective tissue is classified and named according to its structural components: the type, quantity, and organization of its ground substance, protein, and cells. Table 3-1 illustrates and gives the specific functions and locations of the different types of connective tissue.

Bone is a connective tissue which has a rigid, non-compressible ground substance, providing rigid support and bearing the weight of our body.

Cartilage has a stiff but springy ground substance which provides a rigid support, but less so than bone. It is used where good support is needed along with flexibility, lubrication, and shock absorption.

Blood is a connective tissue which is predominantly the ground substance water. Blood provides a fluid transport medium for our body.

Elastic connective tissue has predominantly elastic protein and is used where stretchiness is essential.

Reticular connective tissue has predominantly reticular fibers, providing a light scaffolding for internal organs.

Adipose connective tissue is predominantly cells which can store fat.

• Two words describe the quantity of protein in connective tissue. **Dense** means that there are scads of fibers, making the tissue strong. **Loose** means that there are fewer fibers, when strength is not so important.

Two words describe the protein organization in connective tissue. When all the fibers are oriented in the same direction, the connective tissue is called **regular.** When the fibers are woven in many different directions, the tissue is called **irregular.**

Connective tissues described by these last four terms have a mixture of different types of protein and ground substance.

Any connective tissue can be described by these terms, individually or in combinations, such as elastic cartilage or dense regular connective tissue.

Table 3 - 1 Classification of Connective Tissue

Bone

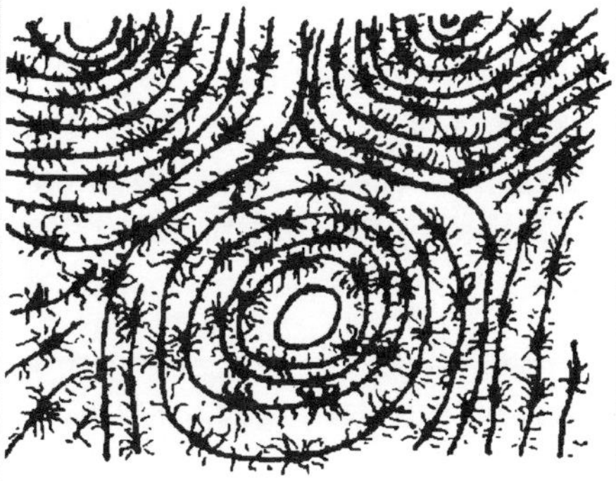

Function: rigid support and protection.
Location: the skeleton.

Cartilage

Function: supports a shape but allows for flexibility, shock absorption, enhances joint mobility, and bone growth.
Location: nose, larynx, trachea, bronchial tubes, anterior rib, intervertebral disc, ear, menisci, joint surface, pubic symphysis, and growing bone.

Blood

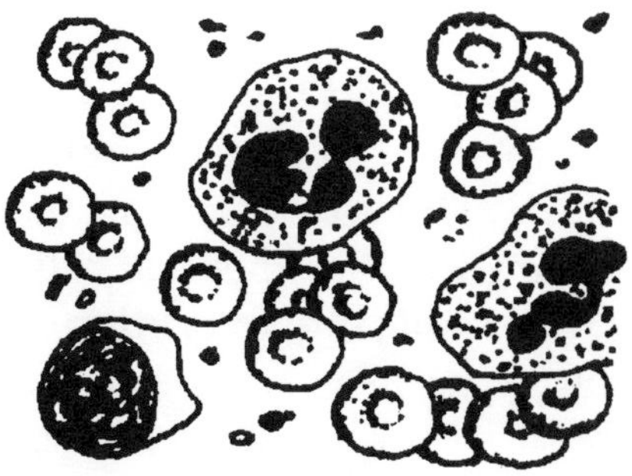

Function: major transport medium of the body, protects against infection, and temperature regulation.
Location: within the blood vessels and heart.

Elastic

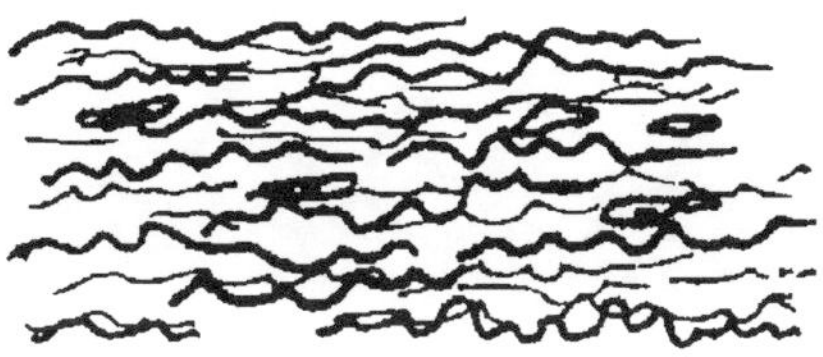

Function: markedly stretchy support.
Location: vocal cords, lungs, some vertebral ligaments and in the walls of blood vessels, trachea, and bronchial tubes.

Reticular

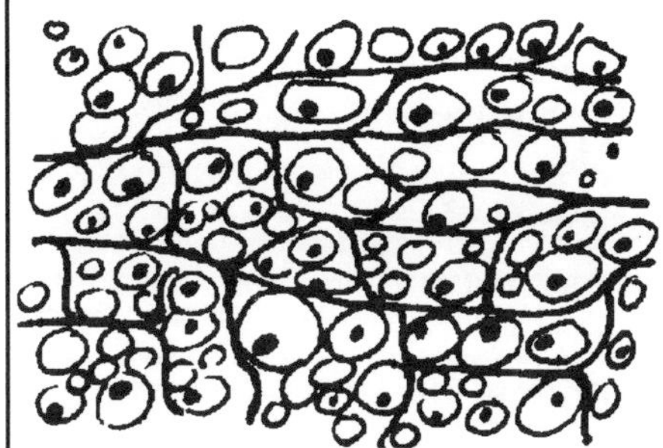

Function: internal supporting scaffold for the cells of solid organs.
Location: in liver, spleen, lymph nodes, adipose tissue, red bone marrow where blood cells reproduce, and visceral muscle.

Adipose

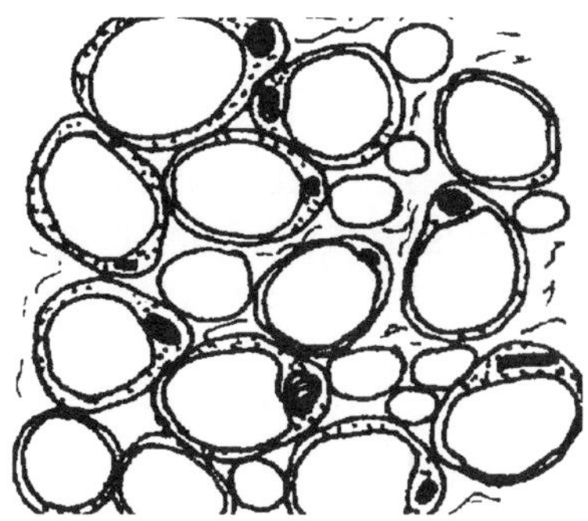

Function: energy reserve, protection, thermal insulation, and packing material.
Location: skin, surrounding the heart, kidney, intestine, and eye, and hollow marrow in long bones.

Dense Regular

Function: withstand great pulling and stretching forces in one direction.
Location: tendons, ligaments, and aponeuroses (broad, flat tendons).

Dense Irregular

Function: withstand moderate pulling and stretching in many directions.
Location: scars, muscle fascia, skin, periosteum, capsules of the kidney, testes, liver, spleen and lymph nodes.

Loose Irregular

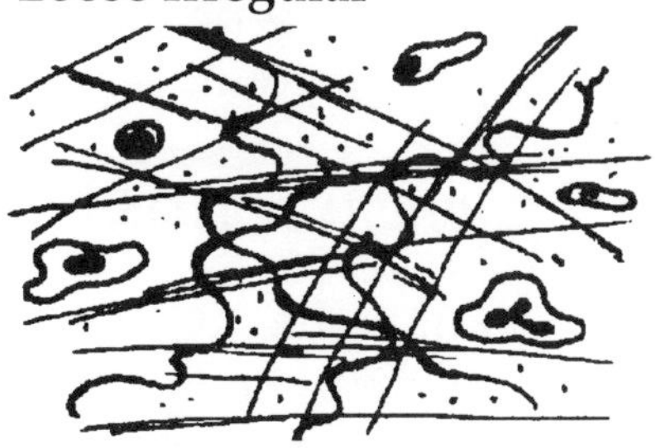

Function: support with elasticity in many directions.
Location: most widespread, between tissues and organs, underlying epithelial tissue in skin and tubular or hollow organs, and surrounding blood and lymph vessels and nerves.

CONTINUOUS WEAVING

• We have listed many individual structures as connective tissue. When we put all these structures together in our body, the connective tissues are all continuous with each other. The fibers of one connective tissue weave into the fibers of adjacent connective tissues, holding everything together. Our muscle fascia is continuous with its tendons, whose fibers, in turn, weave into the periosteum, whose own fibers weave into bone.

Our whole body is one continuous weaving. Each of us is a beautiful tapestry of different threads inlaid with collections of other tissues, cells, water, crystals, and so on.

SOURCE OF NOURISHMENT

Connective tissues generally are considered to be vascular. Many are profusely vascular and support the blood vessels feeding other tissue layers or organs. Some connective tissues are so dense with protein or ground substance that there is not much space for blood vessels. In tendons and ligaments the blood supply is very stingy. Cartilage often is avascular

Color charts of the body show some connective tissues in brighter reds and others in white. The red pigment of our body is in the blood, so these charts tell us which structures are more or less vascular.

• The activity of a connective tissue's cells and that tissue's potential for healing are markedly influenced by its vascularity.

CELL REPRODUCTION

Connective tissue cells can and do reproduce throughout our life.

Since connective tissue cells do not form the integrity of the tissue, cellular replacement is not emphasized as it is in epithelial tissue.

In some connective tissues, like the blood, the cells are constantly being replaced. In most connective tissues, cell reproduction is stimulated by injury to replace damaged cells and to have more cells around to share the work of healing.

TISSUE REPAIR

• With our understanding of connective tissue, what structural component would suffer the most in connective tissue injuries? The supporting proteins. They are cut, torn, or pulled apart by an overwhelming physical stress (Fig. 3-11). Unlike epithelial tissue, cells may or may not be damaged.

Figure 3 - 11
Injuries to connective tissue cut, tear, or pull apart the protein support.

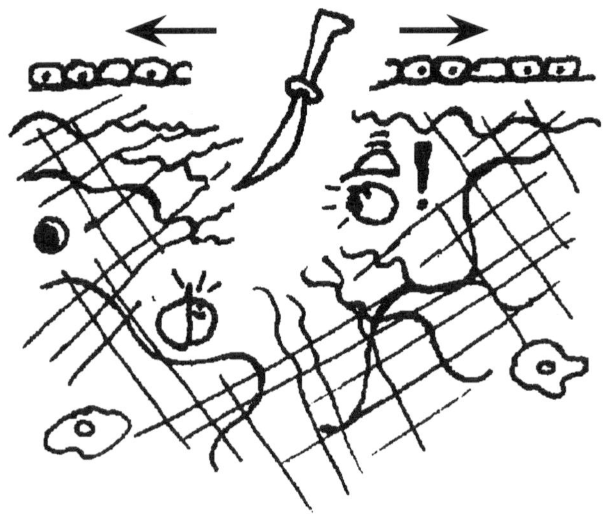

To heal connective tissues then, the big job is to rebuild the protein support system and perhaps to replace a few cells. Regardless, cells adjacent to the injury better get busy and generally they do.

• Healing connective tissue is what we call **scar tissue.** All the basic rules of connective tissue maintenance apply to repair. Diet, hormones, and physical activity always have the same effect on the cell in its work.

• Several problems come into play following an injury. This time it is foul play.

1. Because of the extreme loss of protein support after an injury, even a small stress is scary to the cells so they frequently overbuild, creating an excessively dense tissue (Fig. 3-13). These excess proteins may compress the tissue's blood supply as they mature. The whiteness of dense scars shows their reduced blood flow.

Figure 3 - 12

(**a**) The cells of immobilized connective tissue cannot see the normal stress patterns so they do not know how to lay the protein. (**b**) The result is an enlarged, disorganized, and weak union.

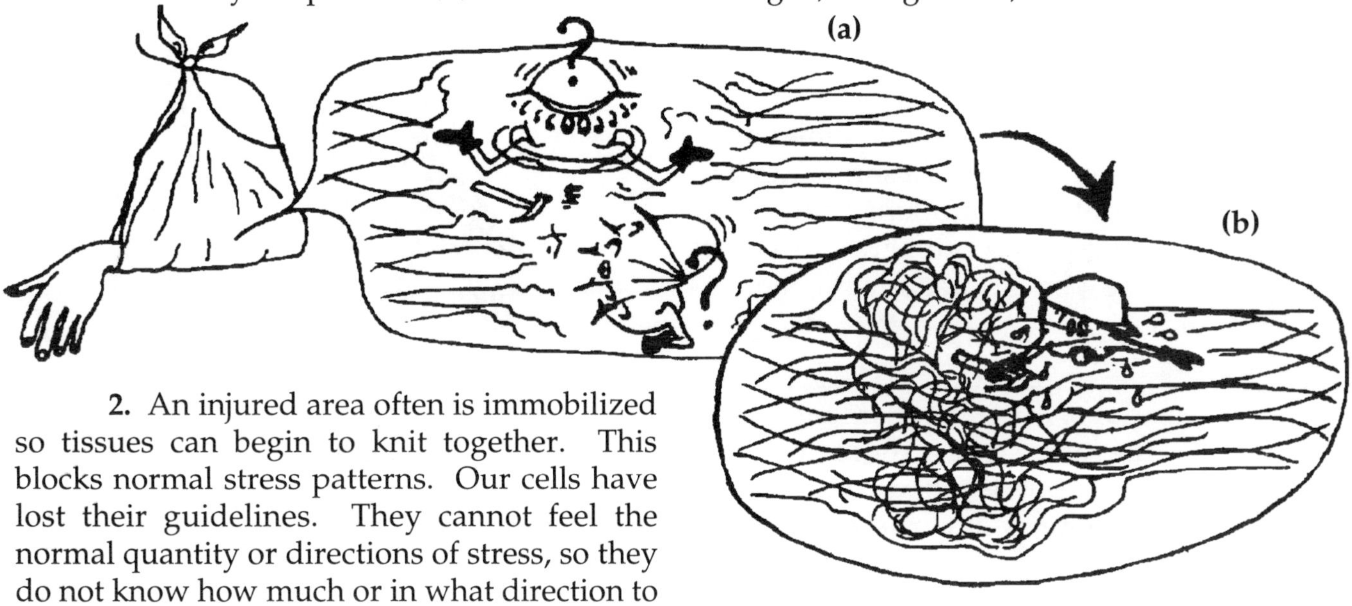

2. An injured area often is immobilized so tissues can begin to knit together. This blocks normal stress patterns. Our cells have lost their guidelines. They cannot feel the normal quantity or directions of stress, so they do not know how much or in what direction to lay new protein (Fig. 3-12a). They cover their bases and go every which way. The scar is a tangle, a mad knot (Fig. 3-12b). It feels like a large, dense connection, but the disorganization may not give strength in the needed direction. In this case there is more tissue but less strength.

Long after a wound is healed and movement is resumed, the internal organization of our tissue is remodeling. Now our cells can see what is going on and nibble away the useless protein and reinforce the appropriate direction(s). This may continue over years, with the scar gradually becoming more subtle and refined.

Figure 3 - 13

Scar tissue frequently has an increased quantity of protein.

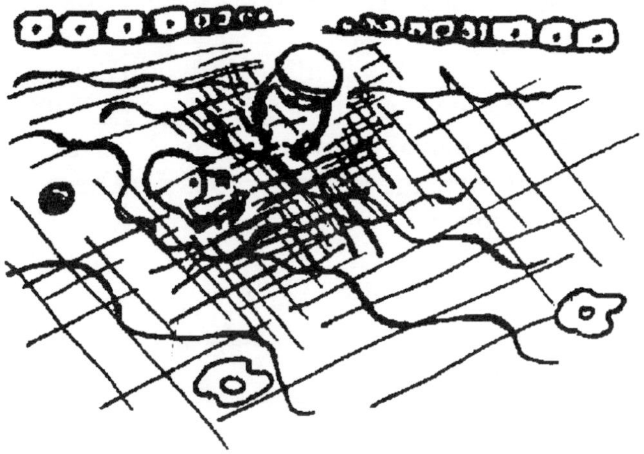

3. Often adjacent tissues, which normally move independently of each other, are injured simultaneously. Reduced movement during healing may allow the new proteins threads of one tissue to weave into neighboring tissue, thus binding the two together. This is called **adhesions.** Adhesions may inhibit movement of our skeleton as well as internal organs (Fig. 3-14).

As adherent tissues try to go their separate ways, the movement may produce pain.

Independent movement only can be re-established by again separating the threads, which is to some degree, another injury.

Figure 3 - 14

Injured loops of intestines may bind to each other as they heal.

4. While shrink happens in all connective tissues, it is strongest in healing connective tissue. This makes sense. When connective tissues are torn or cut, things fall apart and wounds stand open. The shrinking simply pulls things back to their original position.

Deep wounds over joints sometimes will demonstrate how powerful shrink can be. As the wound heals, the shrinking connective tissue may pull the joint so strongly into a bent position that the person cannot straighten it out (Fig. 3-15).

> These days, we take our gaping wounds to the emergency room to be stitched together, but the body was designed before emergency rooms were available.

Figure 3 - 15

The shrink of healing connective tissue may pull a joint into a bent position.

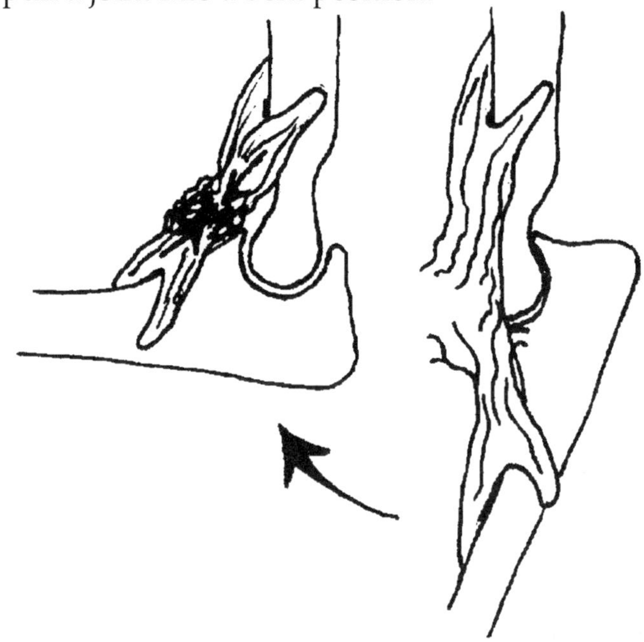

> The advantage of surgically pinning a broken bone, rather than putting it in a cast, is that movement is possible so much earlier. The whole bone and all the surrounding connective tissues do not de-teriorate from immobility and the fracture heals quicker, with stronger organization and less distortion.

- We have emphasized the importance of physical activity for the day-to-day maintenance of connective tissue. In healing connective tissue, activity is even more important. After an injury, the sooner normal movement patterns are integrated with the healing process, the more normal the healing connective tissue will be. There will be increased strength and organization of the scar with reduced shrink, reduced adhesions, and less excessive protein. Of course, activity has to be done with a sense of how strong or fragile the tissue is. Repeated tearing is not a desired effect.

> Water exercises are often therapeutic in unstable or weak conditions. The water supports the weight while producing gentle resistance to movement.

> After extensive knee surgery, the limb is often supported in a contraption that provides a moveable joint at the knee. Small, light movement can begin early and greatly improve the strength and organization of the scar while reducing adhesions.

- The active cells of healing connective tissue have a hearty appetite and need lots of food to do their job. This only is available through an ample blood supply. While we generally say that connective tissues can heal, the healing potential of any specific tissue depends on its vascularity.

> To know the healing potential for any specific tissue, we must ask "How vascular is it?" The answer for tendons, ligaments, and joint cartilage is "Pretty lousy". There is not enough blood in these structures to support a good healing process. Injuries which damage these structures can lead to long-term instability, residual pain, and reduced movement.

Whether connective tissues have become dense, shrunk, or adherent because of a specific, significant injury or just the nonspecific, minute injury of regular use and time, work can be done to reshape them, separating adhesions, reducing shrink, or remodeling scars.

This work is most effective when a therapist's work is combined with daily movements performed by the client to maintain and advance any change. The clients need to do their amazing new movement in order to maintain it. If the clients hold a closed, pinched posture, their changing connective tissues will assume that posture.

So much of how we hold ourself expresses our need to protect or project ourself on a physical and emotional level. Changing our physical, structural stance will go much further as emotional changes accompany it. If we still need a certain posture as our emotional defense when we are out in the world, releasing it from our tissues exposes us to trauma.

On the other hand, freeing our body helps to free our emotions. As our body moves differently, we feel different. Work can be done from both sides.

How aggressive a therapist can be when reshaping connective tissue depends on the client and the situation. Sometimes a client is ready for a quick, intense change. Sometimes the client needs lots of space for processing a change that happens bit by bit.

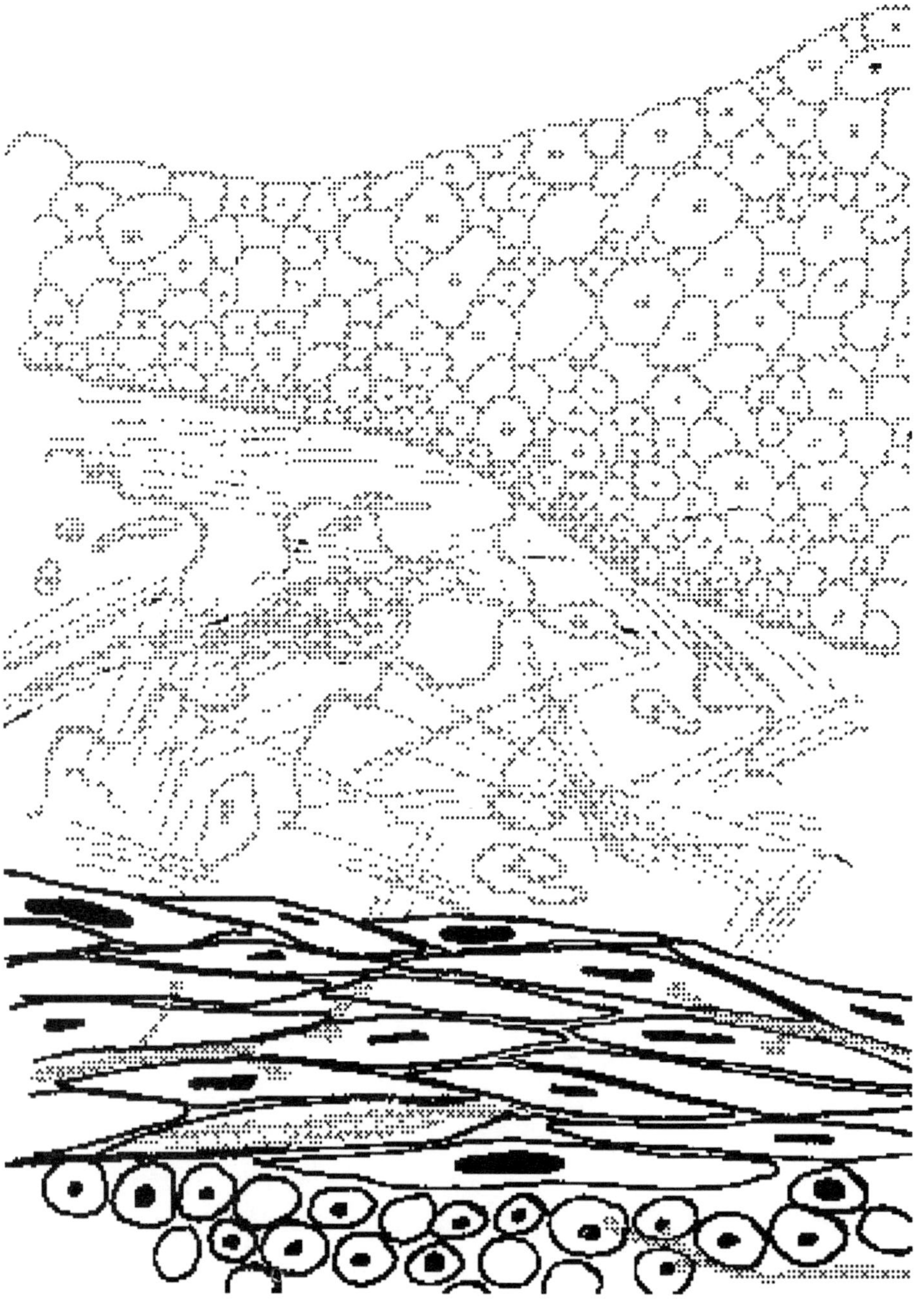

chapter 4: **Muscle Tissue**

In this chapter we only will discuss the features of **muscle tissue** which distinguish it from the other three tissues and divide it into the three types of muscle. Further discussion of any specific muscle is found where that muscle's system is discussed.

GENERAL FUNCTION

• Muscle tissue has one unique function that no other tissues can copy. It can work to create a force of contraction and then quit working or relax (Fig. 4-2). The force of contraction attempts to bring the ends of the muscle tissue together.

If the muscle tissue works successfully, it contracts. It becomes shorter but bulgier. Its two ends pull toward each other and whatever they attach to also will be pulled closer together. This creates physical movement. To work with physical movement is the one and only job of muscle tissue. It can contract to create a movement, or it can contract to oppose or reduce a movement, such as gravity pulling everything down (Fig. 4-1).

The success of a force of contraction depends of the resistance to the movement. Resistance to movement may come from other opposing muscles, other body tissues such as the ligaments, or external objects.

• Relaxation does not change the length of a muscle tissue, but while it is relaxed it may be passively stretched or further compressed by some other process.

Figure 4 - 1
Muscles contract to **(a)** create or **(b)** oppose movement.

Figure 4 - 2
Muscles can contract or relax.

We are often tempted to use the word expansion when describing the lengthening process of a muscle. I caution against this use as the word expand connotes an active working of the muscle as it lengthens. While a small amount of energy is consumed as a muscle lengthens, this is its resting phase. A muscle cannot create a force of pushing or expanding as it lengthens.

LOCATION

Our body puts muscle tissue wherever it needs to physically move something or to hold against physical movement. This includes the visible movement of our body as well as many invisible internal movements such as blood moving through the blood vessels and food moving through the digestive canal. The classification of muscle tissue lists many examples of muscle locations throughout our body (Table 4-1). The only movement that does not use muscle is molecular movement, like transporting a chemical across a cell membrane.

STRUCTURAL CHARACTERISTICS

• The cells of muscle tissue are long and slender in shape, like a finger. They are collected into close groups which vary in size from microscopic to immense. Within the groups, our muscle cells are oriented parallel to each other (Fig. 4-3).

• Muscle tissue is always mixed with connective tissue. Within a group of muscle cells, each cell is individually wrapped in a thin, transparent sheet of connective tissue, like cellophane (Fig. 4-3). This connective tissue has three important jobs: **1.** it shapes our muscle tissue, **2.** it holds our muscle cells together, and **3.** it attaches the two ends of our muscle cells to adjacent structures. As muscle tissue contracts, its connective tissue wrappings are compressed and the adjacent attached structures are pulled together.

• Muscle tissue is very vascular, with many blood vessels supported in the connective tissue wrappings of our muscle cells (Fig. 4-3). Colored pictures usually stain muscle tissue red to indicate its prolific vascularity.

• The connective tissue around our muscle cells supports many nerves.

Some nerves boss our muscles around. They carry instructions from our brain, telling our muscles when, how hard, and how far to contract. These nerves are called **motor** nerves (Fig. 4-4).

Some nerves carry information to our brain about the activity of our muscles. These are called **sensory** nerves. They inform our brain about how a muscle is contracting, if the job is getting done or not. Our muscle's sensory nerves also tattle when something else messes with our muscle to change it, either stretching or compressing it, for example, if gravity pulls us down or someone else pushes our arm. By our sensory nerves we can feel what is happening in muscle tissue. We know that it is moving without looking.

Figure 4 - 3

Muscle cells with their connective tissue wrappings and blood vessels.

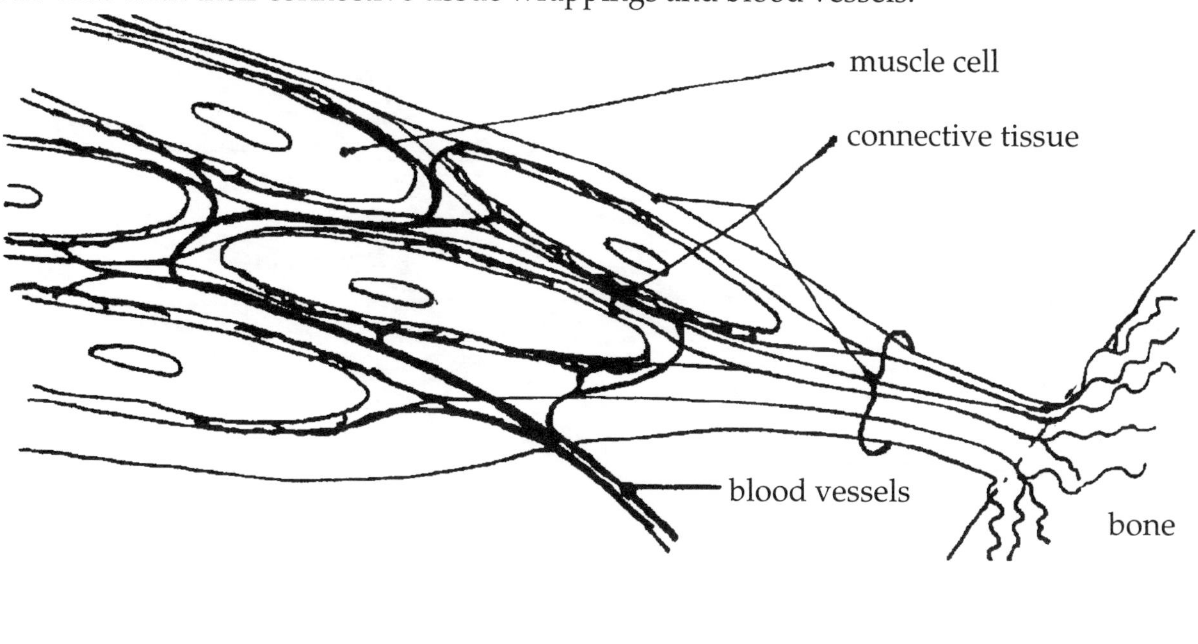

Watching with our eyes as we move, stimulates certain sensory nerves, adding information, making us more stable and confident, but we are not dependent on looking to know if or how we are moving.

Sensory nerves measure and take note of two things about muscle: **1.** a change in the length of our muscle tissue, and **2.** a change in tension or pulling through their connective tissue wrappings onto their attached structures (Fig. 4-4).

Our sensory nerves do not measure an absolute quantity but rather an interval of change. After a certain amount of change in length or tension, the appropriate nerve is stimulated and transmits this information to our brain.

In the muscles of our skeleton, the interval is very small so we feel the tiniest movements. We can know precisely how we are moving. This is explained further in chapter 10, Skeletal Muscle Theory under Sensory Nerves.

In the muscles of our internal organs, the interval is variable, but usually quite large. We feel certain things happening in our internal organs but not so distinctly.

When our urinary bladder is filling with urine, its muscular wall is being stretched. After a certain interval of lengthening its sensory nerves inform our brain. This is when we feel that our bladder is ready to empty.

The intervals in our urinary bladder are quite large, stretching to hold about one cup of urine before it signals our brain. If we postpone our trip to the washroom, we lose the sense of needing to urinate and forget about it until the end of the next interval is reached and another report is sent, usually with a little more urgency.

SOURCE OF NOURISHMENT

Muscle contraction consumes large amounts of energy, so muscle tissue hordes more than its share of the glucose and oxygen available in our body. Muscle cells have an ample storage capacity for these nutrients, but are dependent on a good blood flow for their supply. Blood vessels surround each muscle cell, supported in the connective tissue wrappings as discussed above (Fig. 4-3), so blood should be readily available.

Figure 4 - 4
A muscle cell responding to motor nerve stimulation.

CELL REPRODUCTION

In contrast to epithelial and connective tissue, muscle tissue is not able to reproduce and make new cells after the young age of 2 years old. That sounds crazy. Think of a delicate, wobbly 2-year-old and a bulging, weight lifting 23-year-old. Can you believe that they have the same number of muscle cells? They do. Each cell is just a lot bigger. As we grow or workout, we enlarge the cells we have, we don't make new ones. If one muscle cell dies you will always have one less muscle cell.

During surgery there is almost no damage to muscle tissue. A surgeon never cuts through muscle cells, but separates the muscle cells, going through the connective tissue between them. A surgeon never wants to weaken the muscular support of our body.

TISSUE REPAIR

Fortunately muscle tissue is healthy, with few fatal illnesses. Trauma and oxygen deprivation are its major killers.

- When muscle cells are a little damaged or inflamed, they can heal or repair themselves. When muscle cells are fatally injured, however, there is no cellular replacement. Since muscle cells cannot reproduce, healthy neighbor cells cannot replace the loss. The most they can do is enlarge and strengthen to compensate for it. Losing one cell is not that much to cry over but if it is a larger segment of muscle tissue, it can cause weaker, inefficient muscle contraction.
- When muscle cells have been killed, their surrounding connective tissues clean up the cell debris, fill in the vacant area, and continue to hold the remaining, healthy muscle tissue together. A hard indentation in the area of an old wound indicates where this has happened (Fig. 4-5). The connective tissue prevents things from falling apart, but it cannot produce a force of contraction, so muscular strength is lost.

In a heart attack, cardiac muscle dies from lack of oxygen. Connective tissue fills in the space so there is not a hole, but the heart's contraction may be less efficient, depending on the amount of tissue lost.

A few visceral muscles reproduce new cells in certain circumstances, for example, the pregnant uterus. Correct artificial stimulation also may cause cell reproduction. However, generally when our body is left to its own devices, no muscle cell reproduction is a hard and fast rule.

Figure 4 - 5
When muscle cells are killed, their connective tissues fill in the area and hold the remaining muscle cells together.

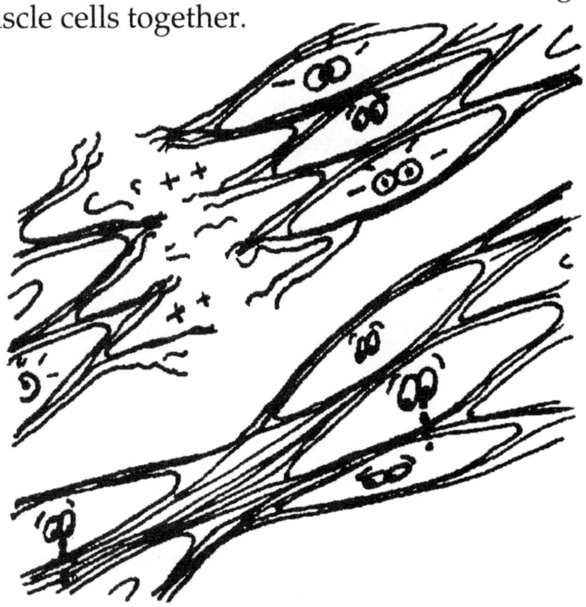

CLASSIFICATION

All the muscles of the body can be divided into three different types: **skeletal muscle, visceral muscle,** and **cardiac muscle.** These three types differ from each other in four ways: **1.** by their location and what they move, **2.** by their microscopic appearance, **3.** by what part of the brain bosses them around, and **4.** by their dependence on instructions from their boss. Each type of muscle is discussed in these four ways. This information is summarized in Table 4-1.

Table 4 - 1 Classification of Muscle Tissue

	Skeletal Muscle	Visceral Muscle	Cardiac Muscle
Location	The skeleton	The internal organs	The heart
Microscopic texture	Striated	Smooth	Latticed & segmented
Main brain control	Conscious	Unconscious	Unconscious
Dependent on brain	Yes	No	No
Examples	Movement of facial expressions, eyeballs, tongue, swallowing, speech, breathing, the abdominal walls, the pelvic floor, and the whole skeleton.	Blood and lymph vessels, hair follicles, bronchi, digestive canal, middle ears, gall bladder, spleen, ureters, urinary bladder, uterus, fallopian tubes, pupil, and lens.	Pumping the heart to circulate the blood.

Skeletal Muscle

1. Skeletal muscles create the obvious movements of our body. Their connective tissue wrappings attach to different bones, so contraction of our muscles causes bending at our joints and movement of our skeleton. Under their power we hop, skip, and jump, or hobble through our day.

Skeletal muscles also move certain structures with respect to our skeleton. For example, skeletal muscles stick out our tongue and rotate our eye balls to look in different directions.

2. Seen through the microscope, skeletal muscle has a distinctive cross stripe, giving it the name **striated** muscle (Fig. 4-6).

Figure 4 - 6

Microscopic view of skeletal muscle.

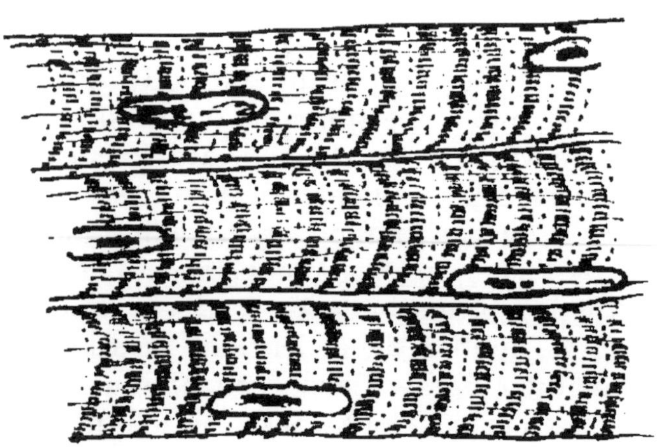

3. The big boss who controls major movements of skeletal muscle resides in conscious parts of our brain. We make a conscious decision, volunteering to work skeletal muscle, so it is called **voluntary** muscle. We do not usually look up and find our hand doing something that we have not decided and told it to do. We may not always want the credit for what happens, but it belongs to us (Fig. 4-7).

On the other hand, it is surprising how much muscle tension and how many habitual movements we do without notice and the small ways that skeletal muscle supports us even if we do not consciously ask. This is discussed further in chapter 10, Skeletal Muscle Theory.

4. Skeletal muscle is completely dependent on our nervous system for instructions. It has no mind of its own. It is very obedient, contracting only when and how much our nervous system tells it to. Without instructions, it does nothing (Fig. 4-8). Most weakness, paralysis, or tremors of skeletal muscle are caused by a problem in our nervous system. The instructions do not get through to the muscle or are inappropriate. The muscle itself is healthy but movement is absent or inappropriate because of the nerve message.

Unit IV of this book elaborates on the general concepts of skeletal muscle and discusses each skeletal muscle individually.

Figure 4 - 7

Skeletal muscle is controlled by the conscious brain whether we admit it or not.

Figure 4 - 8

Skeletal muscle is dependent on instructions from the brain.

While our breathing muscles are skeletal muscles, controlled by our conscious brain, days go by without thinking about taking a single breath. Our brain has reflex interception mechanisms that keep us breathing when we do not think about it.

Water mammals do not have this reflex and need to think about each breath. Even while sleeping, they need to wake up every few minutes to surface and breathe.

Sensory and motor connections from unconscious parts of our brain to our skeletal muscle explain why our leg jumps in response to the doctor's reflex hammer and the occasional jerk of our body just as we fall asleep.

The reflex hammer is tapped briskly on the connective tissue of our thigh muscle just before it attaches to our shin bone. The tap gives a quick stretch to our muscle and our sensory nerve promptly informs our brain. Our brain is surprised. It had not given instructions to stretch that muscle. Our brain does not like changes without its consent. It tells our muscle to contract back to its original length. This muscle contraction causes the quick kick response.

If muscles are still holding contraction as we lie down in bed, some of the tension will leave as we begin to fall asleep. We sag into our mattress. Even these subtle movements will be noticed by our sensory nerves and again, our brain is quickly informed. It is irritated that movement took place without its permission and orders our muscles to take their unsagged length. This quick reposition is the jerk that wakes us up.

Visceral Muscle

1. The function of many body systems besides the skeleton requires physical movement.

We consciously poke food into the entry of our digestive canal, swallow it, and forget about it. The next day it's pushing at the exit door. In one day it has been moved through a canal nearly 30 feet in length.

Blood vessel walls change their diameter in order to change the amount of blood flow to an organ. For example, our skin may be blushed rosy-pink at one moment and blanched ghostly white at another.

As we urinate, our bladder wall squeezes the urine, squirting it out.

Our eye's pupil dilates and constricts.

All these movements are created by muscle tissue within these organs. All physical movement in our internal organs is performed by muscle. A collective name for internal organs is the **viscera**, so this muscle is called visceral muscle. Table 4-1 lists many other examples of visceral muscle. We will discuss these muscles as we go through their systems in Volume Two.

2. Through the microscope, visceral muscle appears amorphous, with no internal structure or texture, so it is called **smooth** muscle (Fig. 4-10).

3 The big boss of visceral muscle is in unconscious parts of our nervous system. We hardly know what happens in our internal organs, let alone make conscious decisions about them, so this muscle is called **involuntary** muscle. When something escapes inadvertently, our conscious mind can rightfully deny responsibility. If we had to consciously organize all our visceral movements, it would take at least two brains for one body or we would have no time for anything else.

We can learn to exert some control over our visceral functions, but it takes a lot of concentration and practice.

Figure 4 - 10

Microscopic view of visceral muscle.

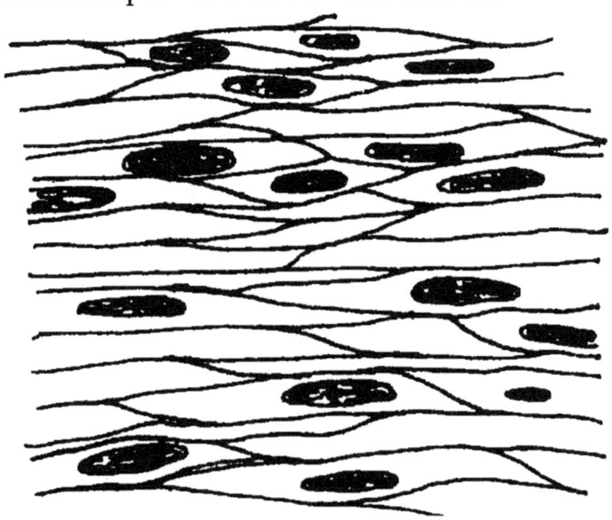

Figure 4 - 9

If the nerves form the brain to the urinary bladder are damaged, the bladder can still contract.

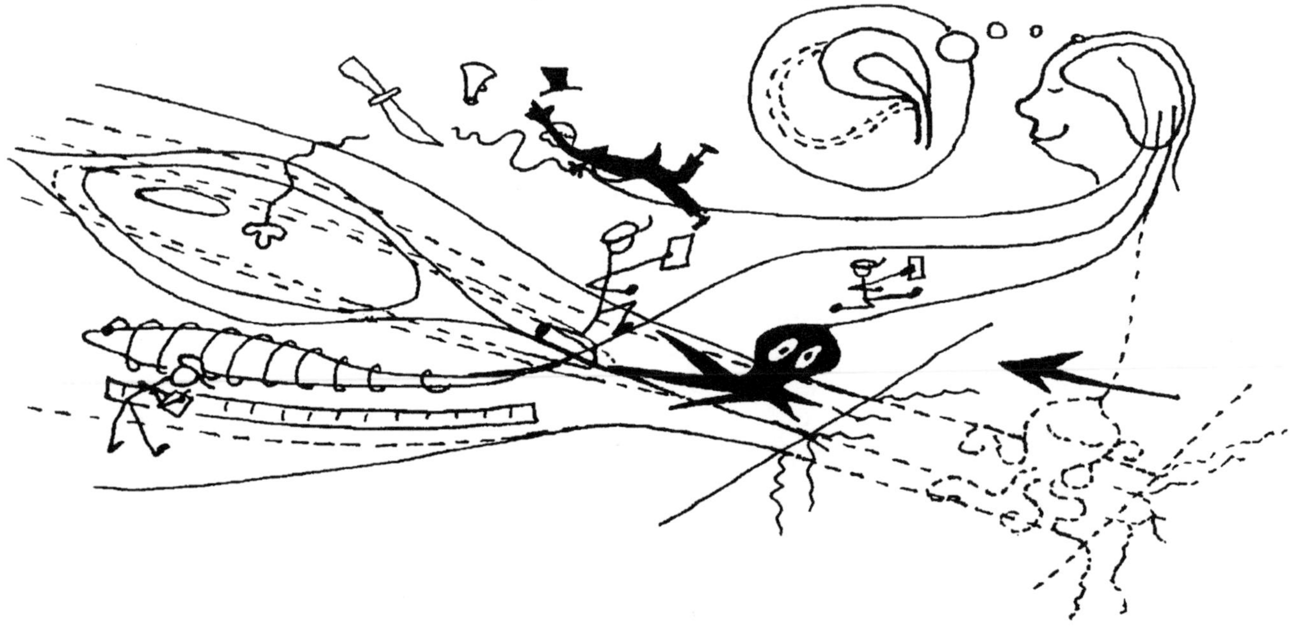

4. While visceral muscle does receive instructions from our nervous system, it can contract without them. Visceral muscle has an inherent ability to initiate its own contraction (Fig. 4-9), and is called **automatic** muscle. This is fortunate. Skeletal muscle impairment due to nervous system damage is bad enough. If all our visceral muscle were so completely affected, many nerve injuries would not be compatible with life. We would swallow food and it would just sit in the back of the throat and go nowhere. Our gall bladder could never empty its bile into our small intestine. Our urinary bladder could never empty, and so on.

The importance of nerve instructions to visceral muscle is to modify the contraction to accommodate fluctuating body needs. Without a nerve message, the muscle cannot change its rate or strength of contraction, but there is at least some function.

The automatic contraction of visceral muscle usually follows a rhythmic pattern. Each muscle has its own specific rate.

Cardiac Muscle

Cardiac muscle is sometimes listed as a specialized visceral muscle or it can be called a type of its own.

1. Cardiac muscle forms the wall of our **heart**, and our heart is the only organ which has cardiac muscle. Even the blood vessels connected to our heart only have visceral muscle.

2. When viewed through the microscope, cardiac muscle has segments like bamboo and contains a subtle latticework, but no standard term is used to define this appearance (Fig. 4-12).

3 and **4.** Cardiac muscle is influenced by the same less-conscious part of our nervous system as visceral muscle and is not dependent on nerve instructions to produce contraction, so it is **involuntary** and **automatic**, the same as visceral muscle (Fig. 4- 1).

The inherent rate of contraction of some cardiac muscle is about 80 beats per minute, the fastest of all automatic muscles.

Figure 4 - 11
(a) As our activity changes, motor instructions change the rate of contraction of cardiac muscle, but **(b)** our heart will contract without instructions.

(a)
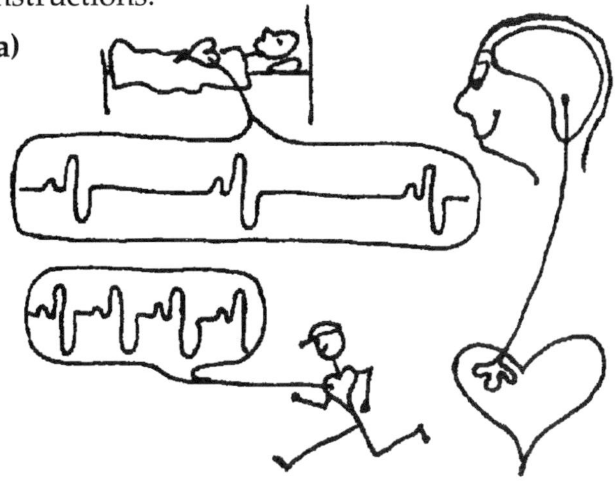

(b)
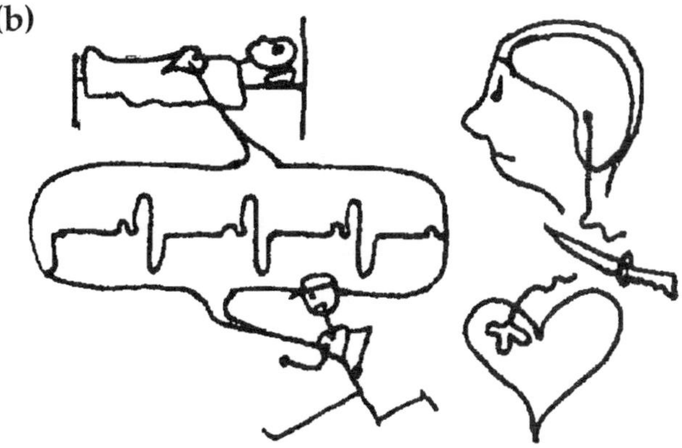

Our whole heart or a single muscle cell from our heart can be entirely removed from our body, no strings attached, and it will continue to contract rhythmically as long as it has adequate nutrients.

Figure 4 - 12
Microscopic view of cardiac muscle.

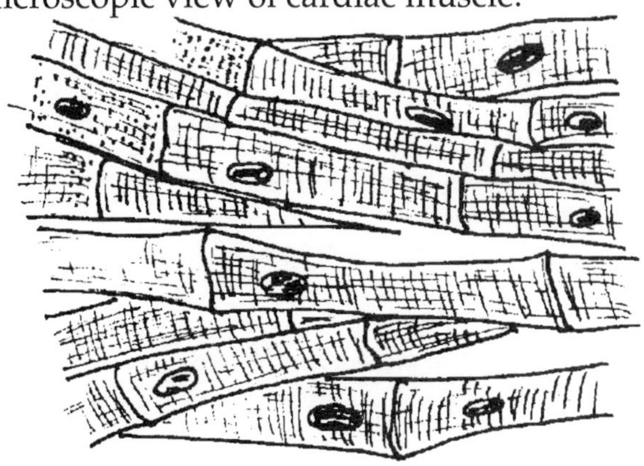

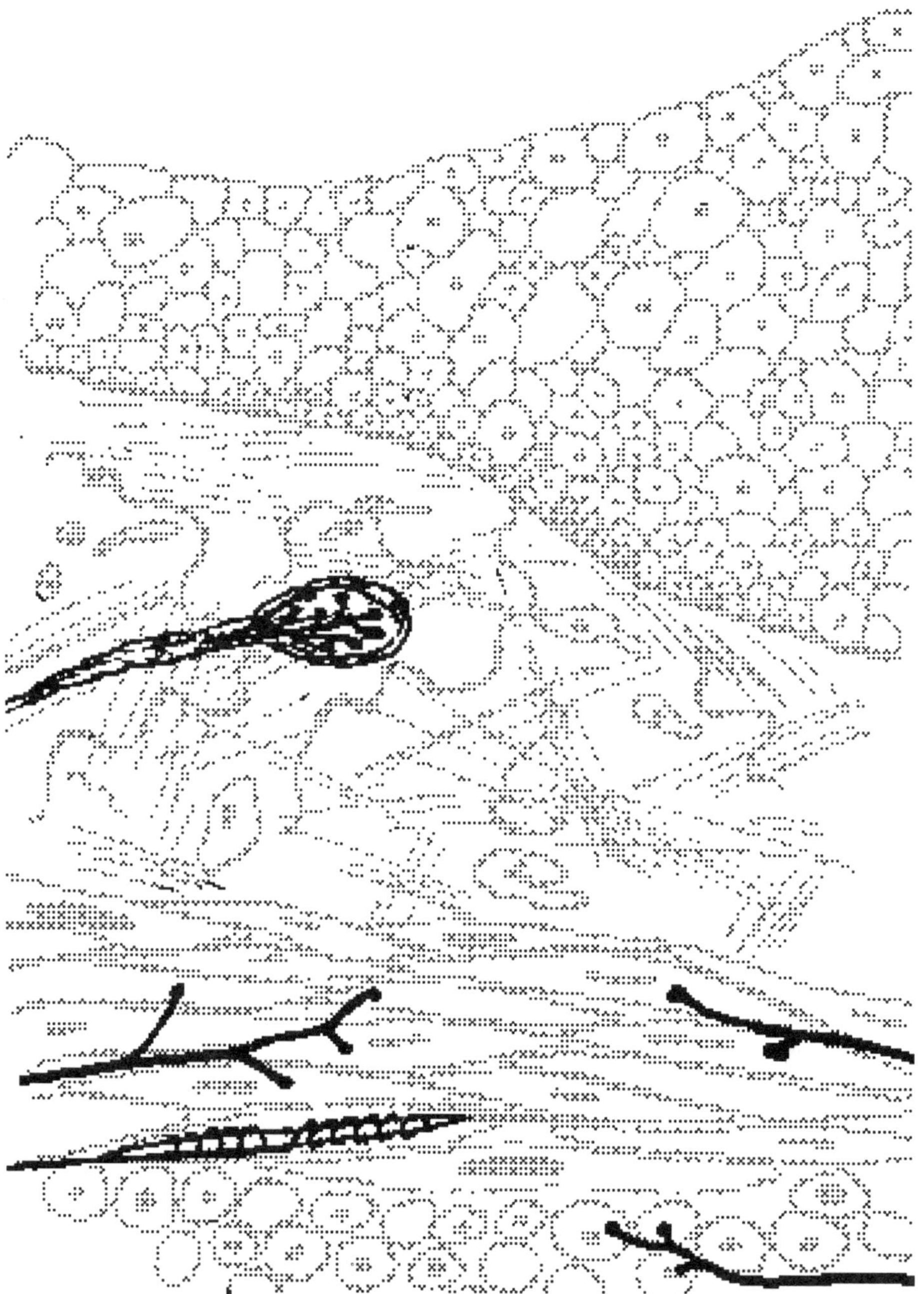

chapter 5: **Nerve Tissue**

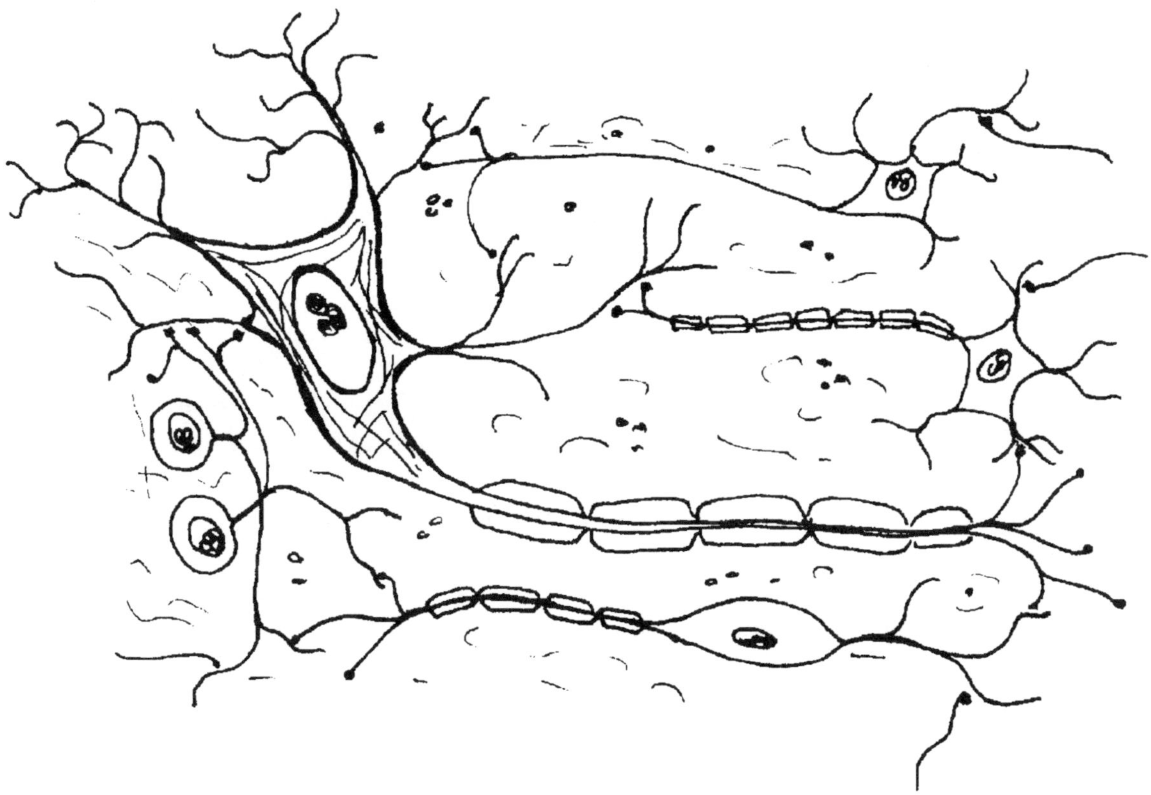

Nerve tissue is used exclusively to build the organs of our nervous system, including our brain, spinal cord, and nerves. The finger tips of nerves touch many other organs, but do not become part of their structure.

In this chapter, we will briefly introduce nerve tissue, leaving its main discussion for the chapters on the nervous system in volume three.

GENERAL FUNCTION

• The general function of nerve tissue and our nervous system is to communicate within our body and with our surrounding environment. This communication has three aspects. **1.** The first step is collecting information about what is happening within and around every part of our body. To this end, our nervous system has an elaborate **sensory network** throughout our body. The sensory information is conducted to our brain, the headquarters of our nervous system, so it can know what is going on. **2.** Our **brain** performs the second aspect of nerve communication. It decides on an appropriate response, what to do about the situation. **3.** Performing the third step, **motor pathways** conduct the instructions for action from our brain to our body to produce the desired response (Fig. 4-4 and 5-2), controlling and coordinating body functions, and interacting with the world around us.

Nerve tissue works like a relay team, passing a message from one nerve cell to the next along the pathways of the nervous system.

• Nerve tissue has two special characteristics which make this function possible. It is **irritable** and **conductive.**

In general conversation, the word irritable has a negative connotation, implying that something is cranky or easily provoked. When applying **irritable** to our nervous system, there is no negative connotation. It simply means that nerve cells can respond to stimuli. The stimulus is said to irritate the receiving nerve cell, whether we interpret it as a delightful or obnoxious experience. For example, light in our eyes irritates the nerve cells of our retina, whether it is reflected from a beautiful rose or garbage in the streets (Fig. 5-1). A gentle caress and a scratch equally irritate the touch receptors of our skin.

The irritation of a nerve cell initiates a nerve message.

To be **conductive** means that any nerve message can be carried along the nerve tissue, traveling and communicating from place to place within our nervous system (Fig. 5-2).

Figure 5 - 1
Light in the eyes is said to irritate the sense receptors of the eye whether it creates a pleasant or obnoxious experience.

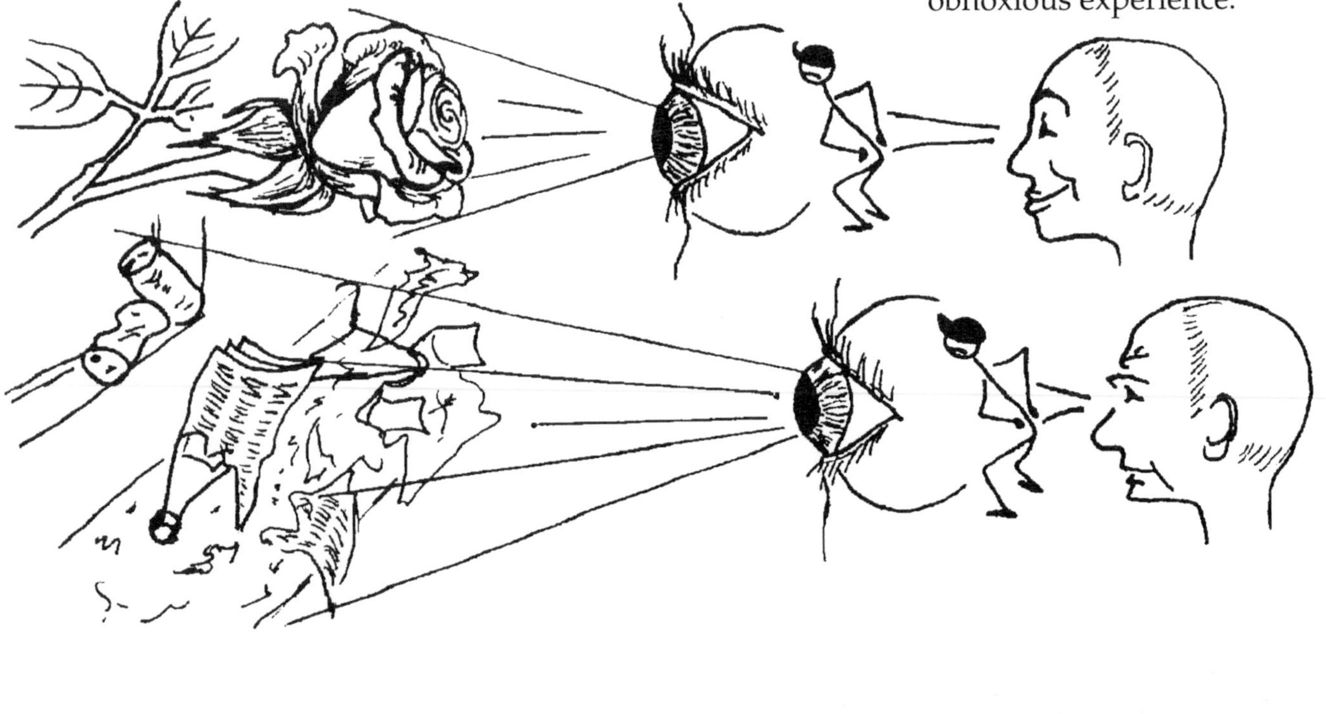

Figure 5 - 2

A schematic showing a thumb tack initiating a sensory message. The message is conducted from the foot, through the spinal cord, to the brain where a decision is made. Instructions for the body's response are sent out along motor nerves, in this case causing the anterior muscle of the leg to lift the foot off the tack. The circles with a central dot represent the nerve cell bodies, while the thickened lines represent the long tentacles of the nerve cells. The short parallel lines crossing the tentacles near the cell bodies are the place where one nerve cell passes its message on to the next. Along the sensory pathway, there are three nerve cells between the sensory stimulus and awareness in the brain (**S 1-S 3**). The motor pathway takes only two nerve cells to go from the brain to the muscle (**M 1-M 2**).

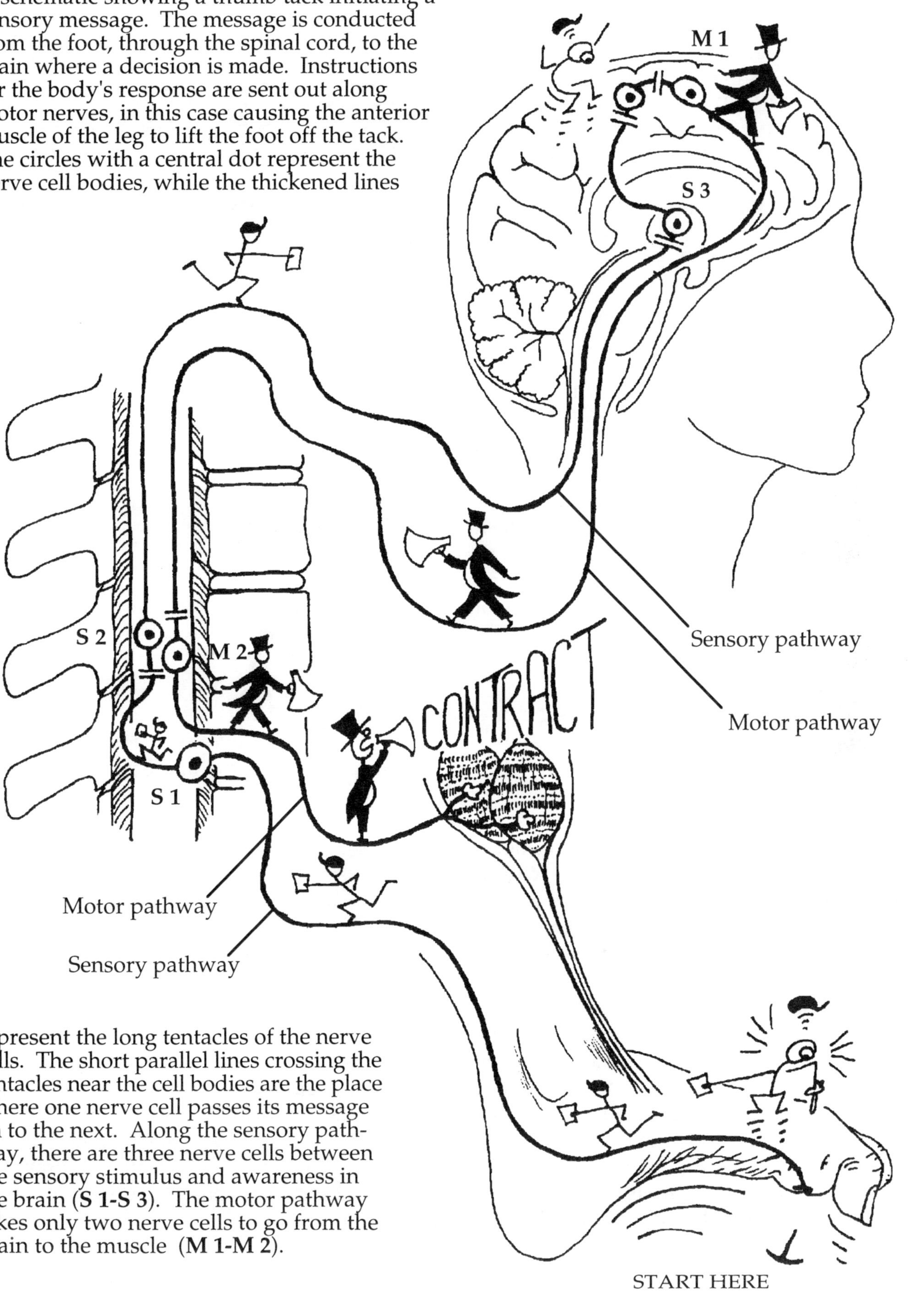

STRUCTURAL CHARACTERISTICS

• Nerve tissue is unique in its structural organization in that all nerve cells throughout our body are in contact with other nerve cells. Since the effect of our nervous system is based on passing a nerve message from cell to cell as much as the activity of a single cell, this structural continuity is vital to its function. Individual cells or clusters of cells which become isolated or cut off from the rest of our nervous system lose their function. When the structural continuity is disrupted, the conducting pathway for a nerve message is interrupted. The message cannot reach its necessary destination (Fig. 4-8). It fizzles and its place in feeling, deciding, coordinating, or commanding our body is lost. Destroying one nerve cell is all it takes, just like removing one domino from a line.

This is distinctly different from our other tissues. Epithelial and muscle tissues have many separated collections scattered throughout our body. Connective tissue has a more continuous weaving, but a proximal lesion does not make the more distal connective tissue fall apart. It does its job wherever it is.

• The cells of nerve tissue have the most amazing shape. Each cell has a central glob, called the **cell body** (Fig. 5-2 and 5-3), which contains most of its organelles and maintains its basic metabolism.

Then, reaching out from its center, a nerve cell has long, thin fingers, like wires or **tentacles.** These fingers or tentacles are like the antennae of the nerve cell, receiving and sending information. One nerve cell may have from two to literally hundreds of tentacles.

The tentacles of a nerve cell are microscopic in width, but may be up to several feet in length, depending on where the cell is in our body and how tall we are. For example (Fig. 5-2), one nerve cell carries the sense of touch from the ball of our foot into the spinal cord in our back. One long tentacle reaches from our foot up through our leg to its cell body which lies against our spinal cord. A second shorter tentacle reaches into our spinal cord to connect

Figure 5 - 3
A nerve cell with its body and tentacles.

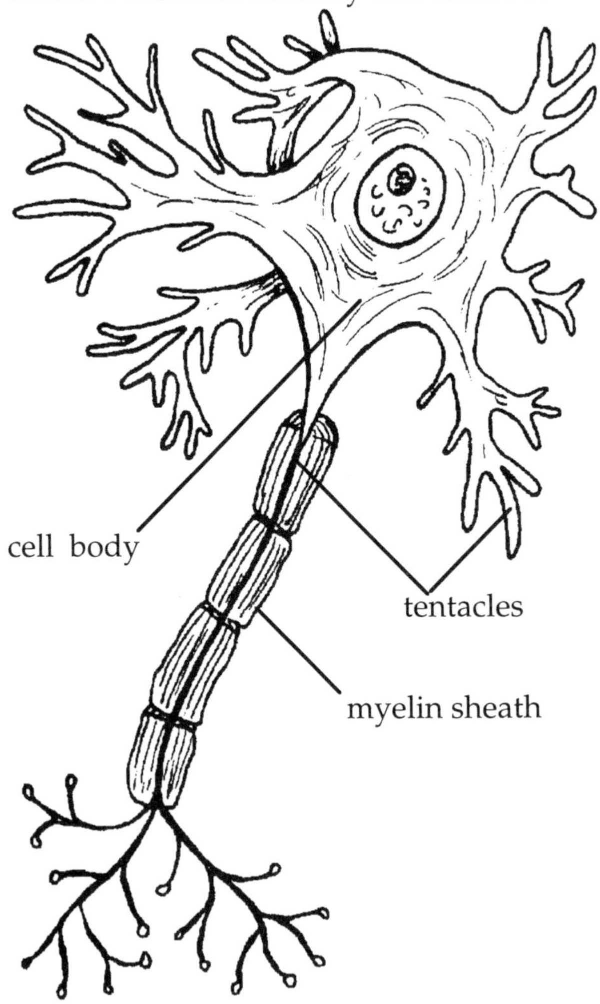

with the ongoing nerve cell. Stabbing the ball of our foot irritates the tip of the first tentacle. The message of being stabbed is conducted up the first tentacle, through the cell's body, to the end of the second tentacle where it is passed on to the next nerve cell and up to our brain. Along a sensory pathway from anywhere in our body, there are always three nerve cells between the sensory stimulus and awareness in our brain.

Instructions to contract a muscle in our leg are carried from our spinal cord to our leg along one nerve cell. This cell has its body within the spinal cord in our back and one long tentacle which reaches down our leg to deliver the message. Any motor pathway uses only two nerve cells to go from our brain to a muscle. Tentacles connecting different parts of our brain may not be so long, but, for their width, they are still fine and delicate.

- Most of what we call our **nerves** are bundles of very long tentacles, while our spinal cord and brain contain most of the nerve cell bodies and shorter tentacles.

SOURCE OF NOURISHMENT

Nerve tissue is very delicate and sensitive. Of all the tissues, it is the least tolerant of fluctuations in body chemistry. Small changes in the concentration of acid-base, oxygen, sodium, potassium, calcium, glucose, and so forth, can cause radical deviations in the function of nerve tissue.

- It is the globular center of a nerve cell which does most of the cell's work and needs an ample blood supply. At the same time, nerve cells are so fussy about their chemical balance, they cannot tolerate even the normal minute fluctuations of blood chemistry. To protect the nerve cells, all their capillaries are wrapped by a cellular barrier which buffers changes in blood chemistry and stores small amounts of essential nutrients.
- Nerve tissue is dependent on glucose as its energy source. During glucose shortage, other tissues adapt, burning fatty or amino acids. Nerve tissue closes shop, leading to muddled thinking, drowsiness, or even coma.

CELL REPRODUCTION , TISSUE REPAIR

- Like muscle tissue, after any part of our nervous system is formed, it cannot make new cells. Many parts of our nervous system are finished making new cells before we are born. The slowest parts are finished by the age of about 2 years old. After this, cell division cannot take place. Cells which die will not be replaced. Injuries which are fatal to nerve cells inflict a permanent loss, even when it happens at or before our birth.
- Nerve cells can change or repair themselves in many ways. Following are four examples. They will enlarge to accommodate for body growth. Interconnections between nerve cells may increase to improve coordination and intellectual associations. New pathways may be found to stimulate new or lost functions. Injured nerve cells may repair themselves or replace severed tentacles.

The part of nerve cells most frequently injured is the long tentacles. These injuries do not usually kill the nerve cell and damaged tentacles can often be regrown. Damage to the globular center of our nerve cell is the most serious and most likely to cause cell death.

When nerve cells are lost, their place is filled in with a connective-like scar tissue.

Figure 5 - 4
A cellular barrier lies between nerve cell bodies and their feeding capillaries.

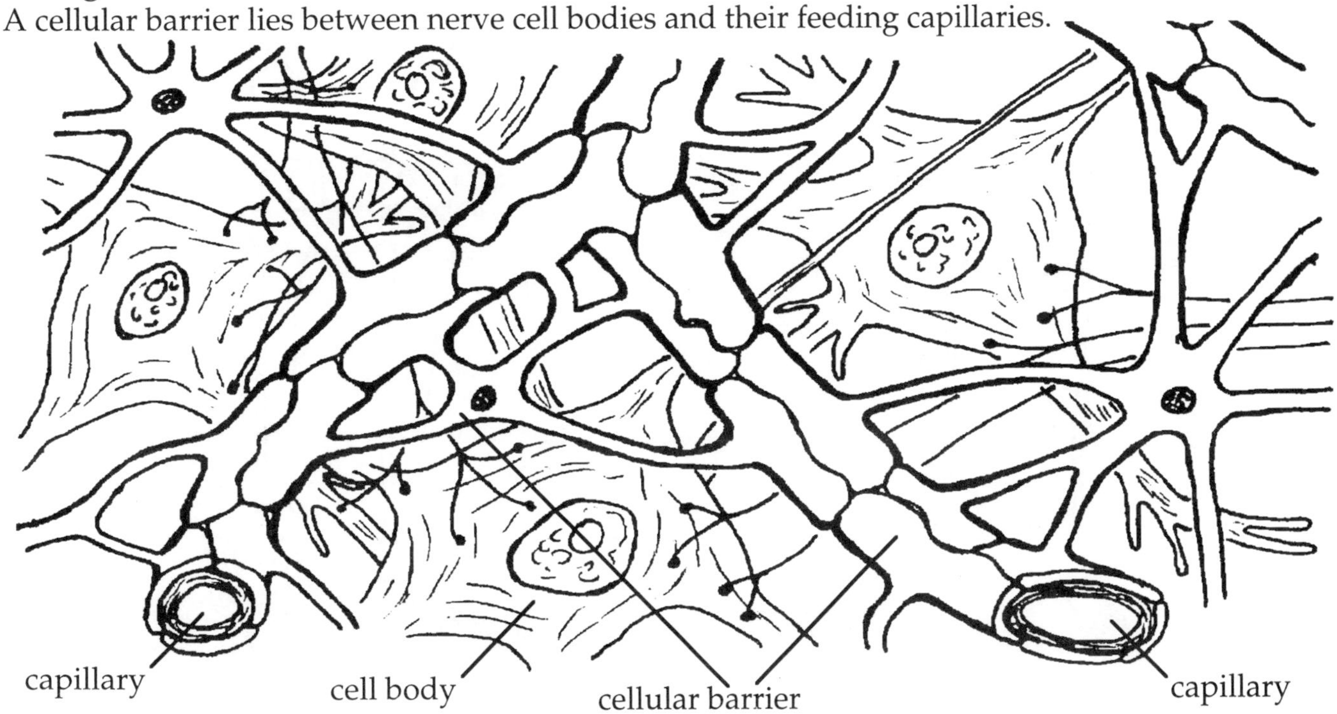

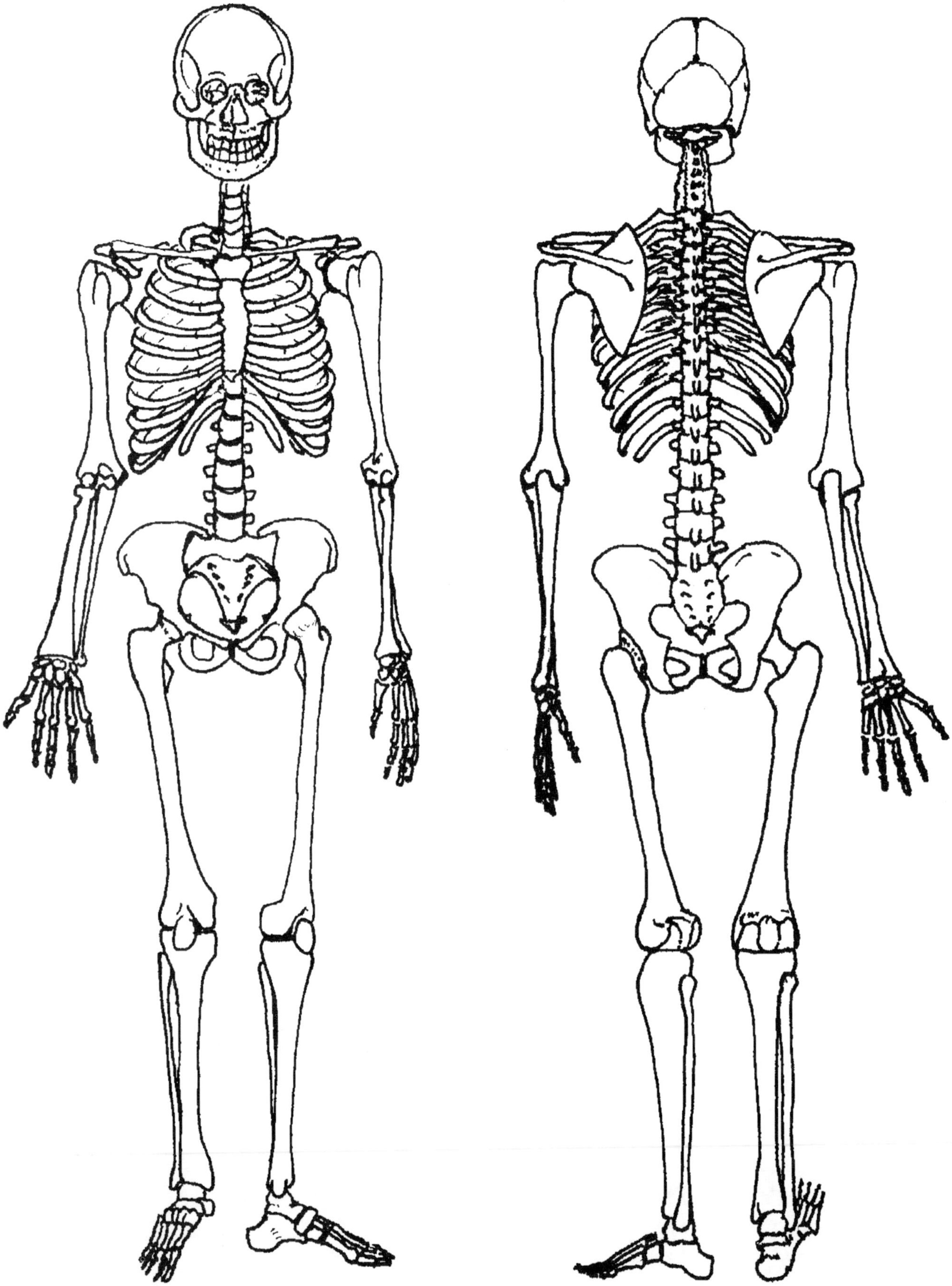

Unit III: **THE SKELETON**

Our skeleton is made exclusively of connective tissue and includes our bones, joints, ligaments, and cartilages. It provides a rigid support structure at the same time as it allows for abundant movement. This brings great pleasure to our life and allows us to respond to changes in our internal and external environment, to maintain conditions essential for life.

While our skeleton is an endoskeleton, lying within the soft tissues of our body, many bones are immediately under our skin and perform protective functions similar to an exoskeleton.

Our skeleton grows as we grow, adapts to changing stresses, and repairs itself with amazing competency.

In this unit we discuss general concepts of the structure, function, formation, and growth of our skeleton as well as the specific details of every bone and joint in our body.

The illustrations which label the individual bones are all drawn to scale - one-half size five-foot-five in an adult. In this way these illustrations are proportionally consistent, depicting the relative size of bones to each other. The exception to this is Fig. 7-10, a lateral view of the vertebral column. It needs to be stretched to be five inches longer. The page simply is not big enough.

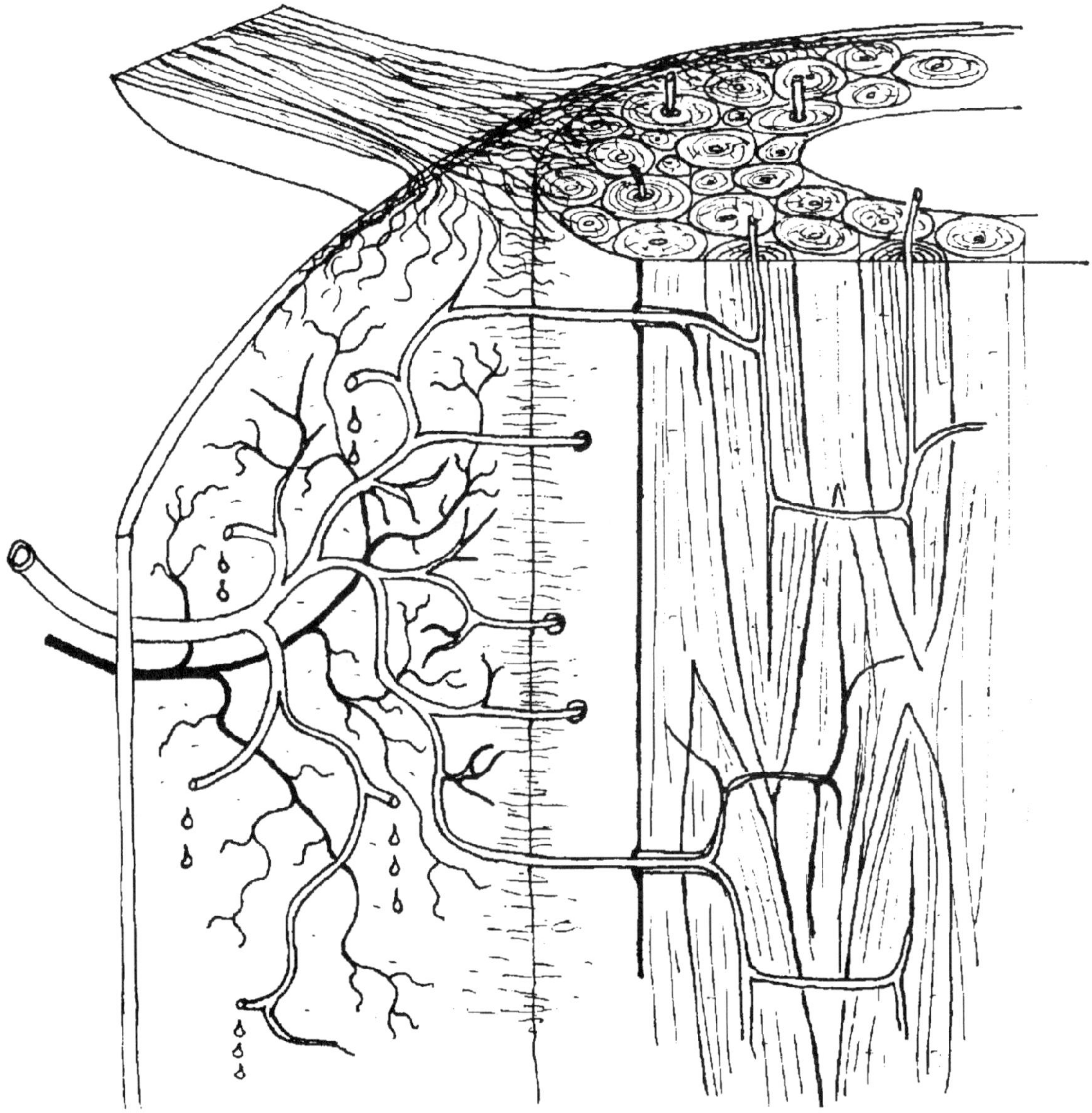

chapter 6: **Skeletal Theory**

Our **skeleton** is the collection of all the **bones** in our body.

• The skeleton and its bones are a symbol of durability. A timeless desert painting portrays a parched rib cage arched against the glaring sun. Walking through a pasture, we may find bony remains which have endured harsh elements and decay long after the skin and flesh succumb. A skeleton hangs in the classroom corner. It never steals a lunch (Fig. 6-1), it never begs a sip of water, it never moves a finger, but at the 25-year class reunion, it's still there with us. We tend to apply this same sense of durability to our own bones. Not so. Bones within a living person or animal are like any tissue, they need oxygen, glucose, water, exercise, and so forth. Without these things they disintegrate.

If bones do not give us immortality, why do we bother with them? A skeleton is so rigid. Who would want one?

Figure 6 - 1

Have you ever caught a skeleton running off with your lunch?

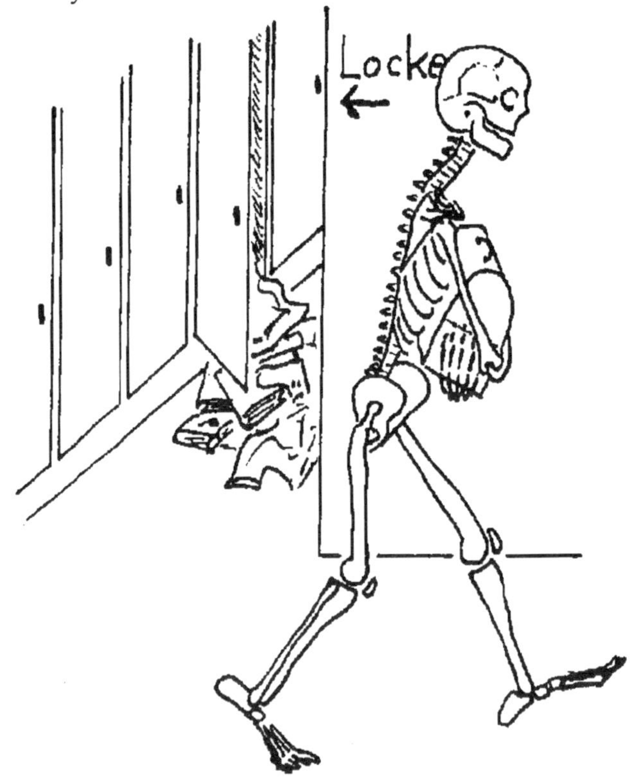

GENERAL FUNCTIONS

Our skeleton performs five functions.

1. The rigidity of our skeleton is precisely its asset to us. It provides a rigid structural support for our body. Without it, we would be a soft blob, like a jelly fish on the ground. This support is a sturdy framework on which everything else in our body hangs, our skin, our muscles, and all our internal organs. Our abdominal cavity is not a bag full of loose odds and ends that fall to the bottom of whatever position we take. Each internal organ is wrapped in connective tissue and suspended from our vertebrae, ribs, or pelvic bones (Fig. 6-2).

Many structures have a rigid supporting framework. The skeleton for a building is a steel or wooden frame on which the dry wall, wires, shelves, or pictures can be secured. The skeleton of a scarecrow is two crossed sticks on which straw and clothes are tied or draped.

By the way, I make no bones about it, I may have pulled some real boners in this presentation, so before swallowing everything I say here, you may want to fish out the bones.

Figure 6 - 2

Internal organs are wrapped in connective tissue and hung from bones.

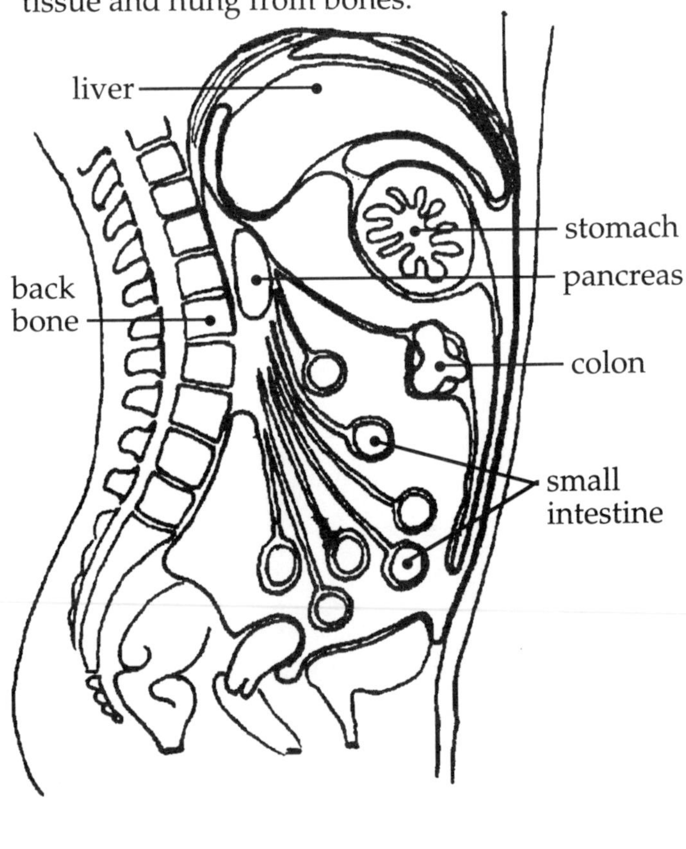

As a structural support, the human skeleton has two unique qualities: we are masters of balance and movement.

The upright human holds a precarious position. It is truly an amazing architectural feat to balance such a tall, slender structure on such a small base. We have no foundation, no roots and no supporting guy wires anchored into the ground (Fig. 6-3). The wind blows and still we stand upright. Arrogantly we lean forward, daring gravity to pull us over. If that is not challenge enough, our skeleton has more than one hundred points of potential, simultaneous movement. A scarecrow's skeleton offers little opportunity for getting around or changing its pose. Buildings have even less. Their only relief comes when the slam of a door vibrates through them, unless they are lucky enough to be on a fault line when the earth shrugs its shoulders.

Our skeleton is a sequence of many small rigid segments called **bones.** The place where any two bones come together is called a **joint.** Our joints are the fixed points or fulcrums around which movement can take place. Our bones act as levers, turning about our joints (Fig. 6-4). Our skeleton does not produce movement, but it makes movement possible, dependent upon the size, shape, and number of bones and joints in our body.

Figure 6 - 4

A lateral view of the bones, joints, and potential movement of the hand.

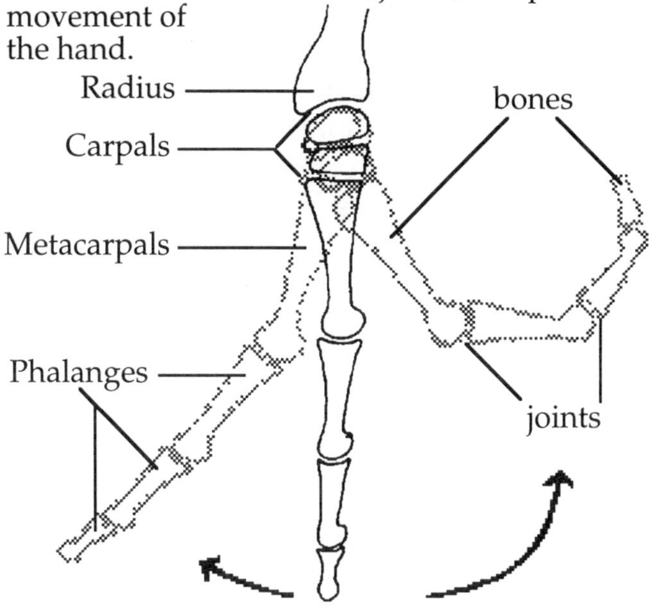

Figure 6 - 3

What if people needed foundations, roots, and guy wires to hold an upright posture and the scare crow went dancing off?

2. A second function of our skeleton is to protect. Most internal organs are protected by bone, much like the juicy parts of an insect are wrapped by its exoskeleton. Bones surround our delicate body structures like shields to ward off or absorb significant bumps and blows. Our skull bones protect our brain, our ribs cage protects our heart, lungs, and upper abdominal organs (Fig. 6-5), our pelvic bones protect our lower abdominal organs, and so on.

In our arms and legs our bones are deep, with the soft tissue unprotected on the outside.

3. A third function of our skeleton is to provide a mineral reserve. Sodium, potassium, magnesium, phosphorus, and calcium are a few of the minerals stored in bones. Of these, **calcium** is the most important.

While calcium renders our bones hard, it also is essential for many other systemic chemical reactions such as muscle contraction, nerve conduction, blood coagulation, and the transfer of energy in every mitochondria. These reactions require a specific blood level of calcium. Too little or too much, both interfere with normal chemical processes. Bones, on the other hand, can give or take a little calcium and still bear our weight. So, if our blood finds its calcium a little low, our bones are a generous neighbor. If our blood level is too high, our bones can always find an empty spot to stuff the extra calcium to relieve our blood (Fig. 6-6). Our skeletal reserve covers us on low days and gets refunded on the high ones.

In spite of the importance of **calcium homeostasis,** it is difficult to absorb from our digestive canal. Its double positive charge (Ca^{++}) pushes it away from cell membranes. This makes our skeletal reservoir even more important. **Vitamin D** enhances the absorption and use of calcium and other minerals in our body. Two hormones, **calcitonin** and **parathyroid hormone,** also influence the absorption, movement, and elimination of calcium. We discuss these hormones in the Endocrine chapter in Volume Three.

Figure 6 - 5

The rib cage protects the heart, lungs, and upper abdominal organs.

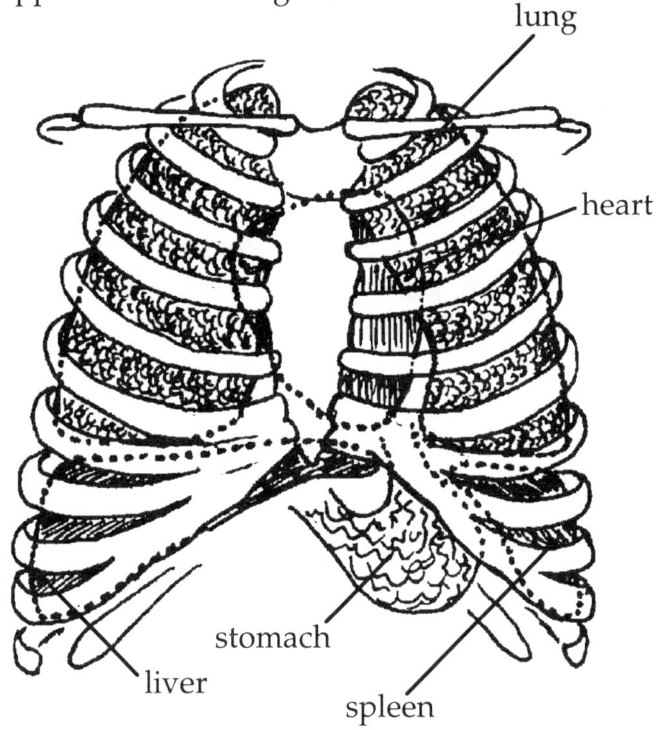

Figure 6 - 6

Bones give or take calcium to keep the blood at the right level.

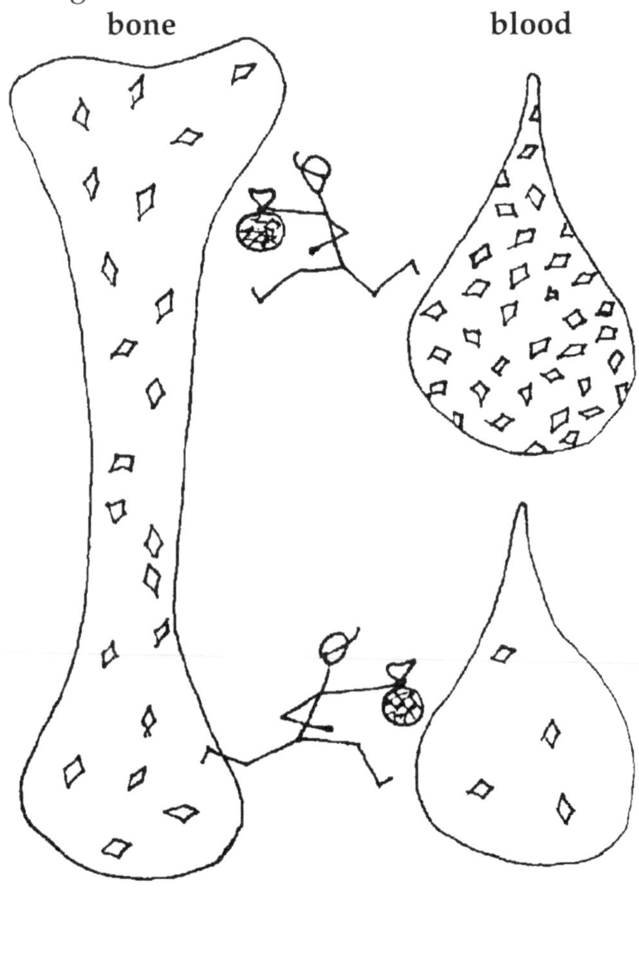

Figure 6 - 7
The blood cell parents actively reproduce to keep the blood supplied with cells.

• The center of bones is filled with a soft, squashy gunk called **marrow.** There are two kinds of marrow, red and yellow. The forth and fifth functions of our skeleton take place in these marrows. As we further differentiate bones, we will locate these two marrows.

4. Our **red** bone marrow contains the parent cells of many of our blood cells**.** The parents stay in our marrow, actively reproduce, and send their mature descendents into our blood to keep it supplied with cells (Fig. 6-7). This is why a person with a blood disease may have a **bone marrow tap** to investigate the disease or a **bone marrow transplant** to introduce new blood cell parents.

5. Our **yellow** bone marrow stores fat as a fuel reserve.

BONE AS CONNECTIVE TISSUE

In the Tissue Unit, bone was listed as an example of connective tissue, and everything we discussed about connective tissue applies to bone.

• Bone is vascular. Blood flows through bone like sap through the limbs of a tree. Microscopic vessels spiral through the length of bone (Fig. 6-8). Next time you grocery shop, visit the meat department. Find a bony piece of meat. It may not be the one you buy, but look for the many red pin points on the cut surface of bone. These points are the cut ends of blood vessels. When bones break, these blood vessels are torn and large amounts of blood may be lost.

• Bones have the three structural components of all connective tissues: ground substance, extracellular protein, and cells.

The ground substance of bone is calcium phosphate crystals and other minerals. These minerals make the bones hard and capable of bearing our body weight.

The extracellular protein of bone is mainly collagen, the strong, non-stretchy fiber. Collagen spirals through the length of bone. It binds to the mineral molecules, holding them in place to maintain the alignment of a bone. The shrink and hardening of bone protein is not significant because of the hardness of the ground substance.

The cells are the active part of bone, so they require good nourishment. They get ringside seats, with several layers of cells encircling each blood vessel. The job of the cells is to build and rejuvenate the protein. Nibblers and builders are both busy, old fibers are digested and new ones built to replace them.

• The three conditions necessary for protein production in any connective tissue are important in bone. To review briefly: **1.** Adequate but not excessive protein must be available through our diet. **2.** Hormones influence the cell's activity, three important ones are growth hormone, the sex hormones, and cortisone. **3.** Bone responds to a test of its strength. Physical activity influences the quantity and organization of bone protein, thus affecting the strength and shape of our bones. Lots of activity, with muscles tugging and body weights pressing, stimulates bone cells to lay down protein. With reduced activity, bones will reduce the density of their protein along with its attached mineral.

In this discussion it is important to differentiate bone density from bone growth. Stimulating cells to build more protein affects the density of bone. The growth of bone is its own special process which is discussed later.

BONE COMPACTION

A long blood vessel with its surrounding cells, proteins, and minerals forms the basic structural unit of bone (Fig. 6-8). Bone is made of many such **filaments** compacted against each other.

Figure 6 - 8
In compact bone, the filaments are tightly packed.

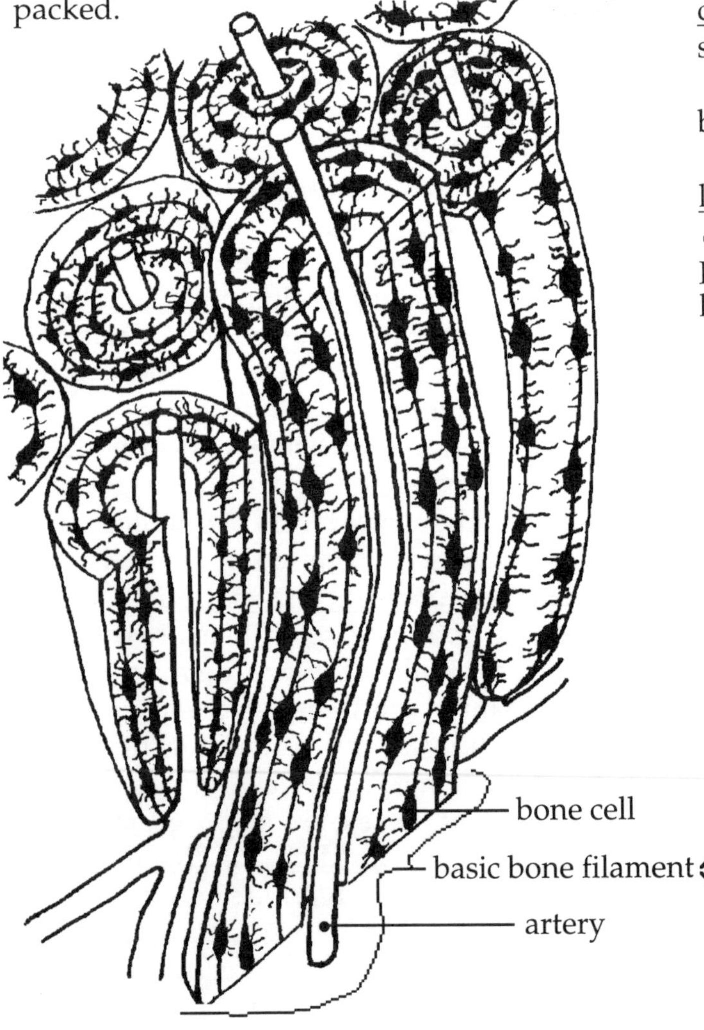

• When the compaction is complete, with the filaments pressed tightly against each other and the marrow all contained in one cylindrical, interior space, it is referred to as **compact bone** (Fig. 6-8 and 6-11). The filaments in compact bone may gently spiral through the bone, but usually are quite uniform in direction, following the length of the bone. The marrow within compact bone is the yellow, fatty type.

• Sometimes the compaction is not so tight. The bone filaments and marrow mingle, dispersed throughout the width of the bone. This is called **spongy bone** (Fig. 6-9 and 6-11). In preserved bones, after the marrow is removed, the many vacant holes give the bone a spongy appearance. But looks are deceiving, this bone is not soft or absorptive like a sponge. It has all the rigidity and strength of compact bone. The same number of filaments are simply spread apart. The filaments of spongy bone generally are scattered into many directions according to the many directions of stress applied through the bone.

Spongy bone has a thin crust of compact bone over its outer surface (Fig. 6-11).

The marrow within spongy bone usually is the red type, reproducing blood cells.

Figure 6 - 9
In spongy bone the filaments are scattered, leaving space for marrow in between.

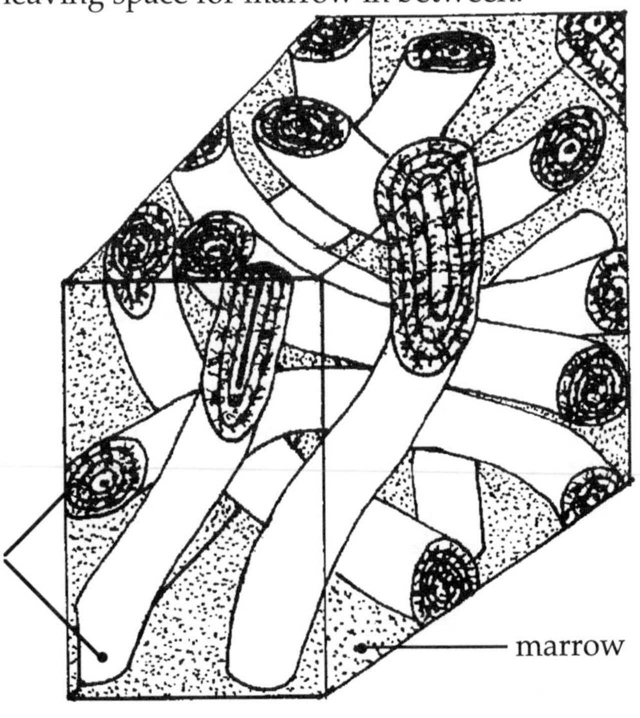

BONE SHAPE

• There is no supreme designer who sculpts the exact size and shape of our bones from a predetermined pattern. Genetic information suggests potentials, but the master stroke of bone shape is in its use. As rigid as they appear, bones are malleable. Their size and shape is formed by the stresses passing through them. Normal fetal movements in utero are the beginning of this process. If our stresses change, the size or shape of our bones also may change. The younger the age, the easier this change takes place.

During my medical education, one radiologist delighted in showing his students an X-ray of the lower leg. It contained only one large bone with a flat top articulating with the thigh bone. The question was, "What bone is this?" and everyone chorused, "The tibia." With sparkling eyes he chuckled, "Wrong. It's the fibula of an adult who had the tibia removed at the age of 6 months." Muscles had been reattached to the fibula. As a toddler and child, this person partook of all the usual activities with no sense of a deformity. Since this fibula took all the stresses usually applied to the tibia, it developed the shape of the tibia (Fig. 6-10). Only upon careful inspection were discrepancies apparent. This is a remarkable example of the formability of bone.

As a rule, there are standard patterns of movement among humans, so human bone shapes follow a standard pattern. Each person molds his/her individual differences into this basic shape.

> When nervous system diseases, such as cerebral palsy, or head injuries produce chronic muscle spasm or paralysis, bones may distort or stay small because of unusual stresses or the lack of normal usage.
>
> Bones deformed by fracture will slowly remodel towards their previous shape, as long as they are used in a normal manner. If the person continues to limp, uses the body in a distorted posture, or abstains from using the injured body part, the continued deformity of bone is encouraged.

Figure 6 - 10
A fibula which bears the stresses of the tibia is reshaped to resemble the tibia.

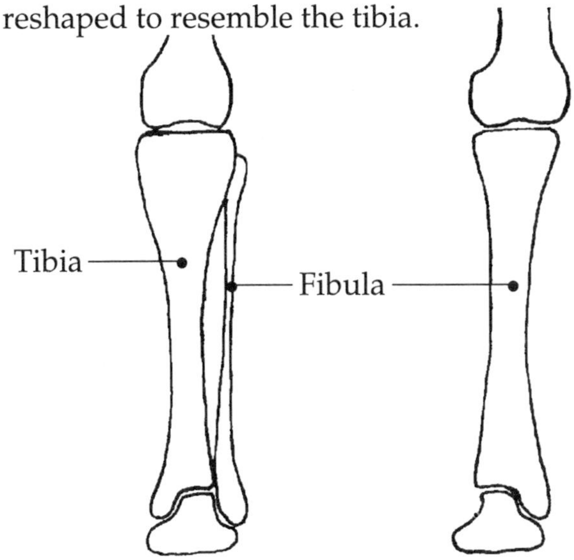

• Bones are sometimes categorized according to their shape.

Most bones in our arms and legs are **long bones** (Fig. 6-11). Their **shafts** are long and cylindrical, increasing their leverage for movement. The shafts are made of compact bone and are filled with fatty, yellow marrow. Both ends are called **epiphyses** (epiphysis in singular). They are rounded and knobby, providing a generous surface for weight bearing and muscle attachments, increasing the stability of the joint. The epiphysis is spongy bone with red marrow making blood cells. It is covered by a thin crust of compact bone.

Our shorter rounded bones, such as those in our wrists and ankles, are called **short bones.** They generally have an outer layer of compact bone with interior spongy bone.

Many of the protective bones in our head and torso are **flat** and plate-like. Flat bones are a sandwich with two outer plates of compact bone filled with spongy bone and red bone marrow (Fig. 6-11).

All our other odd shaped bones are put into one bag called **irregular bones.** Our jaw bone and the posterior part of our vertebrae are examples of irregular bones. They generally have an outer layer of compact bone with spongy bone inside.

As we examine the individual bones of our skeleton, we will see there are many shapes and sizes from which to pick.

Figure 6 - 11
(a) A long bone, cut in half length-wise. **(b)** A piece from the center of a flat bone

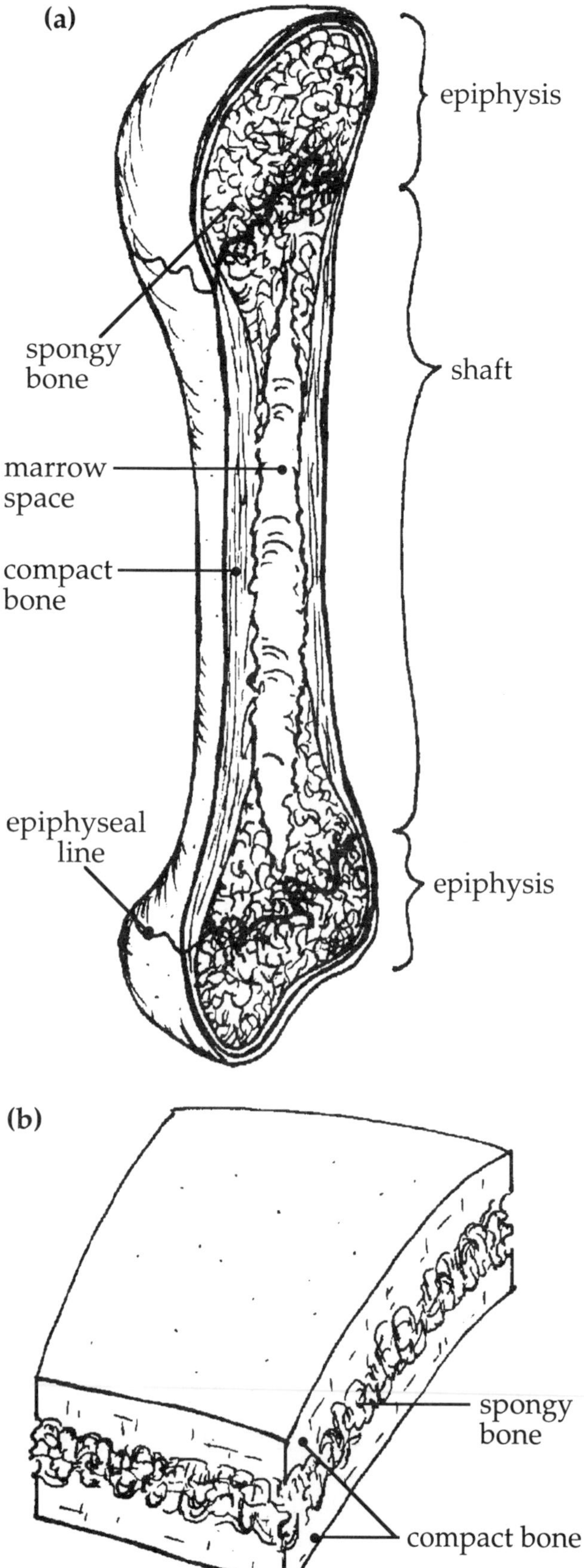

CARTILAGE

Cartilage is connective tissue (Table 3-1) which forms an important part of our structural support. It forms a slippery covering over the ends of many bones (Fig. 6-18). It forms slippery discs within some joints (Fig. 6-17b). It supports structures such as our nose, ears, and wind pipe where flexibility is important. Cartilage ground substance is a stiff gelatinous material. Its protein forms a dense mesh of collagen and elastic fibers. Its cells produce both these materials. Most cartilage is avascular, so its potential to heal is limited.

STRENGTHS OF BONE AND CARTILAGE

Our structural support has two kinds of strength: **compressive** and **tensile.**

To compress something, we press straight into it. Its ability to withstand this pressure measures its compressive strength. The compressive strength of our chair is tested by our weight sitting on it. If its legs do not collapse or squash, we are lucky, it has adequate compressive strength (Fig. 6-12a). If our chair was made out of marshmallows, this would not be true. Marshmallows squash flat under minimal pressure, demonstrating lousy compressive strength (Fig. 6-12b).

Tensile strength is the ability to withstand twisting, stretching, and angular blows or stresses from any direction. The tensile strength of our chair is tested as we lean on it, bang into it, slide across it, slide it across the floor, or tip it over. If the chair does not break, it has adequate tensile strength.

- Wooden blocks stacked squarely on top of each other have good compressive strength, bearing considerable vertical weight (Fig. 6-12d), but a little tap from the side scatters them (Fig. 6-12e). The tap from the side tests their tensile strength, which is very poor. To increase the tensile strength of the blocks, we can attach a thread sequentially to each block. We can repeat this using many threads on all sides of the blocks. The threads hold the blocks in correct alignment during tensile stress (Fig. 6-12c), but they themselves are like wet noodles, with no compressive strength of their own.

Figure 6 - 12

Testing compressive and tensile strength.

(a) Excellent compressive strength.

(b) Lousy compressive strength.

(c) Excellent tensile strength.

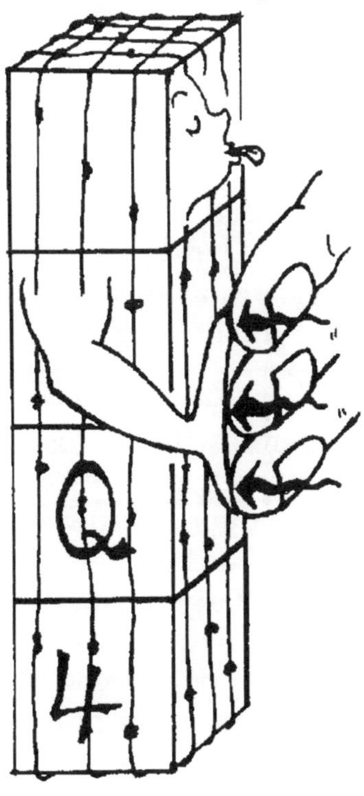

(d) Excellent compressive strength.

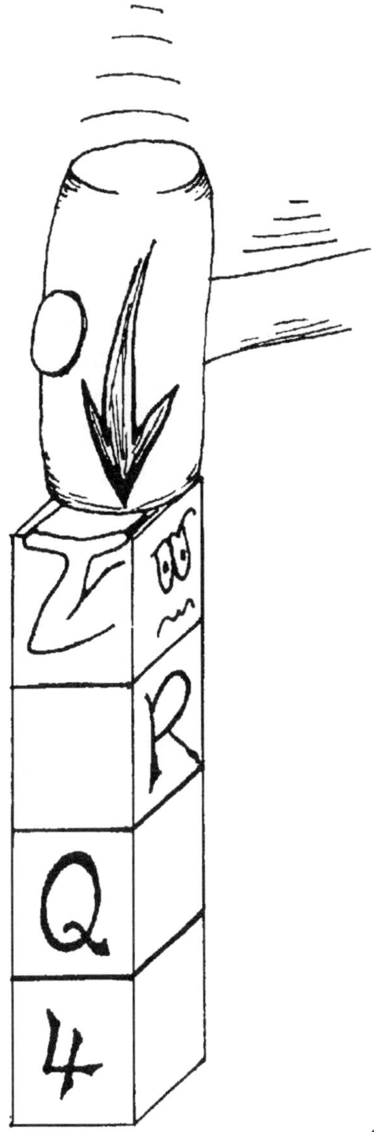

(e) Lousy tensile strength.

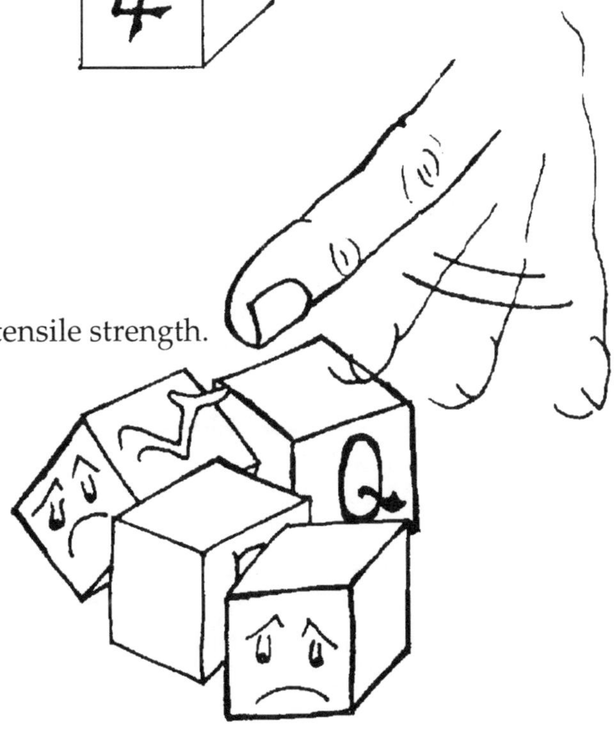

• The structure of bone is very similar to this stack of blocks and threads.

The minerals of bone are like the blocks. They are stacked in bones to bear our weight, maintaining good compressive strength. Bones deficient in mineral are like a marshmallow skeleton. They bow under weight. Their ends enlarge and flatten, producing knobby joints (Fig. 6-13). Rickets is a disease of deficiency of the mineral calcium.

Figure 6 - 13
Bones with poor compressive strength bow under weight and their ends broaden.

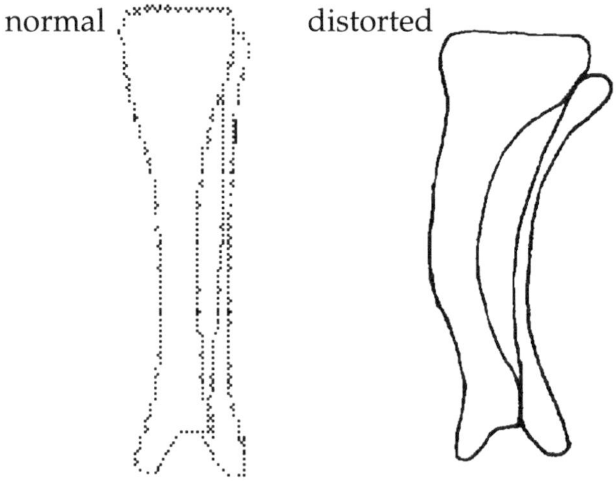

The protein fibers in bone are like the threads, the wet noodles. They have no rigid uprightness of their own, but they bind to the mineral to hold correct alignment, providing tensile strength. Bones deficient in protein become brittle and fracture with the smallest tensile stress, like a stack of blocks without threads. A little stumble or twist might be all it takes to snap the few supporting proteins.

• Comparing cartilage to bone, cartilage has better tensile strength, with less fractures, while bone has better compressive strength, with less distortion under weight. The cartilage in a child's skeleton stands up to all the falls and bumps of developing balance and visual perception. Since children do not weigh much, compression is not a concern. Adults hang their increased weight on a bony skeleton so they get away with it, as long as their coordination skills have improved to reduce the threat of breakage.

> **Osteoporosis** is a controversial disease with many different theories of its cause and treatment. It is apparent that osteoporotic bones have lost their density, being deficient in both protein and calcium. The long bones fracture easily and the vertebrae crumble under the body weight. Public information emphasizes calcium supplementation to prevent osteoporosis. What is often forgotten is that the protein provides the binding sights to hold the calcium in bone. Loading your body with calcium will not replenish protein deficient bones and runs the risk of abnormal calcium deposits and kidney stones.
>
> The first step in preventing osteoporosis is to maintain or rebuild the protein and tensile strength of the bone. Conditions for this are discussed above in Bone As Connective Tissue. During this process it is good to have calcium available to incorporate the lost compressive strength.

PERIOSTEUM

A connective tissue membrane called the **periosteum** covers the outer surface of our bones. It is transparent and thin, but tough, much like cellophane. Short periosteal fibers imbed into the underlying bone like thumbtacks, attaching the periosteum to the bone surface.

Our periosteum performs five functions for bone.

1. Our periosteum distributes nourishing blood to our bones. Periosteum is very vascular. At numerous points its blood vessels penetrate into the underlying bone to form the central artery of our bone filaments (Fig. 6-8 and 6-14). While the outer surface of bone feels smooth, it is peppered with microscopic holes for the passage of these blood vessels.

• Many of the larger marrow spaces have only one big nutrient artery. Its hole through the compact bone is large enough for a stick pin and often is seen on the shafts of preserved long bones. Look for it on the side of a chicken thigh bone the next time you pull one out of the soup.

2. Our periosteum holds a prolific network of sensory nerves which do the feeling for bones (Fig. 6-14). Bones have sensation mainly on their outer surface. There is very little sensation in the deeper parts of bone. Infections or tumors deep inside our bone can progress unfelt until they invade or stretch our periosteum. When this happens, however, the pain may be severe and relentless.

The pain of a bone fracture comes from the accompanied tearing of the periosteum. In fractures with minimal bone displacement, the periosteum may not be significantly torn, so the pain also may be minimal.

The periosteum of our shin bone has very little padding. When banged against a hard edge, its sensory nerves really feel it. They conduct an excited message of pain to our brain.

3. Tendons and ligaments attach to our bones through our periosteum (Fig. 6-14). Their collagen fibers weave into our periosteum and, with the periosteal thumbtacks, imbed into our bones with fierce tenacity.

4. As developing bones grow in length, they also grow in diameter. The inner periosteal cells lay down connective tissue which can be converted into bone for growth in diameter.

5. Periosteal cells clean up the debris following injury to bone and lay down the protein matrix for bone repair.

Figure 6 - 14
A horizontal and vertical cut through bone showing the periosteum partially pealed away.

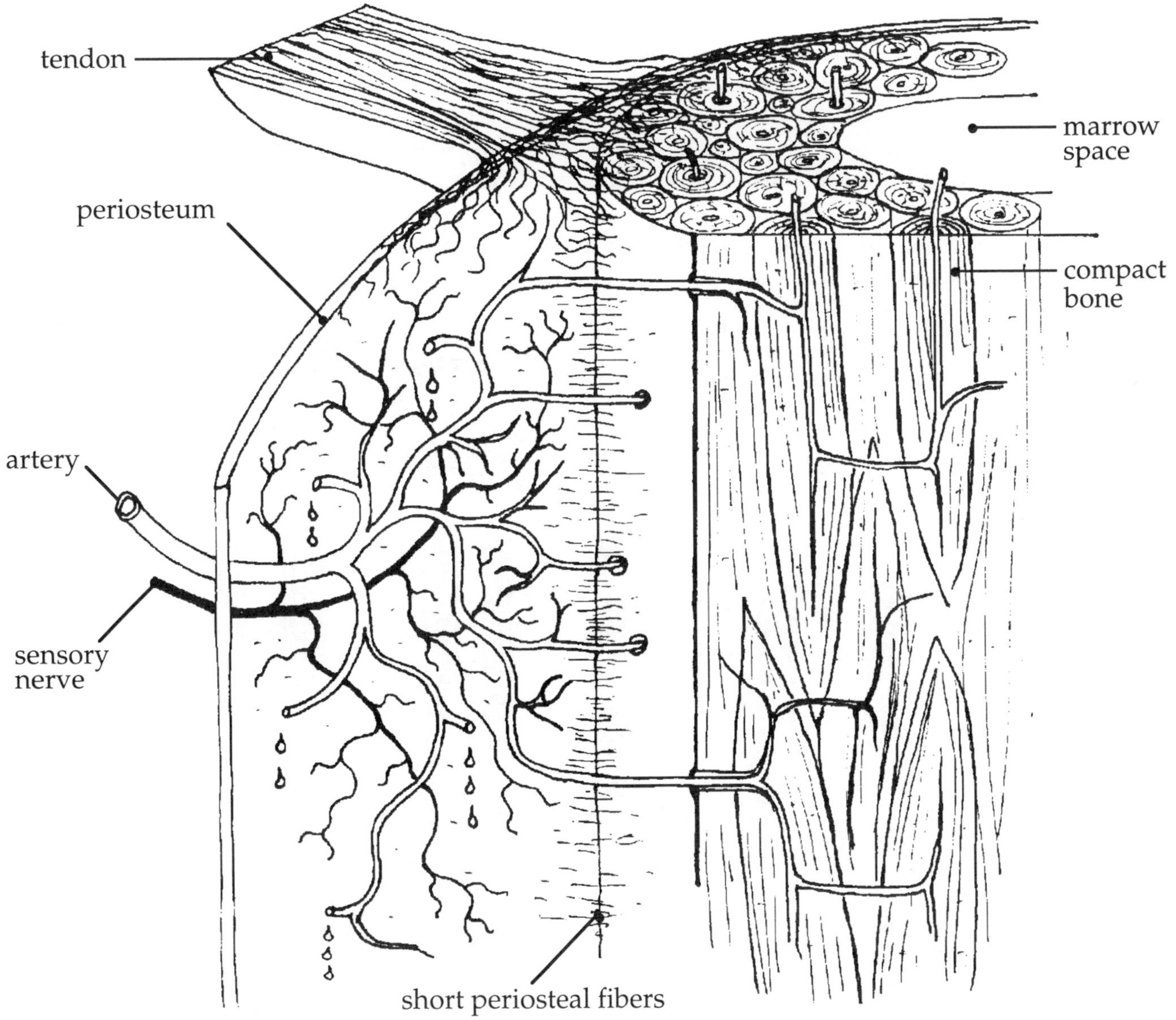

When our bone is **bruised** during trauma, an area of periosteum is pealed off. Its thumbtack fibers and blood vessels are torn. Blood collects between our periosteum and bone. This blood may be reabsorbed by our body or it may be converted into bone, similar to the clot of a healing fracture. The bumps and floating pebbles we feel along our shin are ossified evidence of long forgotten bruises.

BONE FORMATION AND GROWTH

Bones grow in two directions, in length and in width. A different process accommodates each direction.

Unlike most connective tissues, bone cannot form or grow of itself. As an embryo, a template of our skeleton is first built out of a simpler connective tissue and then converted into bone. This conversion process is called **ossification.**

- There are two different types of original connective tissues, fibrous membranes and cartilage. Fibrous membranes are commonly used to make our flat bones Cartilage is used to make our long bones and many of our short and irregular bones.. Regardless of its origin, our mature bone structure is about the same.
- The ossification of each bone begins in the center of its original connective tissue. It begins in a tiny spot, which slowly enlarges, advancing towards the edges. While ossification progresses, the unossified tissue continues to grow, increasing the size of the bone.

Ossification is easiest to see in our long bones (Fig. 6-15). Our tiny embryonic bone is first formed out of cartilage. At a specific time, a band of ossification begins in the center of the shaft of the bone, advancing toward the ends. Some time later, a second ossification begins in the epiphyses at each end of the bone. The cartilage remaining between the two ossifications is called the **epiphyseal plate.** This cartilage continues to reproduce and grow, causing the bone to grow in length. With the passage of time the ossification process over-

Figure 6 - 15
In a long bone, the growth in length takes place in the cartilage of the epiphyseal plate, between the primary and secondary ossification centers. The ossification centers advance toward each other, closing in on the cartilage, reducing the growth potential.

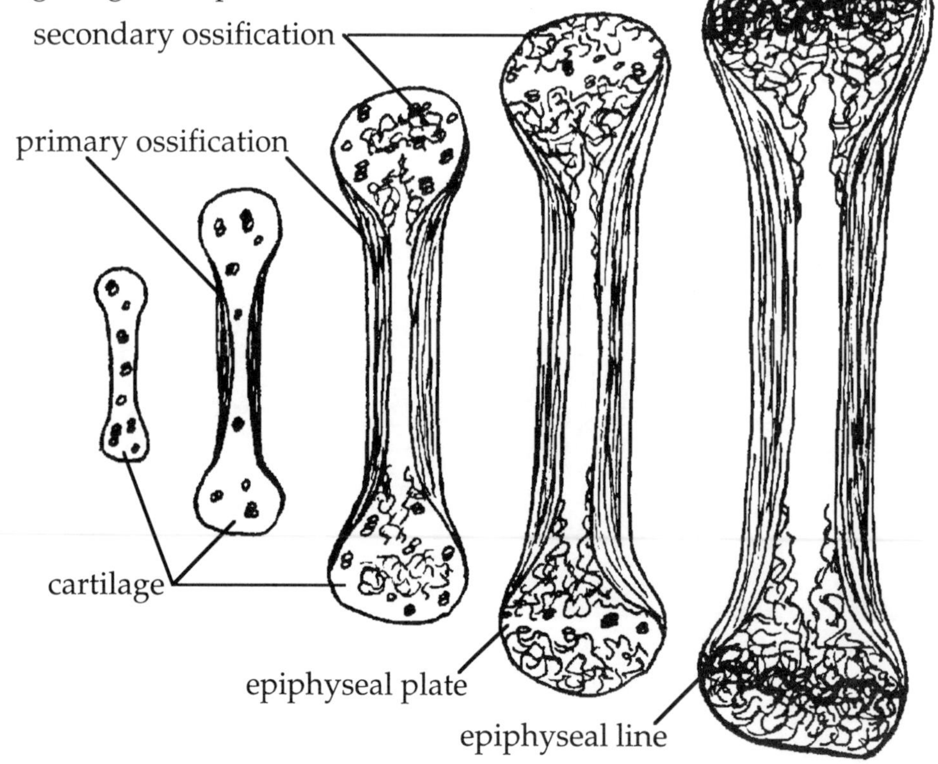

takes the growing cartilage. There is less and less cartilage to reproduce so the growth rate and potential is reduced. As the two ossifications meet, it is called the **closing of the epiphysis.** In mature bones, a dense line indicates where this meeting occurred. This line is called the **epiphyseal line.** It is a jagged line, indicating that the meeting bony surfaces are rough and interlock, like puzzle pieces, to stabilize the shaft on the epiphysis.

• Our skeleton can grow only as long as it has cartilage in its epiphyseal plates. Once completely ossified, bones no longer can grow in length, although their shape may be altered, as we have discussed. The day that every last speck of membrane and cartilage has been converted into bone, our skeleton has reached its end of growth. The ruler on our wall has received its last upward mark. (Although, for better or worse, growth in width is still available to us.) Growth in width of bone is accomplished by the periosteum around the bone and is discussed as the fourth function of the periosteum.

> Every parent has an exceptional child who can do things earlier than expected. It is not good, however, to urge children to bear weight greater than their skeleton can support, whether coaxing a tiny infant to stand or walk, or encouraging younger children to participate in heavy contact sports. Compressive trauma to the cartilaginous skeleton may damage the epiphyseal plate and lead to stunted growth.

> Each bone has its own specific timing for the beginning and finishing of ossification. For any one bone, the timing is the same from person to person. The chronological age of a child can be approximated by examining its skeleton radiographically. Using X-rays to determine the age of a fetus is accurate to within several days.
>
> Some epiphyses only begin to ossify at age 8 or 9 years. Ossification is usually complete by the age of 25 years in women and 30 years in men.

> While our sex hormones stimulate bone growth, they also stimulate ossification, causing the closing of our epiphysis and so, the end of our growth in length.

BONE REPAIR

Our structure has amazing strength, but it is not omnipotent. Sometimes healthy bones fracture because of the massive stresses we put through them. Sometimes a bone fractures under a small stress because some disease process has weakened its structure. This is called a **pathological fracture.** Without pathology, the bone would not have fractured. Diseases such as bone tumors, osteoporosis, or Paget's may lead to pathological fractures. The fracture may be the first indication that something is amiss.

• When a bone fractures, all the protein fibers in a line through the bone have snapped. The disconnected ends of the bone are easily displaced by any tensile stress. The periosteum usually is torn and may be partially separated from the bone.

Blood vessels crossing the fracture are extensively disrupted. In fractures of larger bones, blood loss may be significant. A bulky blood clot accumulates around the fracture. This clot forms the first mending of the bone. It is a dense, fibrous mesh which congeals the area and softly supports the pieces of bone.

Repair of bone utilizes both the nibbler and builder cells of connective tissue, a clean-up crew and a construction crew. Both cells come from the adjacent periosteum. Both crews work at the same time, cleaning up dead tissue and reconstructing the protein continuity, knitting together the pieces of bone.

The fiber of the blood clot is replaced by a fibrous cartilage which assumes the same bulging shape but provides increased support, although normal use would quickly overwhelm its compressive and tensile strength. Fibrous cartilage is abundant within one week after the injury.

The fibrous cartilage has an affinity for calcium and other minerals and is slowly converted into bone. After approximately 6 to 10 weeks, it can support moderate stress and usage.

• The bulging size of the mend disguises its weakness. Immobilization during the repair is essential for the broken ends of bone to knit together but is damaging in two ways. **1.** Immobilization eliminates the normal stress guidelines for the builders so their construction is disorganized and inappropriate. Who could do better when forced to build without plans or a clue of the expected usage of the structure? **2.** Immobilization allows weakening of the rest of the bone as well as all its supporting connective and muscular tissues. This must be kept in mind as usage is resumed so that injury is not repeated.

As normal movement is restored, the nibblers and builders can feel the stresses applied through the bone and remodel it more appropriate to its function. This process happens over many years. The external bulge reduces and the internal strength increases.

> The advantage of surgically pinning a broken bone, rather than casting it, is that movement can be performed so much sooner. The original mend is stronger and more appropriately organized. All the surrounding tissues have less time to deteriorate. This advantage needs to be balanced with all the complications of surgery.

SKELETAL SEXUAL DIFFERENCES

On the average, a male skeleton is larger and heavier than a female skeleton, although there is great variation from person to person and race to race. A male skeleton usually is knobbier, with larger protuberances. The male pelvis is more vertical, taller, and narrower, giving the impression of broader shoulders. The female pelvis is more horizontal, shorter, and wider, with a larger outlet to accommodate the birthing process (Fig. 9-4).

The neck of the femur also is more horizontal in women, giving them a wider stance, with increased lateral stability; this is helpful when pregnancy shifts their center of gravity.

SKELETAL CHANGES WITH AGE

At birth, the most developed system is the nervous system, so an infant's head is relatively large, one quarter of its total body length. In adults, the head is one eighth of the body length (Fig. 6-16). The largest part of an infant's head is the rounded posterior portion which protects the brain. The facial structures are tiny, developing during adolescence.

In a horizontal cross-section, an infant's chest is round compared to the elliptical shape of an adult's. The infant's chest is large and more developed than its pelvis, pelvic organs, and legs.

An infant's arms are more developed than its legs. In adults, both extremities are longer relative to their torso (Fig. 6-16).

Figure 6 - 16
The relative body porportions at four different ages (in years).

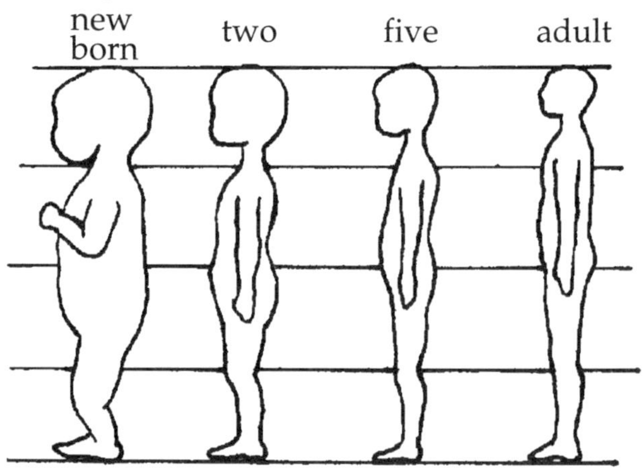

JOINTS

The place where two bones join together is called a **joint**. Joints provide the potential for the movement of our skeleton. They are the fulcrums or pivot points for our bony levers. Another word for joint is **articulation** or articulus, a site of union or junction. Several derivations of articulation are used when talking about joints

Joints present two challenges: **1.** how to lubricate the movement of our bones and all our overlying tissues to prevent friction, wear and tear, and **2.** how to hold our bones together, to stabilize our joint without limiting the range of motion. Generally, the more mobile a joint is, the more unstable it becomes.

All joints can be divided into two categories, non-moveable joints and moveable joints.

Non-moveable Joints

It sounds contradictory to say that the value of a joint is its mobility and yet some joints are non-moveable. Why bother with a joint that cannot move? Actually, all joints are at least slightly moveable. They may move so little that we do not think of them as joints. It may take careful concentration to observe their movement, but that does not discount their importance. It may be essential.

The emphasis in non-moveable joints is that their movement is so small that damage due to friction is not a concern. No lubrication is necessary. The first challenge is met.

Non-moveable joints are built in two ways called **fibrous joints** and **cartilaginous joints.**

Fibrous joints are the simplest. Protein fibers support across the joint, from the edge of one bone to the edge of the other (Fig. 6-17a), holding the two bones together but allowing a small amount of movement. Frequently the bones are somewhat interlocked to increase the joint's stability. Examples of fibrous joints are most joints of our skull and the joint between our distal tibia and fibula.

In **cartilaginous joints,** the ends of the bones are held together with protein fibers

Figure 6 - 17

(a) A fibrous joint. **(b)** A cartilaginous joint.

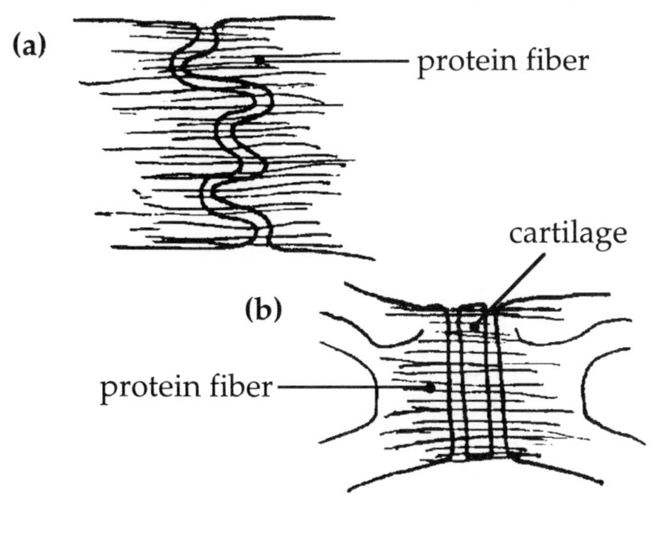

much like fibrous joints, but a flat pad of cartilage is sandwiched between the bones (Fig. 6-17b). Cartilage is dense and smooth. A drop of water makes it slick as a wet bar of soap. The ends of the bones slip over the cartilage with minimal resistance. A pad of cartilage is a simple way to boost the joint's mobility without increasing risks of friction damage. Examples of cartilaginous joints are the joints between the bodies of our vertebrae and the anterior joint where the left and right pubic bones of our pelvis meet.

Moveable Joints

Moveable joints are the ones we usually would point to as joints: our elbows, knees, hips, knuckles, and so on. It is obvious that two bones are coming together and can move extensively. These joints are called **synovial joints** (Fig. 6-18), 'syn' meaning together.

In synovial joints, the ends of the bones which move over each other are covered with a very dense, avascular cartilage called **articular cartilage** (Fig. 6-18). (This cartilage is not part of our initial cartilaginous template of bone and never should ossify.) It provides a slippery contact surface. Our periosteum ends at the edge of our articular cartilage and does not cover its surface.

A loose connective tissue, called the **synovial membrane** (Fig. 6-18), encloses the joint space like a sleeve, attaching into the ends of the periosteum of both bones. The ground substance of the synovial membrane is a viscous, slippery fluid called **synovial fluid.** It oozes between the bones to lubricate their movement. Our synovial membrane is surrounded and supported by an external **joint capsule** (Fig. 6-18) of dense connective tissue, holding the two bones together.

Wherever possible, depending on that joint's range of motion, dense, regular, collagenous bands of connective tissue are woven into or strapped over our joint capsule, binding the two bones together, holding alignment but allowing movement. These bands are called **ligaments** (Fig. 6-18), 'liga' meaning to bind as in ligation. They are tough, non-stretchy, and poorly vascularized. They span the joint, giving it stability, and preventing dislocation in spite of tremendous tensile stress. They limit the range of motion of a joint, preventing excessive movement. If our ligaments are allowed to shrink, our joints lose range of motion. If our ligaments are overstretched, our joints looe stability.

Beside the articular surfaces of joints,

Figure 6 - 18

A synovial joint.

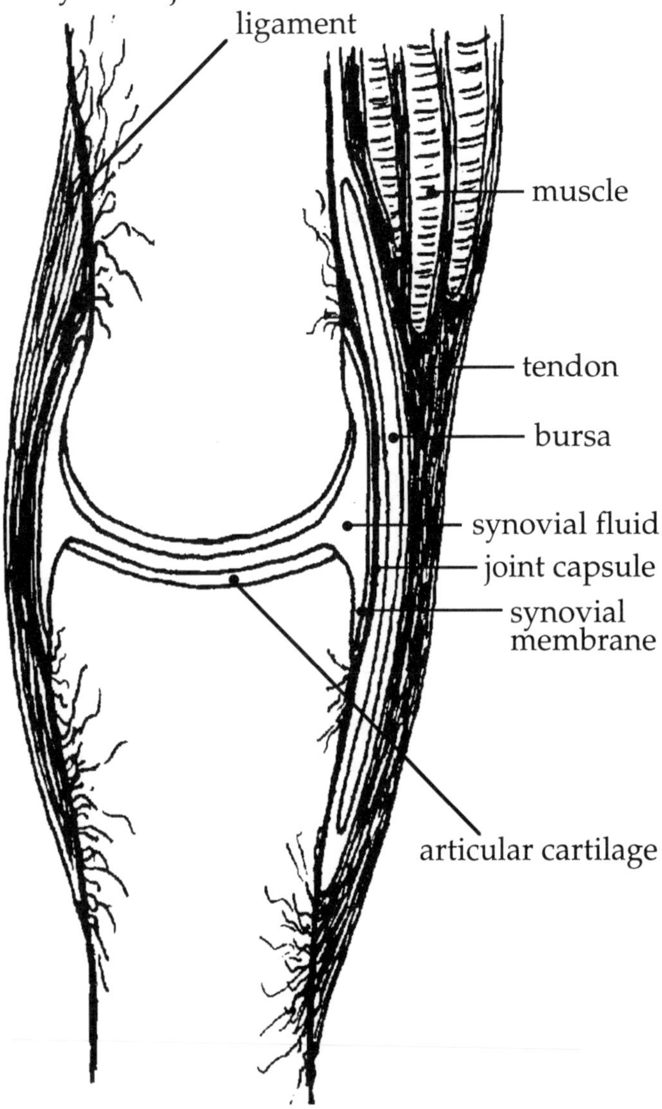

Diagrams often depict a wide, roomy space between the bones of a joint. This is a deceptive exaggeration. If you dip your hands in syrup and press them together, the space between them is similar to that in a joint when bearing weight. The synovial fluid is one or two molecules thick. It is a lubricator, not a cushion. We do not have a water bed mattress in our synovial joints.

there are many tendons and muscles which move over each other or the periosteum as their joint moves. All movement must be lubricated to prevent damage. Connective tissue similar to synovial membrane forms flat pouches which sandwich between the soft, moving structures surrounding a joint. These pouches are called **bursa** (Fig. 6-18) which literally means purse. A tiny bit of slippery ground substance oozes into the pouch so it becomes like a limp water balloon. The bursa walls adhere to their adjacent tissues and move with them, so the actual movement occurs between the two walls of the bursa which are lubricated by their contained fluid.

When lubricating a long tendon, a bursa is too awkward, so the flat pouch is wrapped around the tendon to completely enclose it. It forms a double walled cylinder of connective tissue around the length of the tendon, and is called a **tendon sheath** (Fig. 6-19). This connective tissue is the same stuff that makes synovial membranes and bursa. Its slippery fluid lubricates the tendon as it moves back and forth, like a cord within the cylinder.

Figure 6 - 19

A tendon surrounded by a tendon sheath.

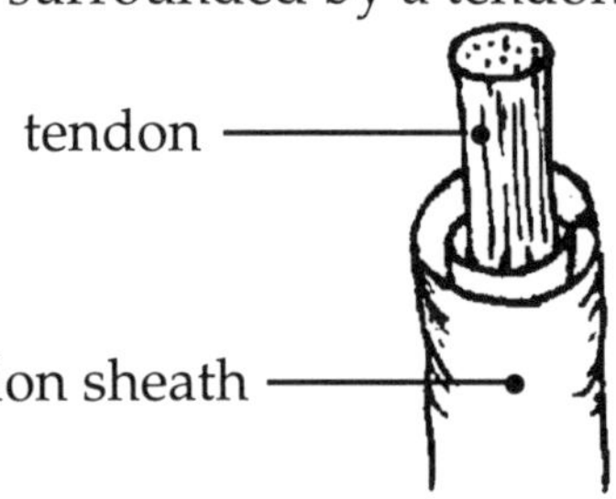

Bursitis happens when movement exceeds the lubrication in the bursa. The inner surface of the bursa is frayed and becomes inflamed and painful. While inflamed, the tissue of the bursa is fragile, so repeated movement of the joint may increase the injury. This movement includes active movements performed by the affected person as well as passive movements done by someone else, hopefully not under the name of therapy.

Tendonitis is like bursitis, only it is the tendon sheath and perhaps the tendon which is frayed and inflamed.

When a joint **dislocates,** supporting structures are inadequate relative to the force applied, so the bones misalign or separate. Joints may dislocate because: **1.** the force is so tremendous, or **2.** their supporting structures are already weak, damaged, or overstretched and provide poor support.

Supporting joint structures may be overstretched or damaged by: **1.** previous trauma, **2.** stressful postures, **3.** overzealous stretching during exercise, and **4.** overachieving therapists. So much emphasis in hands-on therapy is to open, lengthen, or make more space in the body. This may be a great need of people, but it is not the only one. Some joints need more stability, more support, less movement. If joints bear a lot of stress, too much movement may be the cause of damage and pain. Their ligaments need to be allowed to shrink, establishing a limit to movement and increasing support.

When ligaments no longer support a joint, muscles are asked to take over the job. This may create chronic muscle tension. As hands-on therapists, when we feel muscle tension, our first thought is to release it. We see this as our gift to mankind. Not always! Muscle tension may be the only thing holding certain joints together.

My own neck is an example of this. It suffers from not saying "No", not setting its own limits. My neck became so long, flexible, and unstable that a simple turn could dislocate it. Not saying "No" became a pain in the neck. The more therapy I had, the worse it became. Finally I put things together. Now I demand, "No neck lengthening." I say it with my mouth since my neck still goes into spasm trying to and I do exercises to strengthen muscular support.

A **whiplash** injury is like a sprain. It tests the tensile strength of the ligaments supporting the stack of bones in our neck. It may cause considerable damage to our ligaments and may lead to long-term instability of our neck.

When a joint **sprains,** the bones suddenly and forcefully dislocate, at least momentarily. The supporting joint capsule and ligaments are damaged (Fig. 6-20). Until these structures have repaired, the joint is unstable and needs to be supported with muscles or external bandaging.

Because ligaments are poorly vascularized, their cells work slowly. No employer would hire them by choice, but we are stuck with them. Their patch job progresses slowly. It is easy to lose patience, overuse the joint, and repeat the damage. Often, in spite of all the waiting, the repaired tissue is weak and unreliable. Once sprained, joints always may be less stable and dislocate or sprain more easily.

Figure 6 - 20
The damaged ligament and joint capsule of a sprained joint.

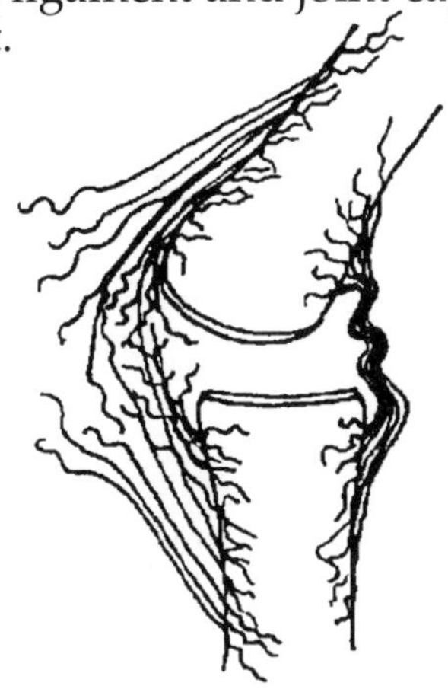

Sometimes a fragment of the synovial membrane type of connective tissue forms a small, isolated capsule or cyst. The fluid content expands until the capsule becomes tense and firm and feels like a marble. This is called a **ganglion.** Ganglia are most common around our wrists.

A layman's therapy is to give the ganglion a good whack with a book. The capsule bursts and the fluid disperses into surrounding tissues.

For better or worse, ganglia are driven to heal like the best of body tissues, so recurrence is common. Surgical removal of a ganglion also is tricky because a few missed cells may again cause recurrence.

SYNOVIAL JOINT MOVEMENTS

The movements of joints are defined according to the three directions in space, assuming our body is in anatomical position. Before learning the movements, it is important to be clear on how the directions are oriented with respect to anatomical position. It also is important to know the ventral and dorsal surfaces of our joints. This information is discussed in the Body Overview.

- In the anterior-posterior direction, as the ventral surfaces on either side of a joint approach each other, the movement is called **flexion** (Fig. 6-21a). As dorsal surfaces approach each other (or ventral surfaces separate) it is called **extension** (Fig. 6-21b). Most synovial joints flex and extend.

Generally speaking, joints flex further than they extend. Our ankle is the one exception to this.

While standing in anatomical position, all our joints are in a moderately extended position. They generally can extend further than anatomical position, which may be called **hyperextension** (Fig. 6-21b).

- In the medial-lateral direction, as body parts move sideways away from the midline, this is called **abduction** (Fig. 6-21c). When someone is abducted, they are taken away. As body parts are moved toward the midline, this is called **adduction** (Fig. 6-21d). They are added to the central mass.
- When a joint spins horizontally around a vertical axis, this is called **rotation.** As the anterior surface turns medially, this is called

Figure 6 - 21

A summary of **(a)** flexion, **(b)** extension, **(c)** abduction, and **(d)** adduction, showing all the joints in the furthest extent of that direction.

(a) flexion

(b) extension or hyperextension

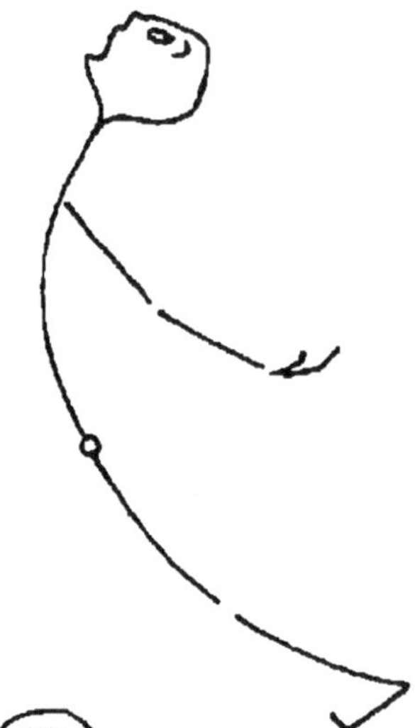

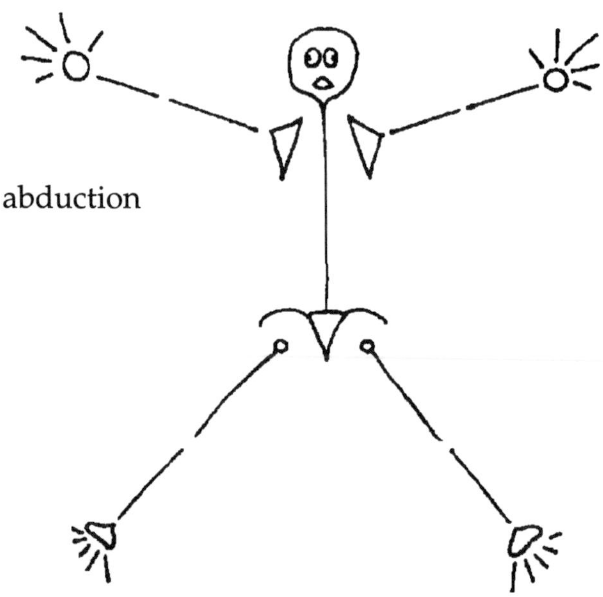

(c) abduction

(d) adduction

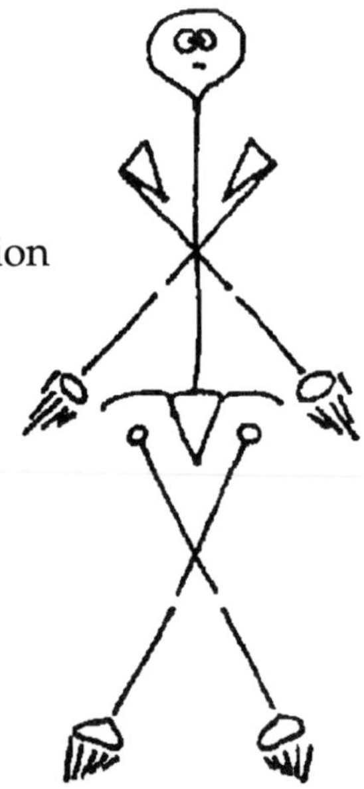

medial or **internal** rotation. As the anterior surface turns laterally, this is called **lateral** or **external** rotation.

These are the basic body movements in the three directions of space. Some joints have unique movements which will be discussed as we examine the specific joints.

• With all of these movements, the name applies to the direction of movement, irrelevant of the beginning or ending position. The movement may be a full sweeping arc or any segment of it. For example, when the elbow is completely bent it is said to be flexed. If the elbow is opened just 2 degrees, its new position still may be called flexed, but since the ventral surfaces have separated, the direction of movement would be called extension.

TYPES OF SYNOVIAL JOINTS

All synovial joints can be divided into six types.

• Two of the types allow movement in only one direction.

a. Hinge joints only allow flexion and extension, much the way a door would swing on its hinges. The cylindrical convex surface of one bone fits into the concave surface of the other (Fig. 6-22). Examples are the humeral-ulnar joint of our elbow and the interphalangeal joints of our fingers and toes.

b. Pivot joints allow rotation, usually in the horizontal plane. A finger-like shape of bone fits into a ring or arch formed by another bone or ligament (Fig. 6-22). Examples of pivot joints are the joint between our proximal ulna, humerus and radius, and the joint between the first two vertebrae in our neck.

• Three of the types of joints allow movement in two directions: flexion/extension and abduction/adduction.

c. A **saddle joint** has two condyles which are concave in one direction and convex in the other, fitting into each other's hollows (Fig. 6-22). Our carpal metacarpal joint at the base of our thumb is a saddle joint.

d. Gliding joints have two opposed flat surfaces which slide over each other (Fig. 6-22). Examples of gliding joints are our intervertebral joints and the joints between all the short bones in our hands and feet.

e. Ellipsoid joints have two oval condyles fitting into each other, one convex, and one concave (Fig. 6-22). Examples of ellipsoid joints are our radial carpal joints and the joint where our head sits on our first vertebra.

• f. The last type of joint has the widest range of motion, moving in all three directions in space. Our **ball-and-socket joints** have a spherical condyle fitted into a concave socket (Fig. 6-22). Examples of ball-and-socket joints are our glenohumeral joint and our hip.

Figure 6 - 22

The six types of synovial joints.

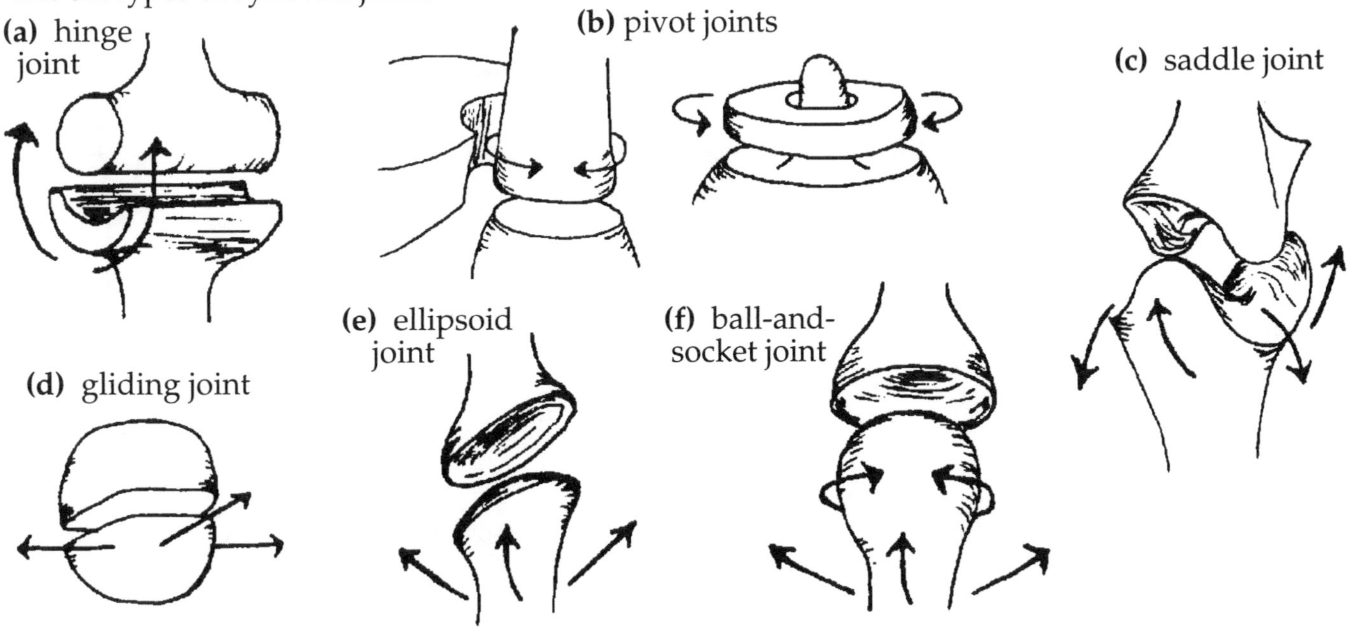

JOINT NAMES

Sometimes joint names do not seem to relate to anything, for example, shoulder, hip, or knee. Sometimes our joints are named after the two bones they join together. For example, the joint where our sternum and clavicle come together is called our sternoclavicular joint. As in this example, it is most common to name the proximal, more stabilized bone first.

BONY TERMS

In discussing and naming individual bones, certain terms are used repeatedly (Fig. 6-23) and have a consistent meaning. Familiarity with these terms will make our discussion of bones easier.

Condyle: the end of the bone which is within the joint capsule and covered with articular cartilage.

Epicondyle: a bony projection adjacent to a condyle.

Foramen (foramina in plural): a hole through a bone.

Fossa (fossae in plural): a shallow depression or concave shape.

Head: the rounded end on a bone, often supported on a narrow **Neck.**

Process: an elongated, finger-like projection of bone.

Spine: a pointed bony prominence.

Tubercle: a smaller rounded bony prominence.

Tuberosity: a larger rounded bony prominence.

Figure 6 - 23
Examples of bones which illustrate the repeated use of specific terms.

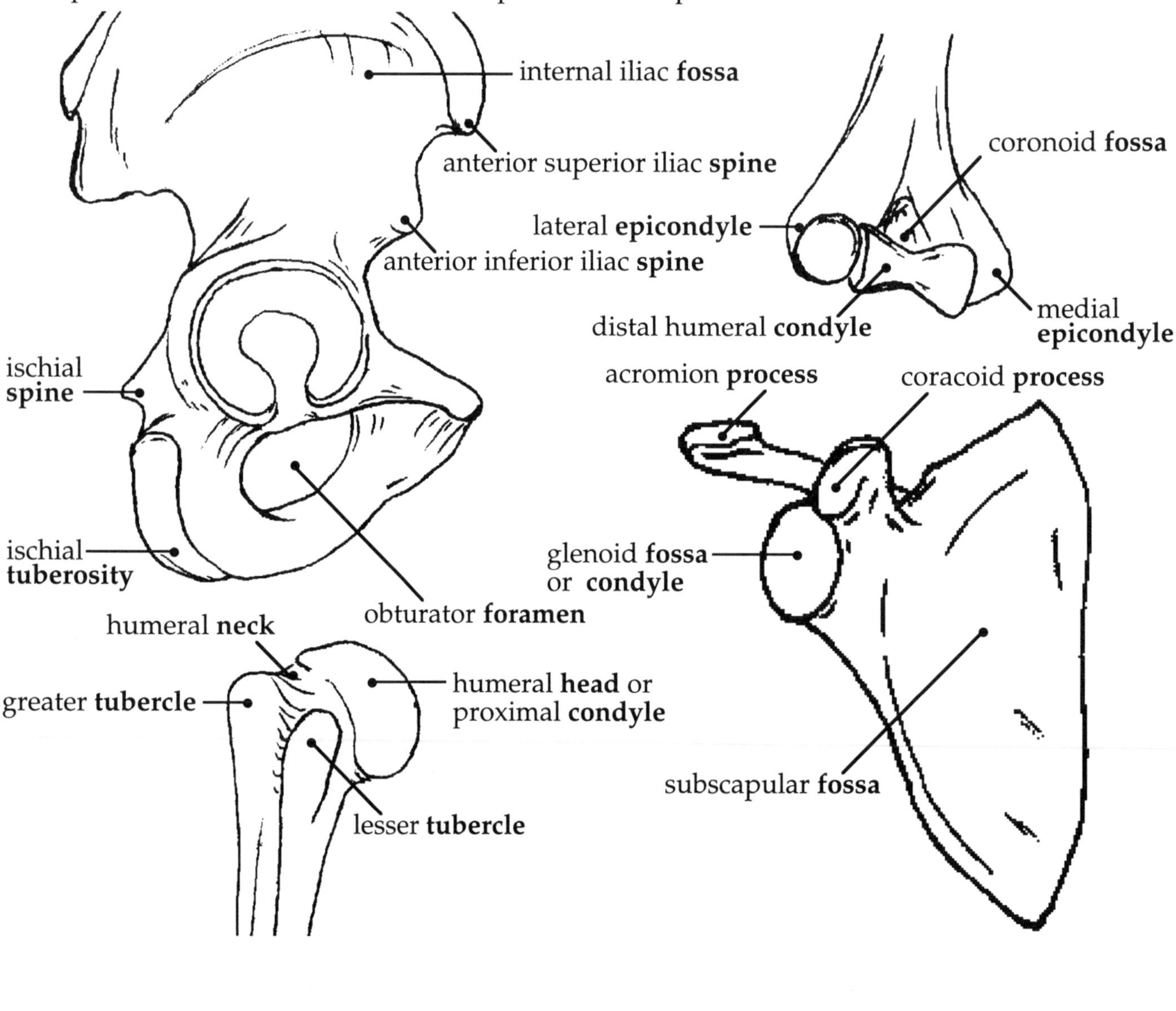

Now we are ready to examine the individual bones and joints of our skeleton. The following three chapters will discuss the location, shape, function, and palpation of each of our bones and the location, structure, and movements of each of our joints. It is for you to discover them inside your own body. The instructions will help you identify the hard knobby projections and follow the subtle edges more deeply. See how much of your internal core structure you can actually feel.

SKELETAL DIVISIONS

All the bones of our skeleton belong to one of two divisions.

1. Our **axial** skeleton includes the bones which support and surround the central vertical axis of our body, including our head, neck, and torso. There are 80 bones in our axial skeleton. When the same bone is repeated on our left and right sides, it counts as two.

2. Our **appendicular** skeleton includes the bones of our arms and legs, and their supporting girdles. An 'appendage' is a beneficial but unnecessary accessory to the main item or object, like the appendix of a book, or the one in our belly. Our arms and legs are helpful, but we can enjoy life without them, as many people have demonstrated.

There are 126 bones in our appendicular skeleton. All of these bones occur on both sides of our skeleton and so are counted twice.

Have fun boning up on all this stuff, so, next time you have a bone to pick with someone, you'll know exactly what you're picking! As you examine all your individual bones, you'll see how many shapes and sizes there are to pick from.

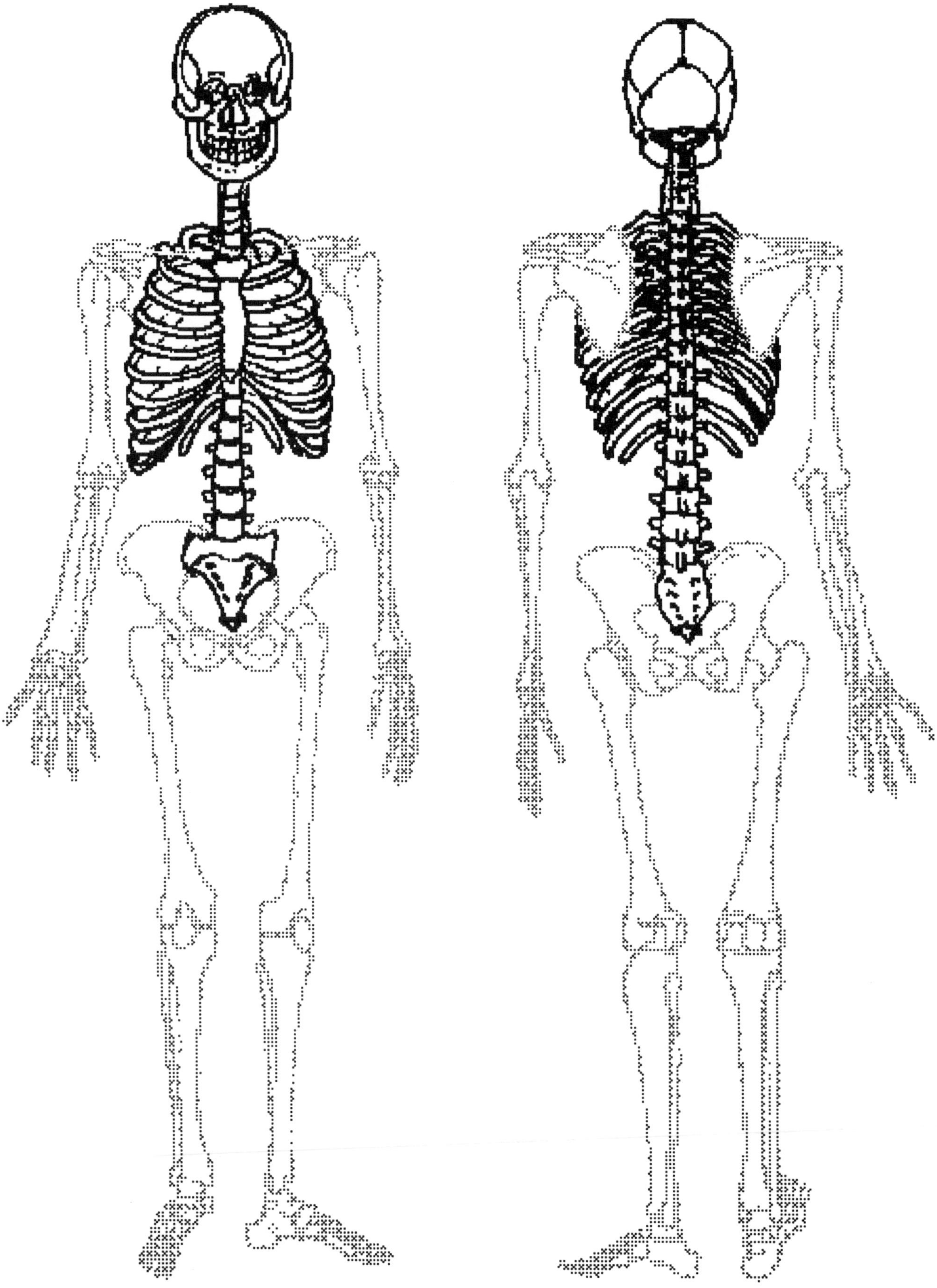

chapter 7: **Axial Skeleton**

Our **axial skeleton** functions to acks as internal organs, and to stabilize and support the central axis of our body. It adds a little movement to augment that of our appendages. Even the most flexible contortionists cannot curl their backs as far as their arms, even though the back has three times as many joints.

The bones of our axial skeleton include our **skull, hyoid bone, vertebral column, sternum,** and **ribs.**

SKULL BONES

There are 28 bones in our skull. Six of these bones conduct sound waves through our middle ears, eight protect our brain, and fourteen form the skeleton of our face. We will emphasize the bones with important functions or landmarks which we can feel.

MIDDLE EAR BONES

Our middle ear is an air-filled chamber within our temporal bone. Three tiny bones, joined by synovial joints, string across each middle ear vibrating the sound waves across the chamber. These bones will be described when we discuss hearing and our ear, under the special senses of our nervous system.

CRANIAL BONES

Collectively, the eight bones which protect our brain are called the **cranium.** Our cranium completely encases our brain, leaving only a few holes for the transmission of blood vessels and nerves through its inferior and anterior aspects.

The joints between our cranial bones are fibrous, non-moveable joints, frequently called **sutures** (Fig. 7-1). The edges of these bones have many tiny projections which interlock to look like stitches holding a cut together. The regular rhythmic movement of these joints is barely perceptible but has profound effects on the function of our nervous system.

In babies, the cranial bones are the most mature bones in the body. The cranium grows rapidly in the first three years of life and has almost reached adult size by five years old.

Our cranium includes four single midline bones: our **frontal, occipital, sphenoid,** and **ethmoid,** and two pairs of bones: our **parietal** and **temporal.**

Frontal Bone

A 'frontal' is a triangular gable ornament over a door or window. Likewise, our **frontal bone** forms the ridge over our eyes and our smooth rounded forehead (Fig. 7-1).

The bony ridge serves as a protective hood over our eyes and contains several air filled pockets called **air sinuses.** Our sinuses are ventilated through little tubes to our nose, keeping their air fresh (Fig. 7-4).

To see how far over the top of your head your frontal bone extends, place the heel of your hand between your eyebrows. Curl your fingers up and over your head. Your finger tips mark the end of your frontal bone and the beginning of your parietal bones.

Parietal Bones

Our two **parietal bones** are simple curved plates, protecting our brain. They lie side by side, and continue backward from our frontal bone about another hand's length. Our parietal bones cover about half the side of our head (Fig. 7-2),'parietal' meaning side wall.

Place the heels of your hands on the top edge of your frontal bone, one hand on each side of the midline. Let your fingers fall behind your head. Now each hand covers one parietal bone. Your finger tips mark the joint between your parietal and occipital bones.

The suture between your two parietal bones lies in the midline of your head, in an anterior-posterior direction between your two hands. It is called the **sagittal suture** and the word 'sagittal' often is used to describe the anterior-posterior direction through our body.

Occipital Bone or Occiput

Palpate your **occiput** as it completes your posterior-inferior cranium and curves forward, out of reach, to form your cranial floor (Fig. 7-1). Our occiput continues forwards to our ears. This inferior portion is referred to as the **base** of our occiput and is covered with muscle attachments.

Figure 7 - 1

Bones of the skull.

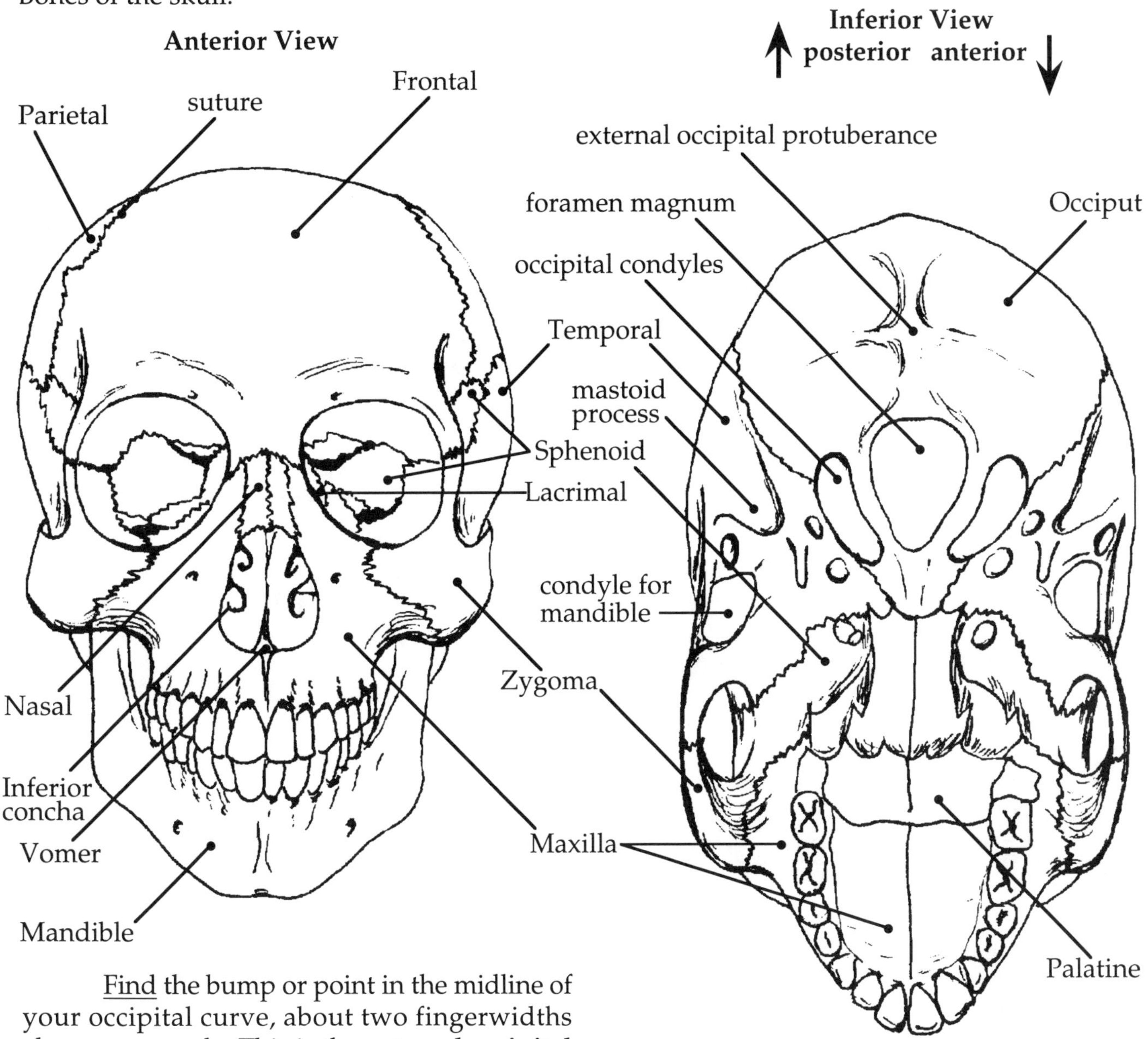

Find the bump or point in the midline of your occipital curve, about two fingerwidths above your neck. This is the **external occipital protuberance** (Fig. 7-1). In some people, it is quite prominent.

A large hole, the **foramen magnum,** consumes the center of the base of our occiput (Fig. 7-1). It is traversed by our spinal cord, arteries, and nerves.

On either side of our foramen magnum, just medial to our ears, our occipital bone is fitted with smooth oval pads called **occipital condyles** (Fig. 7-1). These condyles form synovial joints with our first vertebra, transferring the weight of our head into our neck. The long axis of each condyle is in an anterior-posterior direction, allowing our head to tip forward and backward over our neck. The pivot point of this movement is just below our ears.

Temporal Bones

Our **temporal bones** are covered by our external ears and enclose our middle and internal ears. Spread your hand over the side of your head with your ear in the middle of your palm. Now the palm of your hand covers your temporal bone. The obvious hole on its side is the entry to the **external ear canal** (Fig. 7-2).

Immediately behind and below your ear, lateral to your occiput, find a rounded bump, the **mastoid process** of your temporal

Figure 7 - 2
Bones of the skull.

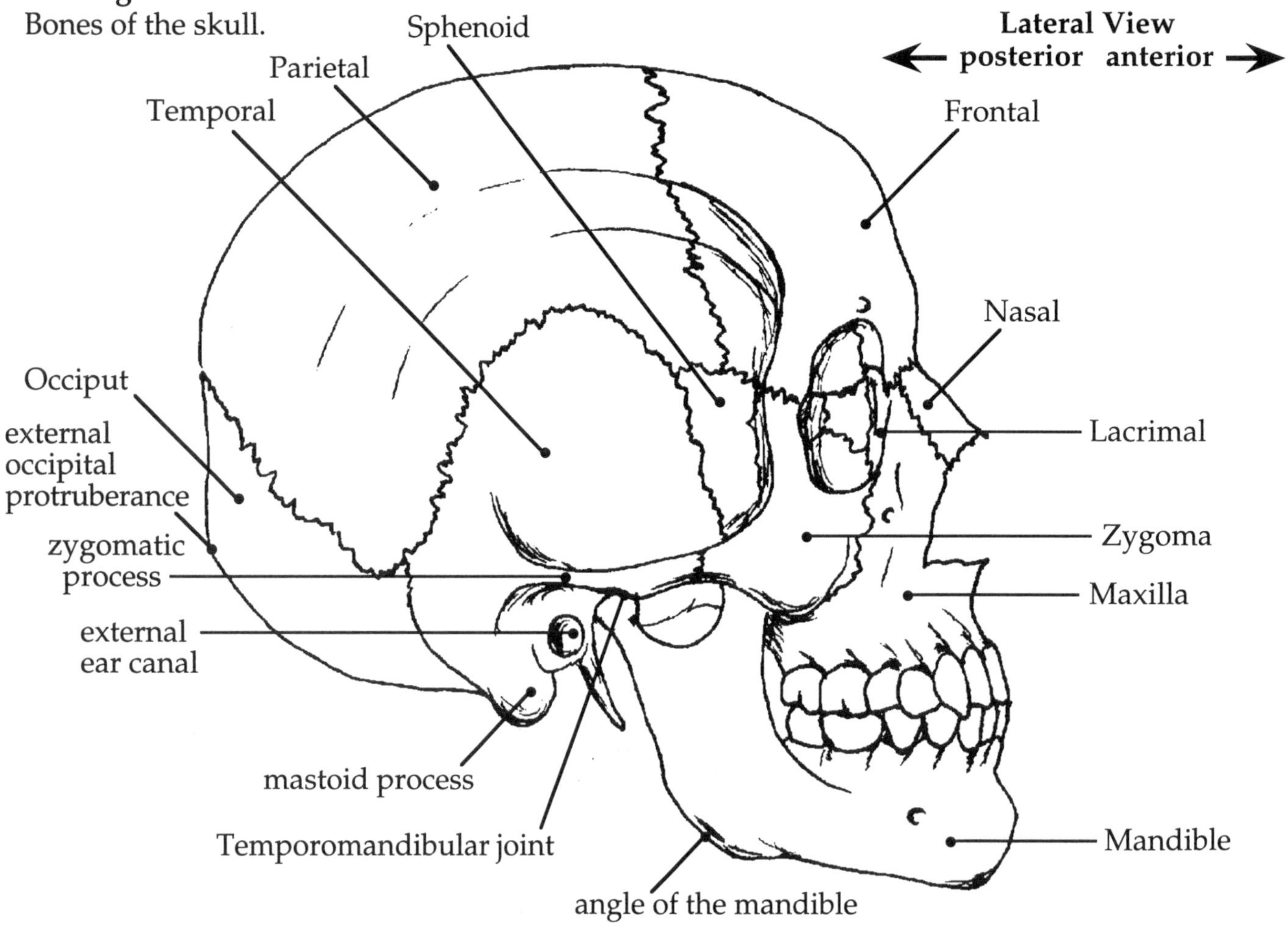

bone (Fig. 7-2). Our mastoid process contains pockets of air ventilated by channels through our ear to our nose (Fig. 7-3).

Feel a horizontal ridge of bone just in front of your ear canal. This ridge is called the **zygomatic arch.** Its posterior portion, called the **zygomatic process** (Fig. 7-2), is a finger of bone protruding outwards and forwards from your temporal bone much like the stylus on a sundial, hence the name temporal for the marking of time.

The inferior surface of our zygomatic process articulates with our mandible, to form our **temporomandibular joint** (Fig. 7-2) often going by its initials **TMJ.**

Sphenoid Bone

'Spheno' means a wedge and describes the shape of our **sphenoid bone.** The point of the wedge begins just in front of the foramen magnum of our occipital bone. It spreads forward and laterally to the back of our eyes, separating our throat and nose from our brain (Fig. 7-1).

The lateral wings of our sphenoid bone are its only palpable part. Find the corner where the prominent lateral brim of your eye socket and your zygomatic arch meet. Press the pad of your thumb into the depression behind and above this corner. Your thumb now covers your entire exposed sphenoid bone (Fig. 7-2).

Our sphenoid bone contains **air sinuses** which ventilate through our nose (Fig. 7-3). This explains why the headache from a stuffy, infected sinus is often felt behind our eyes.

> Our anterior skull is like a 3-story building. Our sphenoid and ethmoid bones divide the top and middle stories, separating our brain and nose. Our maxillary and palatine bones divide the middle and bottom stories, separating our nose and mouth.

Figure 7 - 3

The air sinuses of the skull.

Anterior View

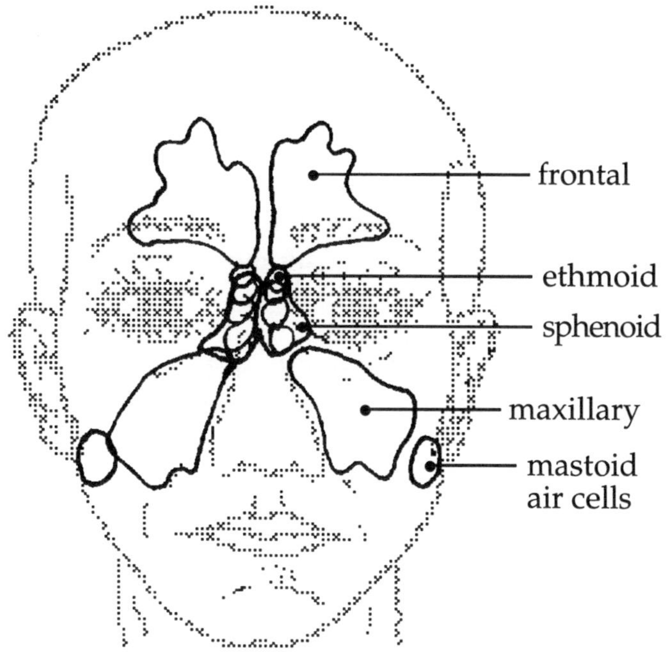

Lateral View

anterior posterior

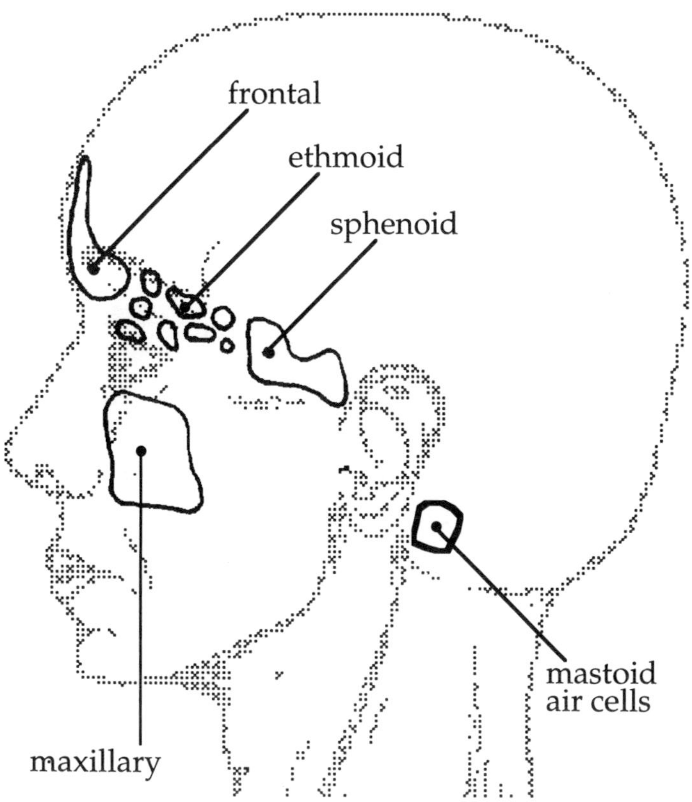

An **air sinus** is an air-filled pocket within a mass of bone. All the air sinuses of our skull have narrow tunnels communicating with our nose (Fig. 7-4). These tunnels allow an exchange of the air in our sinuses and serve as a drain for the fluid and mucus secreted by the epithelium lining the inside of our sinuses.

As drains, the tunnels have one problem. They have not accommodated to upright posture. When we are bent forward, each tunnel is on the bottom of a sinus, but when we stand upright, the tunnel is half way up the side (Fig. 7-4). Imagine emptying a sink with the drain up on the side rather than at the lowest point. This is the challenge facing our sinuses all the time. It is no wonder that they so easily become stagnant ponds, teaming with microscopic life.

We can help our sinuses by bending forward now and then. Feeling fluid flush into our nose as we bend forward tells us that our sinuses are getting behind and could use more of our help. As we lay face down on our stomach, we may notice that our nose gets stuffy and it is hard to breathe. This also may mean that our sinuses are taking advantage of this position to drain.

The five skull bones which contain air sinuses are our frontal, temporal, sphenoid, ethmoid, and maxillary (Fig. 7-3).

Our air-filled **middle ears** follow the same principles and have all the same problems as our air sinuses.

Figure 7 - 4

An air sinus is an air filled pocket within a mass of bone.

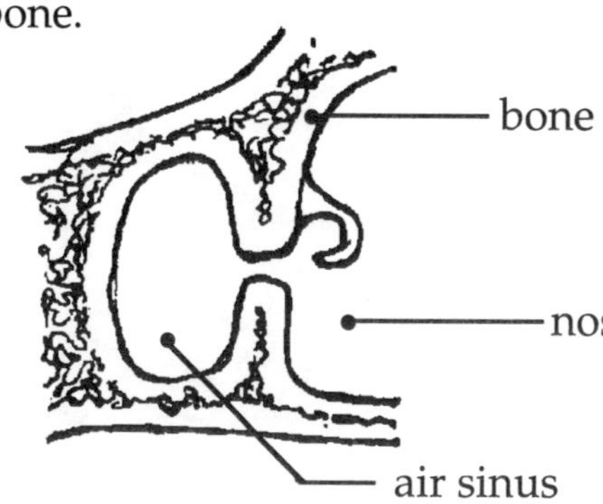

Ethmoid Bone

Our **ethmoid bone** contributes to the roof and walls of our nose and the floor under the front of our brain. It only is palpable from inside our nose.

An 'ethmos' is a sieve and describes the many fine holes which perforate our ethmoid bone, transmitting the nerves which carry smell from our nose to our brain. Our ethmoid bone also contains several air sinuses (Fig. 7-3).

FONTANELS

When a baby is born, the bones of its cranium are not fully formed, leaving soft spaces at the corners where the bones will meet. These are called the soft spots or **fontanels** (Fig. 7- 5). A 'fontanel' is a little fountain, the origin or source of anything, expressing the high regard for the human brain. In the fontanels, the brain is protected only by a dense membrane which is yet to grow and ossify. The fontanels generally are filled with bone or **closed** by two years of age.

At birth there are six fontanels. The largest fontanel is a diamond shape between the frontal and parietal bones, on the top of a baby's head. It is the last to close. A smaller triangular-shaped fontanel is present on the back of the head between the parietal and occipital bones. It is the first to close. There are two irregular fontanels on each side of the head, filling the space where the lateral edges of the cranial bones will meet.

Figure 7 - 5

A lateral view of the skull at birth.

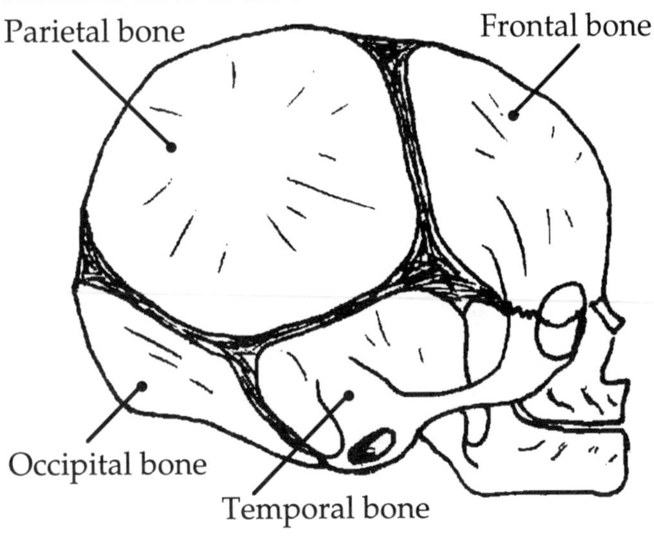

During the birthing process, the incomplete cranium allows the baby's head to be compressed and conform its shape, as needed, to squeeze through the birth canal.

FACIAL SKELETON

Fourteen bones support and shape our face. Our facial bones stay petite throughout childhood but contribute most of our adolescent head growth. Our air sinuses develop in adolescence, increasing the resonance of our voice (and our opportunity for sinus headaches). There also is growth of the tubes which bring fresh air from our nose to our middle ears, decreasing the incidence of obstuction and middle ear infection.

- Seven facial bones are palpable on the external surface of our face, including three pairs of bones: our **nasal, zygomatic,** and **maxillary,** and one single midline bone: our **mandible.**

Nasal Bones

Skeletons could not keep their noses out of other's business, so they lost it. The nose that sticks out on our face mostly is supported by cartilage. It is easy to twist and bend. Slide up the pliable cartilage side of your nose, to feel the ridge where the **nasal bones** start, one fingerwidth anterior to the corner of your eye. Follow the ridge as it slopes down and back toward your cheek. Our nasal bones are like two thickened fingernails sitting side by side at the top of our nose, barely big enough to prop up a pair of glasses (Fig. 7-1).

Zygomatic Bones or Zygomae

A 'zygoma' is a yoke, binding things together. Our **zygoma** is our cheekbone, binding our face to our cranium. Feel its upper edge forming the lateral-inferior corner of your eye socket, follow it down two fingerwidths to the rounded prominence of your cheek, then continue back over your zygomatic arch. Our zygoma forms the anterior half of our zygomatic arch, fusing with the zygomatic process of our temporal bone (Fig. 7-2).

Figure 7 - 6
Surface view of the skull.

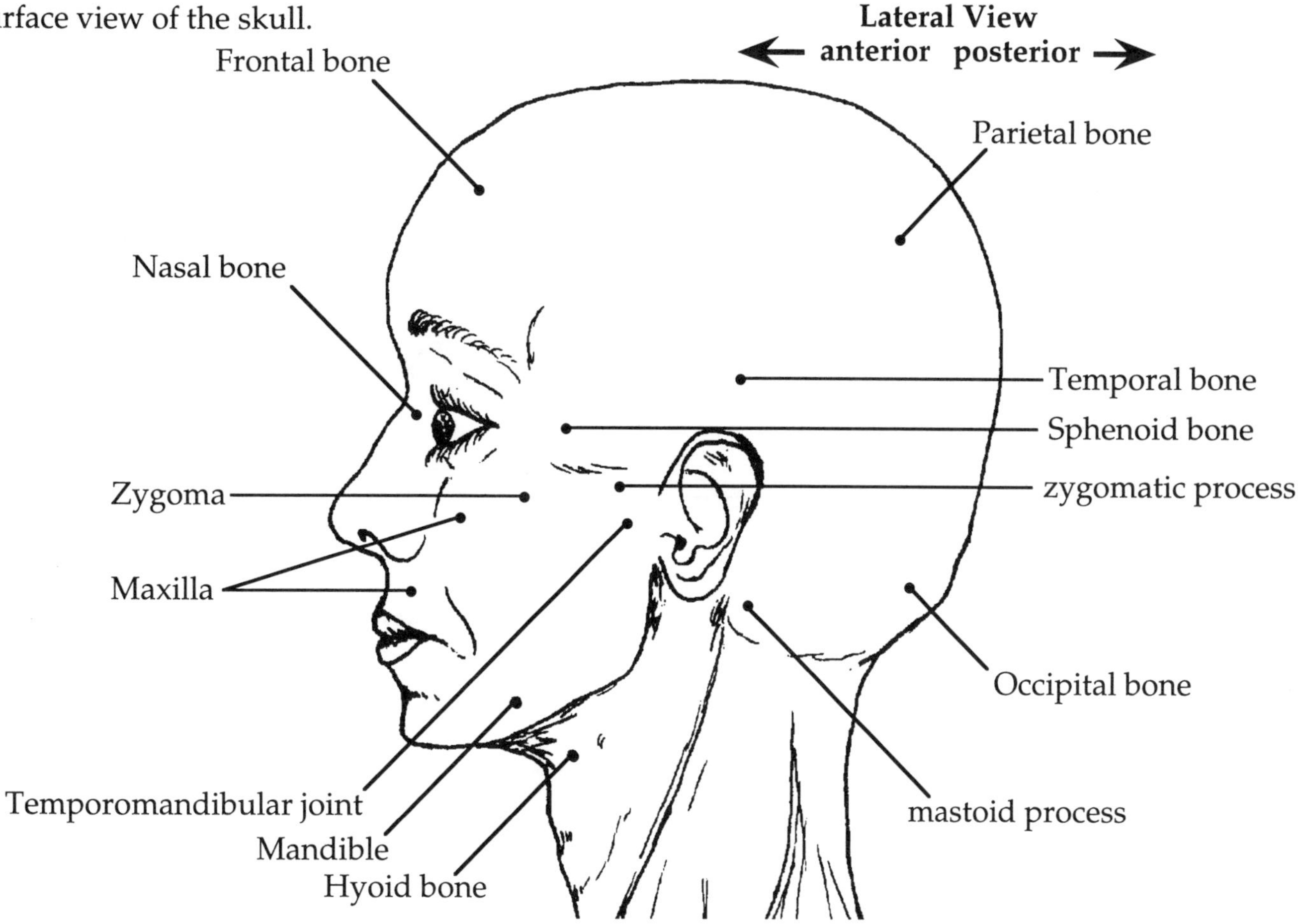

Anterior View

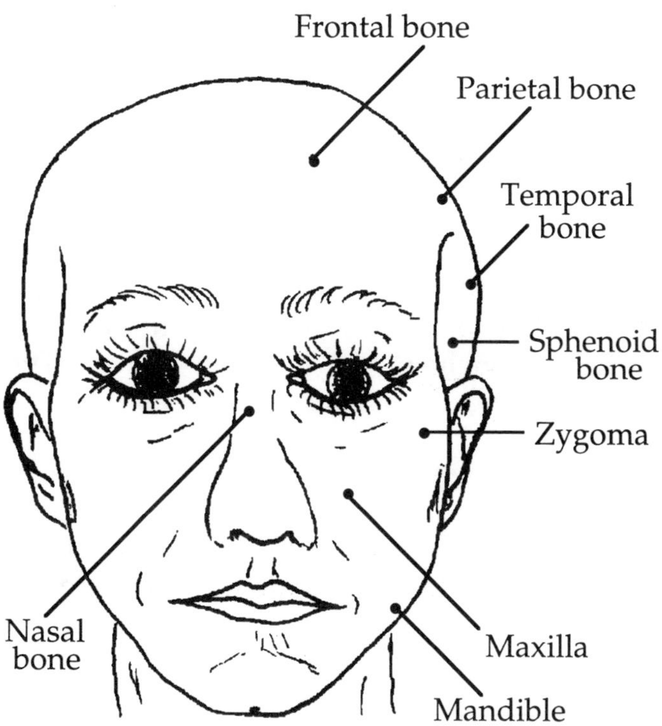

Maxillary Bones or Maxillae

Our **maxillae** are the keystones of our face. All the other bones fit into place around them. They contribute to the floor of our eye sockets and the sides of our nose. They fuse in the midline to form the floor of our nose, the roof of our mouth or **palate** and our upper jaw, holding our upper teeth (Fig. 7-1).

Each maxilla contains a large air sinus, located just lateral to our nose and ventilating through our nose (Fig. 7-3).

Palpate the upper edge of your maxilla as the medial inferior brim of your eye socket. Follow it down into the hollow between your nose and zygoma. Further down, feel it bulge as it sockets your upper teeth. Follow this bulge backward below your zygoma and forward below your nose, where it fuses with the maxilla from the other side.

Our upper jaw and palate are formed by the fusion of our right and left maxilla. Maxillary fusion is one step of fetal development which follows a tight schedule. The fetus allows a brief time for fusion, then immediately covers the bones with epithelium. If the fusion is delayed for any reason, epithelial tissue comes anyway, covering the edges of the bones. This epithelium makes ongoing fusion impossible. The palate, upper jaw, and even the upper lip may remain cleft. The degree of separation depends on how early the delay in fusion happens.

Mandible

Our **mandible** is a single midline bone commonly known as our lower jaw. It outlines the bottom of our face and is made of three parts, one horizontal U-shape and two vertical struts (Fig. 7-2). Our lower teeth imbed into the U-shape and its flaring curve forms our chin. One wide vertical strut is fixed to each end of the U-shape, forming a distinct corner called the **angle** of our mandible.

Follow the bottom edge of your mandible from your chin back to its angle just below your ear. Follow the posterior edge of bone from the angle upward to just in front of your ear. The subtle roundness at the top is the **mandibular condyle,** part of your temporomandibular joint (Fig. 7-2). Hold two fingers just below your zygomatic arch, in front of your ear. Open and close your mouth. Feel your condyle sweeping back and forth over the inferior surface of the zygomatic process of your temporal bone.

Temporomandibular Joint

Our **temporomandibular joint** is the one synovial joint of our facial skeleton. Its movement is complex (Fig. 7-7). As our condyles sweep back and forth, our mandible swings up and down, respectively called **elevation** or flexion, and **depression** or extension. Our mandible slides forward and backward, called respectively **protraction** and **retraction.** Our mandible also waggles from side to side, called **lateral excursion** in technical circles.

Figure 7 - 7
Movements of the temporomandibular joint.

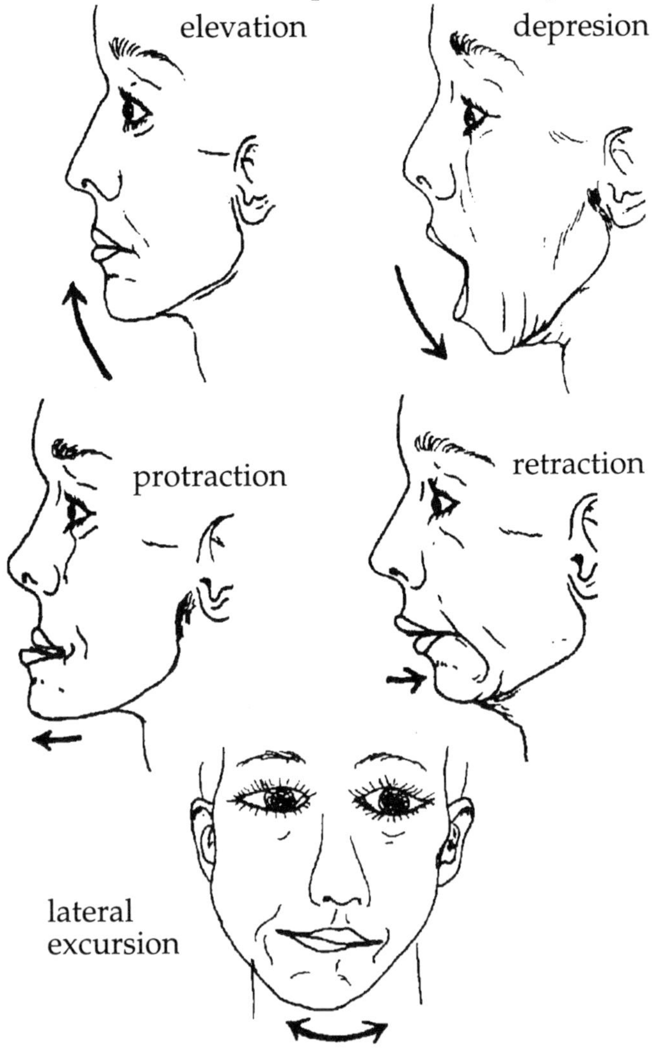

• Seven irregular bones contribute to the internal structures of our face. They only are palpable from inside our nose and mouth, dangerous places to go digging on anyone but yourself, although sometimes appropriate therapeutically. The first six bones are paired, with one on each side. They are our **lacrimal, inferior conchae,** and **palatine** bones. The seventh bone, our **vomer,** is in the midline.

Lacrimal Bones

If you are shy to say that you have been crying, try using the Latin equivalent, lacrimating. Each **lacrimal bone** forms an internal part of the side wall of our nose (Fig. 7-2). It contains a tiny tube which sneaks our tears from the corner of our eye into our nose so they do not have to spill over our zygoma or maxilla. This is why our nose drips when we cry.

Inferior Conchae

The lateral walls of our nose have three bony shelves which create turbulence in the air as it passes through our nose. On this basis, these bones often are called the **turbinates.** Each shelf is coiled into itself like a conch shell and so these bones are also called **conchae.** The bottom shelf on each side of our nose is supported by its own little bone called the **inferior turbinate** or **concha** (Fig. 7-1).

Palatine Bones

Our **palatine bones** are posterior to our maxillae. They do their morning stretch all day long. Each palatine bone stretches a hand upward to support our eyeball and presses a long, flat foot downward to brace the back of our oral palate. Its body contributes to the side of our nose. Over all, each palatine bone is shaped like a capital letter L (Fig. 7-8).

Vomer

Our **vomer** is a single midline bone. 'Vomer' is Latin for plowshare. Have you ever seen a plowshare? If you are curious about what it looks like, the bony plate which separates our nose into two equal sides is a close replica and its namesake (Fig. 7-1).

Figure 7 - 8
Anterior view of the palatine bone showing its relative positiont within the skull.

Figure 7 - 9
Lateral view of the hyoid bone.

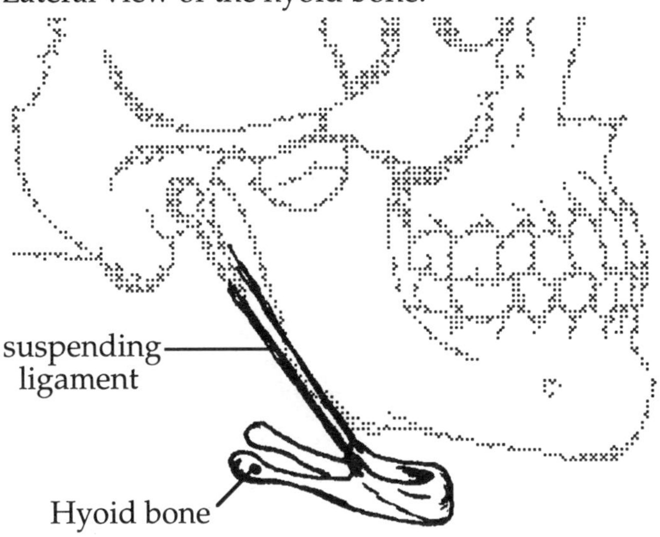

HYOID BONE

We all wear a good luck charm around our necks. Our **hyoid bone** is a thin, horseshoe-shaped bone in the front of our neck. It lies horizontally, with its curved end forward (Fig. 7-9).

To palpate your hyoid bone, go straight back from your chin, to the corner before your neck drops downwards. Your hyoid bone hangs out in this corner (Fig. 7-6). Feel its hard, curved edge just above the prominent cartilage of your voice box or larynx. Feel it bob up and down as you swallow and talk.

Our hyoid bone literally hangs. It is the one bone of our body which does not articulate with any other bone. It dangles by two ligaments, one from the bottom of each temporal bone. It supports our tongue, the floor of our mouth, our larynx, and anterior neck muscles.

VERTEBRAL COLUMN

Our back bone or **vertebral column** is a central, vertical column through the posterior part of our body. It protects our spinal cord and nerve roots and provides the bony support for our head, neck, and torso. It is the will-power and courage that holds us up.

One root of 'vertebral' is 'verto', to turn, or a turning point or vertex, relating to the rotation of one vertebra over the next. We also can say that the vertical orientation of our vertebral column was a major turning point in our evolution. Vertebral also means jointed segments. This describes the structure of our vertebral column. It is a stack of 32 to 34 bones, called **vertebrae** (Fig. 7-10) with their inter-

Figure 7 - 10
Lateral view of the vertebral column.

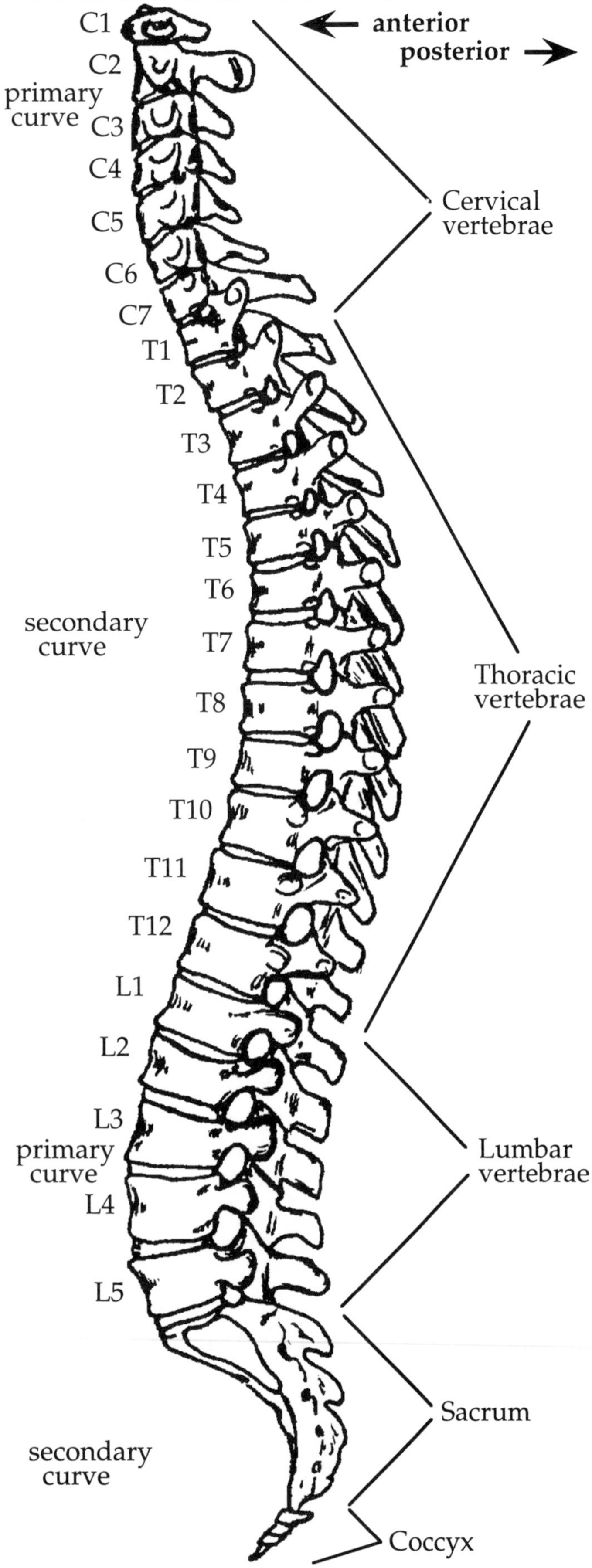

vening joints, the **intervertebral joints,** 'inter' meaning between. As we mature from infants to adults, some vertebrae fuse, so the stack has fewer bones, only 27 to 30.

VERTEBRAL GROUPINGS

All the bones of our vertebral column are divided into five groups.

Cervical Vertebrae

The top seven bones are called our **cervical vertebrae** (Fig. 7-10), usually abbreviated as C. They are numbered 1 through 7 from the top down, so it is common to refer to them as C1, C2, C3, C4, etc. They form the bones and joints of our neck, 'cervical' meaning a narrow column. They are the most flexible and least stable of all our vertebrae.

Thoracic Vertebrae

The next 12 bones are called our **thoracic vertebrae** (Fig. 7-10), often abbreviated as T for treasure chest. They also are numbered from the top down, 1 through 12, and are referred to as T1, T2, T3, T4, etc. They support our chest and articulate with our ribs. They are the third most flexible vertebrae.

Lumbar Vertebrae

The next five bones are called our **lumbar vertebrae** (Fig. 7-10), abbreviated as L. They also are numbered from the top down, 1 through 5, and are referred to as L1, L2, L3, etc. They support our waist area or loin, between our pelvis and ribs. 'Loin' means long, but this area often is quite compressed. Lumbar are the second most flexible vertebrae, being quite unstable for the weight they carry and contributing to many back problems.

Sacrum

The next bone is our **sacrum** (Fig. 7-10) from the word sacred. It begins as five **sacral vertebrae** which fuse. They are abbreviated as S, numbered from the top down, 1 through 5, and are referred to as S1, S2, etc. They support through our lower abdominal and pelvic area. When these bones are fused, they have no mobility and increase the stability of our pelvis.

Figure 7 - 11
Movements of the intervertebral joints.

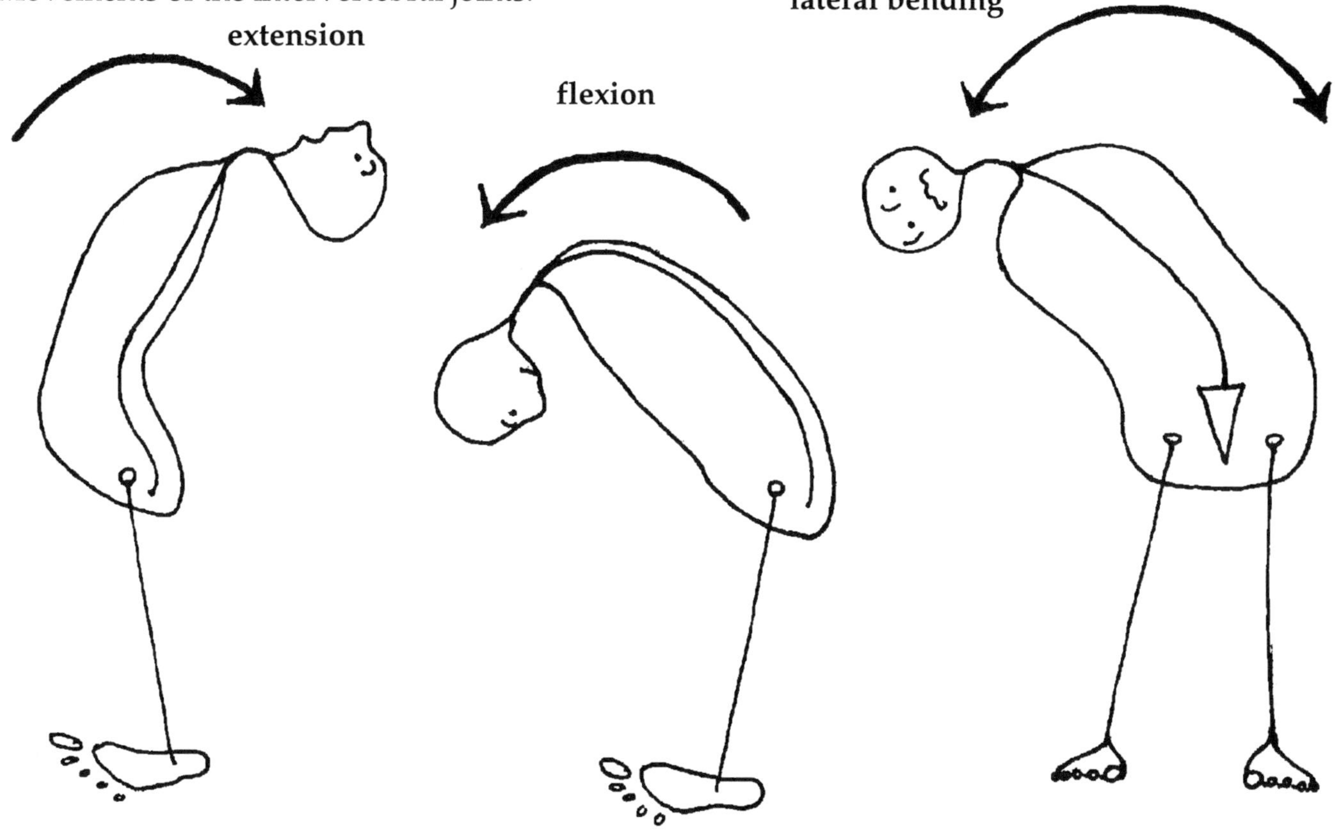

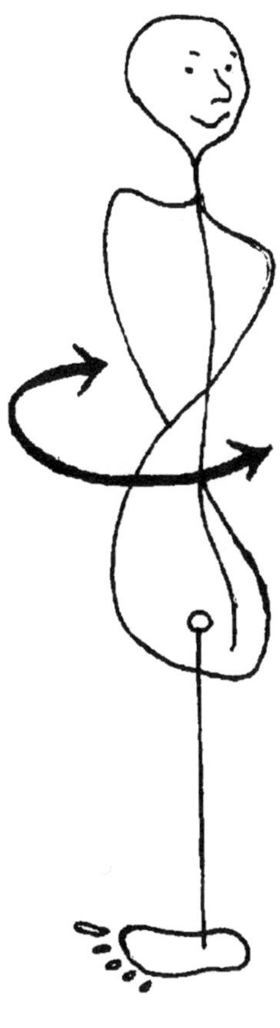

Coccyx

The last group is called our **coccyx** or **coccygeal vertebrae** (Fig. 7-10), and are never abbreviated. They are 3 to 5 in number. Except for the top one, they are often fused. This rigid stub is all we can flaunt for a tail and commonly is called our **tail bone**.

VERTEBRAL MOVEMENT

No one intervertebral joint moves very far, but since there are so many, our torso is quite mobile. We have three vertical movements. Curling forward, so our face approaches our abdomen, is called **flexion** of our vertebral column (Fig. 7-11). Lifting up to stand erect and bending backward is **extension.** Bending to the side, while facing forward, is called **lateral bending** or **lateral flexion.** We have one horizontal movement. Twisting around to face the side or behind is called **rotation** of our vertebral column.

NORMAL CURVES

Viewed from the back, our vertebral column is straight. Any curve from side to side is abnormal. Viewed from the side, no Greek builder would sign his name to a column as curvey as the one supporting our torso. In us, curves in the anterior-posterior direction are appropriate and are called the **normal curves** of our vertebral column (Fig. 7-10).

• Normal curves function in three ways: **1.** We are not immobilized Corinthian temples. Our curvey column is architecturally the strongest support when movement is a must. **2.** Our curving vertebral column helps balance our weight more centrally over our supporting pedestals. If a perfectly straight Greek column supported our back, with all our internal organs hung from its front and sides, can you imagine how long our feet would have to be to keep our nose out of the dirt in front of us (Fig. 7-12)? **3.** Our normal curves absorb the jolt of landing on a hard surface as we walk, sit, or move about. If you jump up and land with your knees straight rather than slightly bent, you can see the difference between the absorbability of a straight or gently curved vertebral column.

• A normal adult vertebral column has four alternating curves. How do these curves form?

At birth, the vertebral column has only one curve, convex posteriorly (Fig. 7-13a).

As a baby becomes curious and starts lifting its head to look about, the curve in its cervical area reverses to be concave posteriorly (Fig. 7-13b). Since this is the second direction of curving in this area, it is called a **secondary curve.**

As an infant grows in curiousity and pulls itself up onto its legs, the curve in its lumbar area reverses to accommodate its upright posture. Its lumbar area now curves concave posteriorly (Fig. 7-13c). Since this is the second direction of curving in this area, it also is called a **secondary curve.**

The curves in our thoracic and sacrococcygeal area maintain their original curve throughout our life (Fig. 7-13c) and so are called **primary curves.**

A TYPICAL VERTEBRA

All our vertebrae have certain aspects in common and each vertebra has certain unique characteristics. First, we will look at the similarities of vertebrae. To begin, we will take any vertebra and look down on it from the top, commonly called a superior view.

Figure 7 - 12

The normal curves of our vertebrae help to balance our weight over our center of gravity.

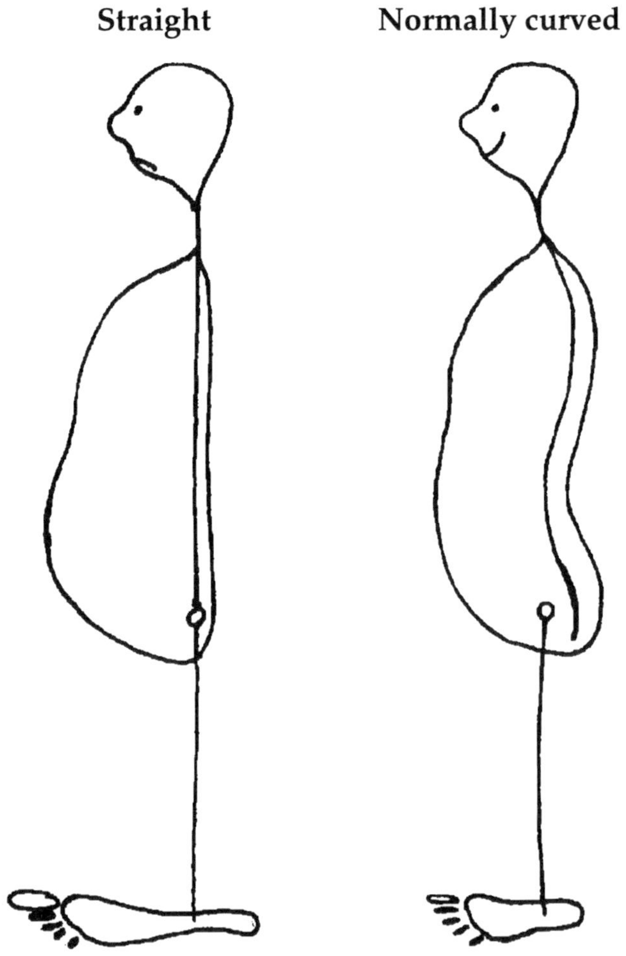

Figure 7 - 13

Formation of the normal vertebral curves.
(a) A new born infant has one curve.
(b) Lifting the head, reverses the cervical curve.
(c) Standing up on the legs, reverses the lumbar curve.

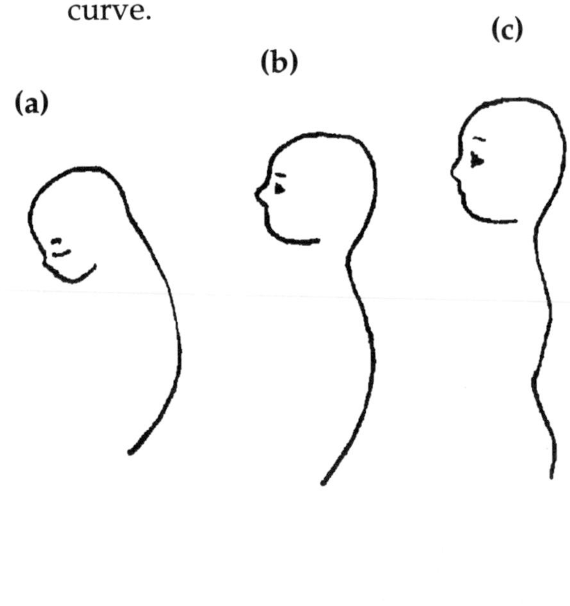

Normal curves are sometimes exaggerated, reducing the advantage of their function, increasing muscle tension, and cramping internal organs.

An excess primary curve in the thoracic area is called **kyphosis.** It leads to a hunched back and caved-in chest.

An excess secondary curve in the lumbar area is called **lordosis.** It pushes the abdomen forward, giving the appearance of a paunch.

Any curve from side to side is abnormal. It is called **scoliosis.** Scoliosis may cause marked twisting and distortion of the trunk of the body.

To maintain balance, one abnormal curve usually leads to a second abnormal curve in another part of the vertebral column. The sequel curve is equal to the first in size, but opposite in direction

Mild abnormal curvature may be corrected with therapy or exercises which re-establish a correct sense of balance, tone, and support in the surrounding muscles. More severe cases often can be improved with similar exercises. This may include releasing or relaxing certain muscles and stimulating and strengthening others.

Body

The most anterior part of each vertebra is a thick, round block called the **body** of our vertebra (Fig. 7-14). It functions to support the weight of everything that is stacked above it. What happens to the size of the vertebral body as we look down our vertebral column in Figure 7-10? The body of each vertebra is progressively larger than the one above it. Why? It is stimulated by the accumulating weight and enlarges to increase its weight-bearing capacity.

The bodies of most vertebrae are deep inside us, covered by organs or other bones such as the intestines, heart, sternum, larynx, and so forth. Only in our mid-abdomen can the anterior surface of the bodies sometimes be felt. The depth of a vertebra from front to back is 2 to 3 inches. At our waist, our front to back

Figure 7 - 14

Superior view of a typical vertebra.

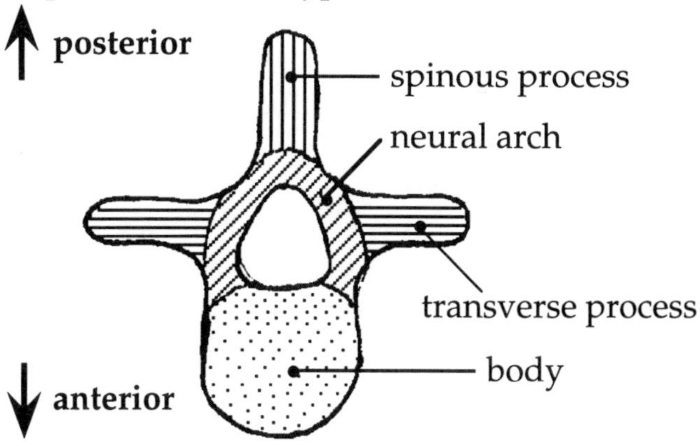

depth may be only 6 to 10 inches, depending on our ability at sit-ups or playing Santa Claus. When we lie down, our abdominal muscles relax, everything sags to the side, and our belly caves in. This leaves our vertebrae even closer to the front.

Have a friend lie down on his/her back. Press gently into the center of his/her abdomen above the belly button. Does your hand feel pushed away by a giant pulse? The main abdominal artery lies just to the left of the vertebral bodies. With gentle, continued pressure, do you run into a rock-hard mass? Don't panic! It is his/her lumbar vertebrae curving forward to meet you. If you do not find either of these things, still don't panic and never create discomfort in your friend.

You may be able to find the vertebral bodies on yourself, although your own muscular effort of pressing may disguise them.

Neural Arch

Attached to the posterior aspect of our vertebral body is an arch of bone called our **neural arch.** It is disguised by all the things poking out of it (Fig. 7-14). The name of this arch implies its function, to protect our spinal cord and nerve roots. Our vertebral body and neural arch together form a complete bony case for our delicate nerve tissue.

Our neural arch is not normally palpable anywhere along our vertebral column.

Processes

Each vertebra has three finger-like projections. Two point laterally and so are called **transverse processes.** One points posteriorly and is called our **spinous process** (Fig. 7-14). A process functions to increase the bony surface area for ligament and muscle attachment and to increase the mechanical advantage or leverage of the muscles' action. The further a muscle is attached from the pivot point of joint movement, the easier it is for the muscle to create movement. It requires less force and energy, and produces less stress and waste.

On a bicycle, short handle bars are designed for showy street cruising. Rugged mountain bikes have long handle bars for increased control and maneuverability. Our vertebral processes are like long handle bars. The distance they add increases the stability of our vertebral column and reduces the effort of movement. Our transverse processes work especially with lateral bending and rotation of our vertebral column, while our spinous processes affect extension.

Transverse processes are normally only palpable in our cervical area. Find the little balls just inferior to your ears and posterior to your jaw. These balls are the tips of the large transverse processes of your first cervical vertebra. Press into the vertical indentation in the posterior-lateral aspect of your neck, just anterior to your upper trapezius (Fig. 7-15). The deep firmness here is the transverse processes of your lower cervical vertebrae. Lower than this, the transverse processes only are palpable if the vertebrae are abnormally rotated.

Figure 7 - 15

Locating the transverse processes of the cervical vertebrae.

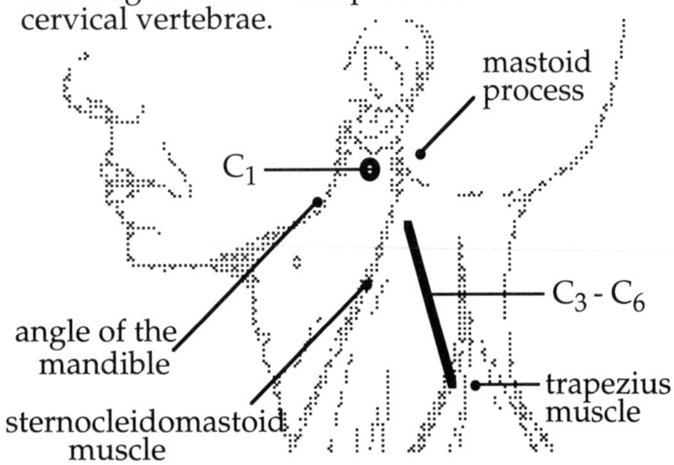

The tips of our spinous processes are palpable as the bumpity ride all the way down the center of our back, from the base of our occiput to our coccyx.

Intervertebral Structures

• Now let us examine a lateral view of our typical vertebra.

Looking from the side, our vertebral body is like a square block, capable of good support. Our neural arch is quite narrow where it takes off from the back of its body. After a short way, the arch flares out to its full height and continues back, around, and forward on the other side where it again dwindles for a narrow attachment to its vertebral body. In this view, our transverse process points out of the page at us and our spinous process points posteriorly from the back of its arch (Fig. 7-16).

Figure 7 - 16

Lateral view of a typical vertebra.

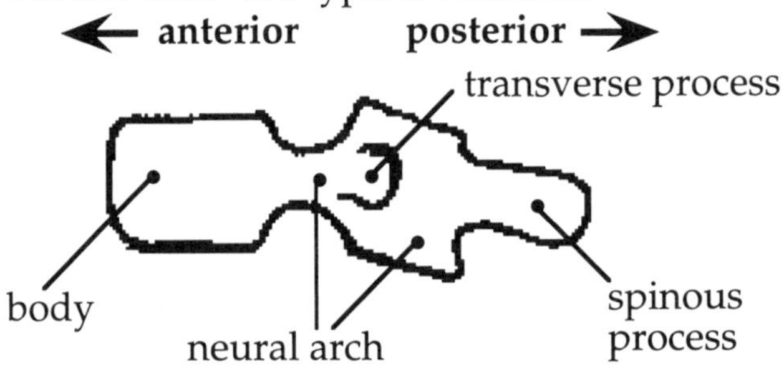

• When we stack a bunch of these vertebrae on top of each other, as we have through our torso, there are several things to notice.

Our vertebral bodies sit squarely over each other, forming a stable weight bearing column. The bodies articulate with each other through cartilaginous joints called **intervertebral joints.** A cartilage disc, called the **intervertebral disc,** lies between adjacent vertebrae to cushion and lubricate each intervertebral joint (Fig. 7-18).

Each intervertebral disc has two parts: **1.** a donut of tough, fibrous cartilage oriented horizontally between our vertebral bodies, and **2.** a springy, gelatinous jelly bean in the donut hole (Fig. 7-17a). Like the planet Saturn and its rings, our jelly bean has greater height than its donut, bulging above and below (Fig. 7-17b). Thus, our jelly bean bears the brunt of our

Figure 7 - 17
Intervertebral disc **(a)** superior view **(b)** lateral view.

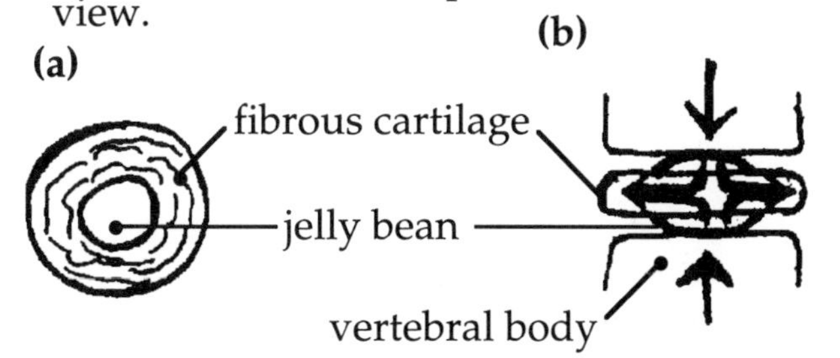

Figure 7 - 18
Lateral view of three stacked vertebrae.

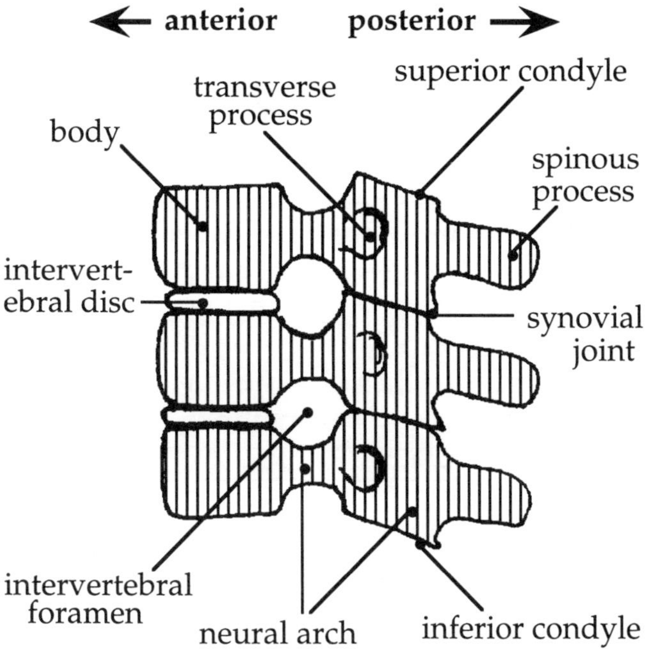

weight and absorbs the shock from one vertebral body to the next.

- Just behind the bodies, where each arch is narrow, the space between adjacent arches leaves a series of convenient holes These holes are called our **intervertebral foramina** (Fig. 7-18).
- If you could peek in through the intervertebral foramen, what would you see? The spinal cord, protected inside the neural arch. But no peeking is allowed. Our intervertebral foramen are stuffed with nerves passing in and out between our spinal cord and our peripheral tissues.

The wider, posterior parts of adjacent neural arches wear **condyles** on their superior and inferior surfaces. Their contact with each other forms a second series of intervertebral joints. These joints are lubricated, synovial joints (Fig. 7-18).

'Jelly bean' is not the anatomical term for the bulging center of an intervertebral disc, so don't tell anyone that that is what we call it here. In case you ever need to know, its real name is **nucleus pulposus.**

In good erect posture, the space between adjacent vertebral bodies has parallel edges. Our weight on our bulging jelly bean is distributed evenly in all directions (Fig. 7-17). When bending our vertebral column in any direction, including the excessive curves of poor posture, our intervertebral space becomes wedged. Our jelly bean is pushed toward the open space (Fig. 7-19a). When heavy weight is lifted from a bent position, there is considerable stress on our jelly bean, however, daily poor posture adds up. Time after time, bit by bit, our jelly bean tears its fibrous support. This may produce pain and spasm in our back muscles.

Finally one day the jelly bean may break the last straw holding the camel's back and pop through its donut (Fig. 7-19b). (Like any jelly donut, biting down on one side squirts the stuffing out the other.) This is called a **ruptured** or **herniated disc.** As the tissue is torn, the back pain may be excruciating and temporarily immobilize the person. If the jelly bean compresses nerve tissue, the pain may radiate or shoot out along a certain path, often down the arm or leg. The exact path depends of the location of the rupture and which nerve is involved. If we know nerve distributions, the location of the pain tells us which disc is involved.

Perhaps the most common error is bending forward to pick up things. This puts a backward pressure on the lumbar jelly beans, pushing them toward the space meant for nerve tissue. This is why we should pick things up with our legs rather than our backs. An ounce of prevention is worth a ton of cure.

Disc problems also may occur in the cervical area, where they are frequently misdiagnosed.

Figure 7 - 19
(a) A bent vertebral column wedges the intervertebral space, placing a unidirectional force on the jelly bean. **(b)** This may cause the disc to herniate.

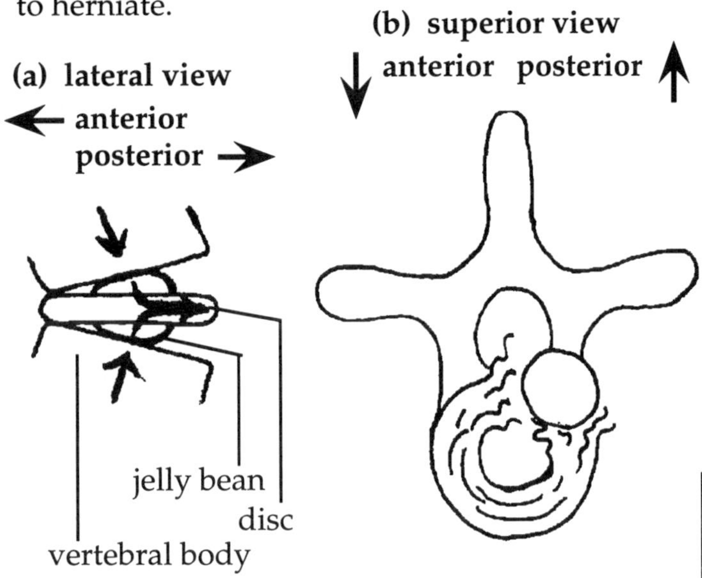

Figure 7 - 20
Superior view of a vertebra showing the ligament support. The numbers correspond with the ligament descriptions from the text.

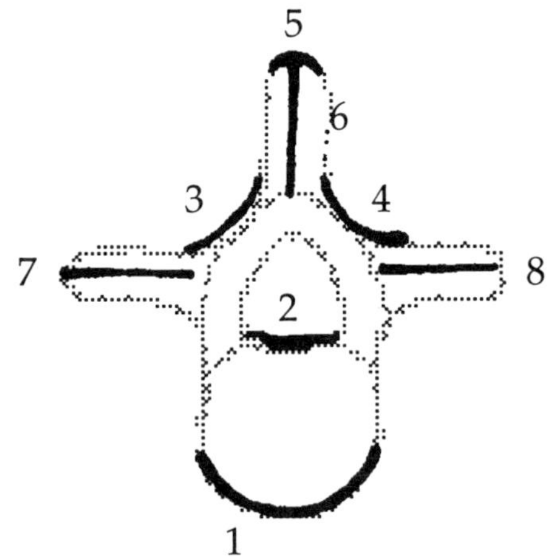

VERTEBRAL LIGAMENTS

Ligament support stabilizing our vertebral column is very extensive. The general positions of our ligaments are described to demonstrate how extensive they are. Their names and details you can find in other books.

Five ligaments form long belts which support the entire length of our vertebral column, attaching to each vertebra (Fig. 7-20) **1.** One surrounds the anterior surface of our vertebral body. **2.** One covers the posterior surface of our vertebral body. **3,** and **4.** One on each side of our spinous process covers our posterior, synovial, intervertebral joints. **5.** Our last long belt strings along the tips of our spinous processes.

6. Another set of short ligaments spans the space between adjacent spinous processes, attaching into the bottom of one spinous process and the top of the next. **7** and **8.** Two similar sets of ligaments span the spaces between adjacent transverse processes (Fig. 7-20).

The only part of our vertebrae which is not linked with ligaments is the arch just behind the body, in front of the transverse processes. Why is this area not strapped in? That is where our nerves are coming and going through their intervertebral foramina.

> Forceful, abrupt movements of our vertebrae may tear its ligament support. This is the basis of **whiplash** injury. Because ligaments have a lousy blood supply, they heal poorly, so these injuries may produce long-term pain and instability.

UNIQUE CHARACTERISTICS

Cervical Vertebrae

• The transverse processes of each cervical vertebra are deep, front to back, and pierced vertically by a hole (Fig. 7-21). An artery on its way from our heart to our brain, threads sequentially through the holes of the stacked vertebrae. These left and right arteries together supply nearly one third of the blood to our brain.

• The tips of our cervical spinous processes may be forked horizontally, so we would feel two bumps side by side (Fig. 7-21). This is normal in our cervical area, but in any other vertebrae it warns of the possibility of deeper bony instability or neurological impairment.

The spinous processes of C2 and C7 are the largest and most prominent (Fig. 7-10).

Our top two cervical vertebra, C1 and C2, are highly specialized.

• The overall impression of C1 is that of a horizontal ring with large lateral masses (Fig.

Figure 7 - 21
Unique characteristics of cervical vertebrae.

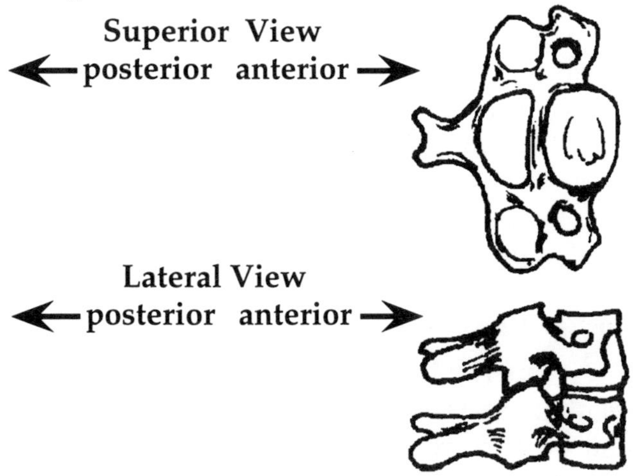

7-22). C1 is named the **atlas,** after the Greek god who balances the earth on his shoulders. Our atlas balances the globe of our head high above our shoulders, receiving the whole weight of our head from our occipital bone. Remember the two occipital condyles lateral to the foramen magnum (Fig. 7-1)? A central weight-bearing vertebral body would fall between these two condyles. Our atlas matches our occipital condyles with two lateral concave pads, leaving vacant the space usually taken by the body (Fig. 7-22).

Our **occipital atlas joint** is the balance point and pivot point for head movement. It is oriented in an anterior-posterior direction to facilitate **flexion** and **extension,** as in nodding "yes". It is a synovial joint so its range of motion exceeds that of the other cervical joints. Our atlas has no spinous process (Fig. 7-22), leaving space for further extension of our head.

Our occipital atlas joint is located deep inside, just under our ears. Long transverse processes poke laterally from our atlas condyles to mark the spot (Fig. 7-15). Feel the tips of these processes as the hard balls under your ears, immediately behind your mandible, inferior and anterior to your mastoid process. Hold these balls and nod agreement to get a sense of the movement of your occipital atlas joint.

Palpate your atlas from behind by sliding your fingers down from the midline of your occipital bone. First you fall into a hole, then you feel a larger bump. The hole is your atlas missing its spinous process. The bump is the large spinous process of C2.

The neural arch of our atlas lines up with the foramen magnum in our occipital bone to accommodate our spinal cord.

- Another name for C2 is **axis.** 'Axis' implies rotation around a central axle. All the special features of our axis facilitate horizontal rotation of our atlas over our axis, as in shaking our head "no". Our axis has a very prominent spinous process (Fig. 7-10). We have just felt it as the bump protruding just below the hole of our atlas. Our axis again receives the weight from our atlas on lateral condyles. Between these condyles, a finger of bone, called the **dens,** sprouts straight upwards. It fits into the space left vacant by the missing body of our atlas (Fig. 7-22), The dens forms the axle around which our atlas rotates. The large transverse processes of our atlas are efficient handlebars for muscles to produce rotation. So the axis itself does not rotate but provides the axle around which our atlas rotates.

Place a finger on the spinous process of your axis and your thumb on the transverse process of your atlas. Look from side to side. Feel the positions shift as your **atlas axis joint rotates**. This is a lubricated synovial joint.

Our axis has a central body under its dens (Fig. 7-22), so from here on down our vertebral column, weight is passed through the bodies and their cartilaginous joints.

Figure 7 - 22
(a) The atlas and **(b)** the axis.

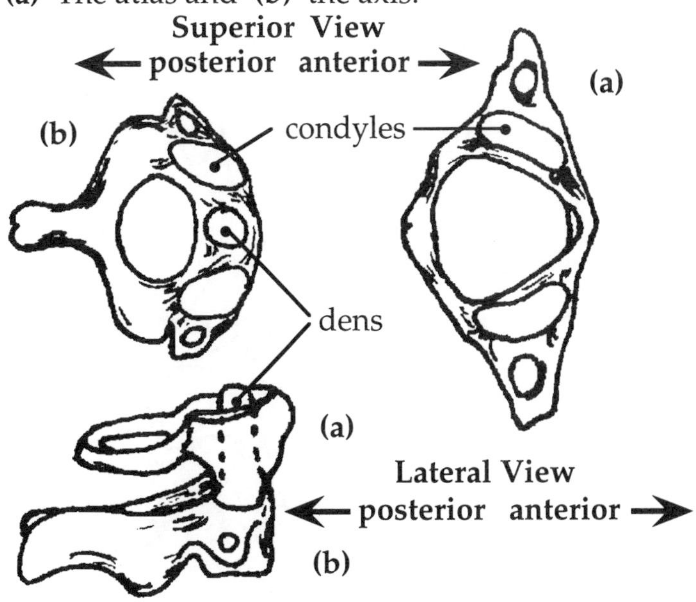

Thoracic Vertebrae

Our thoracic vertebrae are delicate but classic examples of the typical vertebra. One minor modification is made in our middle thoracic area. The spinous processes are long and slope downwards, reducing the appearance of the curve (Fig. 7-23). This is important to remember when counting spinous processes to locate other structures since the slope brings the tip of a specific spinous process well below the rib with the same number. For example, the tip of the T7 spinous process is beside rib 8.

Figure 7 - 23
Thoracic vertebrae.

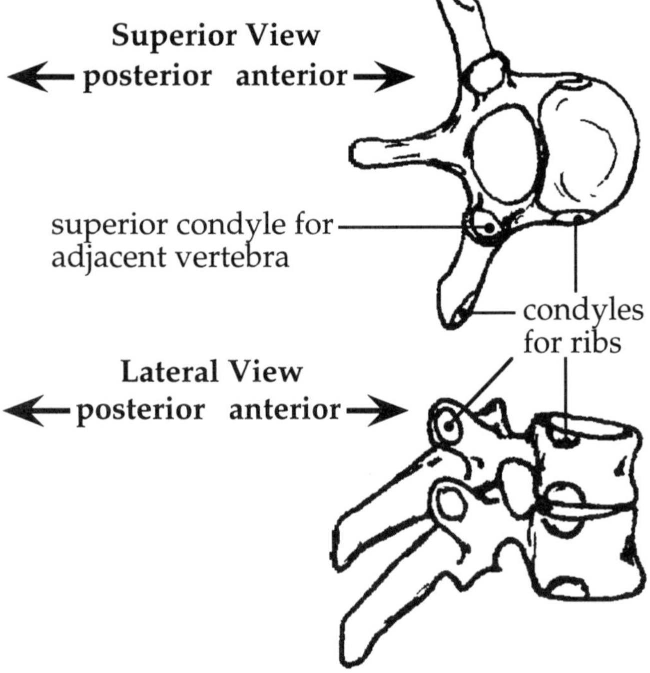

Our thoracic vertebrae articulate with our ribs. For this purpose they have special **condyles** on the sides of their bodies and the fronts of their transverse processes (Fig. 7-23).

Lumbar Vertebrae

Our lumbar vertebrae boldly follow all the characteristics of the typical vertebra. Their bodies are robust with an arch and processes to match (Fig. 7-24).

Figure 7 - 24
Lumbar vertebrae.

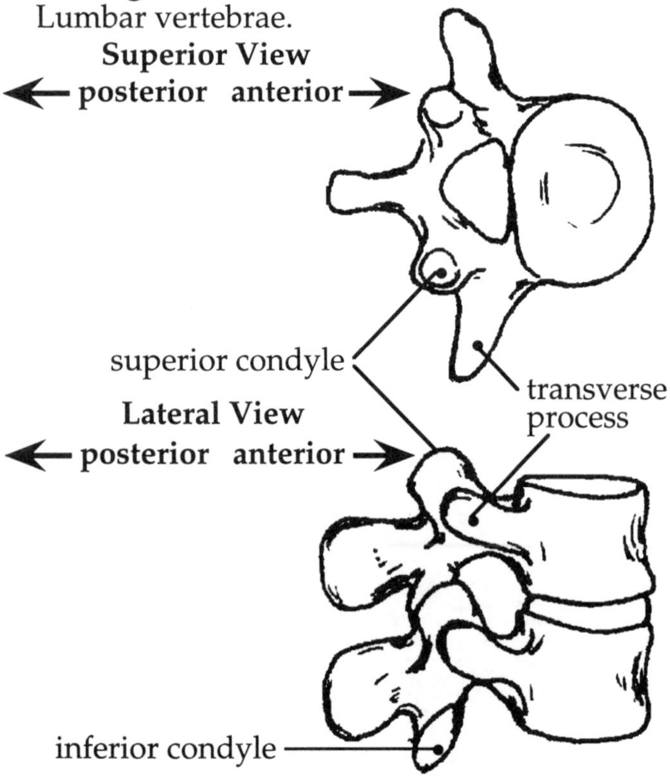

Sacrum

Our **sacrum** is somewhat specialized. In adults, its five segments fuse to form one bone, with no intervertebral discs (Fig. 7-25).

The body of the first segment is substantial, but distal to that, the bodies dwindle to almost nothing (Fig. 7-25). Why can this happen? Recall the function of our vertebral body? It bears our weight. Our first sacral segment receives the weight of the whole central column above it. But where does the weight go from here? It passes laterally into our pelvis. Our lower sacral segments are relieved of their burden and their size diminishes accordingly.

Now look at the transverse processes of the upper sacral segments, they are gigantic (Fig. 7-25). They have a new job, to transmit all that weight laterally into our pelvis, forming the joint with our pelvic bones. They grow to the occasion.

Our oversized transverse processes consume the usual space of our intervertebral foramina, squashing each foramen into two. The larger portion of each foramen moves forward and transmits most of our nerves through the protected anterior surface of our sacrum. The smaller portion opens on the back of our sacrum (Fig. 7-25), transmitting only a few nerves.

The spinous processes of our sacrum are vestigial nubs (Fig. 7-25). Since there is no movement, there are no attaching muscles and no need for handlebars.

Figure 7 - 25

The sacrum and coccyx, with pelvic shadows.

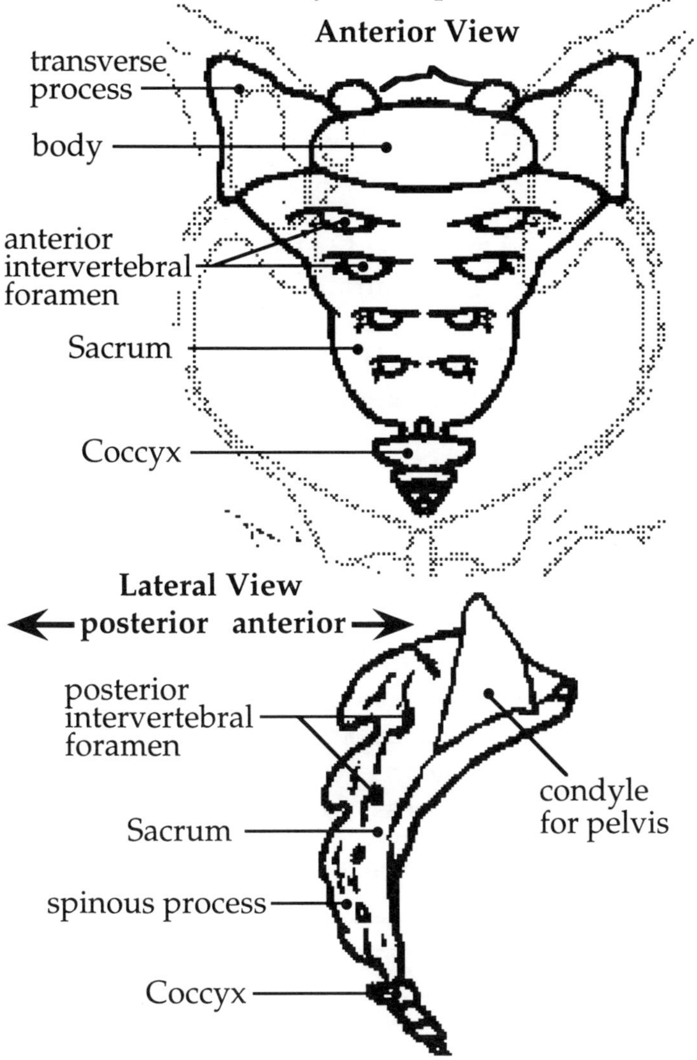

Coccyx

Our coccyx has no typical characteristics. It does few of the typical jobs, so why should it? No weight bearing, no movement, only tiny holes to transmit our last pair of nerves. A 'coccyx' is a berry. Our coccygeal segments are <u>tiny irregular blocks which are frequently fused</u> and hang like a berry from the bottom of our sacrum (Fig. 7-25).

STERNUM

Our **sternum** is the flat, vertical bone in the middle of our anterior chest. A 'stern' is where something is steered from. In boats, it is from the rear. We are steered from our front, led forward rather than pushed from behind. Our sternum is made of three parts, a **manubrium, body,** and **xiphoid process.**

- **Manubrium** means handle, coming from 'manus' or hand. Our manubrium is the handle of our sternum. It is short and wide and sits at the top of our sternum (Fig. 7-27). For fun we can say, "Our manubrium is the little, fat man at our stern, steering our ship."

<u>Feel</u> the superior edge of your sternum as the distinct central ridge at the bottom of your neck, inches below your chin. Move slightly lateral, over a groove, to a bump. This hump is your clavicle, articulating with the top of your manubrium (Fig. 7-27). The groove is the joint space. Right below your clavicle, your first rib cartilage joins onto your manubrium. It is hard to palpate since it is so tucked under your clavicle and covered by muscle.

- Below our first rib, our manubrium swoops medially to a narrow attachment to the body of our sternum (Fig. 7-27). <u>Take</u> a smooth slide three to four fingerwidths down your central manubrium to a bump, marking <u>the bottom of your manubrium and the top of your sternal body</u> (Fig. 7-27). This joint is called our **sternal angle** because from a lateral view, our sternum changes direction at this point. Our manubrium is more sloped, while our sternal body is more vertical (Fig. 7-26). <u>Feel</u> how your sternum changes direction at its angle.

Slide laterally from your sternal angle onto a horizontal ridge. This is <u>the cartilage attachment of your second rib</u>, half on the manubrium and half on the body of your sternum (Fig. 7-27). Dip into the space above and below your second rib cartilage.

Figure 7 - 26

The upper sternum.

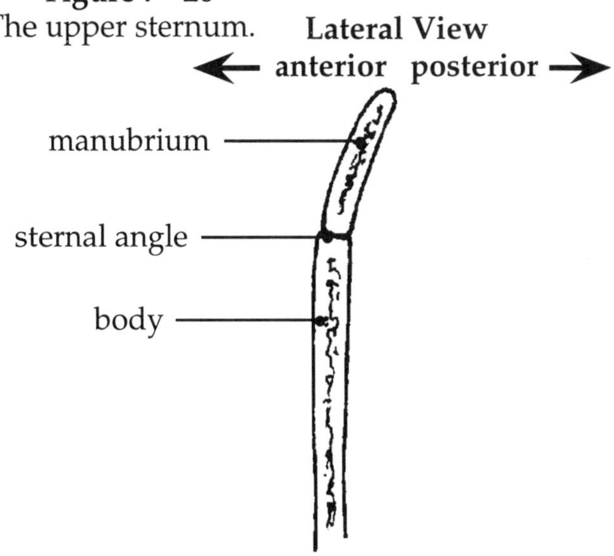

Figure 7 - 27
Anterior view of the sternum with the rib cartilages numbered.

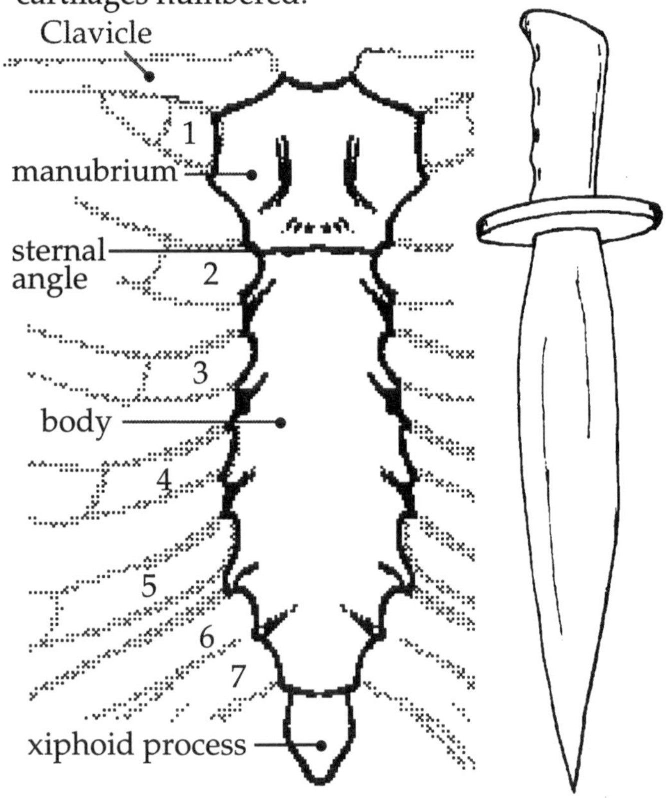

Since our first rib is difficult to feel clearly, our sternal angle is an important landmark for counting our ribs. When we move directly lateral from our sternal angle, we are always on rib 2.

- The **body** is the middle part of our sternum. It is long and thin like the blade of a sword (Fig. 7-27). It lies flat within our chest, turning neither inward against ourself, nor outward against our world. It is planted to establish our boundary. We can lift and project it to extend our boundary and bravely proclaim who we are. We can sink and withdraw it to hide, forsaking our boundary and our statement of who we are.

Our sternal body articulates with our rib cartilages from the bottom half of our second rib through our seventh rib (Fig. 7-27).

Slide your fingers over the smooth center of the body of your sternum. Feel how it swings up and out as you inhale and drifts down and in as your exhale. Move laterally and follow the vertical bump-dip-bump-dip of your rib cartilages and their intervening spaces. How many ribs can you count?

> Our **heart** lies behind the body of our sternum from rib 2 to rib 6 and, in front of the bodies of our thoracic vertebrae. The intention of **cardiopulmonary resuscitation** (CPR) is to use the external compression of the sternum against the vertebrae to pump blood through the heart. It is therefore important, while doing CPR, to place the heel of our hand on the body of the sternum since that is where we can affect the heart most directly.
>
> Even though our heart lies two thirds to the left of the center of our sternum, we do not move our hand to compress more on the left side of the chest. Compressing on the left would push the heart into soft lung tissue with no compressive action. When off center, there also is more chance for panicked pressure to pop the rib cartilages off their bones, hardly a resuscitative measure.
>
> Placing our hand too low on the sternum compresses upper abdominal organs more than the heart. It also risks fracturing the xiphoid process, puncturing the liver, heart, or other adjacent vital organs. If the resuscitee lives to tell the story, he/she may not be as thankful as we had hoped.

- The inferior point of our sternum is called our **xiphoid process.** It expands the space for muscle attachments. Palpate it as a downward point at the peak of the A-frame formed by the inferior edge of your rib cage. It is easier to feel after 30 years of age when it has completely ossified. It may be bent more inward or outward depending on a person's particular muscle tension and posture.

RIBS

The word 'costal' pertains to **rib,** and many terms defining aspects of ribs have costal as a word part. For example, the space between adjacent ribs is called our **intercostal space** (Fig. 7-29).

We each have twelve pair of ribs, men and women alike, with all due respect for the story of God taking one of man's ribs to make woman. Our ribs form a cage which protects the many vital organs in our chest and upper abdomen.

• Our ribs articulate posteriorly with our 12 thoracic vertebrae. The posterior end of each rib has a roundish knob called the **head.** The head articulates with the side of our vertebral body (Fig. 7-28a). After a thinner **neck,** each rib has a small bump which articulates with the front of its vertebral transverse process (Fig. 7-28a).

From here, our ribs become flat, measuring from outside in, although their height also is small. Our ribs scoop first backward, then laterally to our side, then forward and medially, wrapping around our chest and upper abdomen (Fig. 7-28a).

• Anteriorly, each rib ends in a strip of cartilage called its **costal cartilage** (Fig. 7-28a).

Our top seven costal cartilages articulate directly with our sternum. Their ribs are called **true ribs.** Rib 8, 9, and 10 are called **false ribs.** Their cartilages bend upward to join into the cartilage above them, forming the lower A-frame edge of our rib cage. This edge is called our **costal margin** (Fig. 7-29).

Our last two ribs have no cartilage and no anterior attachment (Fig. 7-29). They float along as our rib cage moves, thus explaining their name **floating ribs.**

• The heads and necks of ribs are not palpable, buried like a scared ostrich. We can first feel our ribs at the **angle** where their direction changes from backward to lateral. Find the spinous processes of your thoracic vertebrae. Slide laterally over muscle about one hand width, till you touch the posterior corner of your ribs. Follow your ribs around your sides and chest, interrupted here and there by bones, muscles, or breast. Feel how your ribs slope downward as they move forward.

Anteriorly, your first costal cartilage is tucked under your clavicle. Slide down beside your sternum, bumping over your second through seventh cartilages and spaces. Follow your distinctive costal margin, sloping obliquely to your side, just above your waist. Continue back to find the tips of your floating ribs, the twelfth one an inch or so behind the eleventh.

• At rest, ribs slope inferiorly as they proceed forward around our torso (Fig. 7-29), much like a bucket handle hanging down against the side of the bucket (Fig. 7-28b).

Our ribs move by pivoting up and down on their vertebrae. As their anterior ends swing upward, they arc forward. Our sternum and proximal clavicles move with our ribs, increasing the anterior-posterior dimension of our chest and the space within our thoracic cavity. This is demonstrated by swinging the bucket handle up and out from the side of the bucket (Fig. 7-28b).

Increasing the space inside our chest

Figure 7 - 28

(a) A horizontal section through the thoracic cavity.

(b) A bucket handle demonstrates how the pivoting movement of the ribs increases the depth of the thoracic cavity. The arrows show the depth in each position.

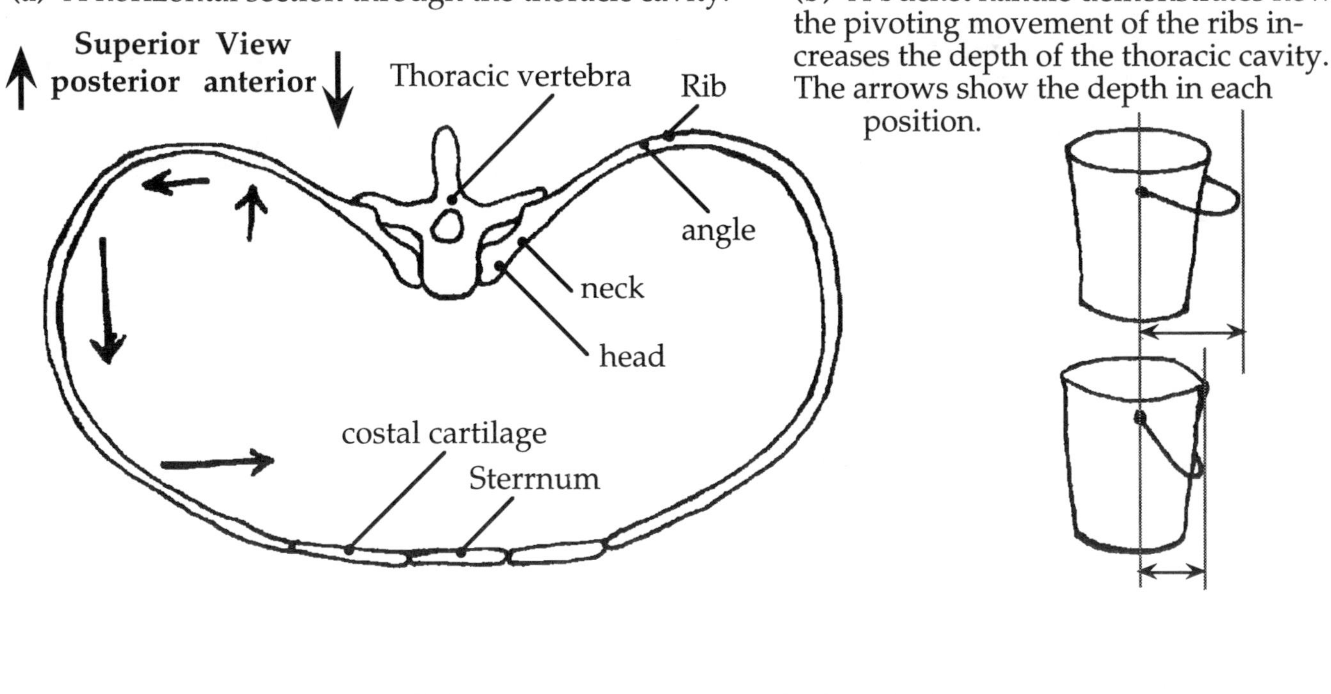

Figure 7 - 29

The rib cage.

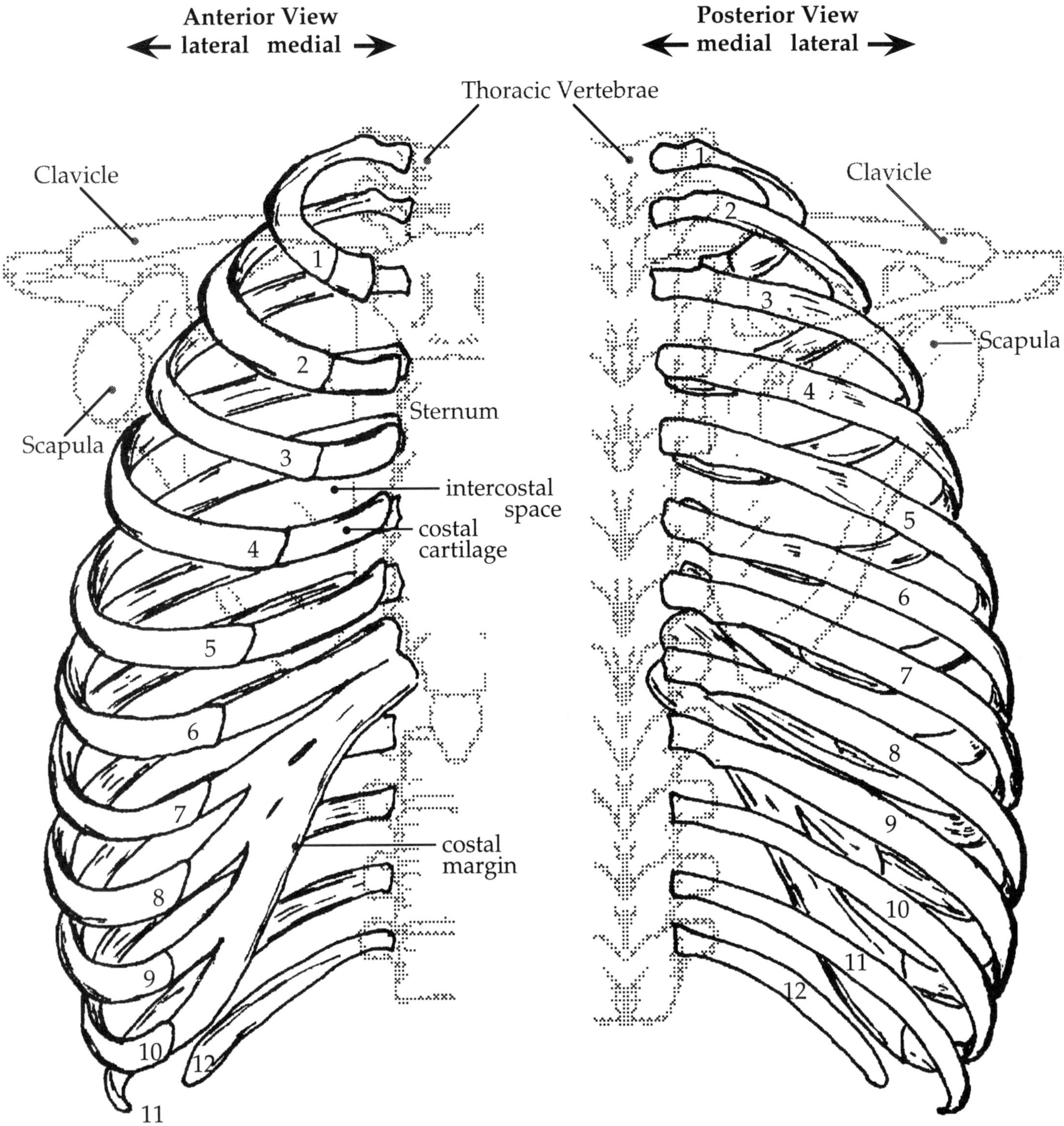

reduces chest pressure and is a major influence on many body systems. It is a force which draws in our breath, moves our blood and lymph, massages our transverse colon, and so forth. As we explore other systems, we will return many times to the changing space in our chest.

Figure 7 - 30

Anterior surface view of the upper chest.

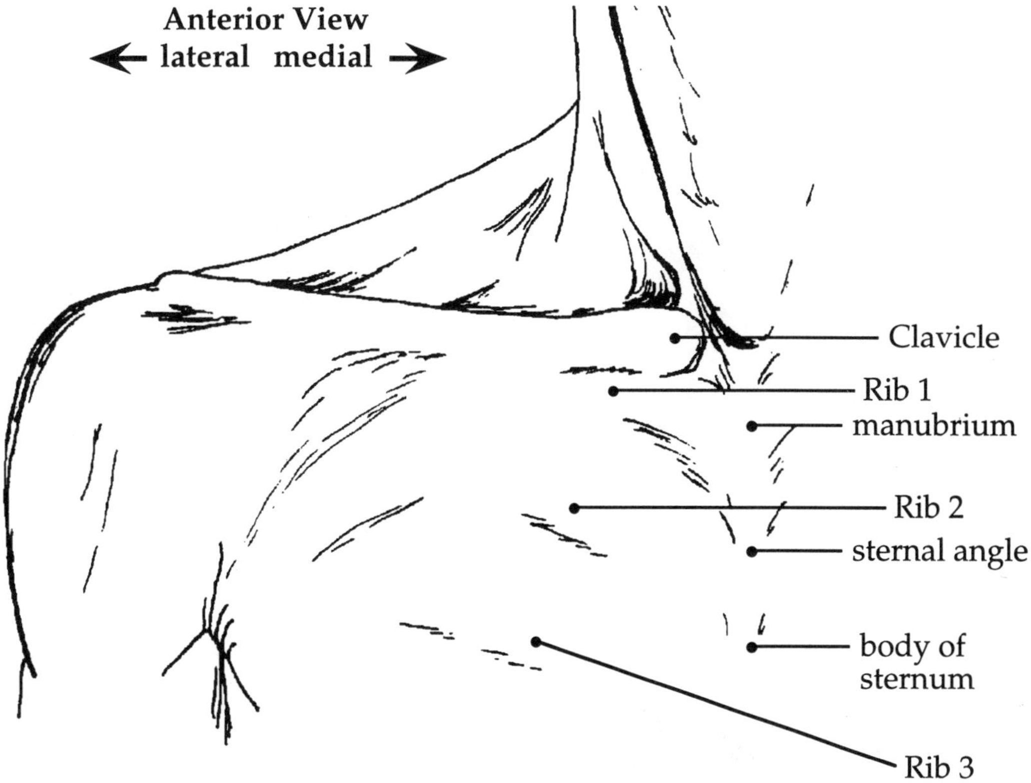

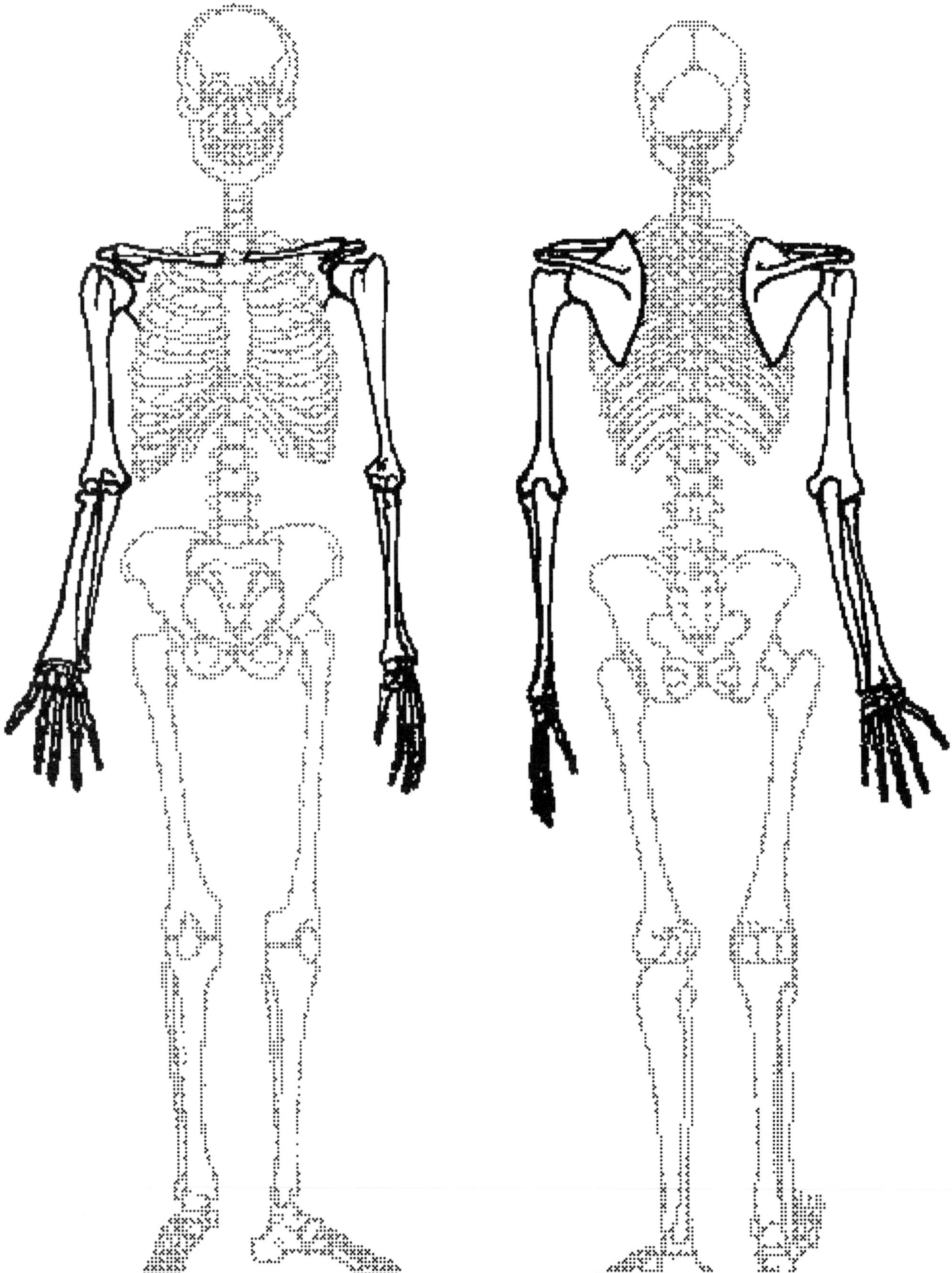

chapter 8: **Upper Appendage Skeleton**

The skeleton of our upper appendage includes the bones of our **arm** and its supporting yoke or girdle called our **pectoral girdle.**

A recurrent theme in our upper appendage is the inseparable partners, decreased stability leading to increased mobility. This already is evident in our pectoral girdle.

PECTORAL GIRDLE

'Pectoral' refers to the area of our chest or breast. A 'girdle' goes all the way around a structure for the purpose of support (Fig. 8-1a). Our pectoral girdle then, encircles the upper part of our chest to support our arms onto our axial skeleton. But our pectoral girdle goofs. It does not go all the way around. It forgets to close things up across our back, leaving our arm support unstable. Yahoo! Exactly what our arms were hoping for. A less stable support leads to more mobile arms (Fig. 8-1b). The whole rest of our body is a pillar of support so our arms can forget support and be out there moving, doing things for us.

Figure 8 - 1
(a) A girdle goes all the way around for the purpose of support. **(b)** The pectoral girdle has a posterior deficit, allowing for movement.

(a)

(b)

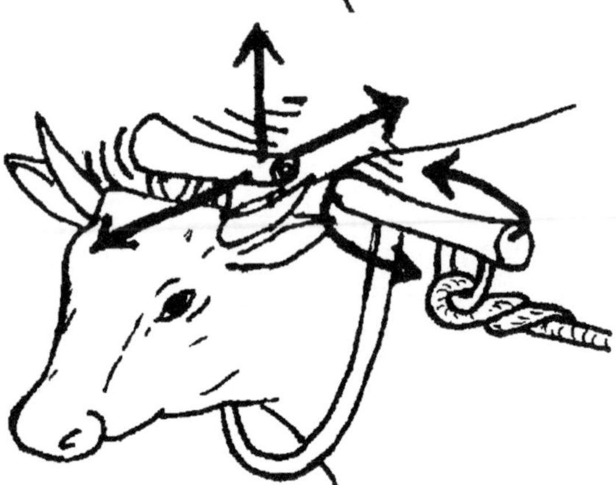

Each side of our pectoral girdle consists of two bones, a **clavicle** and **scapula** (Fig. 8-5). Each side of our pectoral girdle attaches to our axial skeleton by a joint between the manubrium of our sternum and the medial end of our clavicle, called our **sternoclavicular joint.**

> From our sternoclavicular joint, our pectoral girdle defies gravity, reaching out horizontally with no other skeletal connection except our arm dangling from its most lateral aspect. Bones and ligaments by themselves never could maintain this position. On their own, our clavicles, scapulae, and arms would dangle from our sternum straight down in front of us (Fig. 12-1). As we will see, muscles save the day, working continually to support our pectoral girdle onto our axial skeleton.

Clavicle

Our clavicle is commonly known as our collarbone. This collar has a variety of styles.

Figure 8 - 2
As you look from the front, the collar formed by the two clavicles should be a horizontal line, a boat-neck. The collar more frequently worn, however, is a V-neck, with the lateral clavicles elevated by chronic tension.

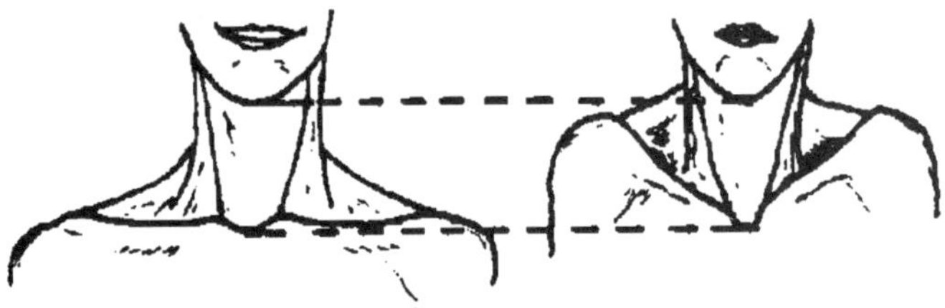

Figure 8 - 3
As you look from above, each clavicle is S-shaped. The rounded medial end curves forward, while the flattened lateral end curves backward.

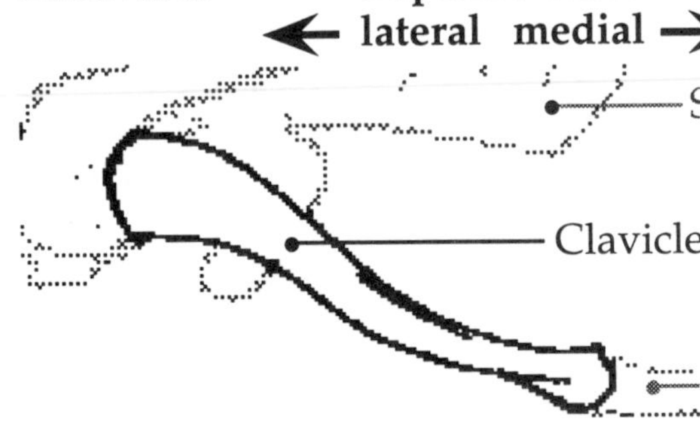

The function of our clavicle is to form a strut between our sternum and scapula, preventing our shoulders from wrapping too far around to the front of our chest. When stalking on all fours, the hanging weight of the body keeps the scapulae back. The clavicle, therefore, is a requirement of upright posture when the arms are free to reach forward. The clavicle is porportionally large in upright creatures such as humans and birds.

'Clavicle' means little key. A big key to success is taking an upright posture. Perhaps the little key that supports success is holding our shoulders back in confidence (Fig. 8-4).

Figure 8 - 4

The clavicle is the little key holding the shoulders back.

Scapula (scapulae in plural)

Our scapula is our shoulder blade, helping us cut through a lot of stuff when we are weary of shouldering the world. Our scapula is a thin, flat bone floating over the upper part of our posterior rib cage like a shield. The basic shape of our scapula is triangular. In geometric language, the sides of a triangle are called borders and the corners are called angles. The naming of our scapula employs this same vocabulary.

- Our scapula is positioned so that its longest border lies medially and parallel to our vertebral column, thus assuming the name **medial** or **vertebral border** (Fig. 8-5). Our vertebral border is the easiest to find. Reach through your arm pit to feel yours as a vertical ridge about two inches lateral to your vertebral spinous processes. Normally it goes from our second through our seventh thoracic vertebra (Fig. 8-11), although it may be higher in people who are into V-neck collars. We can frequently slip under our scapula from the vertebral border. Try this movement on a friend.

Figure 8 - 5

The Pectoral Girdle.

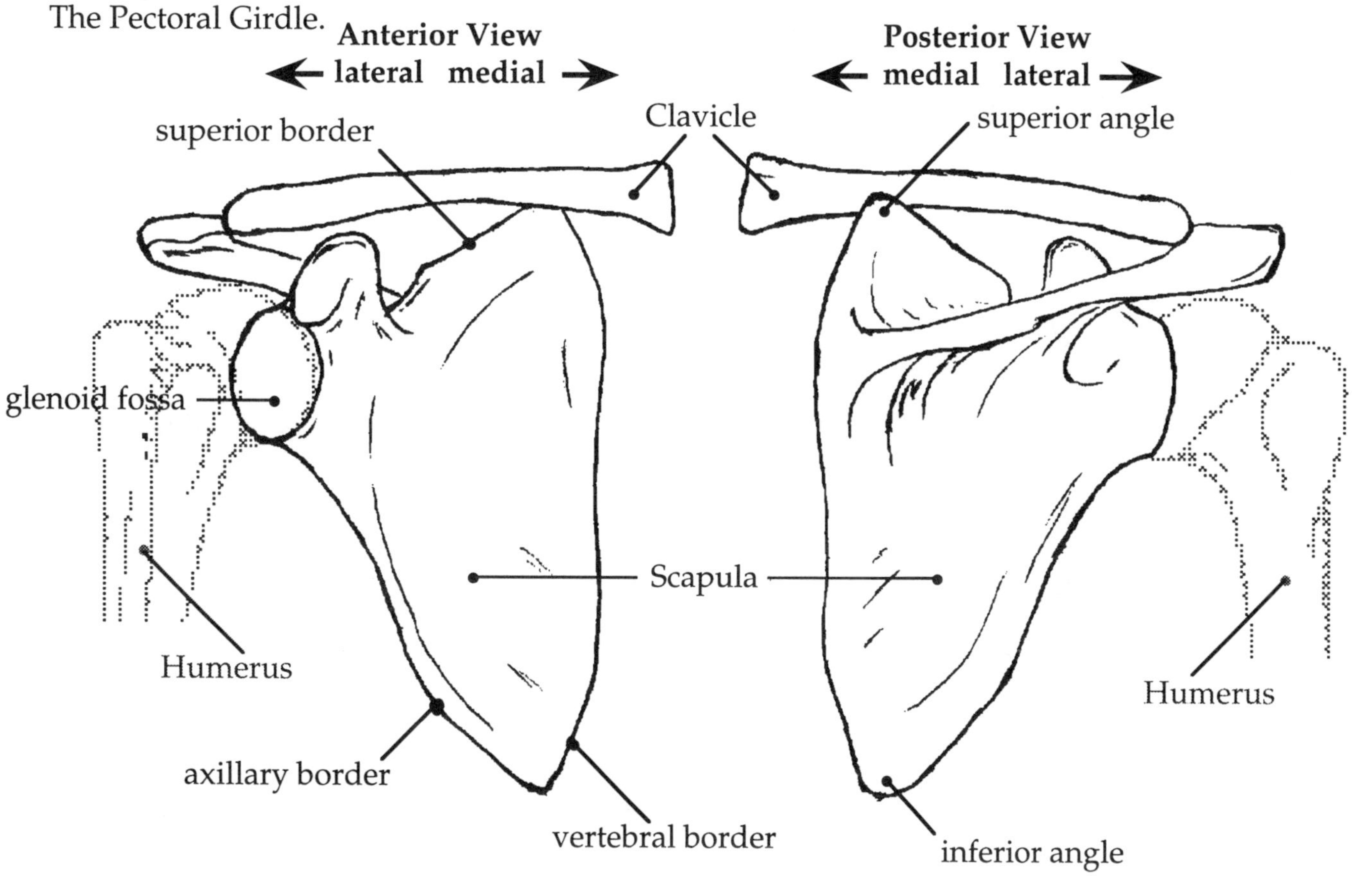

The fancy word for armpit is 'axilla'. The lateral border of our scapula lies parallel to our axilla and is called our **lateral** or **axillary border** (Fig. 8-5). Press from your back forward into the muscle forming the posterior wall of your armpit. The firm edge is the axillary border of your scapula (Fig. 8-6).

Figure 8 - 6
Palpating the axillary border of the scapula.

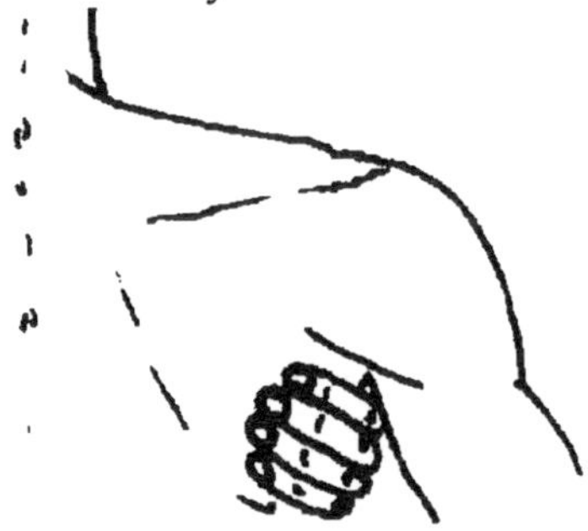

The **superior border,** as we might guess, is at the top of our scapula (Fig. 8-5). It is not the bony ridge felt spreading across our back and over our shoulder. Our superior border lies above this ridge, buried in muscle, the most difficult part of our scapula to feel.

• The names of the angles of our scapula are derived from their orientation. The bottom one, joining our vertebral and axillary borders is called our **inferior angle.** It is the easiest to palpate (Fig. 8-5). Bring your hand behind your back at the level of your waist to make your inferior angle protrude distinctly.

The top angle, joining our vertebral and superior borders, is called our **superior angle.** If you have been lazy about physical exercise, you may be able to find yours. Press into your muscle about two finger widths above the medial end of your scapular spine (Fig. 8-7). Lift your scapula up to make it more obvious.

Figure 8 - 7
Palpating the superior angle of the scapula.

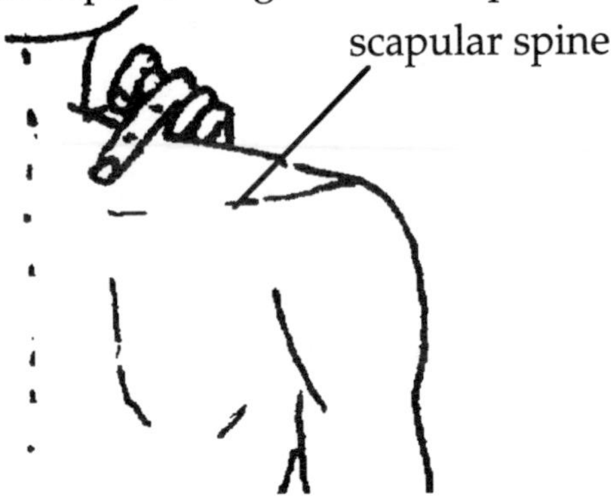

Our lateral angle is impossible to feel, because the area is completely buried in bone, muscle, and tendon and in fact, there is no angle. If we call our scapula a triangle, its lateral corner is in disguise. This corner mischievously flares into a shallow bowl called our **glenoid fossa** (Fig. 8-5). 'Gleno' means socket, and our glenoid fossa becomes the socket which fits the ball of our upper arm bone.

• Our scapula has three structures which protrude outward from its triangle.

One protrusion is called our **coracoid process** (Fig. 8-9). It is like a bent finger which points forward from our glenoid fossa. To locate yours, move one fingerwidth below your clavicle into the hollow between your chest and arm, where your clavicle curves back-ward. If you wrap your fingers around the top of your arm, the tip of your thumb falls on the tip of your coracoid process (Fig. 8-8). If you cannot find it, bring your scapula forward and it will point out further for you. 'Coracoid' means a bird's beak. Our old crow sits behind our shoulder with its head in our arm pit and its beak jutting forward.

Figure 8 - 8
Palpating the coracoid process of the scapula.

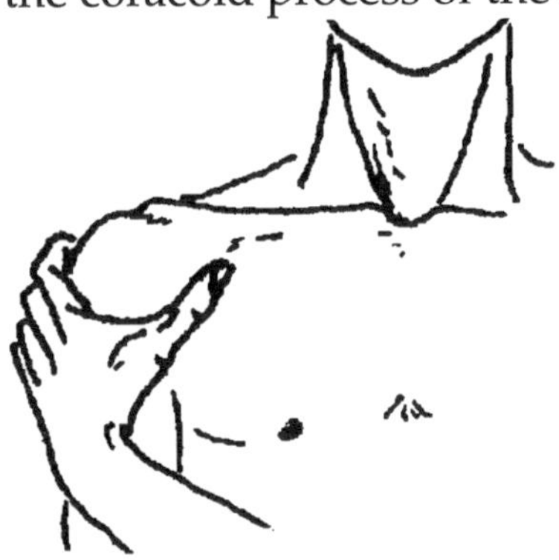

Our scapulae are the eyes of our back, protecting us from behind. Each eye has a brow, a ridge of bone spreading across our back and over our shoulder. This ridge is called the **spine** of our scapula (Fig. 8-9). Each spine begins as a small bump at the vertebral border, about one-third of the way down from our superior angle. The spine lifts prominently away from its scapula as it juts laterally and upward. Finally it leaps free of our scapula to form a shelf over our shoulder, a scout peering forward, to see what we are getting into. This shelf is called our **acromion process** (Fig. 8-9).

The spine of our scapula is only covered by skin, so its outer ridge is usually easy to follow. Follow your spine across your back to its most lateral point. Feel how it turns a right angle and continues forward as your acromion process. Follow your acromion process forward about one and one half inches to its end.

• Our lateral clavicle joins into the medial side of our acromion process to form our **acromion clavicular joint** (Fig. 8-9). Feel your lateral clavicle as a distinct knob, on top of your shoulder, butting into the medial side of your acromial shelf.

• The surfaces of our scapula are divided into three fossae (Fig. 8-9). The anterior surface, hugging against our ribs, is the underside of our scapula or **subscapular fossa.** The posterior surface is divided into two by the spine: a small hollow superior to the spine, the **supraspinous fossa,** and a large one inferior to the spine, the **infraspinous fossa.** These fossae have palpable edges but are filled with muscle.

• Our scapula has four sets of three: three borders, three angles (one still out on a flare), three protrusions, and three surface fossae.

UPPER ARM BONE

Humerus

Our **humerus** is the bone in our upper arm. Too bad the spelling is goofed or we might have made more laughter.

• We will describe our humerus as a funny sort of creature, beginning proximally.

A smooth hemisphere, known as the **head** of our humerus, lies medially with its face buried in our arm pit (Fig. 8-10). The head articulates with the glenoid fossa of our scapula to form our **glenohumeral joint.** This joint is surounded by bone and muscle and difficult to palpate. Our coracoid process is our best clue since it protrudes forward from the top of our glenoid fossa (Fig. 8-11). Where we find our coracoid process, we know that deep inside lies our glenohumeral joint.

A shallow indentation surrounds the head of our humerus. This is all it has for a **neck** (Fig. 8-10). The extensive movement of the head requires support from a thick neck.

Our humerus has two shoulders placed at right angles to each other. The larger one,

Figure 8 - 9
The protrusions and surface fossae of the scapula.

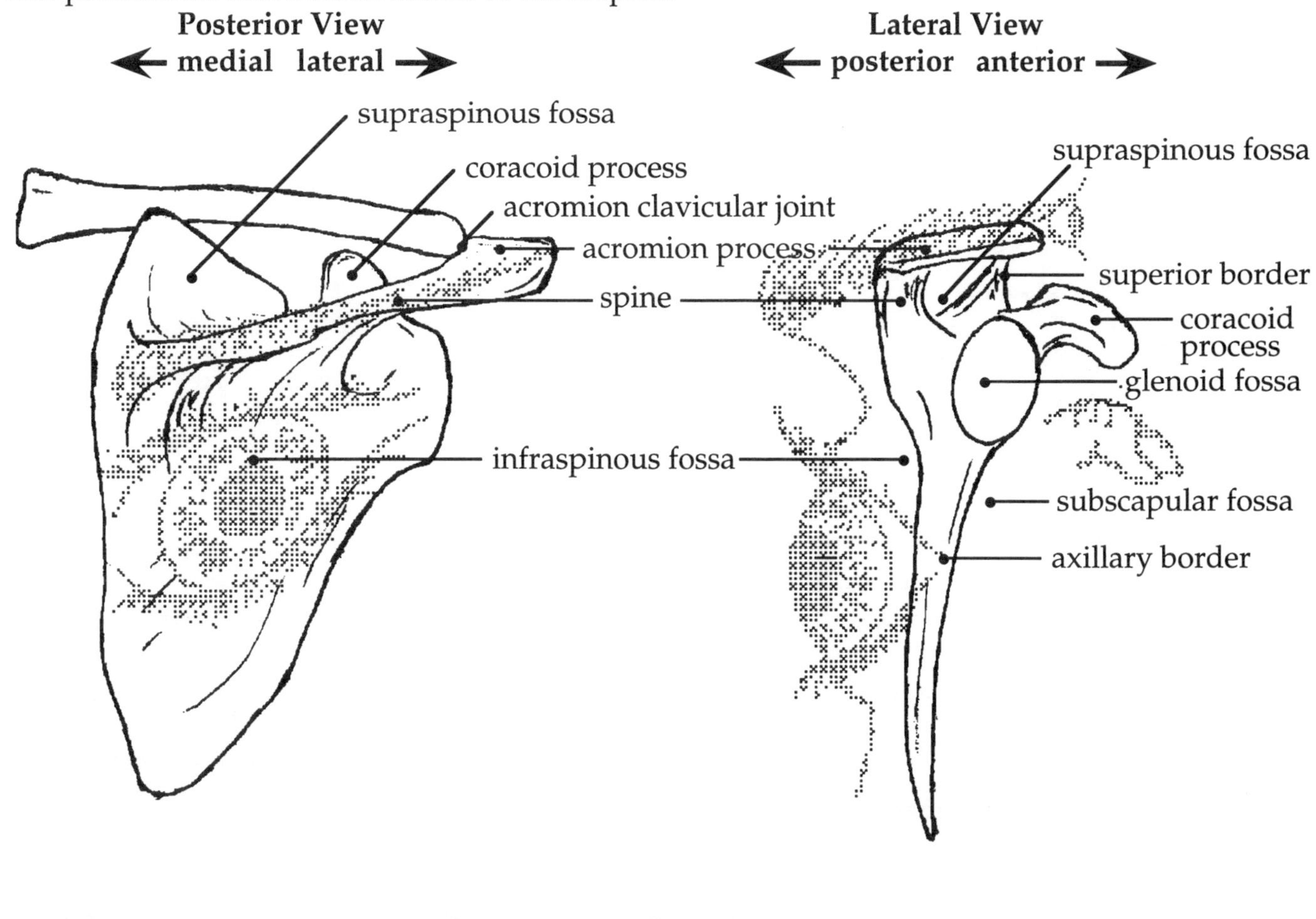

Figure 8 - 10

The Humerus.

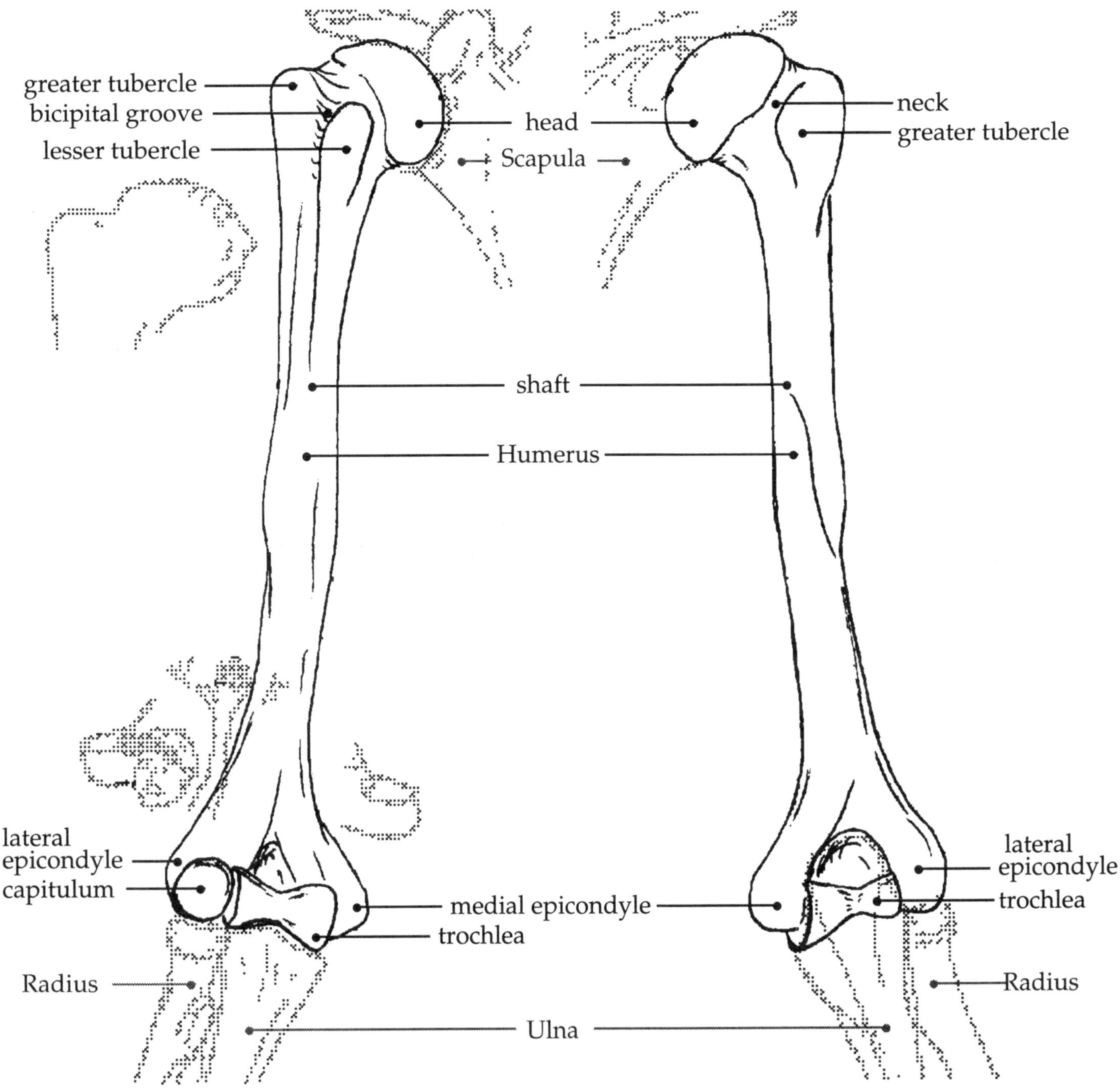

our **greater tubercle,** protrudes laterally (Fig. 8-10). Feel it as the firmness behind your deltoid muscle, one finger width below your acromion shelf. The smaller one, our **lesser tubercle** protrudes anteriorly (Fig. 8-10). A distinct groove lies between the two tubercles. Replacing the word between with its latin equivalent gives us the name of the groove, our **intertubercle groove.** This groove is difficult to feel as it is filled with the dense connective tissue of one of our biceps tendons, giving it the alternate name of **bicipital groove** (Fig. 8-10). To get a sense of all these structures, cup one hand around the top of your other arm immediately below your acromion process and rotate your glenohumeral joint.

Figure 8 - 11

Surface view of the shoulder area, indicating many bony landmarks.

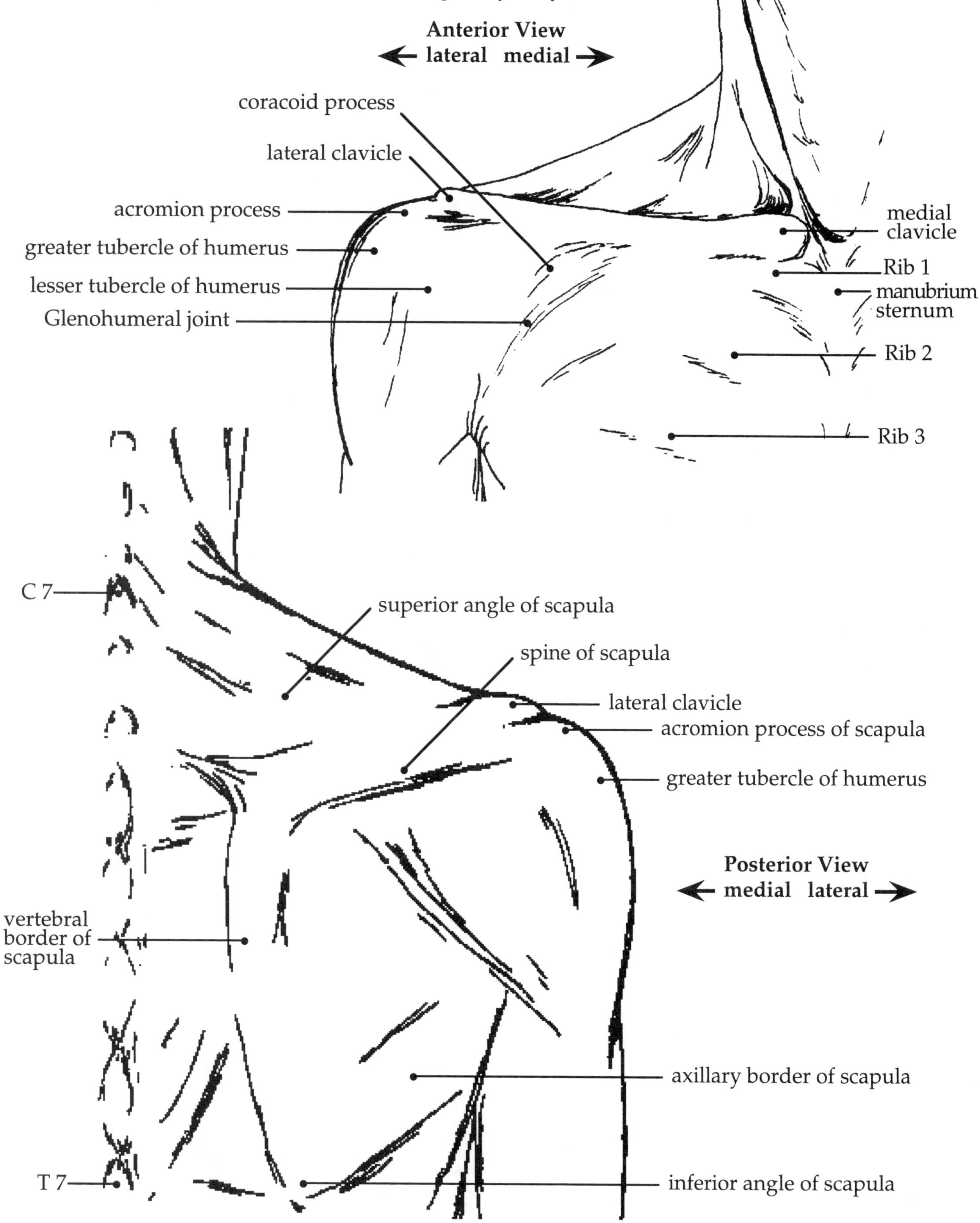

Figure 8 - 12

Superior view of the clavicle, scapula, and proximal humerus.

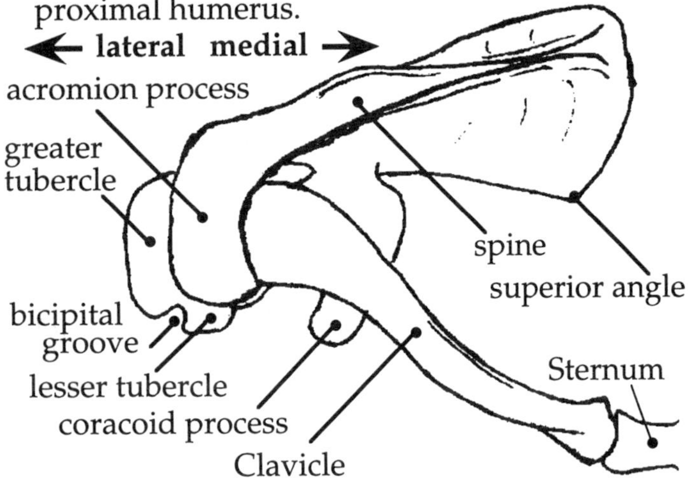

The body of our humerus creature forms the **shaft** of the bone (Fig. 8-10). It is padded by muscle front and back, but its sides are more exposed and palpable, especially medially. The proximal shaft is stout and cylindrical. Distally, it widens from side to side. Press into the side of your upper arm. Feel the firmness of your humeral shaft. Move distally, feel its spreading shins with one stubby toe pointing out on either side of your elbow. These toes are called the **medial** and **lateral epicondyles** of our humerus (Fig. 8-10). Our medial epicondyle is the most prominent. Our lateral epicondyle is most easily felt with our elbow bent to 90 degrees.

> Our medial epicondyle is often referred to as our **funny bone.** There is nothing funny about it until we stub it from behind. Even then, there is not much laughter, but funny as it may seem, the sensation of a thousand needles prickles our medial hand and little finger. The nerve carrying sensation to this area lies just posterior to our medial epicondyle. Banging the bone strongly compresses the nerve, causing our sensory experience.

The distal end of our humerus, between our two epicondyles, holds the condyles which articulate with our lower arm bones. They look like an irregular row of bumps, but are composed of two distinct structures. The medial one is like a spool or hourglass tipped on its side, broad at each end, pinching narrow in its middle. It is called our **trochlea** and articulates with our ulna (Fig. 8-10). Just above the trochlea, our humerus deeply indents to allow our ulna to clasp well around the spool and still have a wide range of motion (Fig. 8-13). The lateral bump is the rounded **capitulum** (Fig. 8-10). 'Caput' is head, the place where we put our cap. 'Ulum' makes it the little head of our humerus. It articulates with our radius. Our trochlea and capitulum are deep within our elbow joint and are not palpable.

> The Greek word 'brachys' means short. Our upper arm is the short part of our shorter appendage, so structures in our upper arm often begin with **brachi** to indicate their location. Two examples are our brachial nerve and our brachialis muscle.

FOREARM BONES

Our forearm contains two bones, our **ulna** and **radius.** In anatomical position they lie parallel to each other, but our radius may rotate to cross over our ulna (Fig. 8-31).

Ulna

Our **ulna** takes the medial position in our forearm. Its proximal end, the **olecranon process,** is large and forms the posterior point of our elbow (Fig. 8-13 and 8-14), good for carving out elbow room. Find yours posteriorly, between your humeral epicondyles. Anterior to the olecranon process, our ulna is scooped out. This scoop is like the open end of a hook or wrench fitting around the waist of our

Figure 8 - 13

Medial view of the elbow.

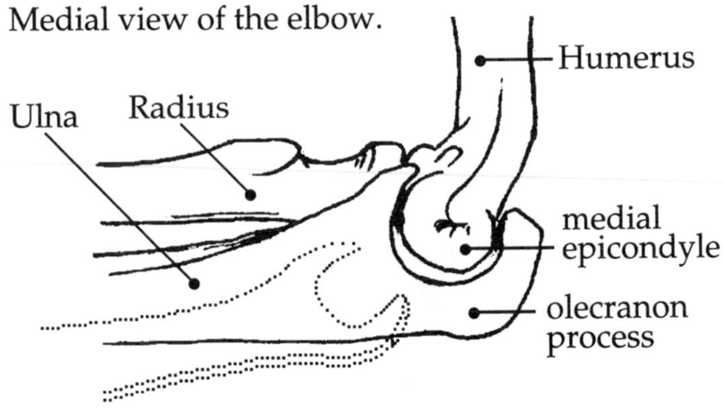

humeral trochlea to form a secure hinge joint.

Moving distally, our ulna tapers to end in a small **head** (Fig. 8-14) seen as the round knob on our medial posterior forearm just before our wrist. A small point, the **styloid process,** juts out from the head, reminding us of a stylus used to write or engrave.

Palpate the entire length of your ulna along the back of your forearm, from the point of your elbow to the point at your wrist. This bony ridge forms the partition between the two muscle groups of our forearm.

Radius

Our **radius** is the lateral of our two forearm bones. The **head** of our raduis is cut flat into a disc and articulates with our capitulum at its end and our ulna on its side (Fig. 8-14). To feel the side of your radial head, flex your elbow to 90 degrees. Find the lateral epicondyle of your humerus. Step one finger width toward your hand. Feel the firm roundness. This is the side of your flat radial head. Rotate your forearm to feel the head roll under your fingers.

Our radius has a long neck and only one small shoulder, our **radial tuberosity.** In anatomical position this shoulder protrudes anteriorly (Fig. 8-14). It is buried under muscle and is not palpable but holds the attachment of an important muscle.

Distal to its small head and shoulder, our radius has one long leg with a heavy ankle and a large foot. The bulkiness of our distal radius emerges through our muscles to be felt in our lateral forearm, just before our wrist.

Figure 8 - 14
Ulna and radius in anatomical position.

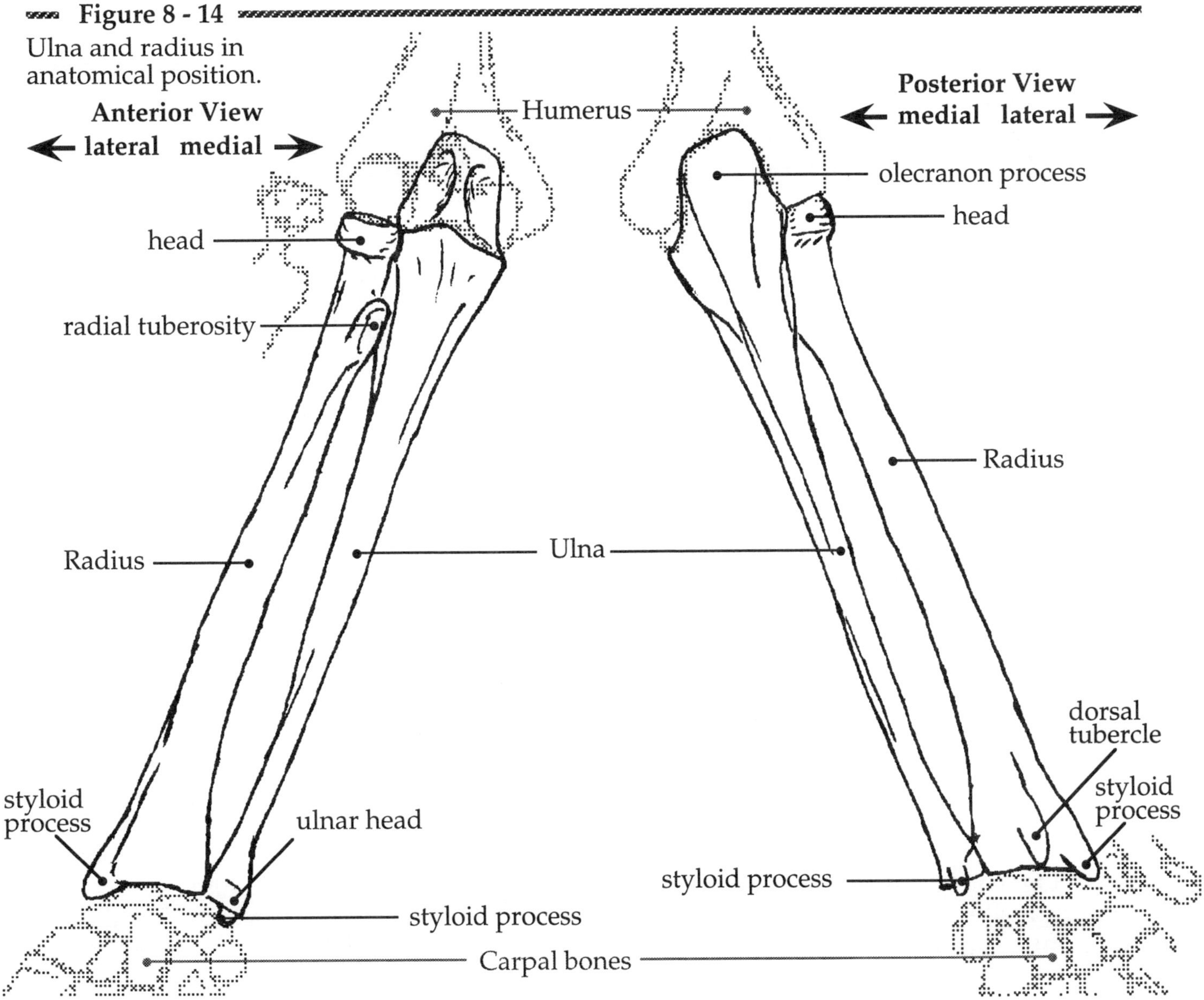

Figure 8 - 15
Surface view of the posterior forearm.

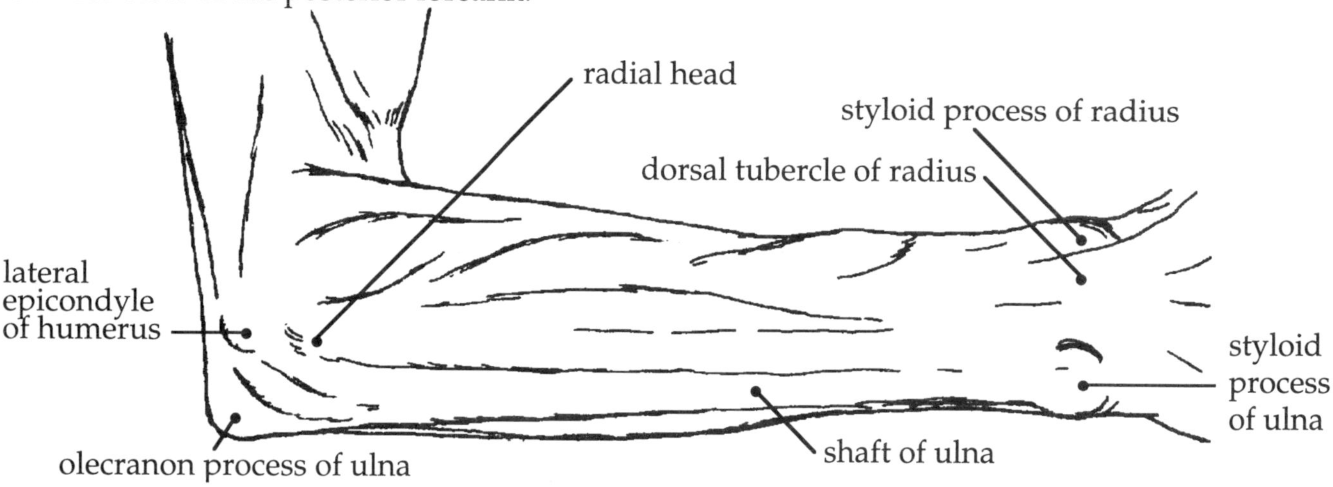

Figure 8 - 16
Surface view of the proximal posterior forearm, with the elbow bent.

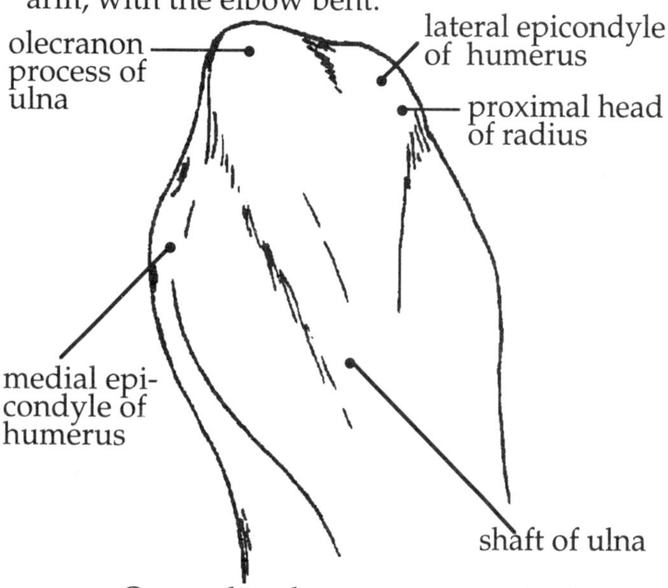

Our radius has two toes. The big toe points laterally and forms another engraving tool, somewhat more blunt than our ulna's, but given the same name, the **styloid process** (Fig. 8-14 and 8-15). Feel yours as the lateral bulge of your distal radius. Bend your hand toward your ulna to make it more prominent. Its little toe is called our **dorsal tubercle**. Feel yours on the back of your radius, half way between your two styoid processes (Fig. 8-14 and 8-15).

HAND BONES

Our hand is the most maneuverable and sophisticated tool in our body. The latin word for hand is manus. When 'man' is part of a word, it often refers to our hand. A 'maneuver' is a skillfull procedure, but translated literally means to work by hand. 'Manicure' means the care of the hand.

So important is the hand to the development of our species that we are named after it. 'Human beings' can be said as having-a-hand beings, or 'mankind' is the kind of creature with a hand.

Each hand contains 27 bones: 8 carpal bones, 5 metacarpal bones, and 14 phalanges.

CARPAL BONES

Our **carpal bones** begin in the subtle hollows just beyond the styloid processes of our ulna and radius. While we have eight carpal bones, they are all so small that two finger widths across our wrist more than covers them. On the ventral surface, our carpal bones are referred to as the **heel** of our hand.

Each carpal bone is delicately carved into a specific shape described by its name. They are placed in two transverse rows of four bones each (Fig. 8-17).

Scaphoid

Our **scaphoid** is the most lateral bone in the proximal row. A Greek 'skaphe' is a small, light boat. Our boat is tiny with our thumb as its giant oar and our index finger its tall sail (Fig. 8-18). All you can feel of the scaphoid is its round bow. Pull your thumb out to the side perpendicular to your hand. See the two tendons which pop out in a V-shape on the

Figure 8 - 17

The carpal bones.

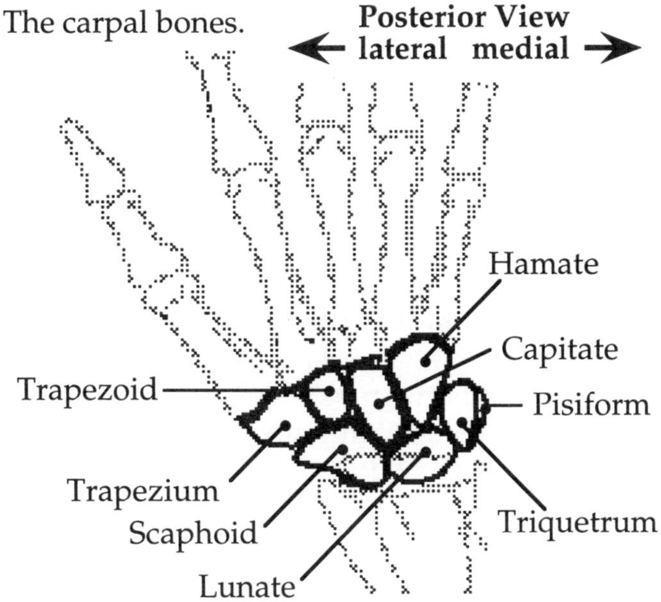

> Our carpal bones may be called our wrist bones but **wrist** also is used to refer to the distal end of our ulna and radius, or the joint between our forearm and hand. When someone says that they fractured their wrist, it is usually their distal radius and ulna.

Figure 8 - 18

The scaphoid.

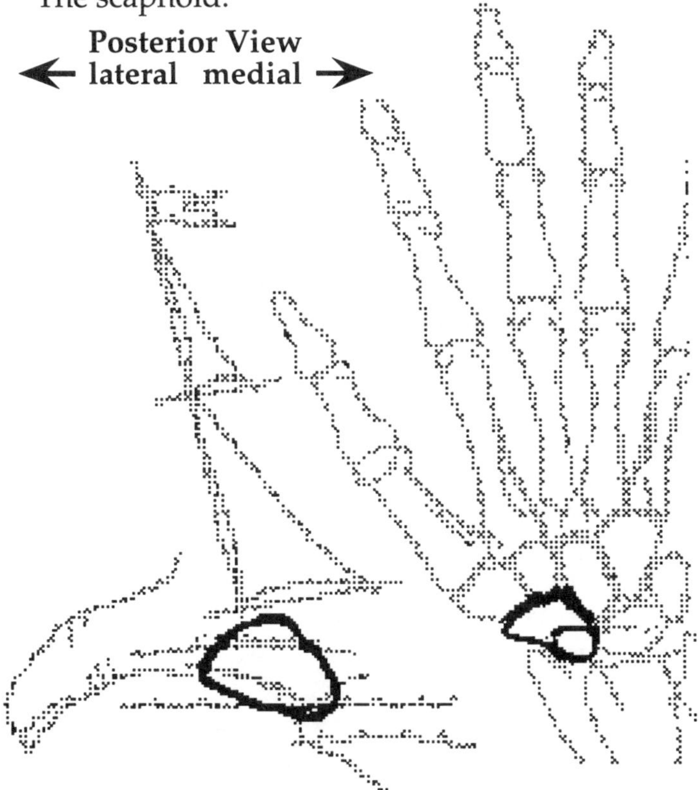

back of your thumb, forming a triangular hollow just distal to your radial styloid process. This hollow is called your **anatomical snuff box.** Press into your snuff box to feel the round bow of your scaphoid. It is easier to feel if you relax your thumb and let your hand droop toward your ulna.

Lunate

Our little boat glides by the light of the silvery moon, with our **lunate** just medial to it. 'Luna' is moon, and in this case refers to the crescent shape of a young moon (Fig. 8-19). Our lunate is our most easily dislocated carpal

Figure 8 - 19

The lunate.

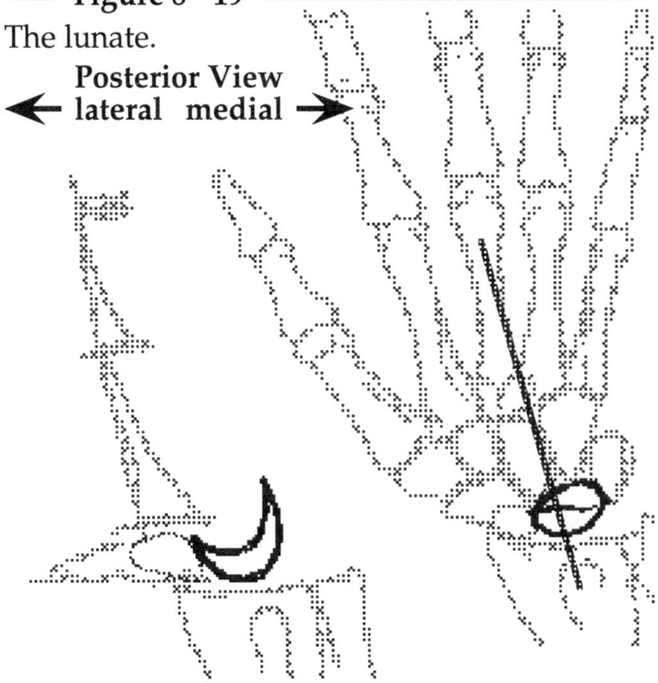

> Our scaphoid is our carpal bone which most often fractures when catching a forward fall with our hands. When we think of a broken bone, we think of extreme pain, swelling, and bruising. When our scaphoid fractures, there is no alarming appearance, so it may be overlooked. The initial symptoms are mild and gone in a week, except for a small persistent pain. Months later, pressure in our snuff box still hurts. Since the bone was not immobilized, it healed as two separate bones. The joint between these bones is not supported or lubricated, so movement continually traumatizes the bones.

bone, popping backward to form a knob in our central, posterior wrist, which is most prominent when our wrist is flexed. Our lunate is difficult to feel, since it is covered in tendons front and back. Draw a line from the dorsal tubercle of your radius to the knuckle of your middle finger. Find your lunate on this line, just distal to your radius. Droop your hand, palm down, to make it easier to feel.

Triquetrum (triquetrium or triquetral bone)

Our **triquetrum** has three corners, a 3-pointed star to guide our boat (Fig. 8-20). It is the most medial bone in the proximal row (Fig. 8-17). Feel yours on the back of your hand, just distal to your ulnar styloid process. Make it more prominent by pulling your hand toward your radius.

Figure 8 - 20
The triquetrum.

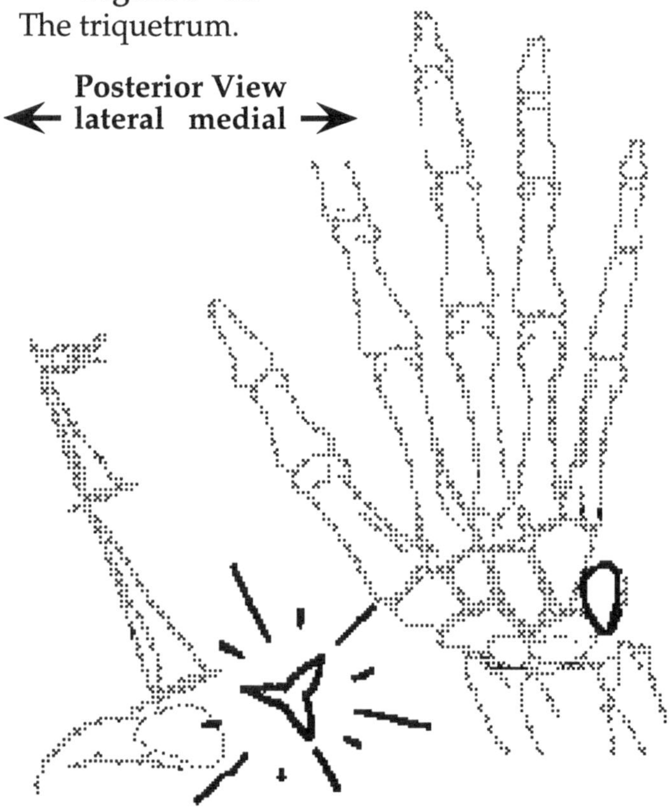

Pisiform

Our **pisiform** is the fourth carpal in the proximal row (Fig. 8-17). It sits directly in front of our triquetrum. 'Pisiform' means pea-shaped. Feel this little, round bone like a hard pea in the proximal medial corner of the palm of your hand. Support and relax your hand. Wiggle your pisiform from side to side.

Trapezium

Our **trapezium** is the most lateral carpal in the distal row (Fig. 8-17). It supports our thumb. It forms an easy trap for our thumb, a special bracket on the side of our little boat to hold its oar. Feel your trapezium between your scaphoid and the base of your thumb.

A 'trapezium' is a little table and refers to a flat plane with four, nonparallel sides. This is our first attempt at table making, and the flat plane warps into a smooth dome. The base of our thumb molds to this dome, giving our thumb the special range of motion denied our other fingers (Fig. 8-21).

Figure 8 - 21
The trapezium.

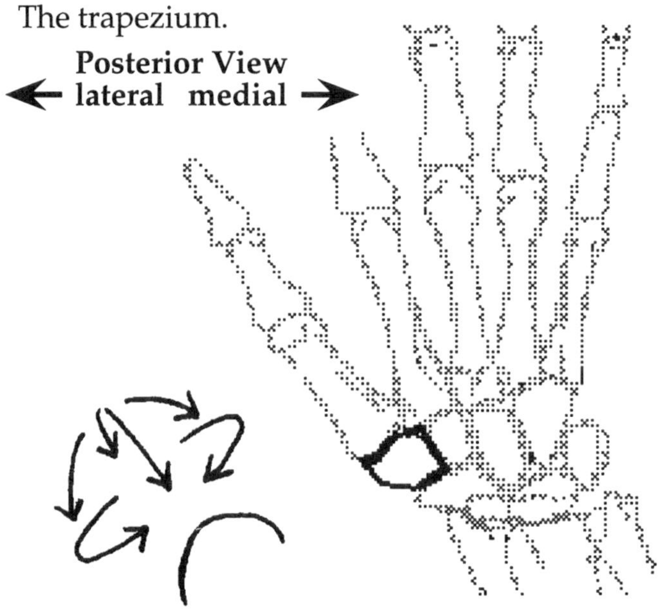

Trapezoid

Our **trapezoid** lies just medial to our trapezium (Fig. 8-17), supporting the base of our index finger, the tall sail of our little boat. Palpate your trapezoid on the back of your hand just distal to your scaphoid in line with your index finger. There are several tendons to push aside. This second attempt at a little table is much improved. It has two sides parallel and the top remains flat. Our index finger is disappointed at this success. A flat surface is much more restricting to movement. At its carpal base, the movement of our index finger is almost imperceptible. See for yourself. Wiggle your index finger while feeling your trapezoid. Where does the movement happen? Not at your trapezoid!

Capitate

Our **capitate** is just medial to our trapezoid in our distal carpal row (Fig. 8-17). It supports the base of our middle finger. Feel yours just distal to your lunate on the line from your dorsal radial tubercle, through your third finger. Flex your wrist to feel it easier.

As we have seen in the term capitulum, 'caput' means head, in this case head-shaped. The Latin who named this bone must have had a lopsided head to equate it to the shape of our capitate. "Cap it ain't." We can, however, still think of our capitate as the head of our carpal bones, since it is the largest of them all and it supports our longest finger.

Hamate

Our **hamate** is our most medial bone in the distal row (Fig. 8-17). On the back of your hand, feel it as the flat piece just distal to your triquetrum, on a line from your ulnar styloid process to between your fourth and fifth fingers. 'Hamate' means hooked. Our hamate got hooked into supporting both our fourth and fifth fingers. Our hamate is the fisherman who fell out of the boat, but he still hangs onto his rod and hook (Fig. 8-22). The **hook** can be felt on the anterior surface of our hand. Visualize the palm of your hand as a square. Draw a line diagonally from your pisiforn to your index finger. Draw a second line from the middle of your fourth finger to your wrist. Palpate the hook of your hamate at the point where these two lines intersect. Press firmly.

Figure 8 - 22

Lateral view of the hamate.

Figure 8 - 23

An end view of a slice made transversely through the carpal bones.

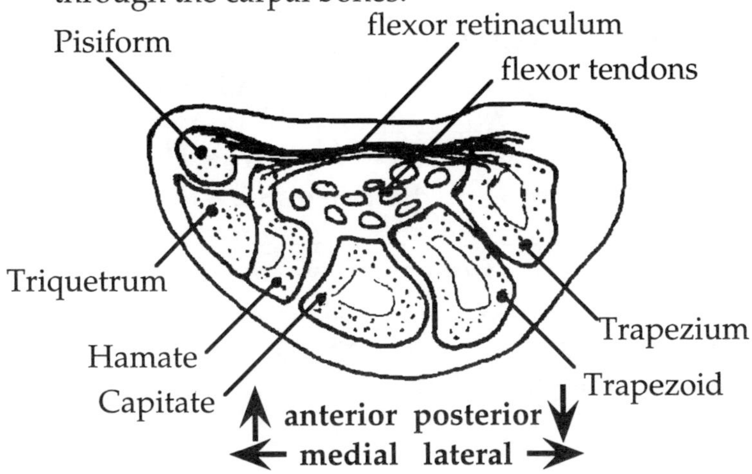

CARPAL TUNNEL

Collectively, our carpal bones are arched transversely, forming a concave or hollow when viewed from the front. A dense ligament, called our **flexor retinaculum,** bridges this hollow, to form a tunnel known as our **carpal tunnel.** Flexing tendons, blood vessels, and nerves go through this tunnel on their way to and from our hand (Fig. 8-23).

Muscles from our anterior forearm insert as far distally as our finger tips. When we contract these muscles to flex our wrist or fingers, they tend to take a shortcut, passing directly from our forearm to their insertion. This could turn our hand into a cumbersome, webbed tool. However, our flexor retinaculum serves like a cinch to hold all our tendons snugly against our carpal bones to prevent this webbing as our wrist flexes (Fig. 8-24).

In a sense, the skin of our entire palm is a continuation of our retinaculum. Pull up the skin on the back of your hand and then on the palm. Feel the difference? The skin of our palm is quite dense and stuck down to our bones, continuing to hold our tendons against our metacarpals.

Figure 8 - 24

The flexor retinaculum gathers the flexor tendons and holds them against the carpal bones.

When our flexor retinaculum is stressed and damaged, the tissues forming our carpal tunnel become inflamed and scarred, resulting in compression of the tunnel contents and adhesions which bind everything together. As the scar matures, its normal shrinkage increases the compression.

Nerves are fussy about pressure. They do not like it and quit effectively conducting their messages when they experience it. Their information regarding sensation and muscle contraction is obscured or lost. Thus, people with carpal tunnel compression may experience burning, prickles, numbness, muscle weakness, or paralysis in their hand, sometimes radiating up their arm. This condition is called **carpal tunnel syndrome.**

Activities most likely to damage our retinaculum are those which combine flexion of our wrist with repetitive movement of our fingers. This presses our tendons into our retinaculum and rubs them back and forth like saws. Our retinaculum is designed to stand up to average everyday use, but when we have a job or hobby which requires the use our hands in forceful or repeated ways, correct use of our hands is of utmost importance to maintain their capabilities. The problem may be obvious where heavy hand use is required such as in construction work, but many so-called light jobs can be very stressful on our hands, for example, pulling big books off a shelf, typing, or detailed work with tiny tools. These movements slowly beat away at the connective tissues of our retinaculum. People who do hands-on therapies also fall into a high-risk category and correct use of their hands cannot be over emphasised.

The first rule of correct hand use is to keep our hands as relaxed as possible and then, even more relaxed than that. Along with our hands, keep our arms and shoulders relaxed. It is too easy to get so caught up in what we are doing that we contract every muscle for a small, simple action. Most of our muscles are just playing tug-of-war amongst themselves, contributing nothing to our intended action and maybe even blocking it. This is fatiguing and stressful on our tissues.

When strength is required of your hands, align the bones is such a way that they balance the weight and force with minimal muscle support. Align your whole body so the strength of your legs and trunk do the work through your arms. If something is too heavy, give yourself the space to use two hands. In light repeated movements, the straighter our joints are, the less our tendons will be pulled out against our retinaculum. Any hand activity should be done with a straight or slightly extended wrist.

All these actions reduce the damage done to our retinaculum and connective tissues through the palm of our hand.

In my experience carpal tunnel syndrome is an overused diagnosis, with nerve compression or other problems at higher levels in the extremity or in the cervical vertebrae, being blamed on carpal tunnel compression.

When doing hands-on therapy, relaxed hands save our structure and so much more. To work effectively, our hands need to be sensitive to what they are working on. Relaxed hands mold into the contour of underlying structures. Tense hands impact only the hard or promiment aspects. Tense hands cannot feel subtle changes, nor can they communicate feelings of relaxation or softness.

METACARPALS

'Meta' means after, beyond or following in a series. After our carpals in our series of arm bones, we have five **metacarpals,** one as the base of each digit (Fig. 8-25). Skeletons appear to have long fingers sprouting immediately from their wrists. In creatures with more tissue, our metacarpals are padded with muscles and connective tissues to form the palm of our hand.

Figure 8 - 25
The bones of the hand.

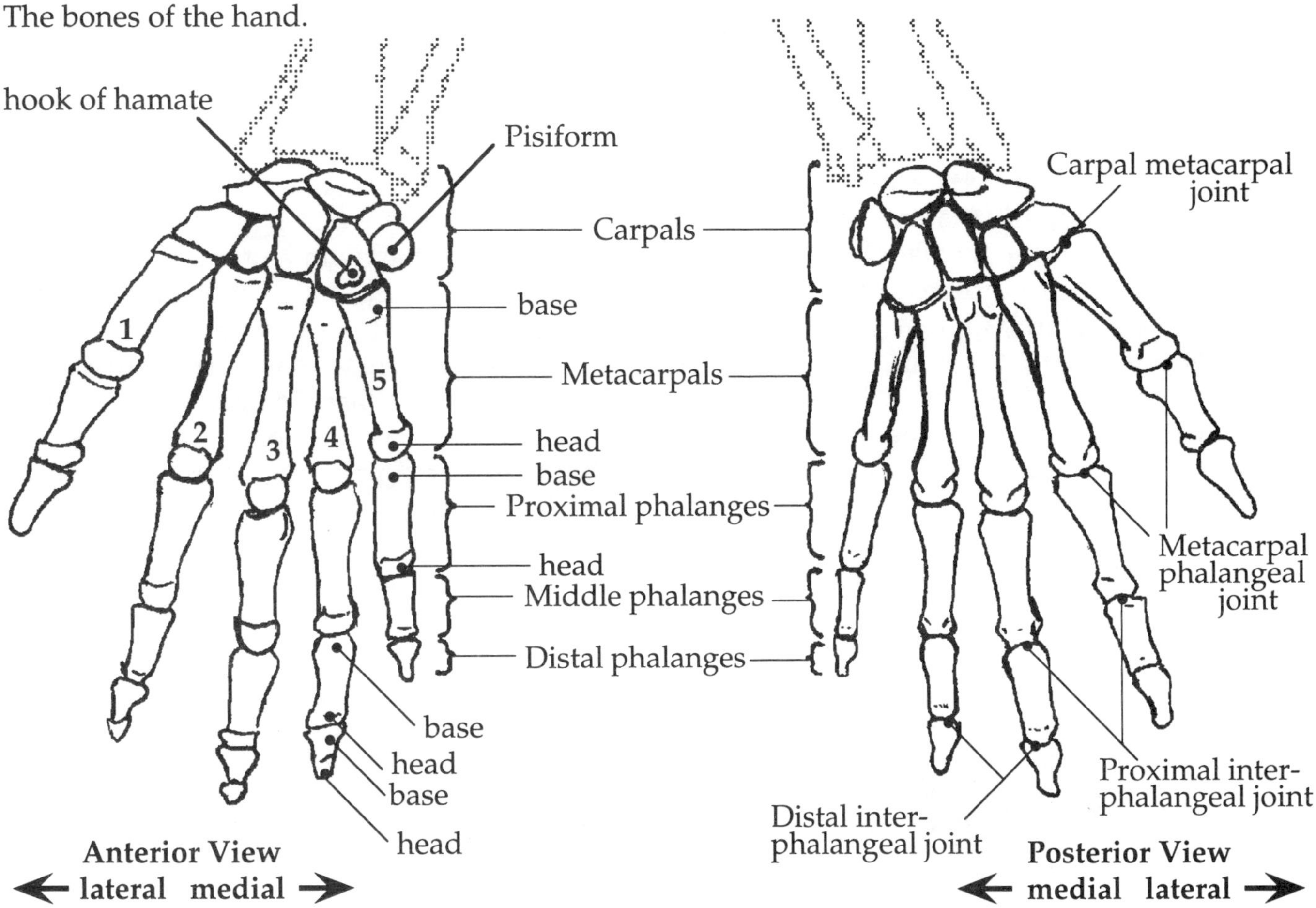

Our metacarpals are the classic shape of a long bone, a straight cylindrical shaft with enlarged ends. The proximal end is stabilized and called the **base** (Fig. 8-25). (We can refer to the broad, stabilized end of anything as its base.) Feel the bases of your metacarpals on the back of your hand, the bumps about two fingerwidths distal to your ulnar and radial styloid processes, just beyond your distal carpals.

The distal end of a metacarpal is called its **head** (Fig. 8-25). It is rounded to enhance the range of motion of our fingers. When making a fist, the knuckles at the base of our fingers are the bald heads of our metacarpals.

To designate one specific metacarpal, we simply number from one to five, our thumb being supported by our first metacarpal, and our little finger by our fifth (Fig. 8-25).

Our first metacarpal is short and stout relative to the others, setting our thumb off to a short start.

PHALANGES

All the bones of our fingers and thumb are called **phalanges** (**phalanx** in singular). 'Phalanges' means a closely knit row. Three of these bones are knit into each finger, but only two into our thumb, which is another contributor to our thumb's stunted stature. Each phalanx has a rounded **head** at its distal end and a proximal **base** which glides over the head of the preceding bone (Fig. 8-25).

Since our phalanges all have the same name, we have to think up words to designate any one specifically. We already know these words. Our **proximal phalanx** follows our metacarpal. Our **distal phalanx** is the tip of our finger. The bone in between is simply called our **middle phalanx.** Our thumb only has a proximal and distal phalanx (Fig. 8-25).

As we curl our fingers, their knuckles are the heads of our proximal and middle phalanx. Grasp the sides of a finger between your thumb and finger. Slide up and down feeling the shape of each phalanx.

UPPER APPENDAGE JOINTS

Sternoclavicular Joint

Our **sternoclavicular joint** is the only bony attachment of our upper appendage to our axial skeleton. It is a synovial joint between the superior-lateral corner of our manubrium and the medial end of our clavicle. The bones are separated by a little disc of cartilage. Ligaments support our clavicle onto our sternum and first rib (Fig. 8-26).

Feel your sternoclavicular joint as the indentation just lateral to the midpoint of the top of your sternum, accenting your sternal notch. Draw a circle with your shoulder and feel your clavicle pivot on your sternum, much like your finger pivots on your hand when drawing a circle with your fingertip (Fig. 9-27).

Figure 8 - 26
The sternoclavicular joint.

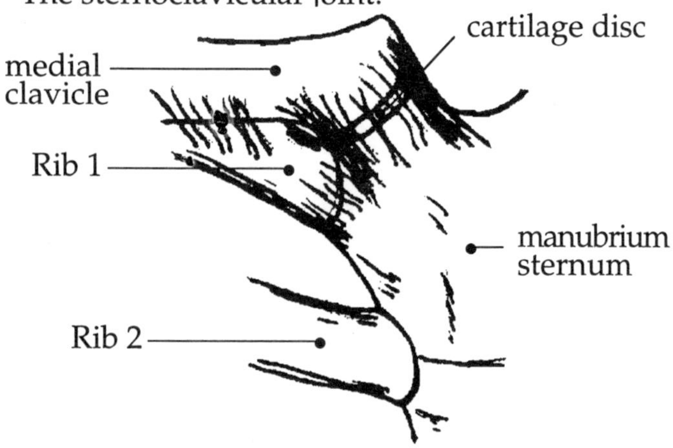

Figure 8 - 27
The clavicle pivots on the sternum much like the finger pivots on the hand when drawing a circle with the fingertip.

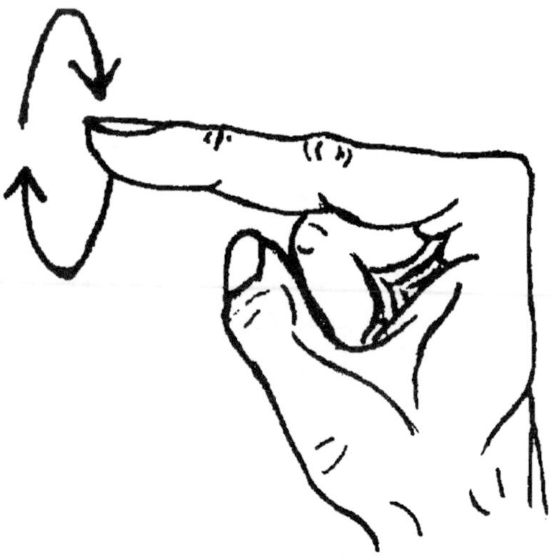

Acromion Clavicular Joint

Our **acromion clavicular joint** is a synovial joint between the lateral end of our clavicle and the medial side of the acromion process of our scapula. This joint allows subtle adjustments between our clavicle and scapula as our pectoral girdle moves. Follow your clavicle laterally until it ends in a bump on the top of your shoulder. Step down onto your acromion shelf just beyond the end of your clavicle.

No muscles cross our acromion clavicular joint. When its ligament support is damaged, we have no ability to compensate with muscular support or to use antagonistic muscles to bring back correct alignment.

If injury to this joint causes separation or dislocation, the structural misalignment is usually permanent unless it is surgically repaired. The function of the pectoral girdle, however, may not be significantly impaired except in severe injuries.

• Most body movements are described as a particular joint bending in a particular direction, for example, our elbow flexes. Our pectoral girdle is an exception to this. Its movements are named by the direction our scapulae move rather than a direction of movement at our sternoclavicular or acromion clavicular joints.

When our scapulae are lifted upward, it is called **elevation** and usually shows that we are up in the sense of uptight rather than happy. When our scapulae are brought downward it is call **depression** and may accompany that mood (Fig. 8-28a).

To 'protract' is to draw out or prolong. As we pull our scapulae out and forward, away from our vertebrae, it is called **protraction.** This position often is a protective stance. We duck behind our shoulders, using them to ward off what is directed at us. Taking our scapulae back toward our vertebrae is called **retraction.** When we retract a statement we take it back (Fig. 8-28b).

Our scapulae also rotate. Imagine a single nail fixing the center of our scapula to an underlying rib. Our scapula still could spin.

Our inferior angle or acromion process are our reference points as rotation is defined.

As our inferior angle moves laterally and our acromion process moves up, it is called **lateral** or **upward rotation.** As our inferior angle moves medially and our acromion process moves down, it is called **medial** or **downward rotation** (Fig. 8-86c). The illustration of scapular rotation also shows the glenohumeral joint abducting and adducting since these movements usually happen simultaneously.

Figure 8 - 28
Movements of the pectoral girdle.

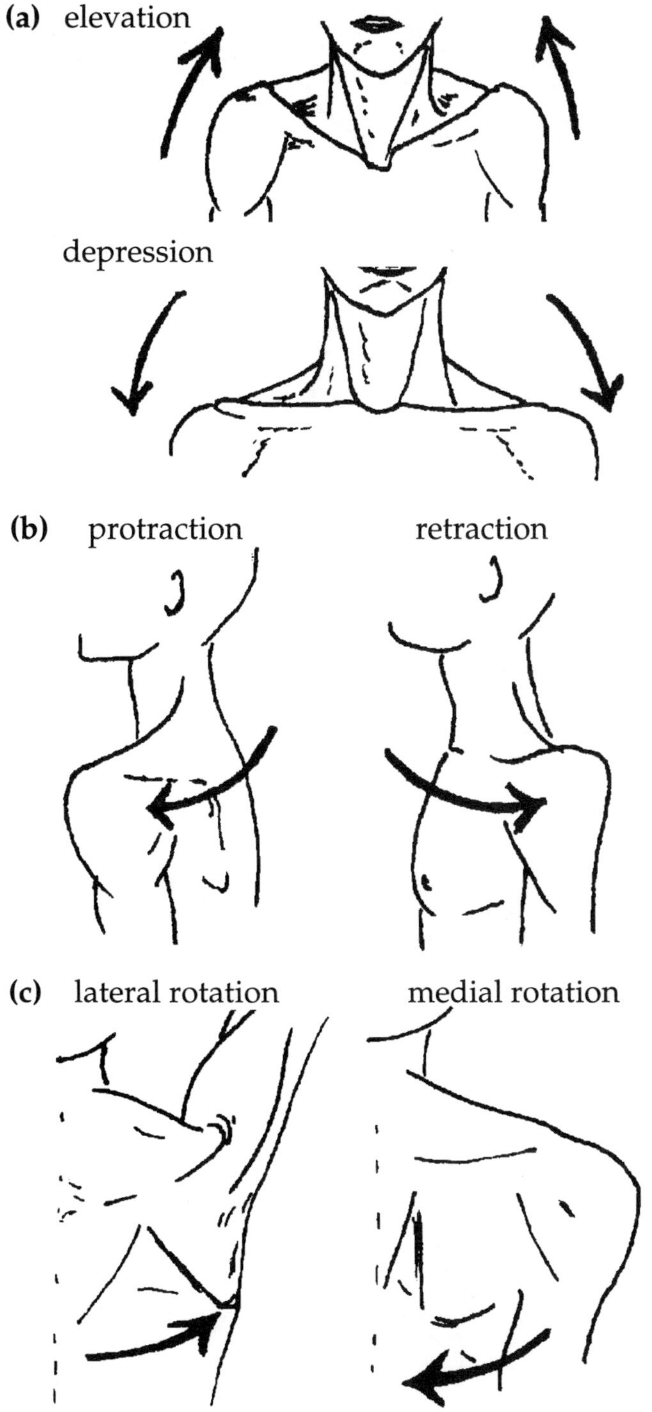

Glenohumeral Joint

Our **glenohumeral joint** is the synovial joint between the glenoid fossa of our scapula and the proximal head of our humerus. It is buried in the tendons and muscles which produce its movement and so is difficult to pal-pate. The bony landmark which is closest to our glenohumeral joint is the coracoid process of our scapula. Our glenohumeral joint lies deep and inferior to our coracoid process. Find your coracoid process to estimate where your glenohumeral joint would be. Press your fingers into the groove just below your coracoid process and move the joint around to feel it.

Our glenohumeral joint is classified as a ball-and-socket joint, but the socket is very shy and shallow in its clasp of the ball. Even with an extra lip of cartilage, the socket stands open and away from the contour of the head (Fig. 8-29b). This loose fit leaves lots of room for movement, making our glenohumeral joint the most moveable joint in our body.

As anywhere else, movement has the inevitable partner of instability and our gleno-humeral joint too often becomes dislocated. Another difficulty with wonderful movement is lubricating the soft tissues surrounding the joint. Many bursa separate and lubricate be-tween the soft tissues of our shoulder area, but it still is the most common place for movement to exceed lubrication and develop bursitis.

Figure 8 - 29
The glenohumeral joint in **(a)** external view and **(b)** vertical section.

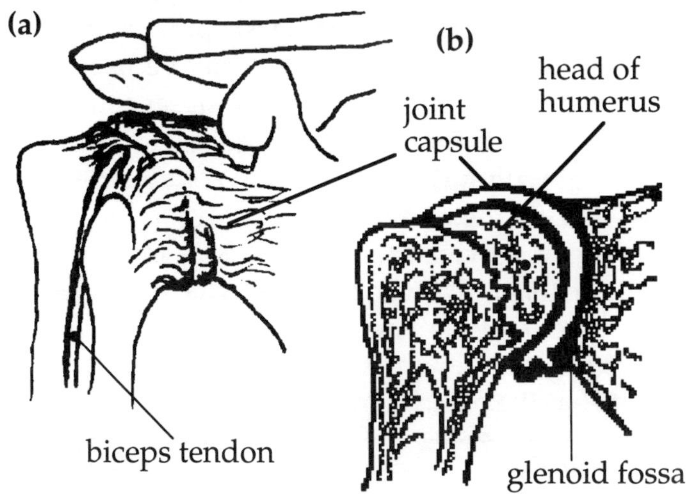

• Movements of our glenohumeral joint can be separated into four pairs of movements. In the <u>anterior-posterior</u> direction, the joint **flexes** as our arm is brought up in front of us, and **extends** as our arm is brought down and back as far as it goes (Fig. 8-30a).

In the <u>sideways</u> direction, the joint **abducts** as our arm is brought away from our body, and **adducts** as our arm is brought across our body in front or behind (Fig. 8-30b).

When our arm is held horizontally carried out to our side and backward, it is called **horizontal abduction.** As it is carried forward in a horizontal plane it is called **horizontal adduction** (Fig. 8-30c).

In any of these positions, our humerus can spin over our glenoid fossa. This spinning is called **rotation,** and is either in a **medial (internal)** direction or a **lateral (external)** direction (Fig. 8-30d).

When we flow all these individual movements into a smooth sequence, drawing a large circle with our arm, the complete movement is called **circumduction.**

• In daily activity we seldom isolate the movement of our glenohumeral joint from that of our pectoral girdle but rather use our girdle to further all our glenohumeral movements. For example, abduction of our glenohumeral joint ends when our arm is about perpendicular to our body, but we can continue the movement of our arm much further by upwardly rotating our scapula.

> When the word **shoulder** is used in this book, it refers to the whole prominent area of our distal clavicle, scapula, and the top of our arm. Shoulder may be used to mean glenohumeral joint or acromion clavicular joint, or to designate any other spot within this whole area, so it loses any specific meaning. When trying to localize specifically, it is clearer to use a more precise term. When clients describe a problem effecting their shoulder, I always ask them to show me where, rather than imagining what they might mean.

Figure 8 - 30

Movements of the glenohumeral joint.

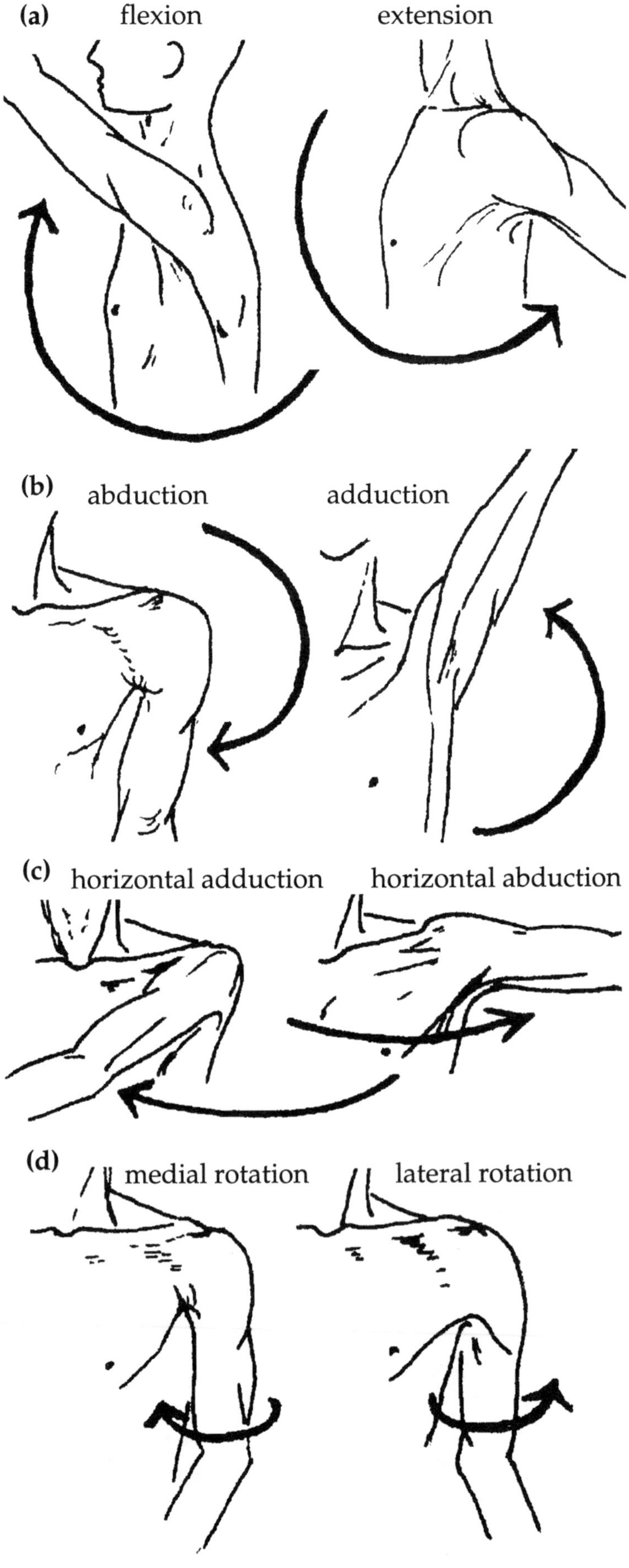

Elbow Joint

During many daily activities our **elbows** are flexed, producing an L-shaped bow in our arm, thus the joint's name. Our elbow includes two parts. In one, the trochlea of our humerus articulates with our proximal ulna. In another, the capitulum of our humerus and the side of our ulna articulates with our radial head.

- The joint between our trochlea and ulna is the larger, weight-bearing part of our elbow joint. Our ulna clasps around the waist of our trochlea, much like a wrench around a nut, to form a hinge joint. Movements are only in the anterior-posterior direction, producing **flexion** and **extension** of our elbow (Fig. 8-32a).
- The joint between our rounded capitulum and flat-topped radius is much like a loose-fitting ball-and-socket joint. This joint cannot express its full range of motion, however, because our ulna keeps a noose around our radial neck and a dense connective tissue membrane binds the two bones together throughout their length. This noose is called our **annular ligament** (Fig. 8-31a). It is an annual event that the earth completes its circle around the sun. Our annular ligament completely encircles our radial neck with both its ends attaching firmly to the side of our ulna. The dense connective tissue membrane goes by the latin words for between the bones, the **interosseus membrane** (Fig. 8-31a). Thus encumbered, our radius accepts the limitations of our ulna and glides over the capitulum as our ulna flexes and extends. The one special dance our radius can perform actually is quite a stunt, pirouetting on its head. This spinning happens in any position of flexion or extension. As our radial head spins back and forth, its distal end sweeps an arc of a circle, filling the space from the center to the circumference, like the windshield wipers on our car (Fig. 8-31b), giving our radius its name. The central, stationary pivot point of the arc is our ulna.

Press your fingertips against the styloid process of your ulna. Flop your hand from palm up to palm down, back and forth. See how your ulna is nearly motionless while the end of your radius sweeps around it from lateral to medial and back again. While the most obvious movement is in your hand, the origin of this movement is the spinning of the head of your radius over your capitulum. Perform this movement while holding the side of your radial head. Feel it roll under your fingers.

Flopping our hand palm up is called **supination** (Fig. 8-32b). When we lie on our back, exposing our vulnerable bellies, we are supine. This is anatomical postion, with our ulna and radius lying parallel (Fig. 8-31a). Flopping our hand palm down is called **pronation** (Fig. 8-32b). When we are prone, we lie face down. In pronation, our radius crosses over our ulna (Fig. 8-31a).

There is a totally non-anatomical way to remember which is which. When we are supinated, we can hold a cup of soup. If we pronate, the cup of soup drops to the ground.

Figure 8 - 31

In supination and pronation, **(a)** the radius sweeps an arc over the ulna, much like **(b)** the movement of a windsheild wiper.

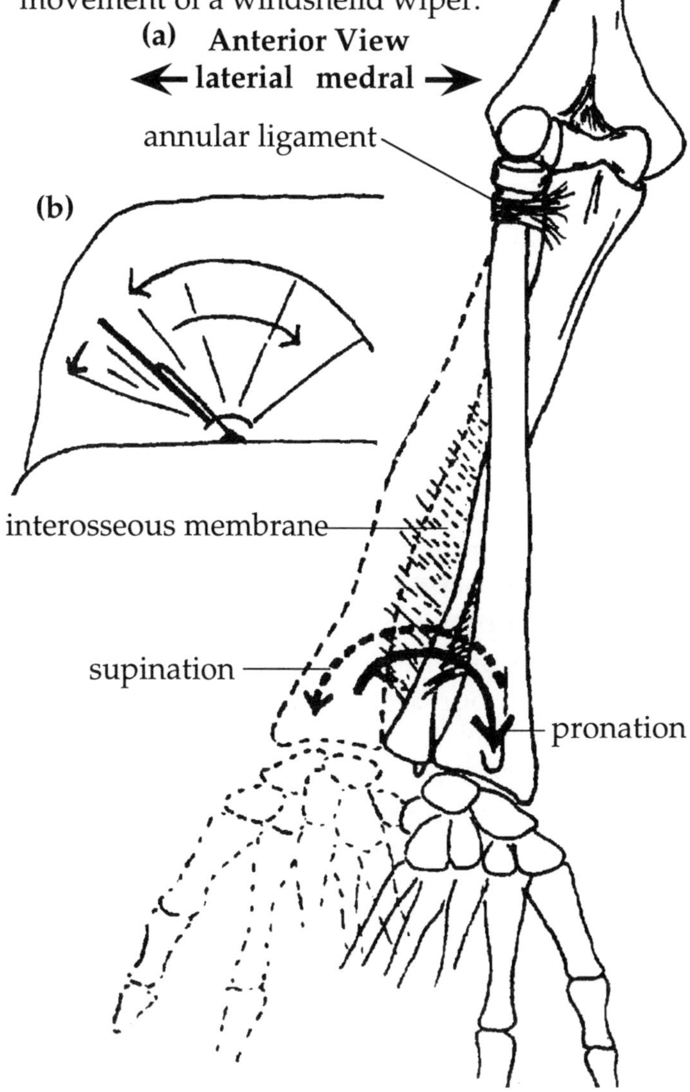

Figure 8 - 32

The movements of the elbow.

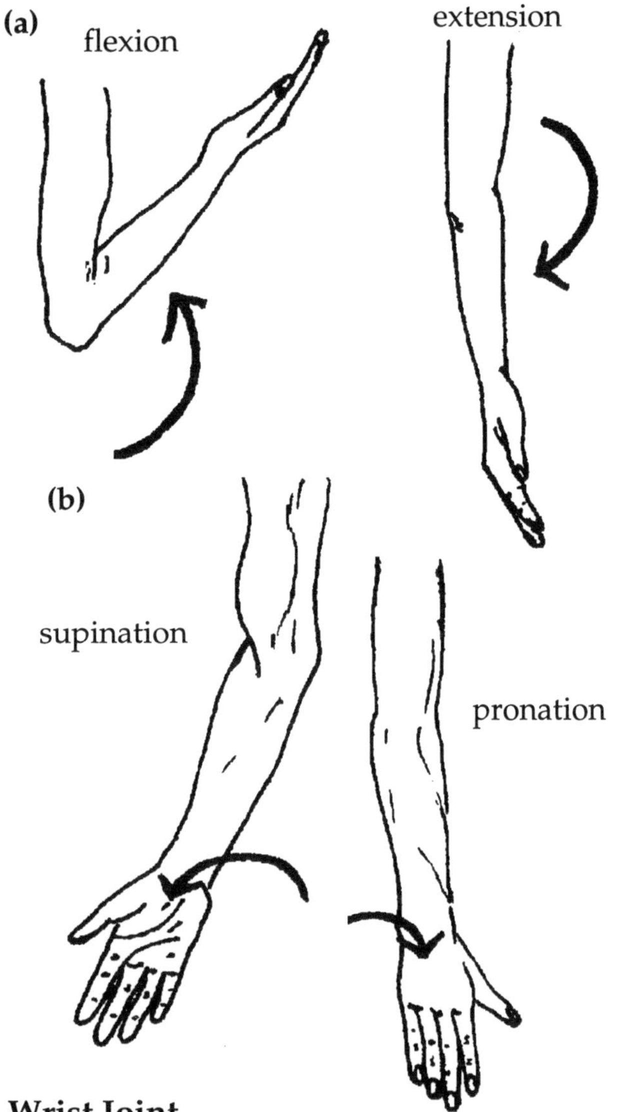

Figure 8 - 33

The movements of the wrist.

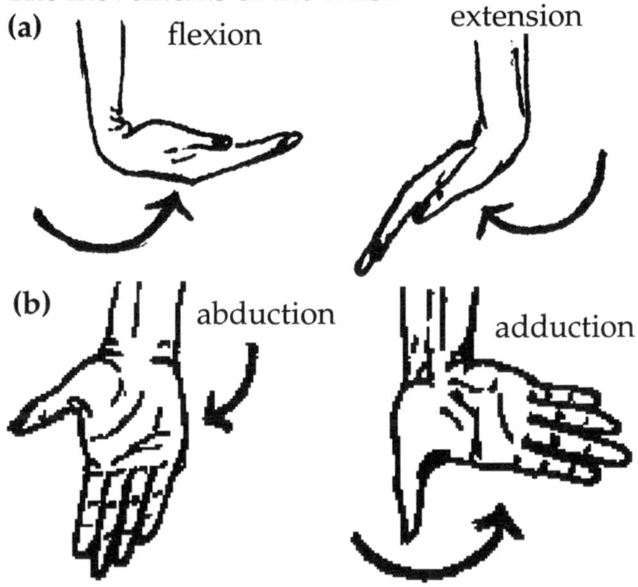

> If we conbine all our wrist movements sequentially, our hand draws quite a nice circle at our wrist. Watch closely. Are you enlarging the circle with supination and pronation from your elbow?

Wrist Joint

Our **wrist** is the joint where our hand moves with respect to our forearm. While our radius is small and accepts restrictions at our elbow, it dominates at our wrist. Its shaft becomes heavy and broad and captures the entire articulation with our hand. That is why our hand sweeps with our radius, and in no way is limited by our ulna. Our carpal bones often are collectively called our wrist bones, but only two of them take part in our wrist joint. Our scaphoid and lunate form the complete attachment of our hand to our radius (Fig. 8-25).

Our wrist **flexes** and **extends** in the anterior-posterior direction. as well as **abducts** as it is pulled sideways away from our body, and **adducts** as it is pulled back toward our body (Fig. 8-33).

Intercarpal Joints

There are many articular surfaces between our eight carpal bones. They slide over each other in short, almost imperceptible steps to further the movements of our hand.

Carpal Metacarpal Joints

- Generally, the base of each metacarpal is well stabilized on its carpal, with little allowance for movement (Fig. 8-25). Our fouth and fifth metacarpals slide a short distance over our hamate allowing us to curl up the medial edge of our palm. This movement is called **opposition** (Fig. 8-34a).
- Our thumb is the renegade, moving to join the opposition. Our first metacarpal forms a saddle joint with our trapezium. Its domed surface glides in a variety of directions (Fig. 8-21). Hold your trapezium and feel your first metacarpal move over it. Curl your first metacarpal around to face the fourth and fifth metacarpals, shaping your hand into a pocket like a baseball mit. This curling of our thumb toward our fourth and fifth finger also is called

opposition (Fig. 8-34a). These opponents work together, greatly enhancing our ability to be 'on the ball' and hang onto things.

Pulling our first metacarpal forwards, away from our hand is **flexion**. Pressing it back beside our hand is **extension** (Fig. 8-34b). Pulling our thunb sideways away from our hand is **abduction,** while pressing it into the side of our hand is **adduction** (Fig. 34c).

Circumduction combines all of these moves to sweep a large circle with our thumb.

Figure 8 - 34
Movements of the carpal metacarpal joints.

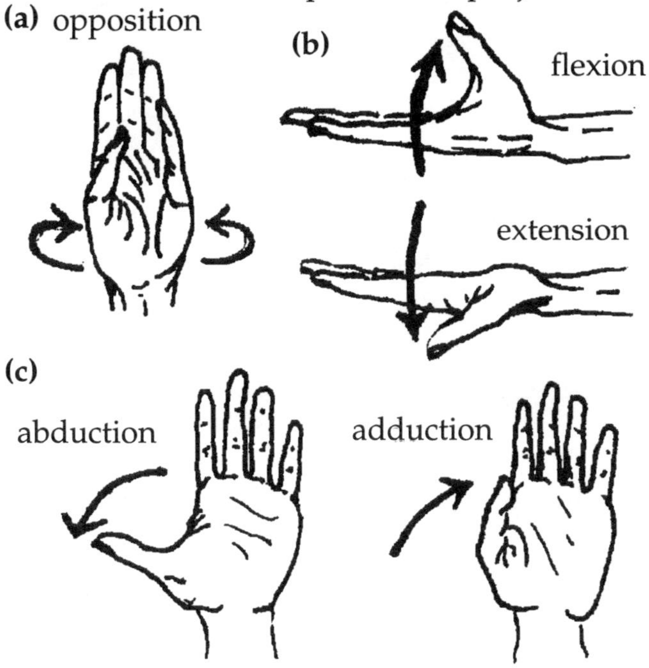

Metacarpal Phalangeal Joints

Each joint where <u>the bald head of a distal metacarpal articulates with the base of a proximal phalanx</u> is simply named by the two interacting bones, our **metacarpal phalangeal joint** (Fig. 8-2). These joints often go by their initials, **MP** joints.

Most MP joints allow two pairs of movements. In the <u>anterior-posterior</u> direction, they **flex** and **extend.** In the <u>sideways</u> direction, as our fingers are pulled apart, our MP joints **abduct.** As our fingers are pressed against each other, our MP joints **adduct** (Fig. 8-35). If we sequentially combine these movements, we can draw a circle with our finger.

• Our thumb is the exception. Our thumb does its fancy foot work at its carpal metacarpal joint. Its MP joint only flexes and extends.

<u>Palpate</u> your MP joints just beyond the knuckles on the back of your hand. Squeeze the sides of a proximal phalanx between your thumb and finger. Slide proximally toward your hand. Feel the flaired base of the phalanx, then an indentation. This is your MP joint.

Interphalangeal Joints

There are two joints in the row of phalanges that form each finger. The latin for 'between phalanges' is interphalangeal, and both these joints use this as their family name. Two familiar words are given as first names to distinguish the two joints. Our **proximal interphalangeal joints** lie <u>between our proximal and middle phalanges</u> (Fig. 8-25). They are marked by a double transverse crease on the palm surface of our fingers. Our **distal interphalangeal joints** lie <u>between our middle and distal phalanges</u> and are marked by a single transverse crease (Fig. 8-36). These joints (and everyone else!) are happy to use their initials, **PIP** and **DIP**, respectively. <u>Press</u> into the knuckles of your fingers to feel these joints.

In both our PIP and DIP joints , the base of each phalanx slides only forward and backward over the head of the preceding phalanx to produce **flexion** and **extension** (Fig. 8-34a).

• Again, our thumb is the exception. Having only two phalanges, it has only one interphalangeal joint which flexes and extends.

Figure 8 - 35
Movements of the metacarpal phalangeal and interphalangeal joints.

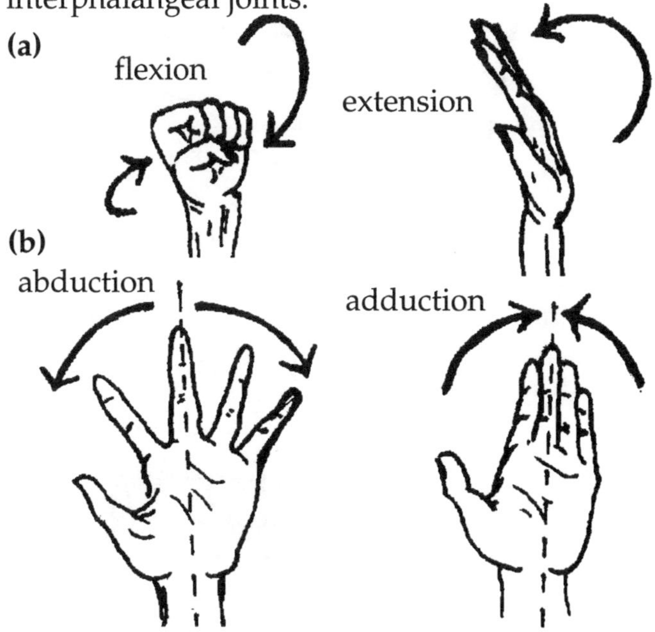

Figure 8 - 36

Surface view of the hand.

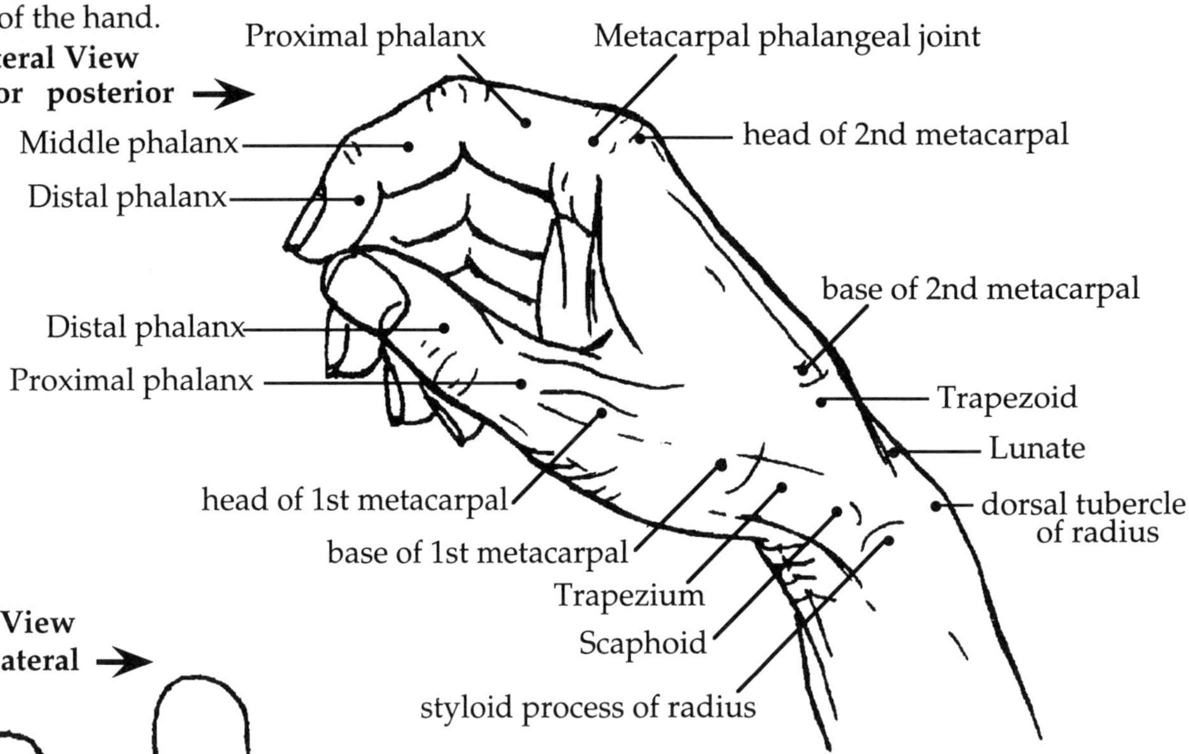

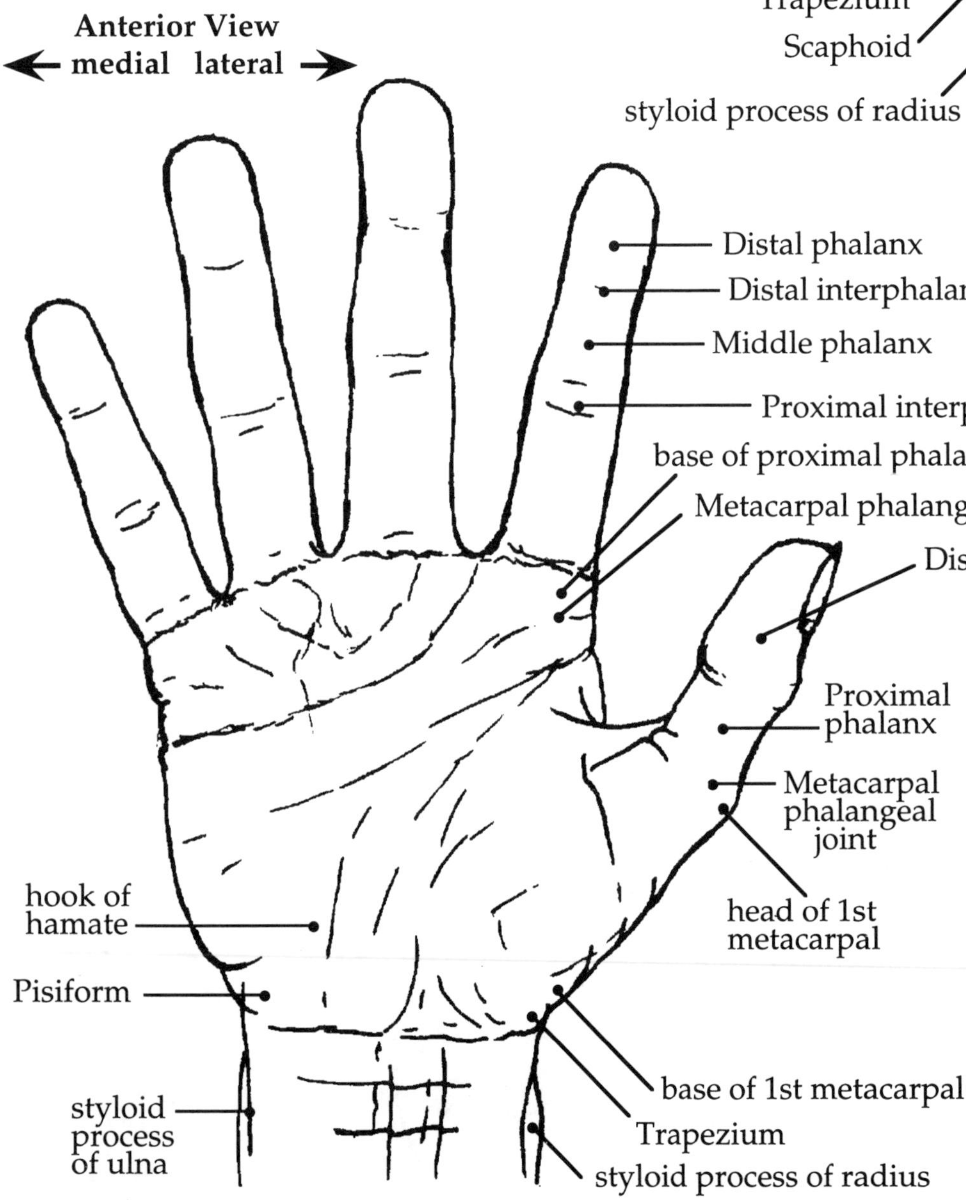

If anyone asked me to name the bones and joints of our hand, I would do it a bit differently. I would give our thumb three phalanges. It is our first metacarpal that is missing, not our middle phalanx. What we now call the metacarpal of our thumb actually is its proximal phalanx. It is shaped like a proximal phalanx, its movement over our tra-

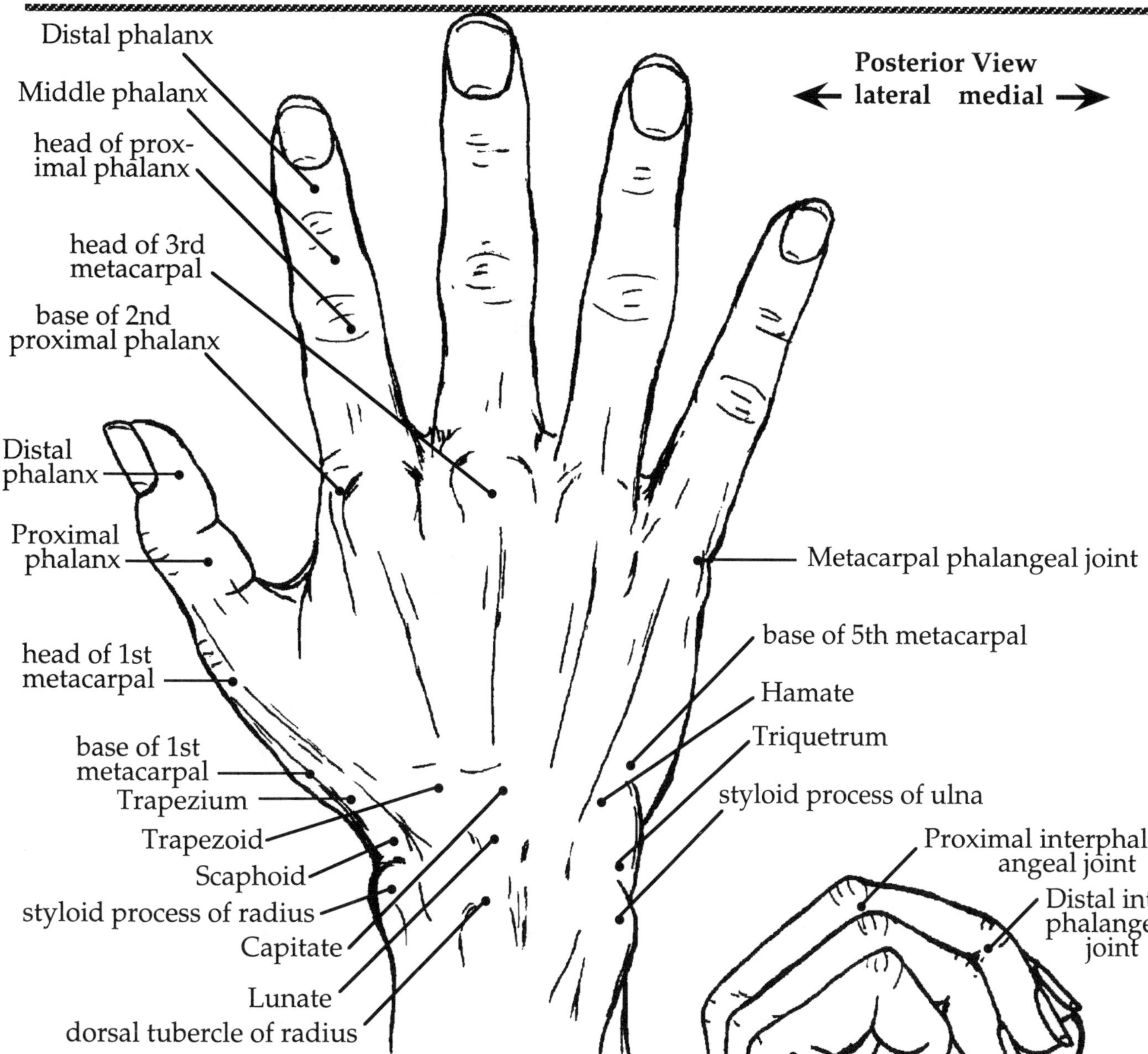

pezium is similar to the movement of a proximal phalanx over a metacarpal and the movement at its distal end is the same as an interphalangeal joint. Metacarpals are given to our four fingers to space them out from our thumb, enhancing the effect of opposition. Shall we revolutionize the naming of the anatomy of our hand?

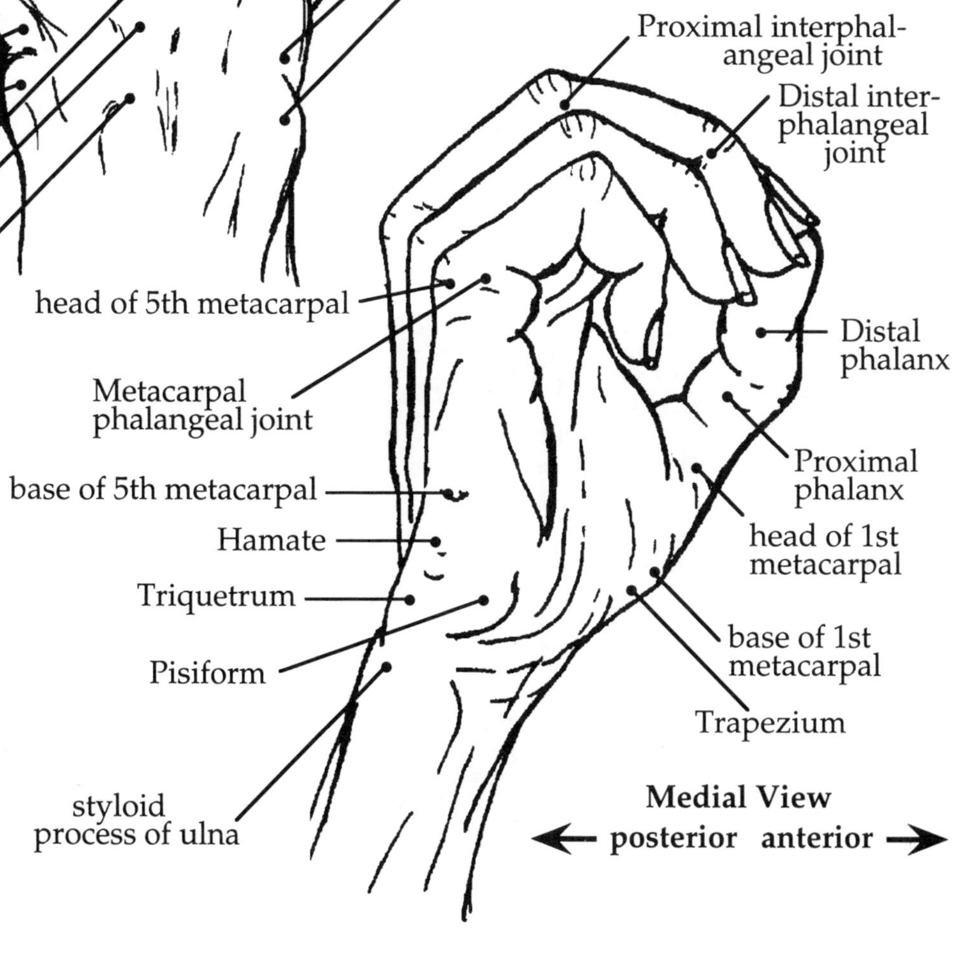

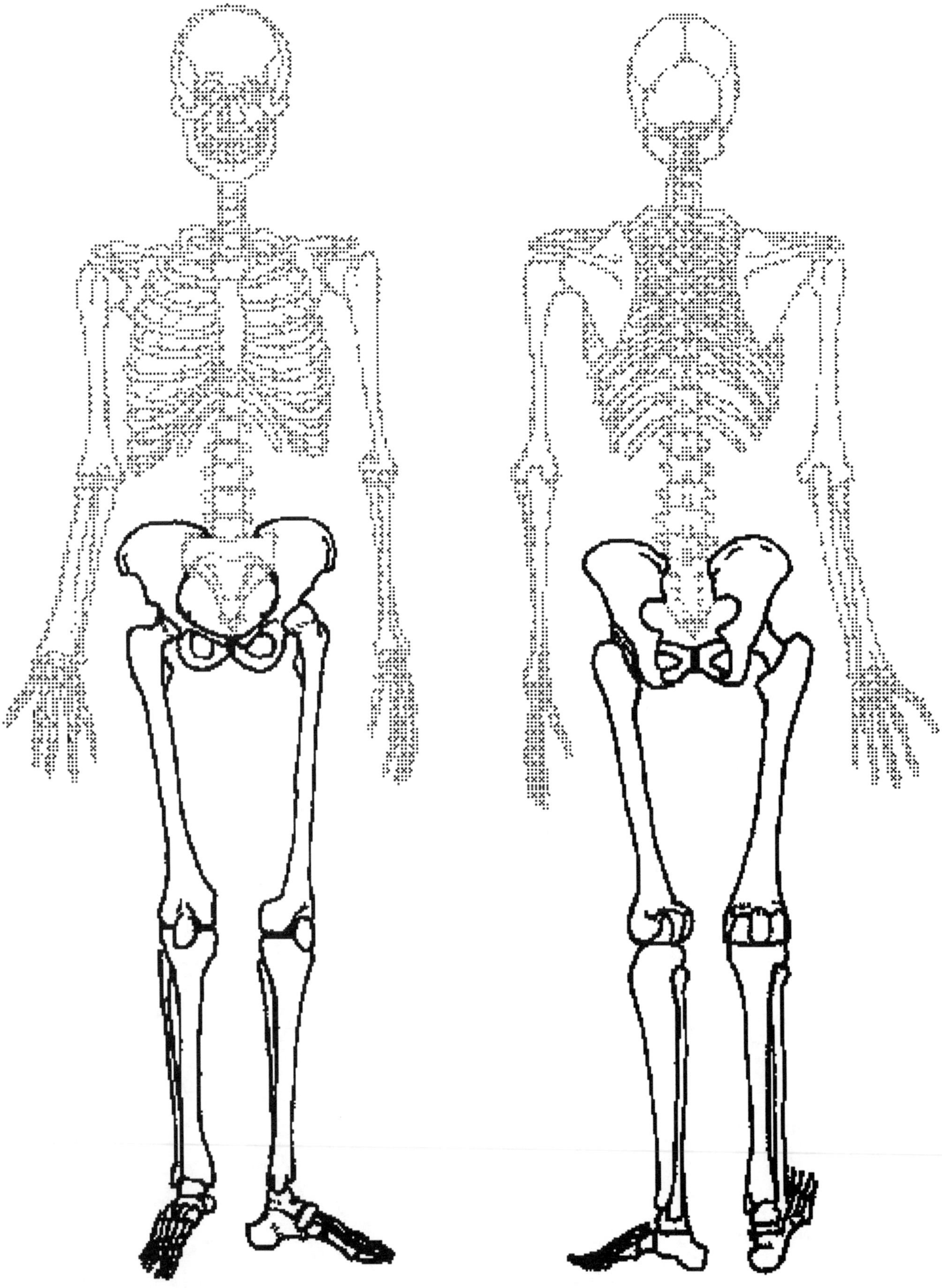

chapter 9: **Lower Appendage Skeleton**

The skeleton of our lower appendage includes the bones of our **pelvic girdle** and **legs.** Our legs form twin pedestals which support our body and move it about. While the emphasis in our upper appendage is movement, the design of our lower appendage incorporates more stability, thus sacrificing some of its mobility.

PELVIC GIRDLE or PELVIS

As we have seen in our upper appendage, a girdle is something that goes all the way around for the purpose of support. Our **pelvic bones,** along with their sacral attachment, form a complete ring of bone (Fig. 9-1). They provide a supporting interface between our torso and our legs, a mechanism for dispersing the weight of our torso into our legs and keeping our legs attached to our axial skeleton. Responsible for the whole weight of our body, this girdle makes no mistakes, it leaves no gaps. From the beginning, our lower appendage chooses increased stability rather than movement. Our pelvic girdle has limited movement under our axial skeleton and the left and right sides of our girdle move with very little independence of each other.

Figure 9 - 1
The pelvic bones and sacrum form an unbroken ring.

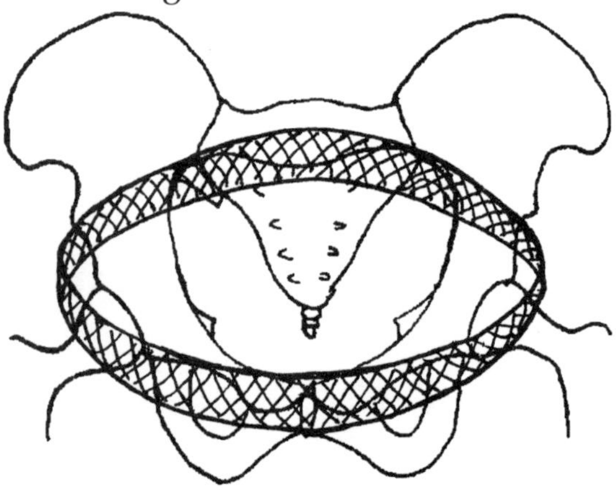

What we call a **pelvic tilt** is actually movement within our lumbar vertebrae, indirectly changing the position of our sacrum and pelvis, rather than adjusting our pelvis over our sacrum (Fig. 9-3).

'Pelvis' means basin. Our basin supports and balances our upper body and is filled with our lower abdominal organs, protecting them from the sides and back (Fig. 9-2). The front rim of our basin has been carved away to allow greater anterior expansibility of our abdomen. This design gives space for a bigger breath, but often is abused by over eating.

Figure 9 - 2
The pelvic basin.

Figure 9 - 3
Changing the tilt of the pelvis is accomplished by flexing and extending the lumbar vertebrae.

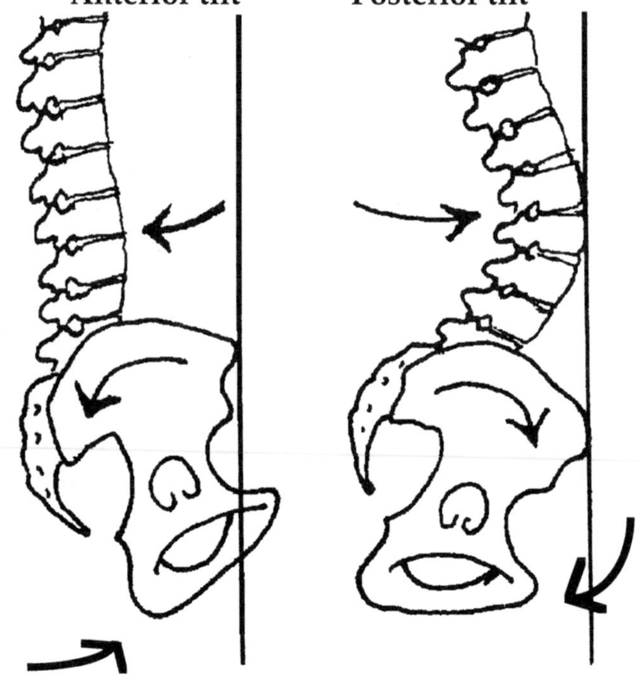

Each side of our pelvic girdle consists of three bones: our **ilium, ischium,** and **pubis.** On each side of our pelvis, these three bones all meet, clasping each other at a central point. Each bone lends a hand, so together they cradle a little saucer (Fig. 9-5a). This central union and saucer is buried deep under muscle and not palpable. From this point of union, each bone reaches out in a different direction, supporting and protecting. These extremities of our pelvis are exposed and can be touched.

In men the ilium is more vertical, so its crest is higher and narrower. Many men think of their waist as below the iliac crest and also wear their belts below it. In women the ilium flares out horzontally, bringing it lower and wider (Fig. 9-4). Women usually find their waist above their iliac crest. The sexual differences in the pelvis are recognizable by the fourth month of fetal life.

Figure 9 - 4
The sexual differences in the pelvis and femur.

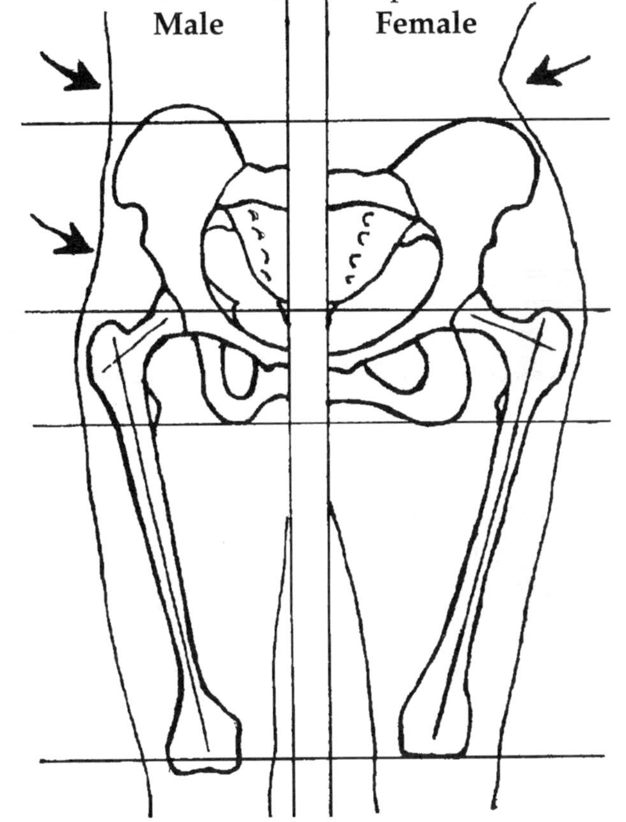

Ilium

Our **ilium** is the largest of the three bones and forms the superior and lateral third of our pelvis. It receives the weight of our torso from our sacrum. Our ilium cradles the superior rim of the saucer and reaches up toward our ribs with a broad, flat shield (Fig. 9-6c), protecting our lower abdominal organs from the side and back. The top edge of the shield is called the **crest** of our ilium or **iliac crest** (Fig. 9-6b). It is the only palpable part of our ilium. Spread apart your index finger and thumb. Press them into your sides, just below your ribs. Direct the pressure downward. Feel your iliac crest all the way from the tip of your finger to the tip of your thumb.

Follow your iliac crest anteriorly, until it ends in a distinct point. This is called your **anterior superior iliac spine.** An **anterior inferior iliac spine** snuggles into your muscle about two finger widths below your anterior superior spine. While not palpable, it is important for its muscle attachment. Posteriorly, our iliac crest ends in a vertical ridge. Feel yours about two inches lateral to your vertebral spinous processes. Feel the top and bottom points of this ridge. They are called your **posterior superior iliac spine** and **posterior inferior iliac spine,** respectively (Fig. 9-5b and 9-6b). This vertical ridge covers the joint between your axial skeleton and pelvic girdle.

Ischium (ischia in plural)

'Ischo' means to hold back, restrain, or suppress, so our ischium kneels in the lower back position of our pelvis. Our ischium cradles the inferior-posterior rim of the saucer from a corner of its bulky, squarish **body.**

Each of the two ischia has one heavy, bent knee which protrudes downward as our **ischial tuberosity** (Fig. 9-5 and 9-6). They are the round knobs which bore holes through our buttocks when we are sitting on hard benches. They are nicknamed **sit bones.** When sitting, our body weight transfers from our ilium into our ischium, and then out our ischial tuberosities into our supporting surface. Sit on your hands and wiggle around to feel them. They are closer together in men than in women.

From our ischial tuberosity, the lower leg of our ischium extends forward and its foot is grasp by our pubis (Fig. 9-6c).

Figure 9 - 5

The pelvic girdle.

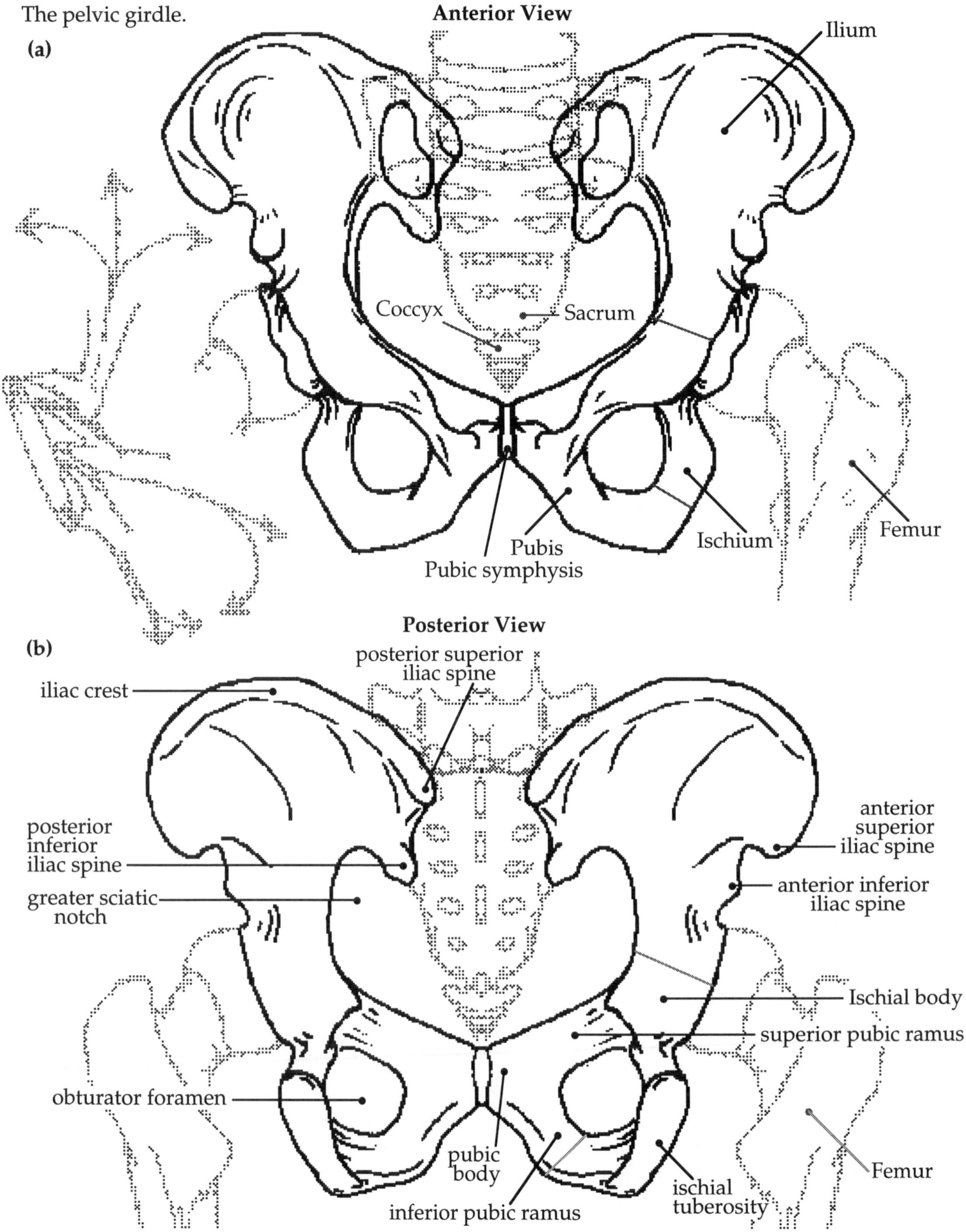

Figure 9 - 6

The pelvic girdle. Lines A and B are pelvic orientation lines.

(a) **Medial View**

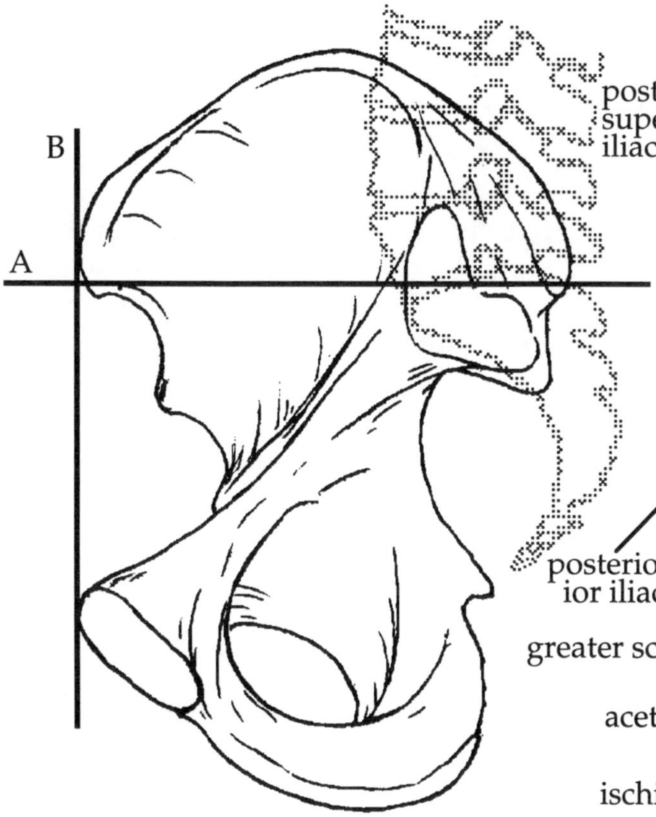

(b) **Lateral View**

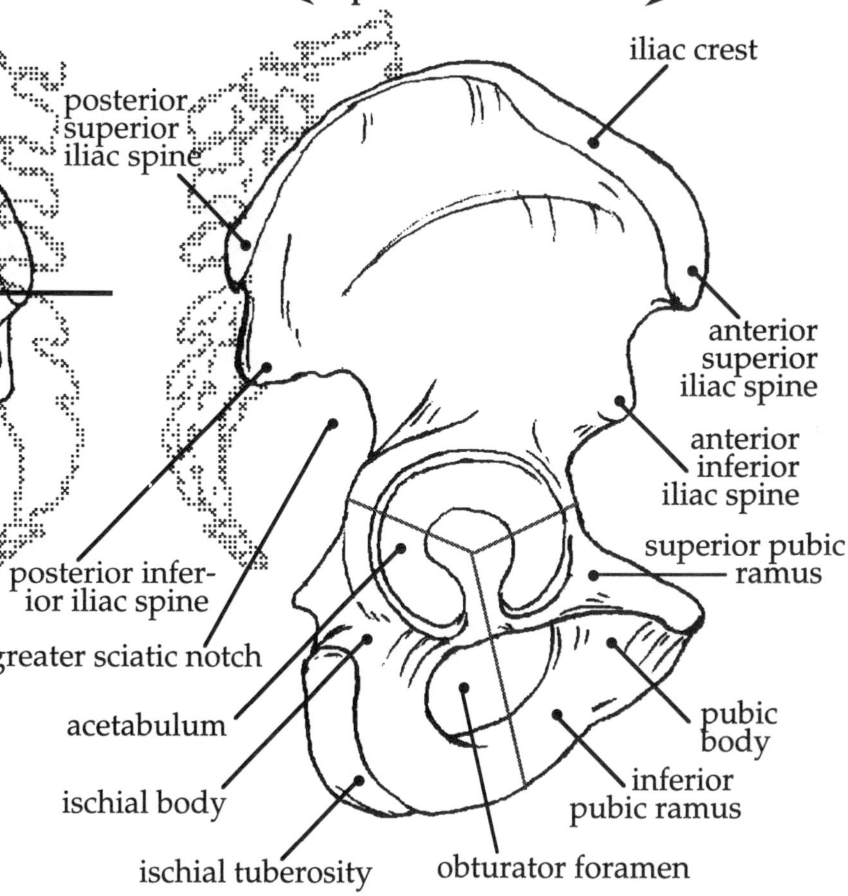

(c)

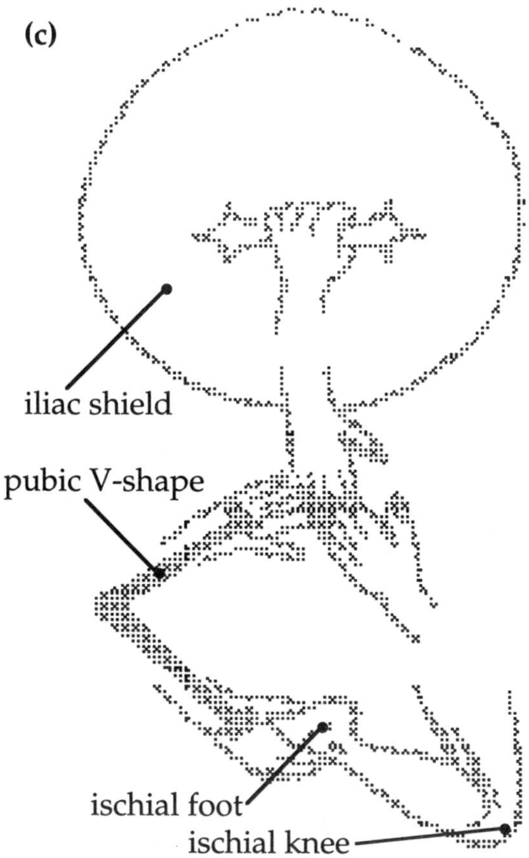

There are two methods of measuring the **orientation of our pelvis.** The first method imagines a line passing through our anterior superior iliac spine and our posterior superior iliac spine. In a well-positioned pelvis this line will be horizontal. The second method imagines a line passing through our anterior superior iliac spine and our pubic body. This line should be vertical in a medial view (Fig. 9-6a, lines A and B).

A large bite, called our **greater sciatic notch,** is taken from our ilium just below our posterior inferior iliac spine (Fig. 9-6b). Our sciatic nerve passes through this notch on its way from our abdominal cavity to our buttock. The notch is not palpable, but to locate it and the entry of your sciatic nerve into your buttock, find your posterior inferior iliac spine and move three fingerwidths diagonally, down and lateral.

Pubis

Our **pubis** or **pubic bone** forms the lower front part of our pelvis. 'Pubis' means the bone of the groin. A 'groin' is where something branches or divides. In our case, our torso is dividing into its two supporting legs.

Each pubic bone is a V-shape that is tipped on its side. The **body** of our pubis is at the point of the V (Fig. 9-5b and 9-6b). Feel your pubic bodies in the center at the very bottom of your belly, in front of your legs.

The bodies of our left and right pubic bones embrace each other at our anterior midline to complete our pelvic ring. This joint is called our **pubic symphysis** (Fig. 9-5a).

The two arms of each reclining pubis are thrust out laterally and posteriorly. One arm, the **superior ramus** of our pubis, reaches upward to hold the inferior-anterior rim of our saucer (Fig. 9-6b). Find yours by moving laterally from your pubic body into the depression between your torso and thigh. After a very short way it is buried in muscle. The other arm, the **inferior ramus** of our pubis, reaches downward to clasp the foot of our ischium (Fig. 9-6b). Find yours by moving posteriorly from your pubic body between your thigh and external genitalia. Follow the ridge of bone all the way to your ischium.

The big hole between our two pubic arms and ischium is our **obturator foramen** (Fig. 9-6b). To 'obturate' means to obstruct. This large hole is almost completely obstructed by a connective tissue membrane, traversed only by one small artery, vein, and nerve.

• The little saucer which is supported externally, at the union of all three pelvic bones is called our **acetabulum** (Fig. 9-6b). 'Acetabulum' specifically refers to a little vinegar saucer, acetic acid being an ingredient of vinegar. Why they preferred vinegar to ketchup or relish, I have no idea. Our acetabulum forms the socket for our hip, articulating with our femoral head.

> Skeletons are purposefully knocked kneed. Our femur does not stand perfectly vertical, but moves medially at its distal end, bringing our knees closer together than our hips (Fig. 9-7). Muscled beings hide this with bulky muscles in our upper medial thigh.

THIGH BONES

Femur

Our **femur** is our main thigh bone. It is the longest and strongest bone of our body.

• Our femur begins with a smooth round **head,** cupped over by our acetabulum to form our **hip joint** (Fig. 9-7). Our femur has a well-defined **neck** which is bent toward the horizontal from the main shaft of the bone, leaving the head cocked medially (Fig. 9-7). The head and neck of our femur are buried deep in muscle and are not palpable.

Our femur has two shoulders. The larger one, our **greater trochanter,** protrudes laterally (Fig. 9-7). It is the first palpable part of our femur. To find it, slide down about one hand length below your iliac crest, on your side, to the level of your pubic bodies. The lateral knob is your greater trochanter. It is the only distinct bony protruberance on your side between your iliac crest and knee.

The smaller shoulder, our **lesser trochanter,** protrudes medially and posterior (Fig. 9-7). It is palpable only with deep pressure into our upper medial thigh muscles, just below our inferior pubic ramus.

The shaft of our femur is a smooth cylinder. It is not directly palpable, although the sense of deeper bony support is apparent when pressing into any of our thigh muscles.

> The proximal growth center of our femur is in its neck, so the angle of the neck determines the direction of growth. In men the neck is more vertical, adding length to the legs, pushing their greater trochanter further from their iliac crest. In women the neck is more horizontal, adding width to their lower pelvic and upper thigh area (Fig. 9-4). This growth center closes in our mid twenties. Women having children before this age may notice permanent growth due to the high levels of estrogen during pregnancy.

Figure 9 - 7

The femur.

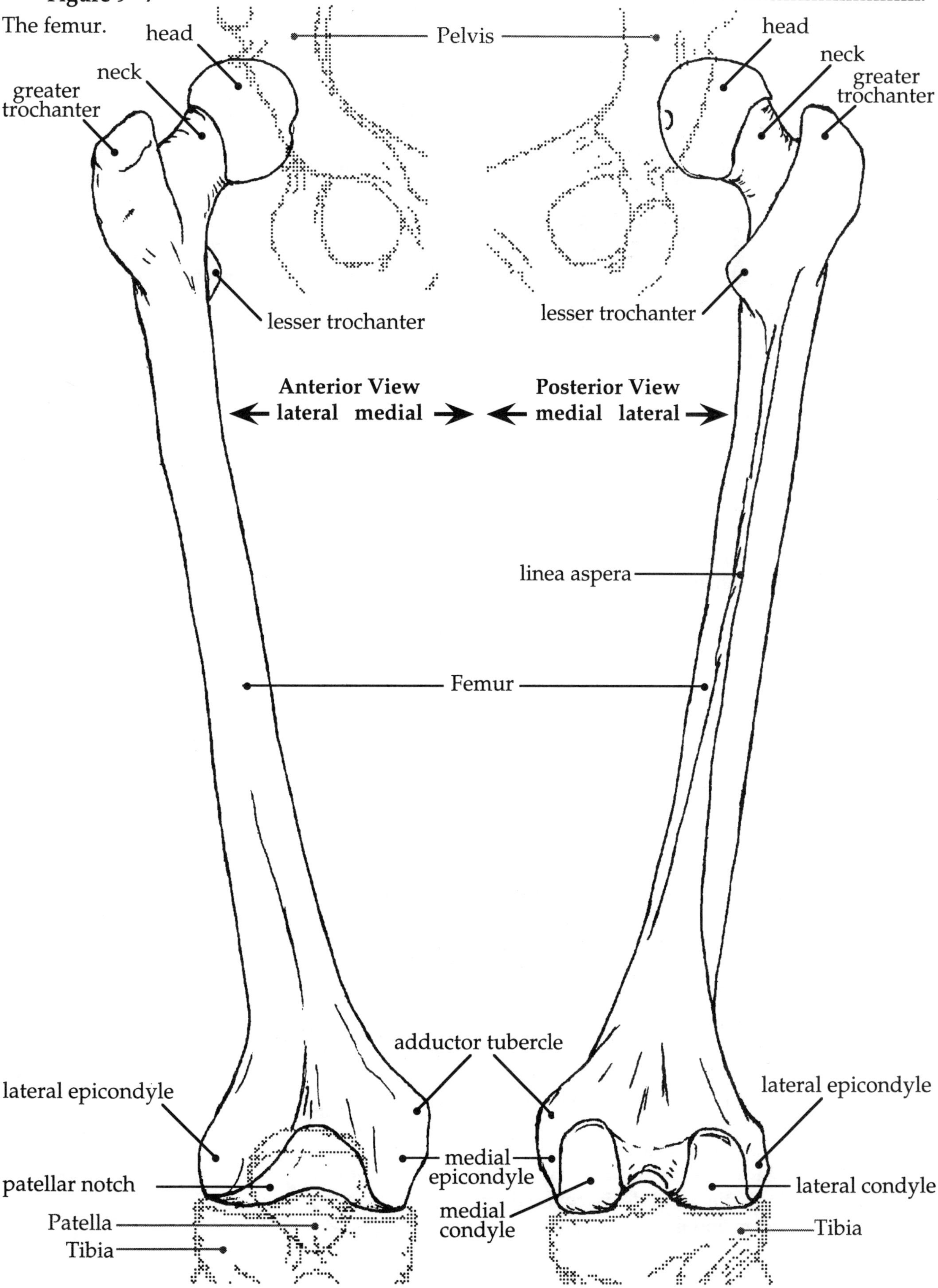

Figure 9 - 8

An end view of the distal condyle of the femur.

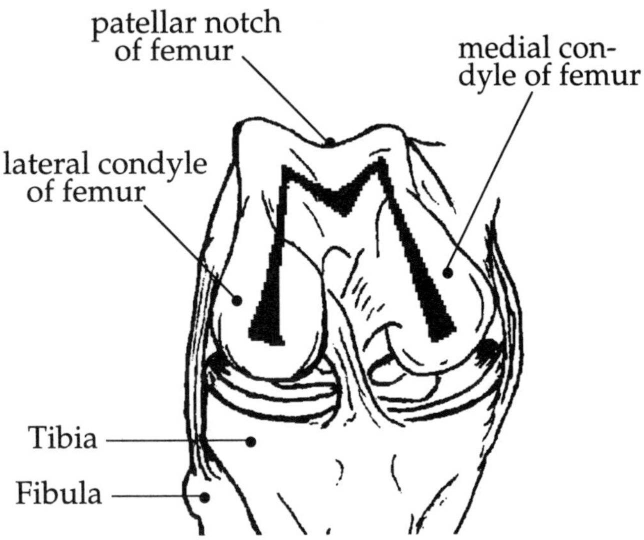

Posteriorly, the smooth shaft is interrupted by a rough line running up and down the length of our femur. It is called the **linea aspera** (Fig. 9-7), 'asperity' meaning roughness.

- As the muscle bellies of our thigh dwindle near our knee, the distal end of our femur emerges. Just like our humerus, our femur spreads from side to side. Slide down the muscles along your medial and lateral thigh till you feel the hard, rounded bumps on each side, just above your knee. These are called **epicondyles,** being located just over our knee condyles. Our inner and outer epicondyles can be distinquished using the familiar words, **medial** and **lateral** (Fig. 9-7). Our medial epicondyle usually is the most prominent. The distinct bump on its upper edge is called our **adductor tubercle** (Fig. 9-7).

The distal joint surface or condyle of our femur is rounded from front to back, and is M-shaped. The top end of the M lies anterior and provides a notch for the nose of our knee cap. The two sides of the M spread apart and curl posteriorly (Fig. 9-8).

When our knee is straight, our lower leg covers the anterior part of the condyle. When our knee bends, our lower leg swings out of the way. Palpate the outer edges of your femoral condyle around your kneecap. The posterior part of the condyle is covered by muscle and tendon and is not palpable.

Patella

A small round bone sits like a little skull cap in front of our knee joint. It is commonly known as our kneecap but in anatomical circles, its proper name is **patella.** 'Patella' means a small pan. Whether we use them or not, our knees are provided with little round pans in which to kneel. If I had a choice, I'd prefer a cushion, but ...

The circumference and anterior surface of our patella are nondescriptively curved, while its posterior surface has a purposeful protrusion, like a stubby nose (Fig. 9-9c).

What is the value of our patella? Straighten your knee and relax your thigh muscles. Feel how moveable your patella is in this position and how you can get under the edges of it. Still holding your patella, slowly bend your knee. Feel your patella sink out of your grasp into the notch between your left and right femoral condyles (Fig. 9-10). Your

Figure 9 - 9

The patella.

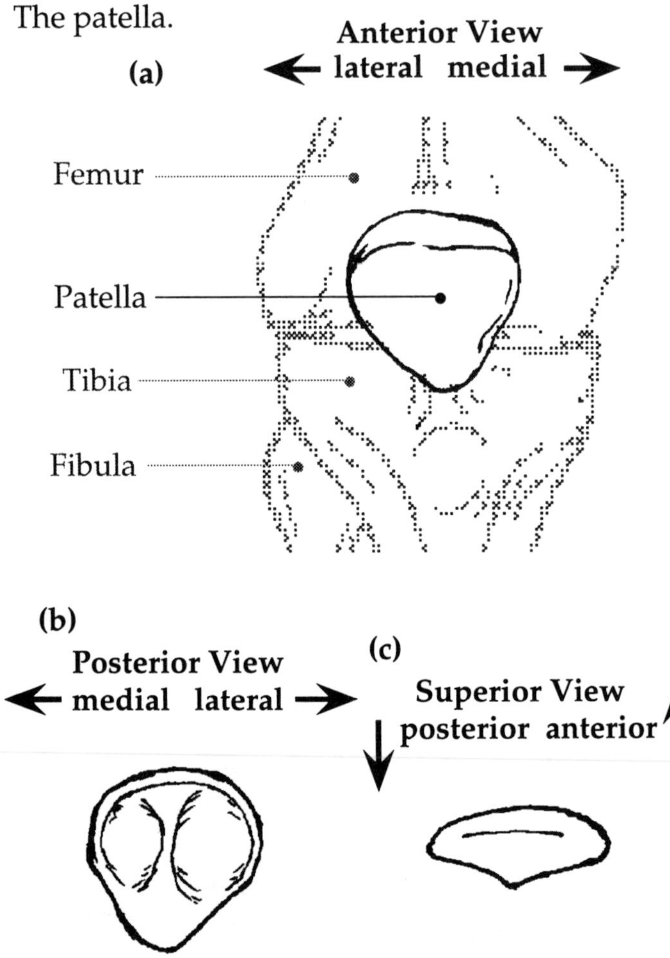

patella is immobilized as its nose wedges into this notch. It cannot move from side to side

The tendon of our large anterior thigh muscles crosses directly in front of our knee. Correct muscle function depends on this position, but as our knee bends, the natural tendency of the tendon is to roll off to the side. To prevent this, our tendon's protein fibers pass right through the substance of our patella. As our patella is immobilized, the correct tendon postion is maintained.

Figure 9 - 10

As the knee bends, the patella sinks into the notch on the anterior femur and is immobilized.

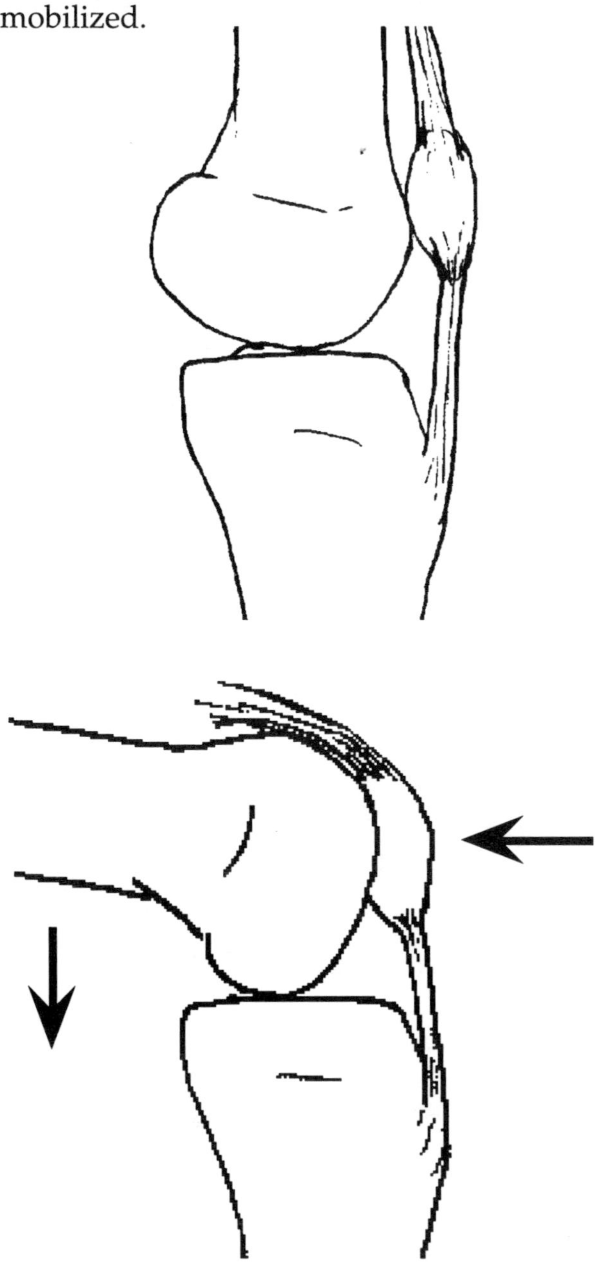

LOWER LEG BONES

Our lower leg is fitted with two bones, our **tibia** and **fibula.**

Tibia

Of the two bones in our lower leg, our **tibia** is the largest and takes the anterior-medial position. Our tibia alone atriculates with our femur to form our knee joint (Fig. 9-11). It bears the entire weight of our body from our knee to our ankle.

- Except for a central bump, the proximal condyle of our tibia is a broad, flat surface like a table top on which our femur sits. To feel your proximal tibia, find the medial and lateral epicondyles of your femur. Slide distally. A small indentation marks the joint space. The next pair of bulges are the ears of the tibia on both its medial and lateral aspect (Fig. 9-11).

What does our tibia have between its ears? Anteriorly, two finger widths below the patella, the nose of our tibia pokes forward. This prominence is called our **tibial tuberosity** (Fig. 9-11).

Our tibia gradually tapers into a stout, three-sided shaft. The medial side of its shaft is a flat surface about two finger widths wide and palpable the entire length of the bone. Follow yours all the way from your knee to your ankle. The two edges of this surface hint at the triangular shape of its shaft. Press into the soft tissues just behind and lateral to the shaft of your tibia.

The anterior edge of our tibia forms a sharp corner. It is commonly called our shin and splints against much abuse. Follow your shin straight downward from your tibial tuberosity. Feel how it flattens near your ankle.

- Distally, our tibia sits squarely over our foot, transferring the weight of our body into the top of our foot's arches (Fig. 9-16). Our tibia falls to the medial side of our foot to form a prominence known as our **medial malleolus** (Fig. 9-11). The root of 'malleolus' is mallet. Some anatomist saw a mallet in our lower leg. Its handle forms our shin and its head forms the bulges of our ankle bones.

Figure 9 - 11

The tibia and fibula.

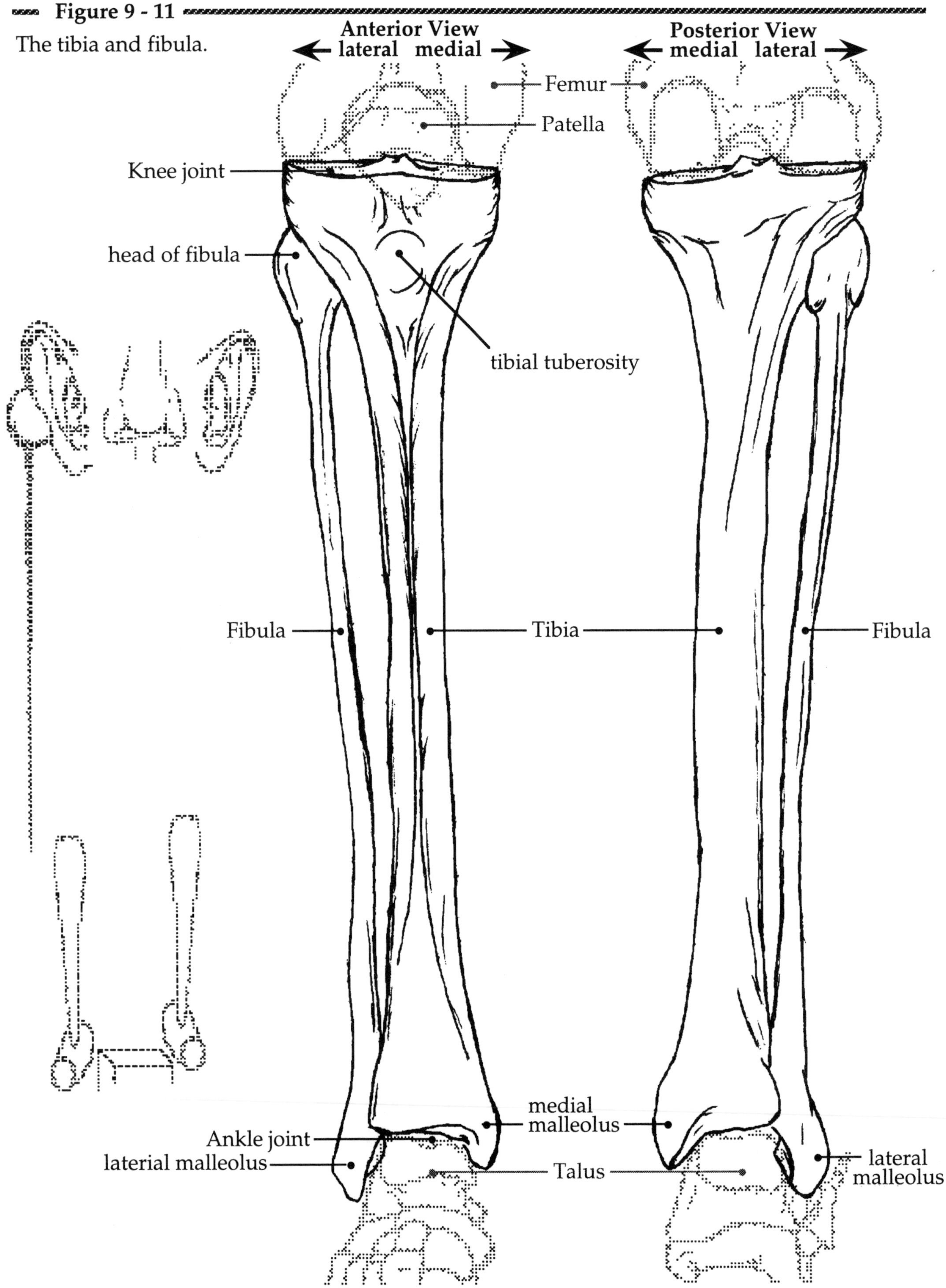

Figure 9 - 12
Surface view of the lower appendage, from the pelvis to the mid calf.

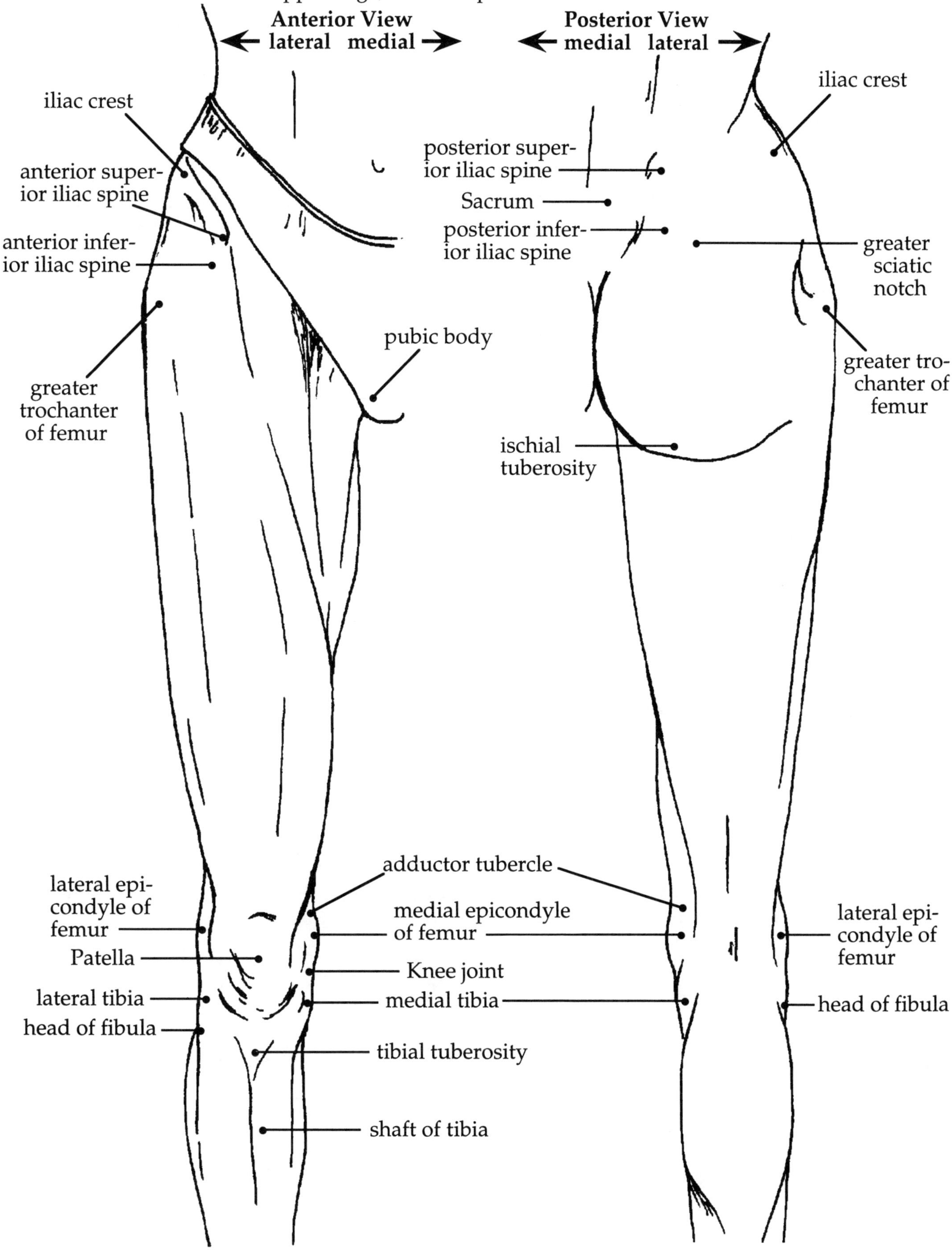

Fibula

A 'fibula' is a pin. Our **fibula** is a stick pin. Relative to its length, it is the thinnest bone in our body. It takes the lateral-posterior position in our lower leg.

Feel the little round **head** of your stick pin just behind the lateral ear of your tibia (Fig. 9-11). The descending shaft of our fibula is surrounded by many muscles and difficult to palpate. It emerges near our ankle where it forms a lateral prominence known as our **lateral malleolus** (Fig. 9-11). Find the lateral head of the mallet in your lower leg.

Our fibula is a sharp pin. It knows how to take life easy. It does not get involved with our knee joint, and bears no body weight.

Still, our fibula maintains two jobs. Its shaft provides a place for many muscle attachments and survives their incessant tugging while our lateral malleolus supports the outside of our ankle joint.

- Our lateral and medial malleoli together with the distal tibia form a box which fits over the top bone of our foot to form our ankle joint (Fig. 9-11).

> Each foot provides a strong and stable weight-bearing mechanism. It has its longest axis from front to back, increasing forward-backward stability. We plant two feet side by side to increase lateral stability, compensating for the narrowness of one foot.
>
> We can increase our front-to-back stability by placing our feet one in front of the other, increasing the size of our base in the front-to-back direction. We can increase our side-to-side stability by placing our feet further apart to increase our lateral base (Fig. 9-13).
>
> Stand with your feet together and have a friend push you gently from the side. How much does it take to throw you off balance? Not much. Spread your feet further apart and let someone push you again. (It does not matter so much now if it is a friend.) Is it easier for you to resist the force?
>
> People who are unsteady on their feet can feel a lot more secure if they widen their base, and often will unconsciously do this. If a client has a wide stance, that is a clue that they may feel unsteady. A beginning toddler naturally takes a wide stance.

FOOT BONES

Like our hand, our foot contains many small bones, but unlike our hand, our foot provides better stability with less mobility. When our feet form the base of our entire support mechanism, they had better incorporate stability.

Each foot has 26 bones; 7 tarsals, 5 metatarsals, and 14 phalanges (Fig. 9-17).

> The stability of our feet is partly by design and partly by usage. Use can shape the structure and function of our skeleton. People who have no arms or hands teach us about this. Anything you can imagine doing with your hands, these people are doing with their feet, all their daily maintenance, plus sewing, typing, playing musical instruments, and so forth. With their feet out in front of them doing things, the feet are stood on much less. Their feet are asked for mobility more often than stability and the structure changes to accommodate. So, part of the reason why our feet have more stability is because we do stand on them.

Figure 9 - 13

We increase our base and stability by standing on two feet and by spreading our feet apart.

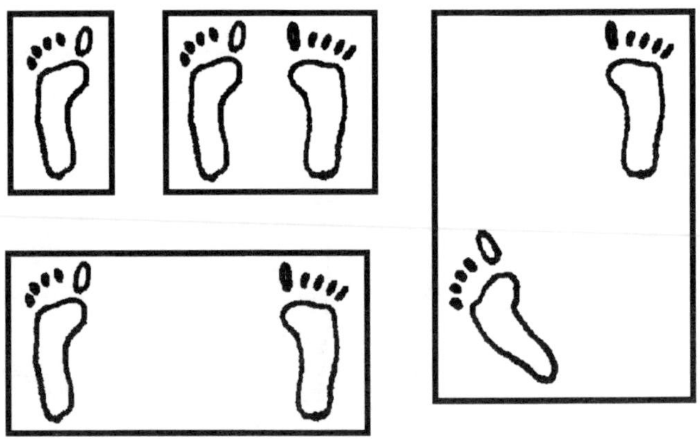

TARSAL BONES

A 'tarsal' is a broad flat surface. In humans, our **tarsal bones** form more than half the sole of our foot (Fig. 9-17), taking up a large space compared to the carpal bones in our hand. We will see, however, that they are more arched than flat (Fig. 9-24).

Talus

A 'talus' is the slope of anything, and refers specifically to the sloping heap of broken rocks at the foot of a cliff. This is a good description of the whole collection of irregular tarsal and metatarsal bones, as our foot slopes forward from our abruptly vertical shin (Fig. 9-16). Only the top bone of this slope, however, takes the name talus.

> Most four-legged animals are ballerinas by birth. They walk on their tiptoes, while their tarsal and metatarsal bones form the shank of their leg (Fig. 9-14). However, they balance on four points rather just two.

Figure 9 - 14
Many four-legged animals normally stand and walk on their distal phalanx.

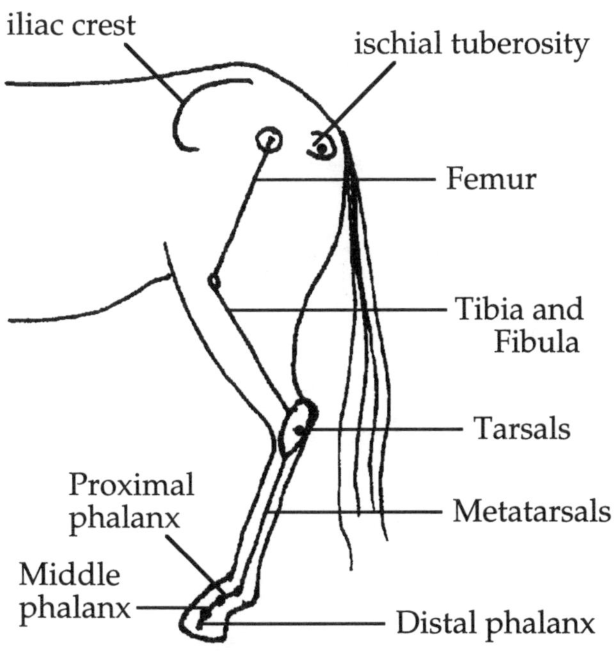

Figure 9 - 15
The dome of the talus fits into the box formed by the distal tibia and fibula.

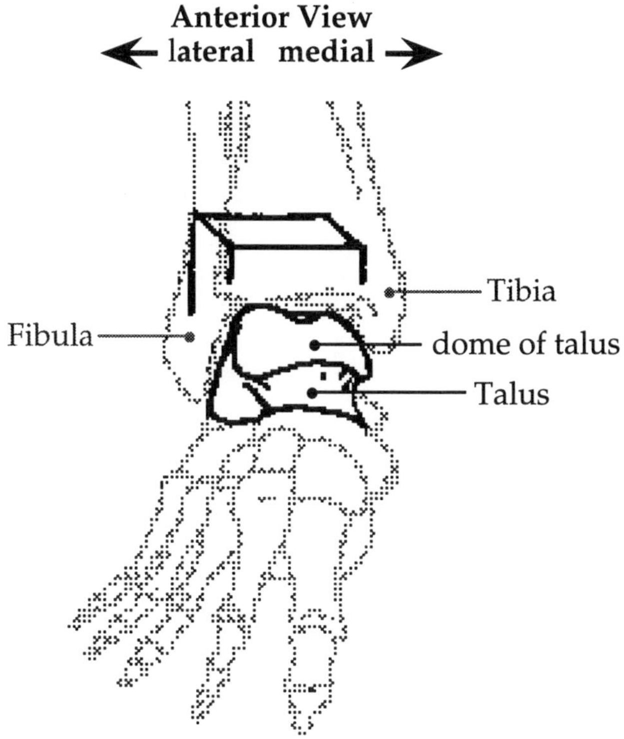

Our **talus** stands at the top of the arches of our foot (Fig. 9-16). Its **dome** fits into the box formed by our distal tibia and fibula (Fig. 9-15), and these three bones together form what is called our **ankle joint.** Our talus receives the weight of our body from our tibia, distributing it downward into our calcaneus and forward into our navicular (Fig. 9-16).

Palpate your talus immediately below and in front of the tip of your medial malleolus, in front of your lateral malleaolus, and on top of your foot between the two malleoli. Straighten your ankle to feel the increased prominence of your talus as it swings forward and downward.

Calcaneus

Our **calcaneus** is our largest tarsal bone. It sits directly under our talus, receiving a large portion of our body weight. Our calcaneus has a large posterior protrusion forming the ball-like **heel** of our foot. This protrusion carries part of the weight of our body backward and

Figure 9 - 16

A medial view of the foot showing the talus bone at the top of the arch and the geographic talus. The arrows show how the talus receives the body weight from the tibia and transfers it into the calcaneus and navicular.

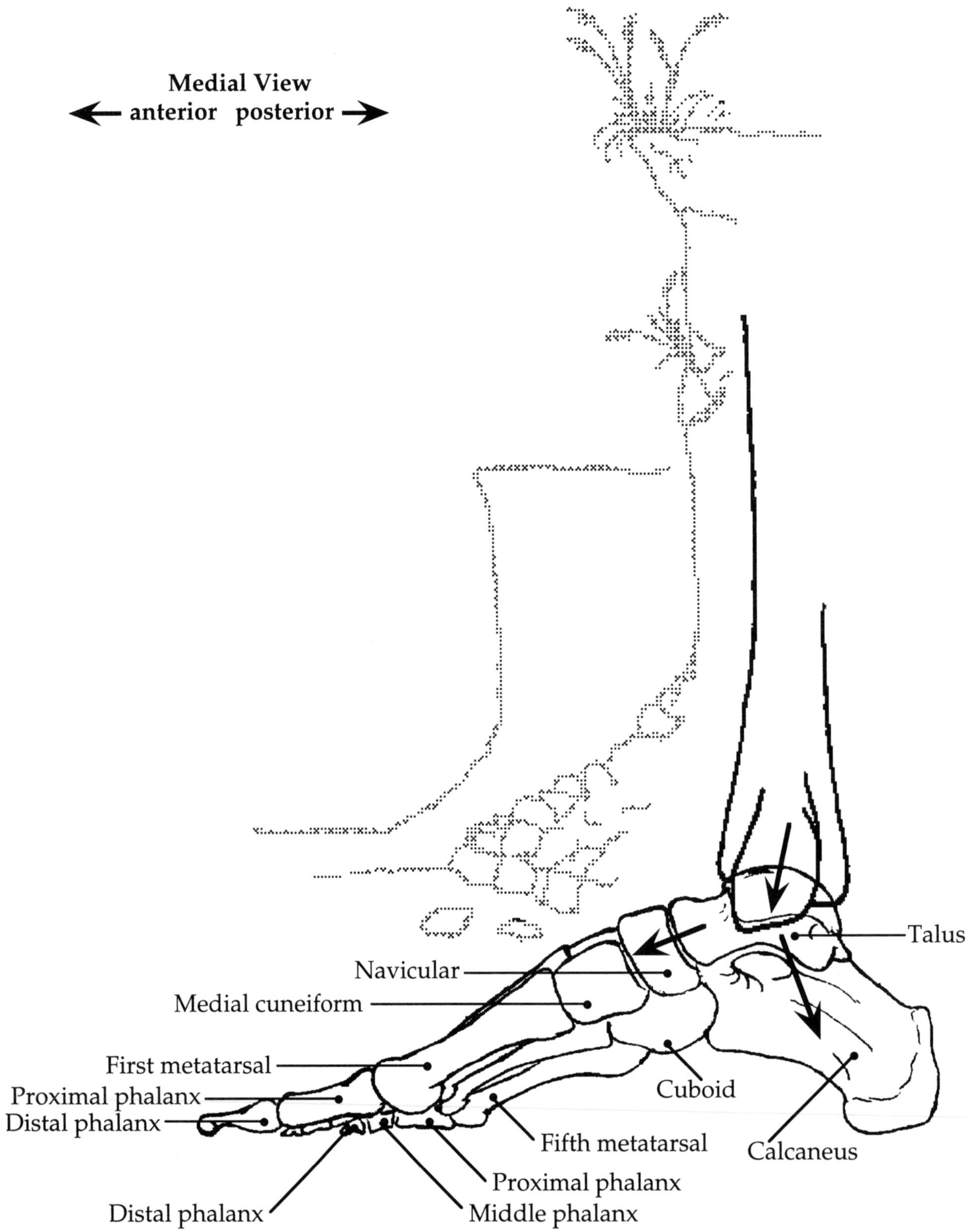

Figure 9 - 17

The foot.

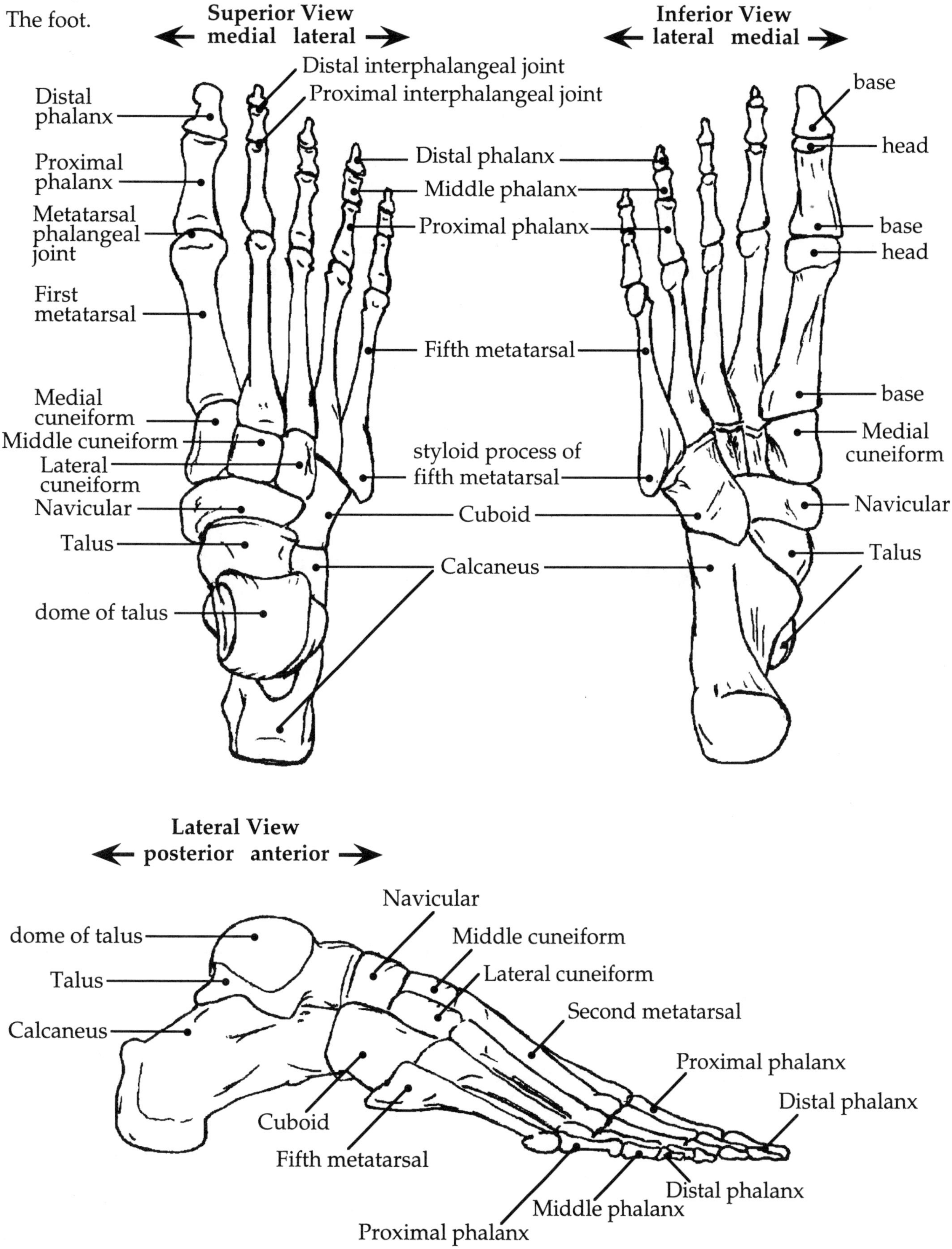

downward through our foot (Fig. 9-18). Feel the ball in your heel on four sides: medial, posterior, lateral, and inferior. It flares out inferiorly to increase the size of its weight-bearing surface.

Our calcaneus is hollowed medially but more prominent on the lateral side of our foot (Fig. 9-20). Slide along its lateral surface all the way from your heel to two finger widths in front of your lateral malleolus.

Our calcaneus participates in the arches of our foot with only its posterior ball in contact with the floor. Anteriorly, the calcaneus passes part of our body weight forward into our cuboid (Fig. 9-18).

Figure 9 - 18

A lateral view of the calcaneus. The arrows show how the weight is transferred.

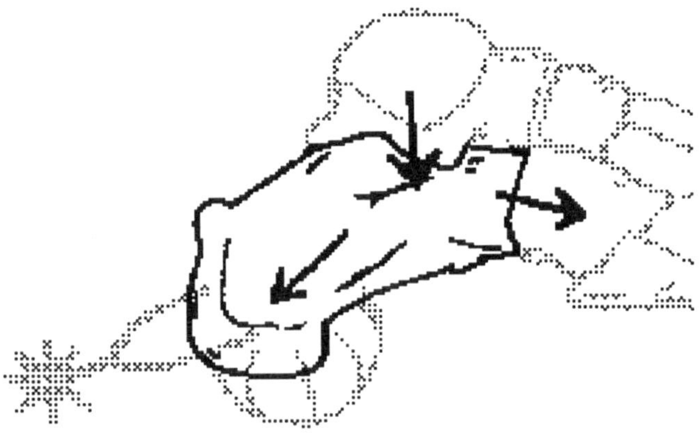

Figure 9 - 19

The cuboid. The arrows show how the weight is transferred.

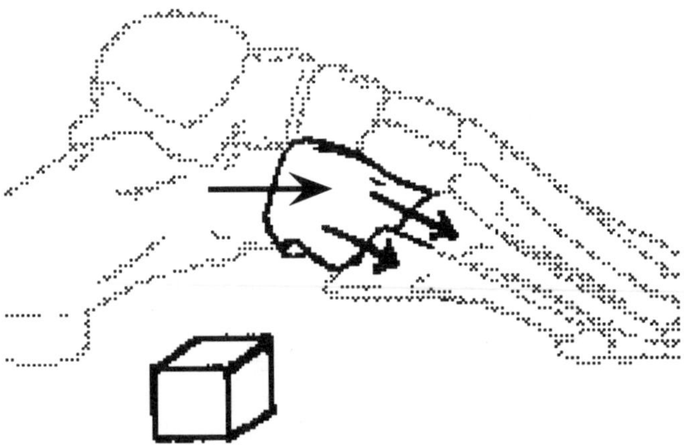

Figure 9 - 20

Inferior view of the calcaneus showing its medial hollowness.

The origin of 'calcaneus' is related to spur. A 'spur' is a projecting appendage. In botany, a spur is the principle root of a tree. In geology, a spur is a mountainous mass that extends for some distance. In poultry, a spur is the hard, pointed projection on a cock's leg, used in offensive and defensive tactics. For horses, a spur digs into the flank to incite action, hastening their pace.

We have our own built-in spur and we have our choice in its use. We can plant our heels to become rooted and immobilized like a tree. We can lay by the heels to confine, squelch, and fetter the world around us. We can dig in our heels to impel us along our path. We can take to the heels and run from it all. Or we can kick up our heels and have some fun. Each case has its appropriate time.

Cuboid

As you might have guessed, our **cuboid bone** is the shape of a cube (Fig. 9-19). It lies anterior to our calcaneus on the lateral side of our foot. It has no prominence, but lies deep in the hollow part of the side of our foot. It forms the base for our fourth and fifth toes and pushes our body weight forward through the lateral side of our foot.

Navicular

Our **navicular bone** is shaped like a ship. Its navigation came to a halt when it jammed sideways across our foot. Our navicular sits in front of our talus and carries about one third of our body weight forward through the medial side of our foot. It relates to the base of our first three toes (Fig. 9-21). The medial end of our navicular bone is often prominent. Find yours about one finger width below and in front of your medial malleolus. Follow it laterally to about halfway across the top of your foot. Several tendons will slip under your finger as you slide along.

Figure 9 - 21
Superior veiw of the navicular bone with arrows showing the transfer of weight.

Cuneiform

The ship of our foot has three wedge-shaped sails in our three **cuneiform bones** which articulate with its distal surface. They are distinquished from each other by their first names **medial, middle,** and **lateral,** or **first, second,** and **third** (Fig. 9-22a).

'Cuneiform' means wedge-shape. All three cuneiform bones are wedged into our foot from above. They have a wide dorsal surface and point into the floor like little ballerina feet, so light they never quite touch the floor (Fig. 9-22b). Each cuneiform bone carries our body weight forward to the base of one of our first three toes (Fig. 9-22a).

Palpate your cuneiform bones across the medial half of your foot, between your navicular and the most prominent bumps on the top of your foot.

Figure 9 - 22
(a) Superior view of the cuneiform bones with arrows showing the pattern of weight transfer.

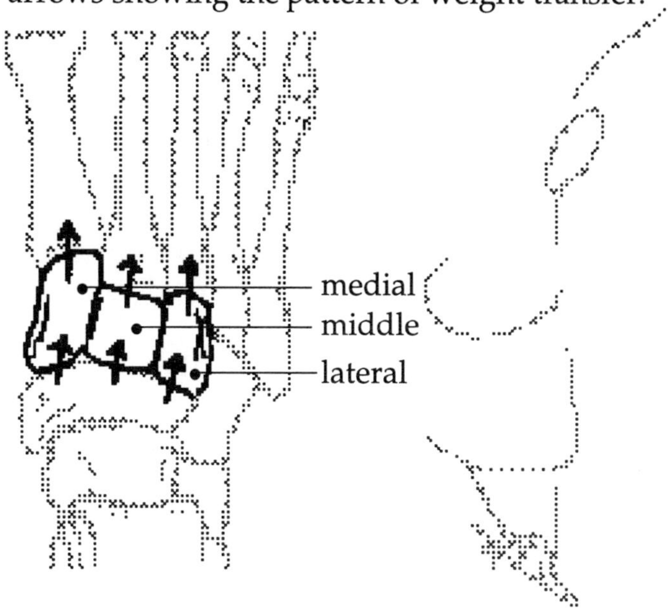

(b) A cross-section through the foot.

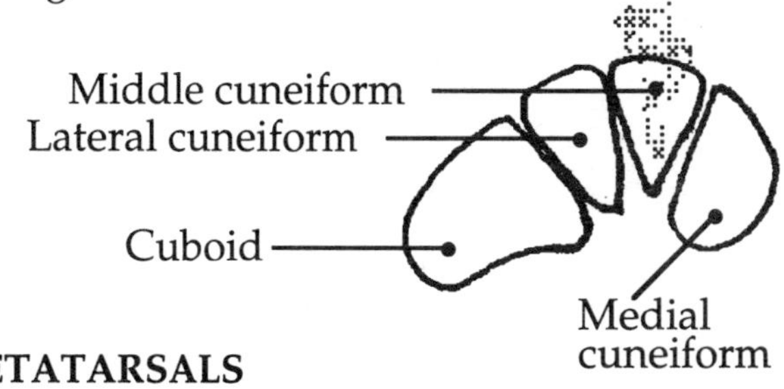

METATARSALS

'Meta' means after, beyond or following in a series. After the tarsals in our series of foot bones, we have five **metatarsals,** one at the base of each toe. To designate any one metatarsal specifically, they are simply numbered from one to five. The first one is shorter and stouter and supports our big toe. The fifth one supports our little toe (Fig. 9-23). Like the hands, the feet of skeletons have long toes, they are missing the soft tissues that web between the metatarsals in more fleshy beings.

• Our metatarsals are straight cylindrical shafts with enlarged ends. The proximal end of each of the five metatarsals is stabilized and called its **base.** Palpate the bases of your metatarsals as small bumps on the top of your foot about halfway between your heel and the tip of your toes (Fig. 9-43).

The base of our fifth metatarsal has a lateral prominence called the **styloid process** (Fig. 9-23a). This is the third engraving tool we have found in our skeleton. Find your styloid process as the prominence in the middle of the lateral edge of your foot, just distal to your cuboid hollow.

• Follow the shafts of your metatarsals along the top of your foot, gently sloping toward the floor. The distal end of our metatarsal is called its **head.** It is rounded to accommodate the swing of our toes (Fig. 9-23b). Curl your toes under. Feel your metatarsal heads as little bumps ending the dorsum of your foot. Swing your toes upward. Feel the heads on the sole of your foot. The first one is the most prominent and commonly is referred to as the ball of our foot.

Figure 9 - 23

(a) Superior view of the metatarsals.

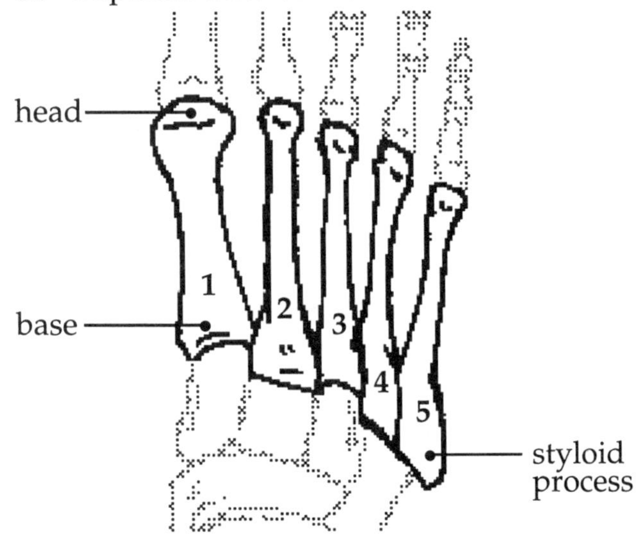

(b) Medial view of the metatarsals.

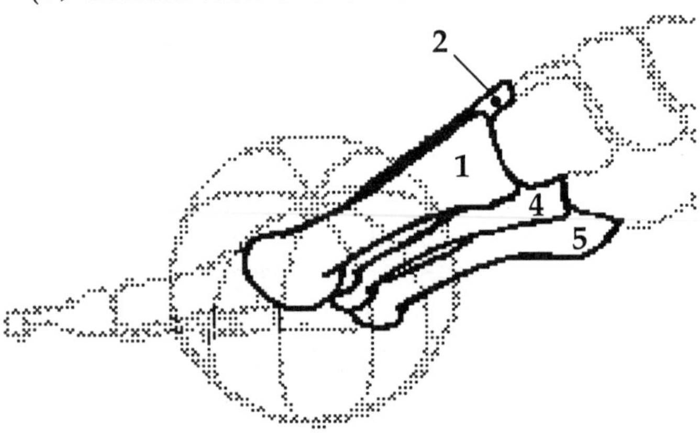

PHALANGES (phalanx in singular)

'Phalanges' means a closely knit row, and all the bones of our toes are appropriately called **phalanges**. Their shape and position are similar to the phalanges of our hand. There are 14 in all, two large ones in our big toe, and three in each of our other toes. To designate any one phalanx specifically, the words **proximal, middle,** and **distal** are used (Fig. 9-17).

Each phalanx has a rounded **head** at its distal end and a proximal **base** which glides over the head of the preceding bone.

As we curl our toes, their knuckles are the heads of the proximal and middle phalanges (Fig. 9-43). Feel yours.

ARCHES OF OUR FEET

An arch is a strong and flexible structure for bearing weight. Each foot has three arches, leaving three points of maximun weight-bearing contact with the ground (Fig. 9-24). Essentially, each foot is a tripod or a three-legged stool which can be planted with stability, without wobbling on any sloped or irregular surface. Our ankle, intertarsal, and tarsal-metatarsal joints all accommodate to keep our leg vertical over any foot position.

The arches of our foot also act as shock absorbers, absorbing the impact of our foot against hard surfaces before it reflects upward through our body.

• Two of our arches parallel the length of our foot and are called **longitudinal arches.** They provide flexibility in front-to-back balance.

Our **medial longitudinal arch** is the most pronounced and the most often referred to. It often is called our instep. Our talus is at the height of this arch and transfers weight downward into our calcaneus and forward through a gradual slope into our navicular, three cuneiforms, and first three metatarsals (Fig. 9-24). Its two contact points are the ball of our calcaneus and the heads of our metatarsals, especially the first one. This arch should be visible on the medial side of our foot even when bearing weight on our feet. Watch what happens to your foot's shape as you step down.

Our **lateral longitudinal arch** also uses the ball of our calcaneus as its posterior contact point. Our weight is pushed forward through our cuboid to our fourth and fifth metatarsals. The heads of these metatarsals, especially the fifth, are the anterior contact points (Fig. 9-24).

This arch should be visible in a relaxed foot but it often flattens when bearing weight. Still the extremities of the arch should take most of our weight and the pads on the sole of our foot should be heaviest over our heel and head of our fifth metatarsal.

- Our **transverse arch** provides side-to-side balance. It links the anterior aspects of our two longitudinal arches from the heads of our metatarsals to our navicular and cuboid.

Feel your transverse arch across the top of your foot. Relax your foot and look at its sole. The heads of your metatarsals should also arch, with the second one at the height of the arch.

When bearing weight, this arch flattens, but our first and fifth metatarsals take most of our weight. The thickened pads on the soles of our feet should demonstrate this with increased thickness over our first and fifth metatarsal heads and minimal thickness over our second. A thickened pad over our second metatarsal head indicates a fallen transverse arch.

Figure 9 - 25
A superior view of the foot showing the three arches and the three points of maximum weight bearing

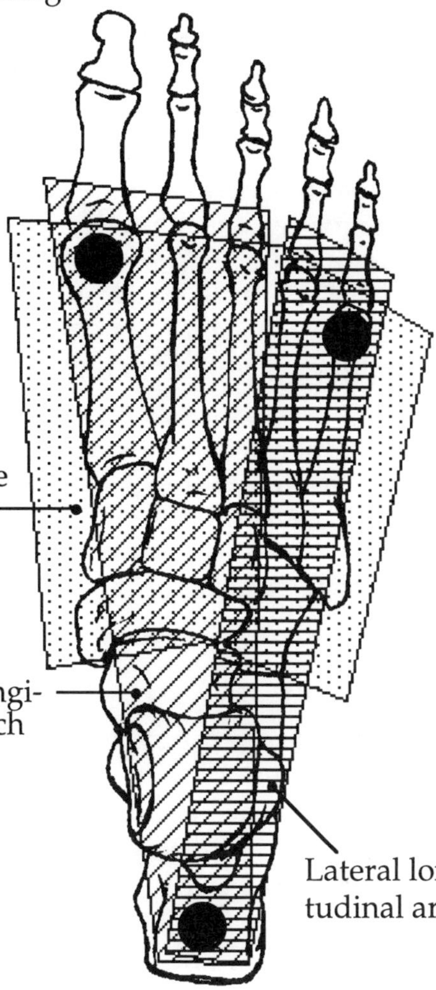

Figure 9 - 24
A medial view of the foot showing the three arches and the three points of maximum weight bearing.

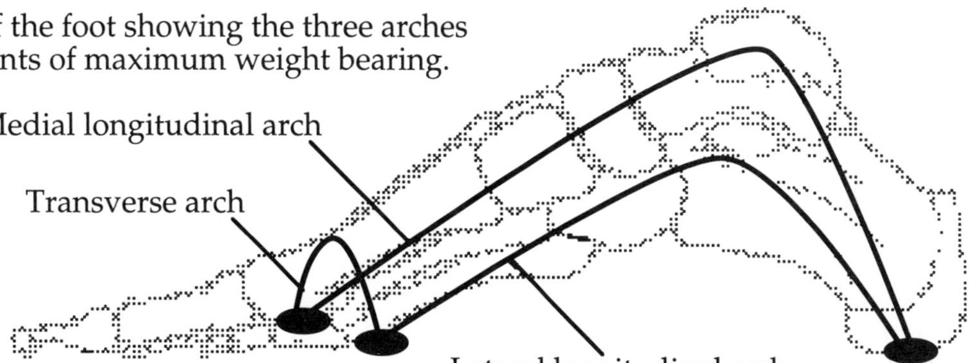

Weak or flattened arches produce **flat feet** which may lead to poor balance and pain in bones that bear weight inappropriately. Supporting our arch by placing a mound in our shoe and forcing our foot to bend over it perhaps gives an arched shape to our foot, but it forces our foot to bear weight through the bones in contact with the mound rather than at the tips of the arch, so it does not necessarily help the function of the arch. To help the arch, the foot needs to be supported so that it bears its weight on its three points.

Our arches are formed by the shape of our bones, their ligament support, and their muscular support. When our arches are weak and flattened, the easiest way to rebuild them is by strengthening their muscular support. As we discuss our muscles, we will indicate which ones support our arches. As a general rule, muscles which flex our toes support our longitudinal arches, while muscles which invert and evert our foot support our transverse arch. Exercising these muscles will increase the support of the arches of our feet. Appropriate exercises might be to pick things up with our toes or push off with our toes when walking. Both of these emphasize toe flexion. Rolling a tennis ball from side to side with the sole of our foot uses eversion and inversion.

LOWER APPENDAGE JOINTS

Sacroiliac Joint

Our **sacroiliac joint** joins our axial skeleton onto our lower appendage. The giant transverse process of our sacrum and our posterior ilium interlock broad, irregular surfaces to transfer the weight of our torso laterally from our vertebrae into our pelvis while siting or standing (Fig. 9-26).

Our sacroiliac joint cannot be palpated directly, since it lies in front of or deep to our posterior iliac spines.

Figure 9 - 26
Posterior view of the sacroiliac joint with arrows showing the pattern of weight transfer.

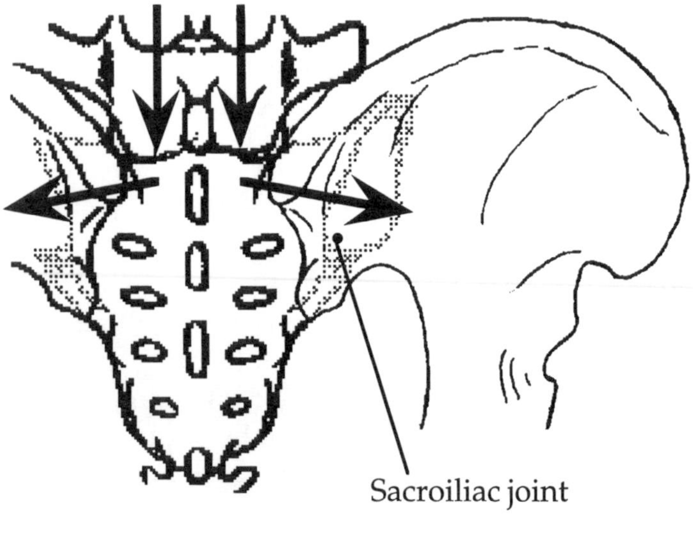

Our sacroiliac joints are synovial in early life, but may become fibrous with time. In spite of being synovial, movement at our sacroiliac joint is minimal, indirect, and never clearly defined. Minimal movement requires minimal musculature. This joint is dependent on ligament support with no muscles which directly support or move it and, therefore, no muscles to compensate in case of ligament injury, leaving this joint more vulnerable to chronic problems.

• Stress on this joint may be evidenced by tenderness between our posterior iliac spines and our sacrum. Bending forward, especially when lifting or holding weight and pregnancy and childbirth are stressful on this joint.

Pubic Symphysis

The ring of our pelvis is completed anteriorly as the bodies of our left and right pubic bones embrace each other at our **pubic symphysis** (Fig. 9-27). 'Symphysis' means growing together, like all the instruments in a symphony. While these bones never fuse, their independent movement is minimal.

Figure 9 - 27
Anterior view of the left and right pubic bones and the pubic symphysis.

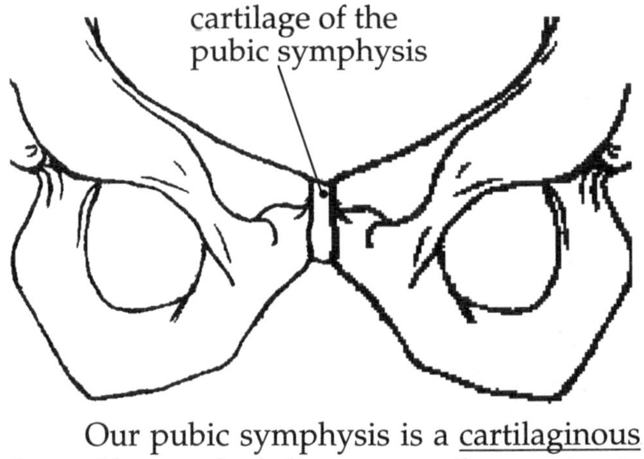

Our pubic symphysis is a cartilaginous joint, with a pad of slippery cartilage separating the two pubic bodies. There are no muscles which cross our pubic symphysis to move it, but indirectly it performs tiny figure eights as we walk. This joint also allows our pelvis to enlarge during childbirth.

Palpate your pubic symphysis in the midline, at the bottom of your belly.

Hip Joint

Hip hip hip hurrah! No matter how down homely we might feel, we have two joints that are 'hip' and give up a sense of fashion and sophistication.

Our hip is a synovial ball-and-socket joint, formed as our round femoral head fits into the acetabulum of our pelvis (Fig. 9-28). The socket is deepened by a lip of cartilage, to make a relatively secure but mobile joint.

Our hip joint is located in our lateral pelvis about a hands width below our iliac crest. It is just above our greater trochanter, but buried deep in muscle and not palpable.

• Our hip uses all its surrounding muscles to create its many movements. It performs three distinct pairs of movements, plus all the degrees of movements in between. In the anterior-posterior direction, our hip **flexes** as our femur swings forward, and **extends** as our femur swings back. In the side-to-side direction, our hip **abducts** as our femur swings out from our side, and **adducts** as it swings in toward our other leg. Our hip can rotate in any position. **Medial rotation** turns our kneecaps toward each other, and **lateral rotation** turns them away from each other (Fig. 9-29).

Figure 9 - 28

A cross section through the hip.

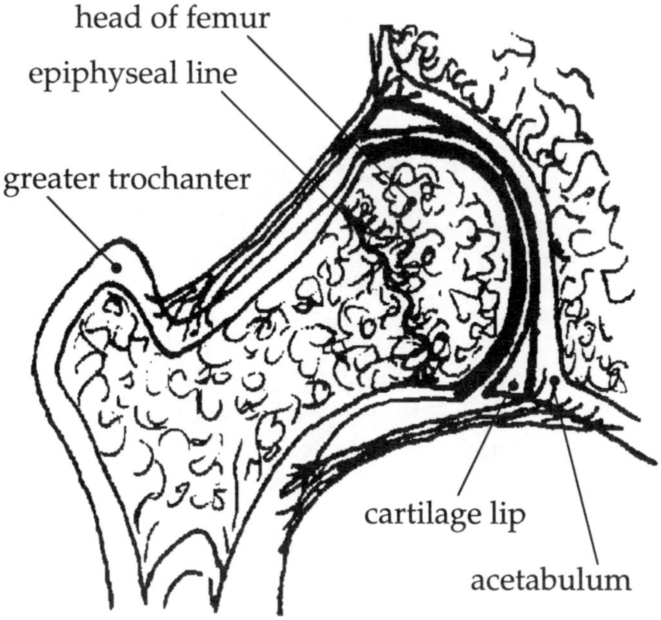

Figure 9 - 29

Movements of the hip.

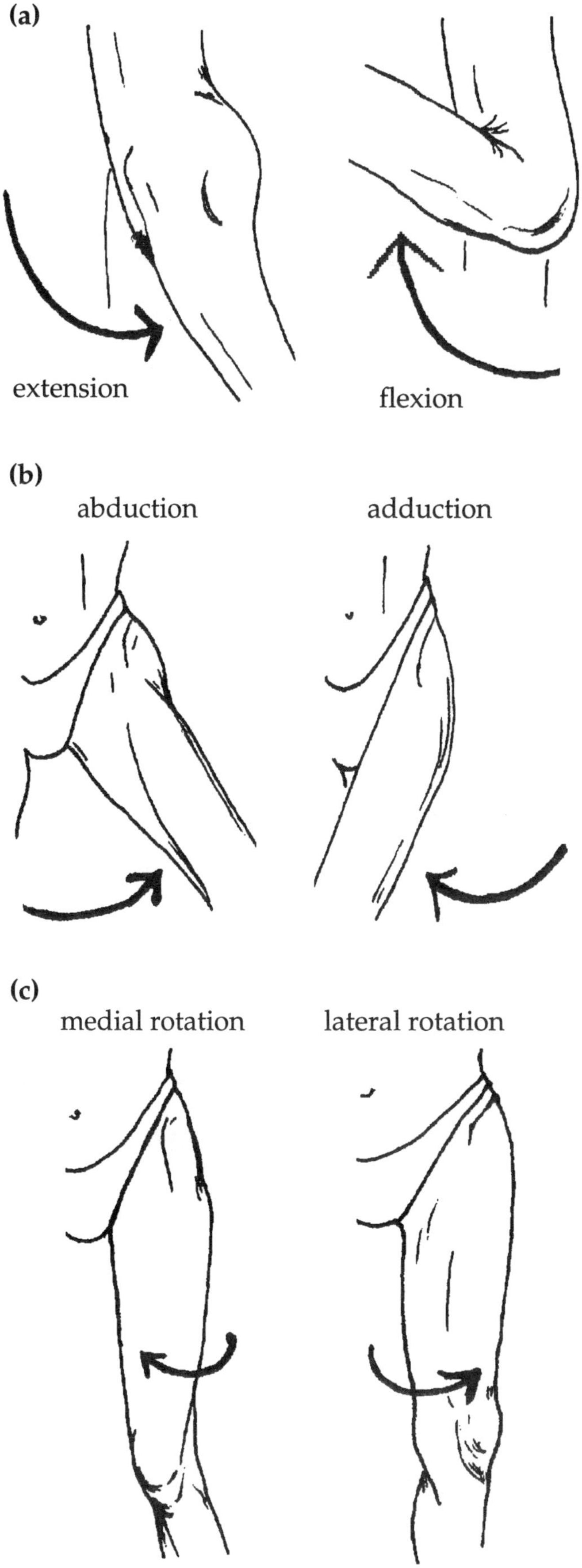

Knee Joint

Our **knee joint** is a synovial joint formed as our distal femur meets with our proximal tibia (Fig. 9-11, 9-12, and 9-35).

For its size and weight-bearing capacity, our knee is the most unstable joint in our body. With our curved femoral condyle and the flat top of our tibia, it is like balancing a ball on top of a moving table. To increase its stability, each knee has two wedged cartilages and two pairs of ligaments. Still, they suffer many injuries.

• When a ball sits on a table, all of its weight presses on a very small point. To increase the area of the weight-bearing surface, each knee is fitted with two wedged cartilages, called **menisci** (meniscus in singular). 'Meniscus' means cresent-shaped. Each cartilage is C-shaped and cups around one of the femoral condyles (Fig. 9-30, 9-31, and 9-32). It is thick on the outer edge of its curve and thin on the inner edge, blocking the front, side, and back of its condyle much like one would block the wheels of a car to prevent it from rolling. The weight-bearing surface area between our femur and tibia increases from two small dots to two long fossae. Menisci are in addition to, not a replacement for, the articular cartilage covering the end of our femur and fibia.

Figure 9 - 30

The distal end of the femur with each condyle cupped by a meniscus.

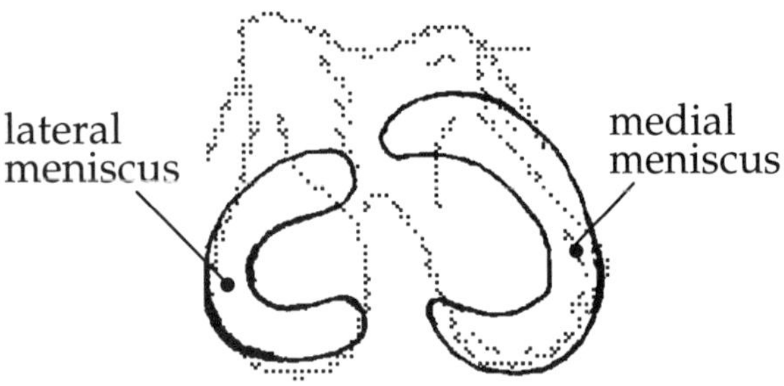

Figure 9 - 31

A lateral view of the knee showing how the menisci increase the weight bearing surface.

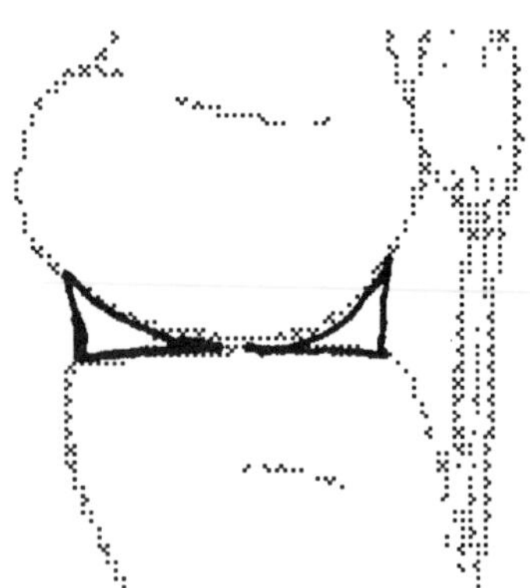

Our menisci float within our knee joint. They are lightly attached to the synovial membrane on their thick external edge, but they are not adherent to our tibia nor our femur. If trauma breaks off a central piece of cartilage, it may float freely wherever it wants within the joint space. It may wedge into unwanted places causing pain and/or limiting movement. When cartilage has to be removed from a knee joint, it is frequently a broken piece of meniscus.

• One pair of ligaments is called our **colateral ligaments.** They hold the sides of our femur and tibia together (Fig. 9-32), preventing what would be abduction and adduction of our knee. When our knee experiences a heavy blow from the medial or lateral side, these ligaments may be torn. With a blow from the lateral side, our medial knee is separated. With a blow from the medial side, our lateral knee is separated (Fig. 9-34).

'Cruciate' means cross-shaped, like a crucifix. Our **cruciate ligaments** form a criss-cross in the center of our knee, within the

Figure 9 - 32

Anterior view of the bent knee showing the ligaments and menisci.

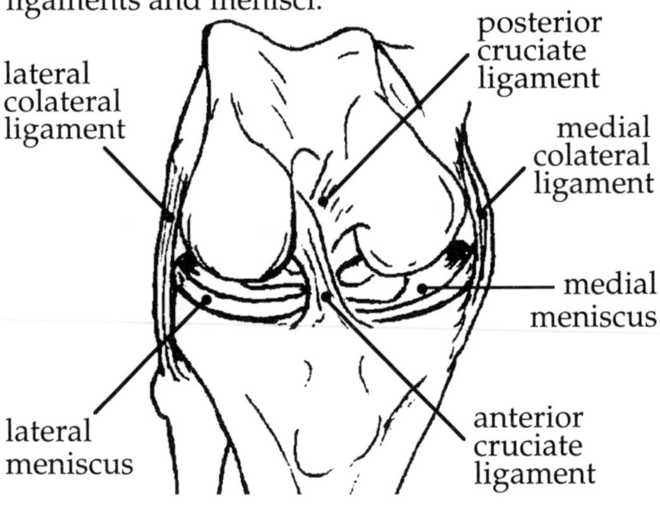

synovial cavity. These ligaments go from the notch between our femoral condyles to the top of our tibia (Fig. 9-32). These ligaments stabilize against excessive forward and backward movement between our tibia and femur. Stress from the front or back may tear them, leaving our knee with excessive anterior or posterior movement (Fig. 9-33).

Figure 9 - 33

A lateral view of **(a)** a normal knee and **(b)** a knee where the anterior cruciate ligament has been torn. allowing increased anterior movement of the tibia.

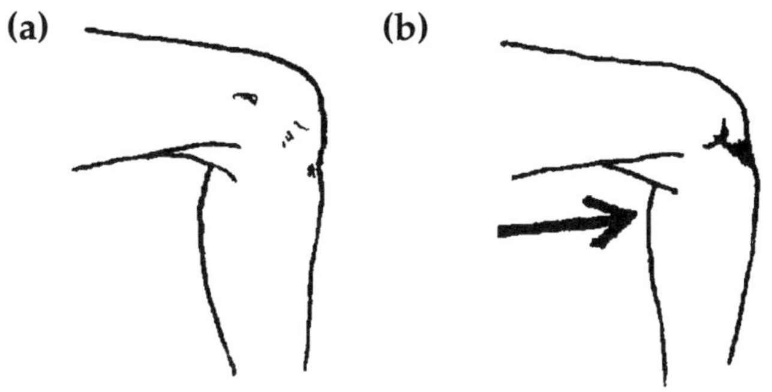

Figure 9 - 34

A longitudinal section through the knee showing how a force from the medial side might damage the medial meniscus and tear the lateral colateral ligament.

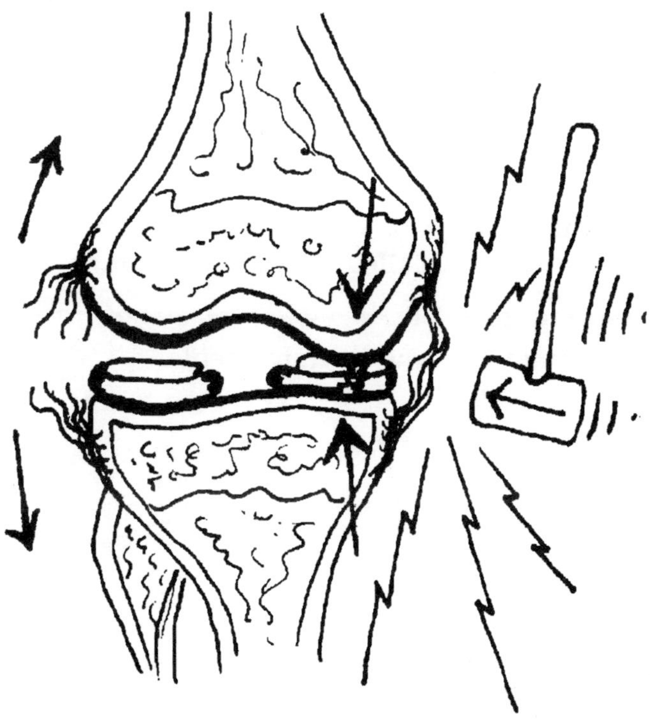

• Palpate the space of your knee joint as a small indentation on both the medial and lateral side, between the epicondyles of your femur and your tibia (Fig. 9-35).

Figure 9 - 35

Surface view of the lateral knee.

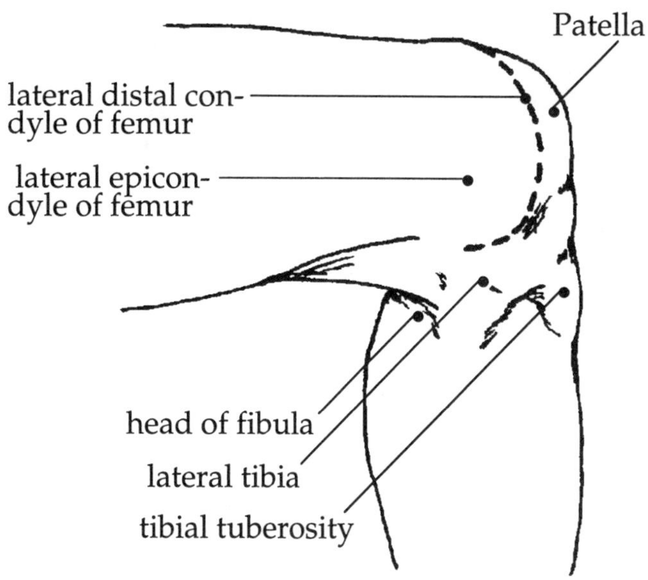

• The most obvious movement of our knee joint is in the anterior-posterior direction. Remember that the ventral and dorsal surfaces at our knee have rotated, bringing our ventral surface posterior and our dorsal surface anterior. To flex any joint is to bring the ventral surfaces together, so to **flex** our knee, we swing our tibia backward. To **extend** our knee, we separate the ventral surfaces, swinging our tibia forward until our leg is straight (Fig. 9-36a).

Except when fully extended, our knee also can perform a few degrees of **medial** and **lateral rotation** (Fig. 9-36b). With your knee partially flexed, watch or lay your fingers across the indentation of your joint space. Twist your foot inward and outward. Watch or feel for this subtle but essential movement at your knee joint, between your femur and tibia.

Another important movement of our knee joint is the sinking of our patella as our knee flexes. This is discussed under our patella (Fig. 9-10).

Figure 9 - 36

Movements of the knee, shown with a lateral view of the left knee.

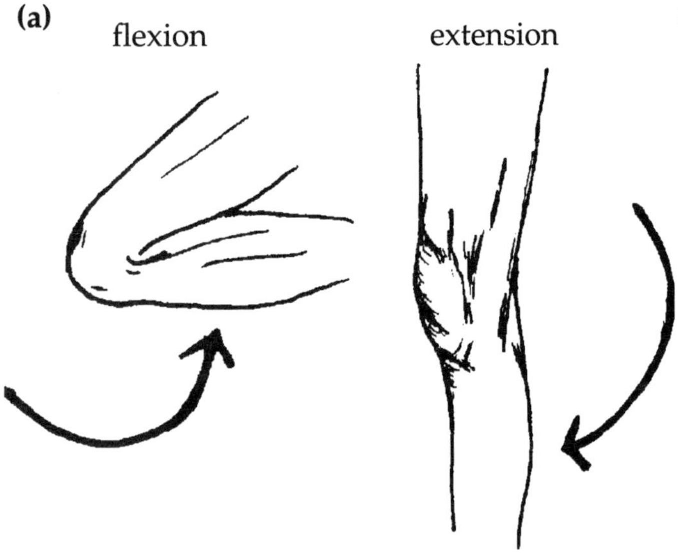

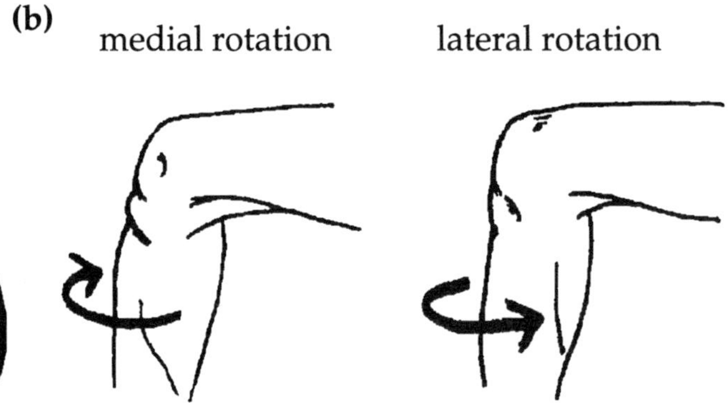

Tibia Fibular Joints

Our tibia and fibula are strongly attached to each other. Proximally, the head of our fibula forms a synovial joint with our lateral-posterior tibia. Small, gliding movements take place at this joint. Throughout their length, these two bones continue to be strongly bound together by a dense connective tissue called the **interosseous membrane,** literally translated, a 'between bones membrane'. Distally, strong proteins bind the bones together in a fibrous joint (Fig. 9-37).

Figure 9 - 37

The tibia fibular joints.

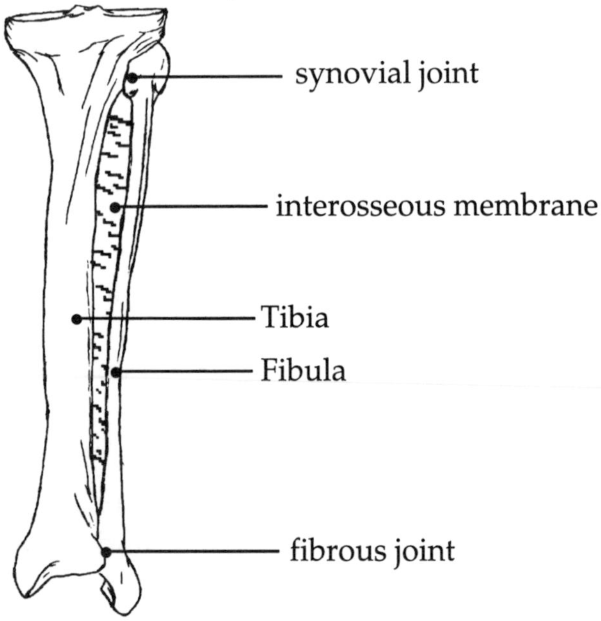

Ankle Joint

Our lateral and medial malleoli together with our distal tibia form a box which fits over our talus to form our ankle joint (Fig. 9-39). It is a hinged, synovial joint. Our body weight passes from our tibia to our talus to be distributed through our foot (Fig. 9-16).

The only movement of our ankle is in the anterior-posterior direction. This is the only joint that extends further than it flexes. The sole of our foot is its ventral surface. Swinging our foot down to point our toe brings it closer to the ventral surface on the back of our calf, so this movement is **flexion.** The top of our foot and our shin are dorsal surfaces, so lifting our foot up toward our shin is **extension** (Fig. 9-38). Since extension of our ankle appears very much like flexion, some people just had to call it flexion, so new names were developed for ankle movement. Extension also is called **dorsiflexion** and flexion also is called **plantar flexion.**

Figure 9 - 38

The ankle joint and its movements.

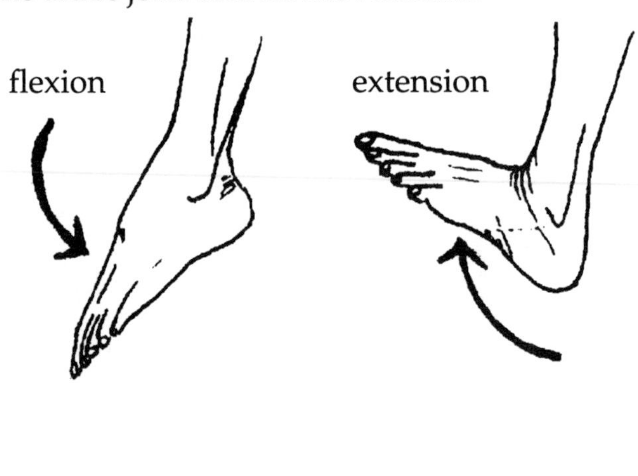

Figure 9 - 39

The ankle joint.

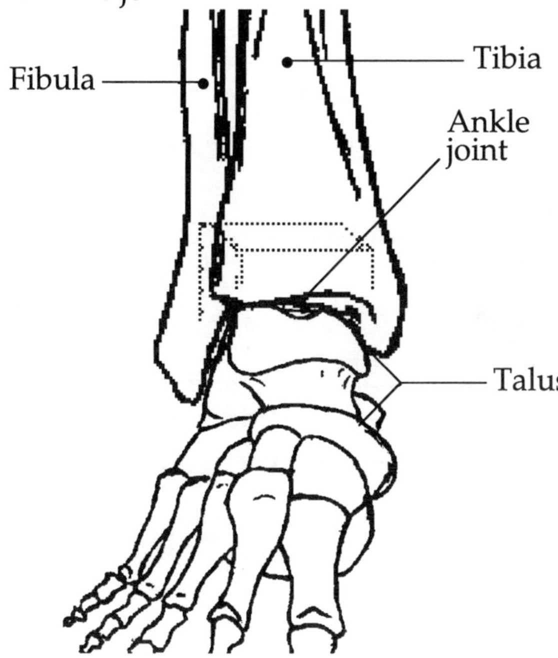

Intertarsal and Tarsal Metatarsal Joints

The numerous articulating surfaces between our seven tarsals and five metatarsals slide slightly over each other. Their combined efforts carry the effect of ankle **flexion** and **extension** into our foot (Fig. 9-38), or twist our foot from side to side.

Twisting our foot outward so its lateral edge is lifted is called **eversion.** The 'e' is for exit or out. Twisting our foot inward so its medial edge is lifted is called **inversion** (Fig. 9-40).

Figure 9 - 40

Inversion and eversion of the foot.

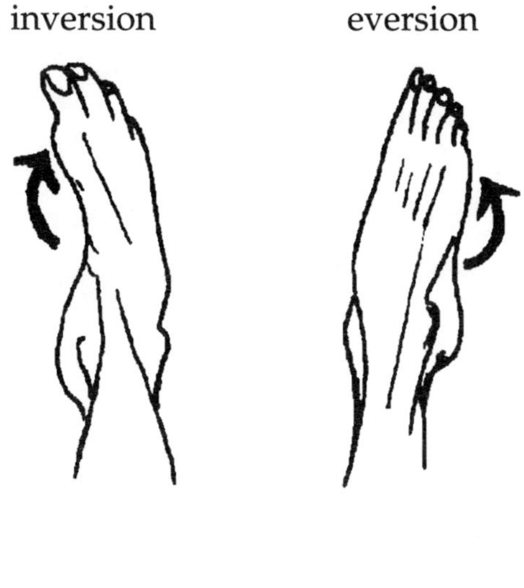

Metatarsal Phalangeal Joints

Our **metatarsal phalangeal joints** are located where the bald head of each metatarsal is caressed by the base of its proximal phalanx (Fig. 9-17). These joints simplify life by using their initials, **MP.** Palpate your MP joints just beyond the knuckles that appear on the top of your foot when your toes are curled under (Fig. 9-43). As our toes curl under, our MP joints **flex.** As our toes swing upward, our MP joints **extend. Abducting** our MP joints spreads our toes apart. **Adducting** our MP joints pulls our toes together (Fig. 9-41).

Figure 9 - 41

Movements of the metatarsal phalangeal, proximal interphalangeal and distal interphalangeal joints.

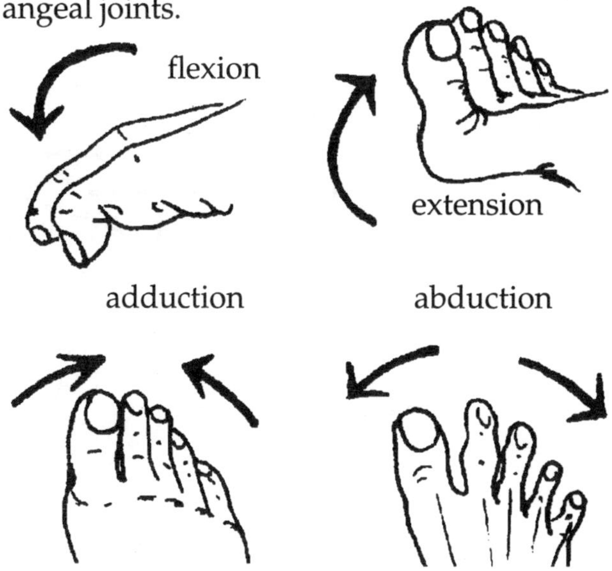

Interphalangeal Joints

'Interphalangeal' is latin for between phalanges, the location of these joints. Our big toe has only two phalanges with one **interphalangeal joint**. Our second through fifth toes each have three phalanges, with two interphalangeal joints. Our **proximal interphalangeal joint** lies between our proximal and middle phalanges. Our **distal interphalangeal joint** lies between our middle and distal phalanges (Fig. 9-17 and 9-43).

Our interphalangeal joints only move in the anterior-posterior direction, **flexing** as our toes curl under and **extending** as our toes radiate upward (Fig. 9-41).

Figure 9 - 42

Surface view of the top or dorsum of the foot.

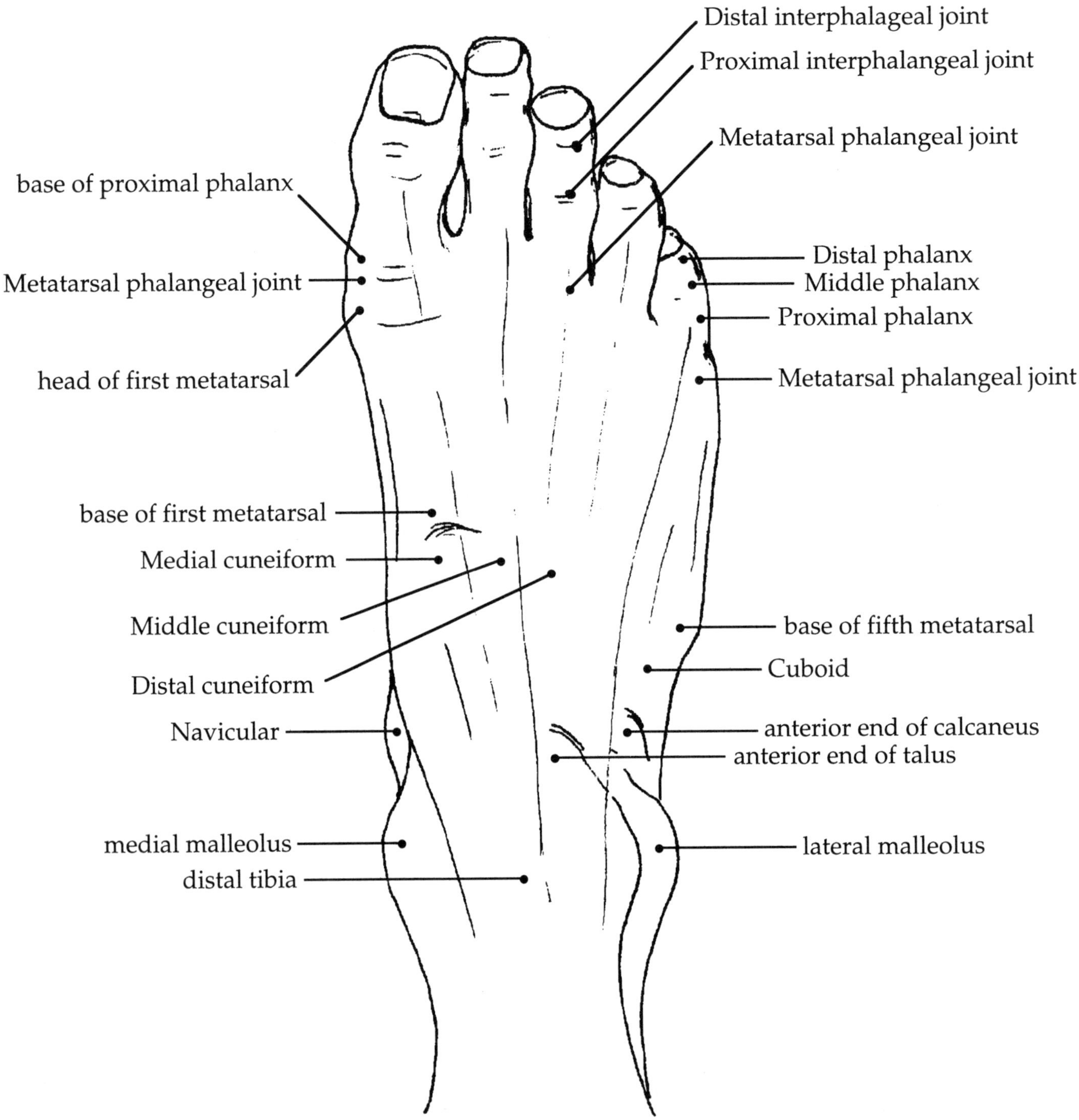

Figure 9 - 43

Surface view of the medial foot.

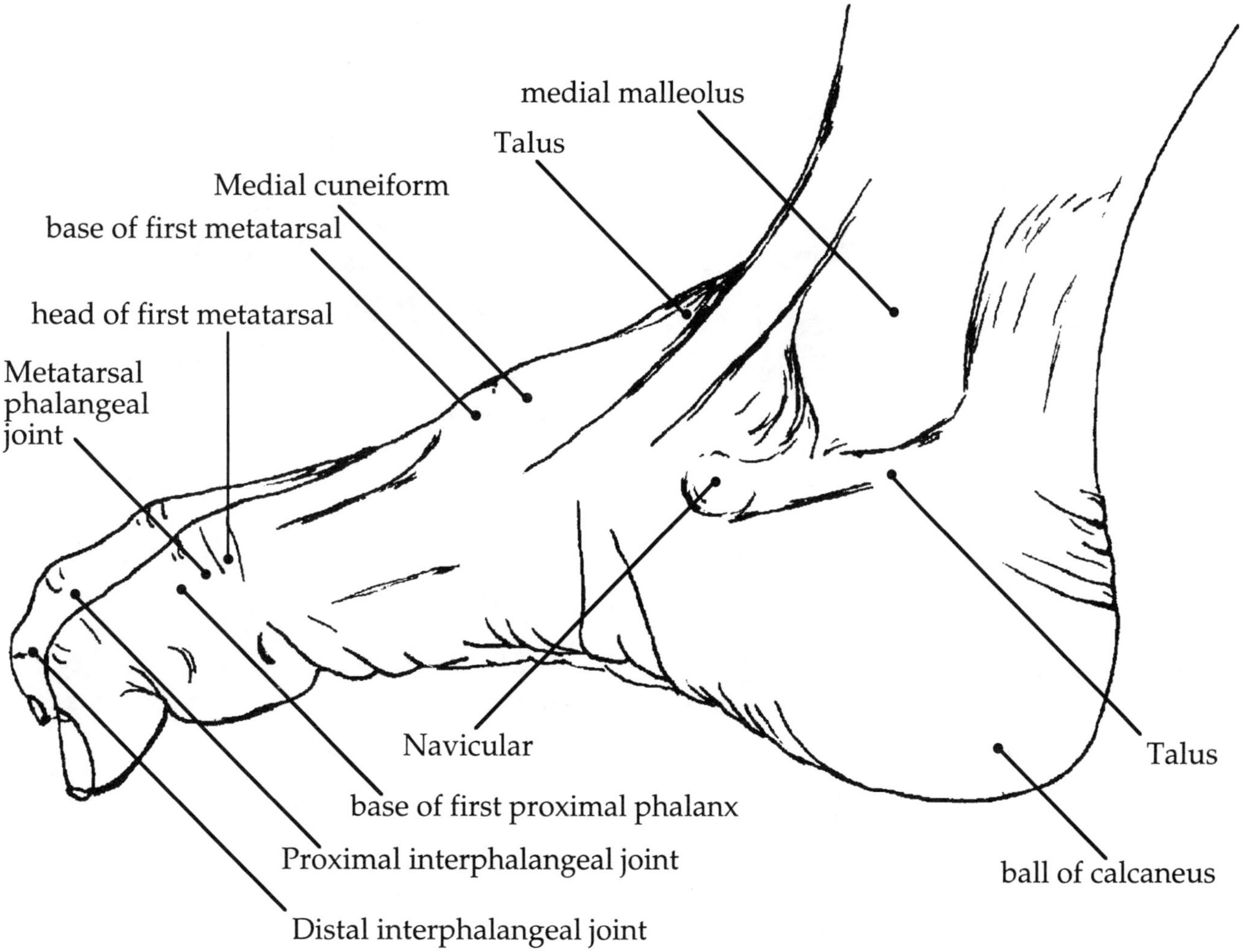

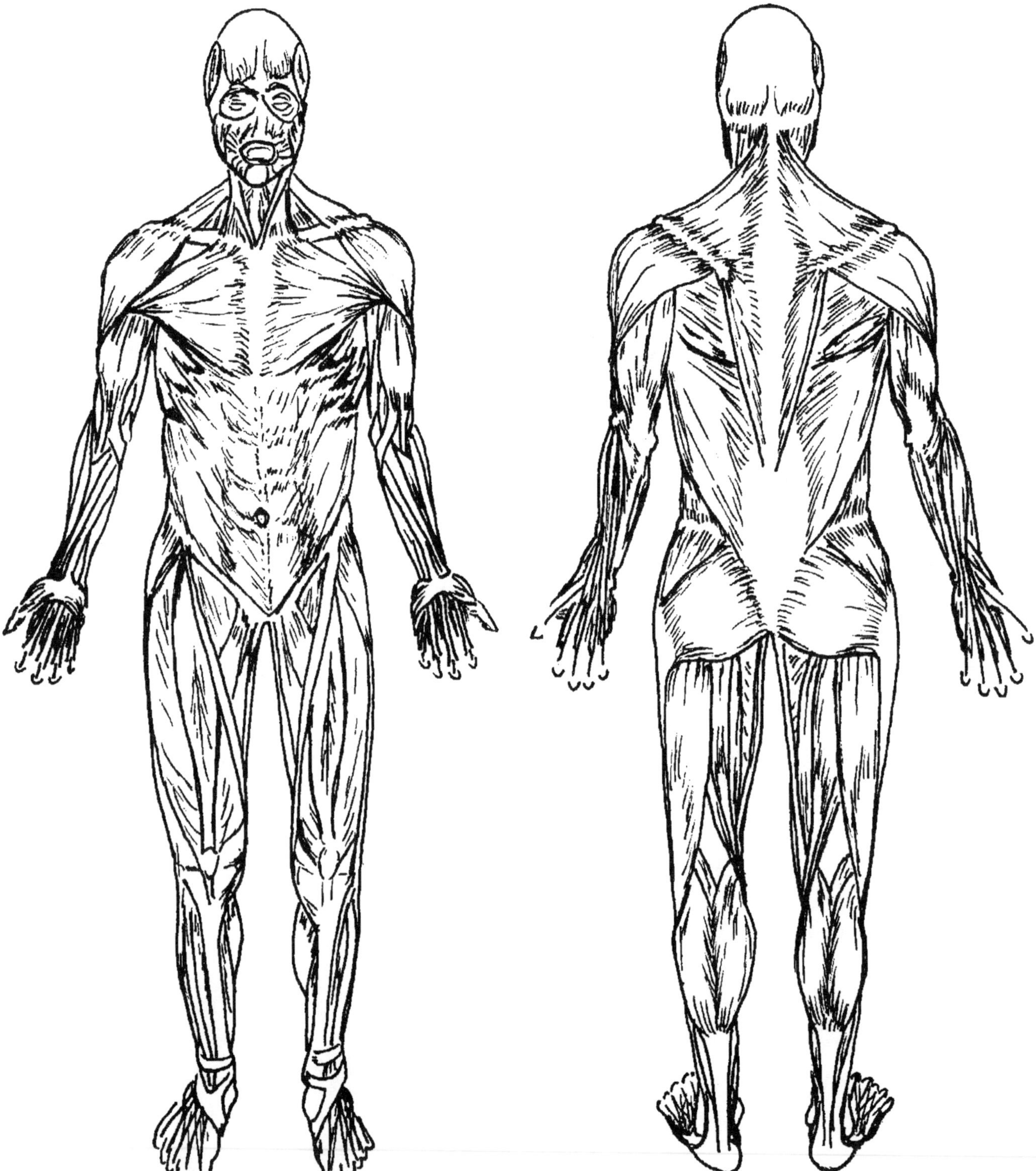

Unit IV: THE SKELETAL MUSCLES

While our skeleton provides a framework and leverage for the movement of our body, it is our skeletal muscles who exert the forces, doing the work of movement. This movement is a major aspect of our survival mechanisms.

Our skeletal muscles make up about 40 to 50 perccent of the weight of our body and consume enormous amounts of energy. Our muscles are composed of muscle tissue and the connective tissue which attaches them to their bones. Each muscle is considered to be an organ.

In this unit we discuss general concepts of skeletal muscle structure, function, and nerve innervation, principles of muscle contraction, as well as the specific details of the attachments, actions, and palpation of each muscle.

The scale of one-half adult five-foot-five is not maintained in the illustrations of specific muscles. Since muscles often cross several joints and span the length of more than one bone, too many pages are not big enough. There is no relative proportion amongst the muscle illustrations.

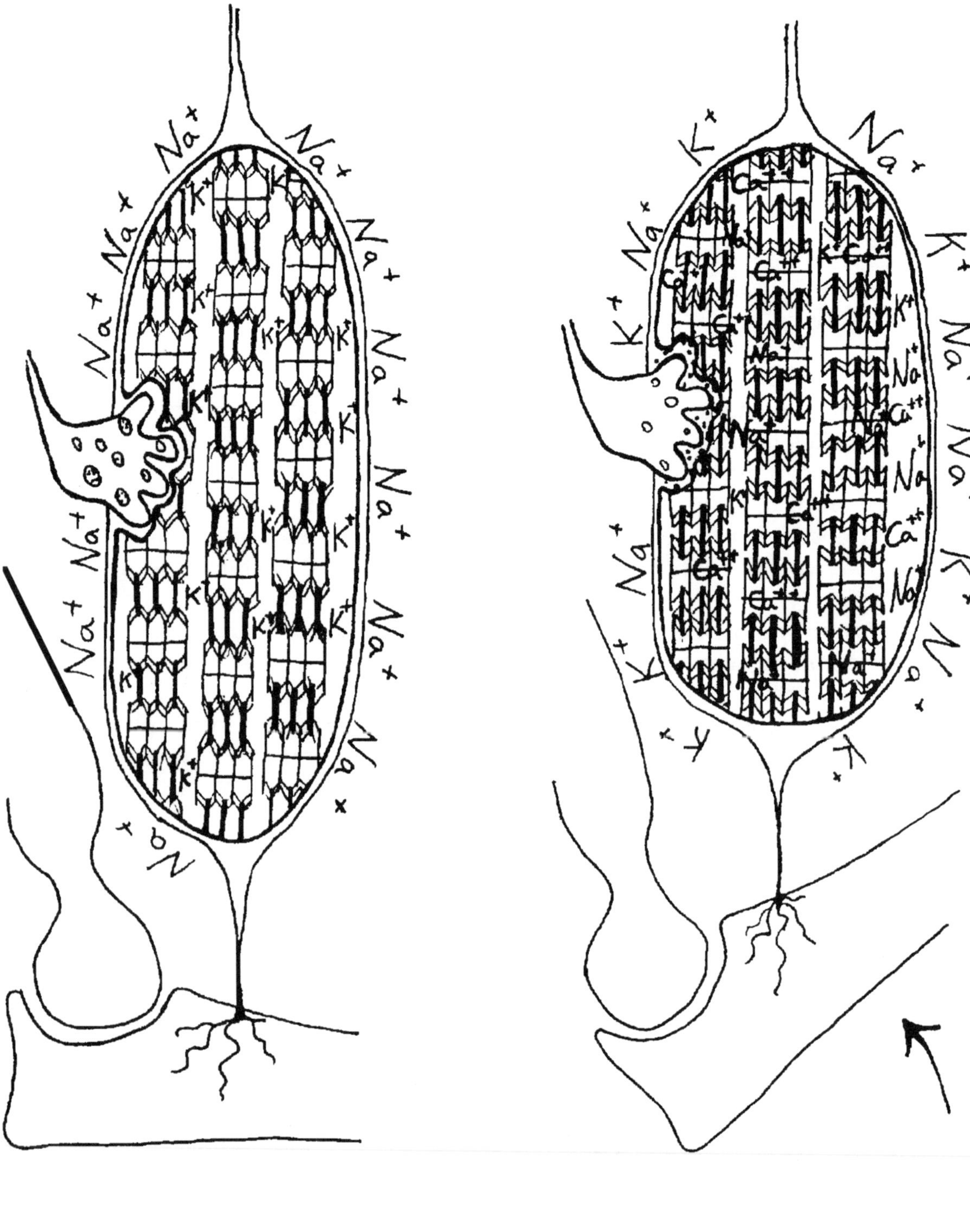
Na+
K+
Ca++

chapter 10: **Skeletal Muscle Theory**

Our discussion of skeletal muscle began in chapter 4 under Muscle Tissue. The special ability of muscle tissue is to exert a contracting force which may shorten our muscle and cause movement. Our skeleton is the structural support of our body, providing a system of levers and fulcrums which allow for movement. Our skeletal muscles attach to our skeleton to create the movement, to exert the force of action, to accomplish the lever's work.

GENERAL FUNCTIONS

Briefly outlined, skeletal muscle performs seven general functions for our body:

1. To create the physical movement of our skeleton.

2. To assist in joint support.

3. To pump blood and lymph from our tissues back to our heart.

4. To support upright posture or to hold a particular position.

5. To express our emotional feelings.

6. To protect against emotional and physical trauma.

7. To produce heat.

1. The most obvious job of skeletal muscle is to create the physical movement of our skeleton. The structure of our skeleton provides joints which allow for movement. Muscles attach to our bones to produce the movement.

Each end of any muscle attaches to different bones, so our muscle always crosses at least one joint. As our muscle shortens, our bones are pulled toward each other (Fig. 10-1). Our joint must bend. It is that simple. This is the only way voluntary movement of our body happens. Any muscle that crosses a joint may move it, depending on the mechanical advantage. Whether we dance, swim, shuffle about our house, speak, or wink an eye, it is all accomplished through the work of skeletal muscles. If you do not feel like moving a muscle, you are left to the world of day dreams.

2. The bones, joints, and ligaments of our skeleton serve as the structural support of our body. While ligaments are the unchanging stalwart of joint support, certain positions or abuse may make them slack and unsupportive.

Figure 10 - 1

As muscles contract, bones are pulled toward each other to produce physical movement.

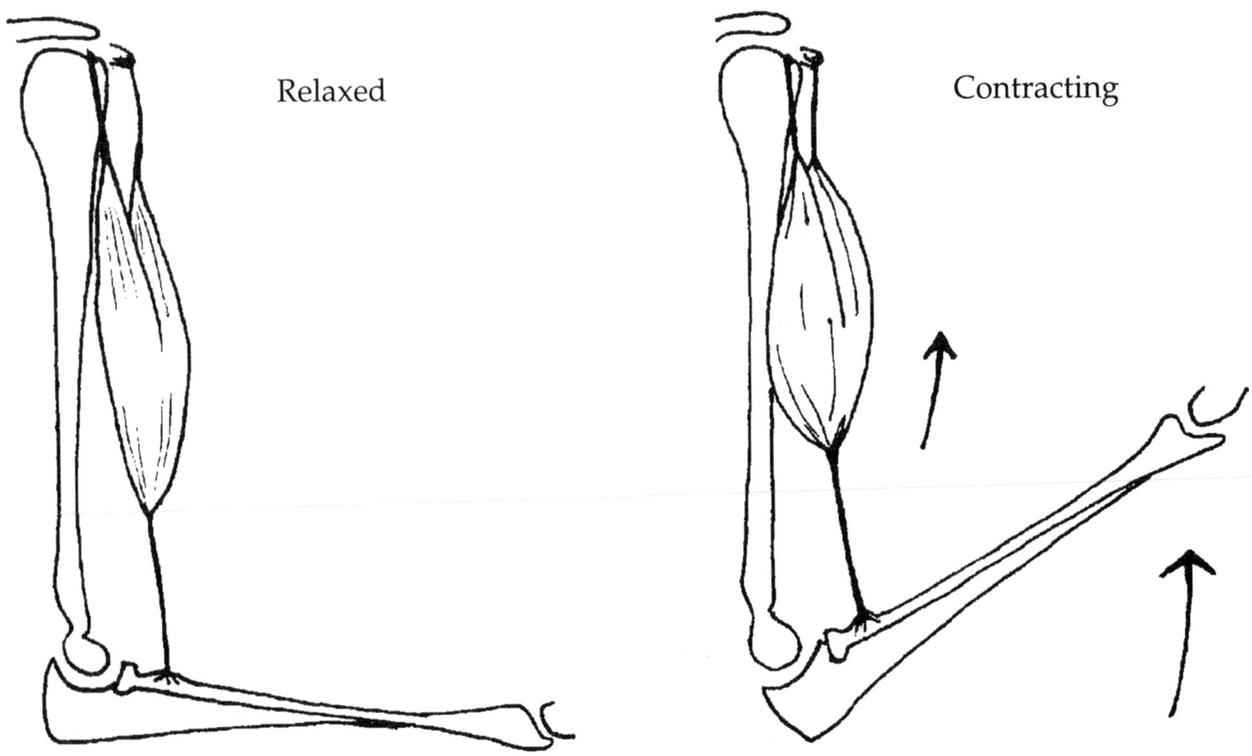

Muscles can take up the slack to keep us together and aligned. Muscles assist in joint support. Strengthening and toning muscles will increase the stability of the joints they cross. This is especially useful when normal ligament support has been damaged.

In very mobile joints, like our glenohumeral joint, it is difficult to find an available place to put a nonstretchy ligament so it does not limit the joint mobility. Consequently, the connective tissue support of the joint may be skimpy. These joints are even more dependent on muscular support. If the muscles which support our glenohumeral joint lose all their strength, the weight of our dangling arm will, itself, dislocate the joint.

Some joints have little or no muscular support. If their ligaments are traumatized or weakened, there is no backup support. These joints may develop chronic instability or even remain permanently dislocated. Examples of such joints are our acromion clavicular joints and our sacroiliac joints.

3. As skeletal muscles alternately contract and relax to produce repetitive movements, they indirectly pump blood and lymph from our tissues toward our heart. A contracted muscle is hard and compresses surrounding tissues, pushing our blood or lymph along within its vessel (Fig. 10-2). As our muscle relaxes, it softens. The compression is removed from our vessel so it fills with more fluid from the distal tissues, ready to be pushed along by the next contraction.

It is important to notice that our flow of blood and lymph benefits when muscles alternately contract and relax. When muscles hold continuous contraction, the opposite effect occurs. Our flow of blood and lymph is reduced. This effect is discussed further in the tid bit box on page 206.

Figure 10 - 2

(a) Relaxed skeletal muscles allow blood and lymph to stagnate. **(b)** As skeletal muscles contract, they compress adjacent blood or lymph vessels, pushing the fluid along within the vessel.

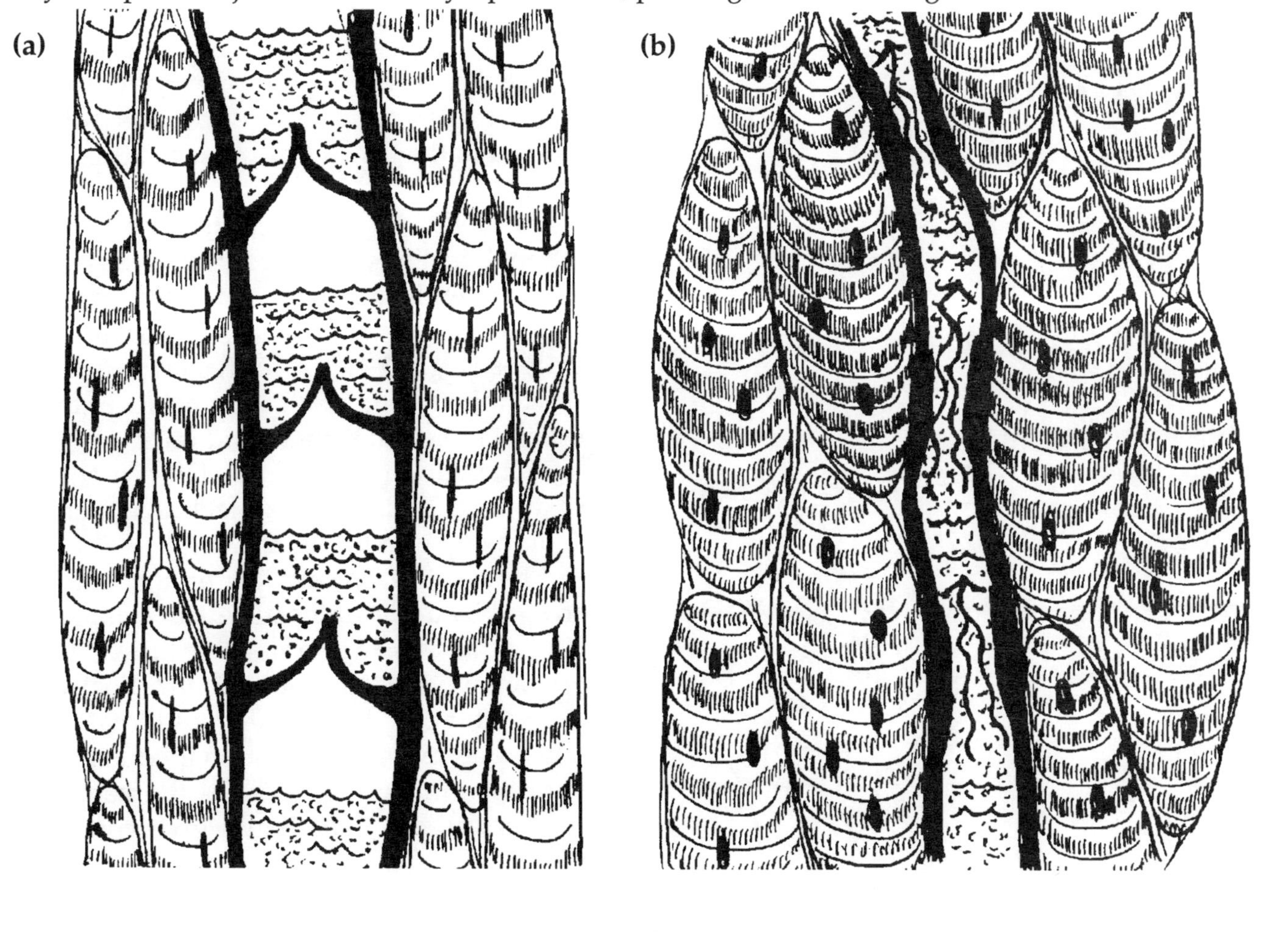

4. The earth has an invisible grasp on all its inhabitants, pulling them down toward its center. With all of our effort, we humans can only leap upward a few feet before gravity pulls us back down to the earth's surface. Indeed, lifting and holding any part of our body upward requires continuous muscular effort against the force of gravity (Fig. 10-3). One of the monotonous functions of skeletal muscle, then, is to support upright posture or to hold any particular position. Only when our body is completely supported against gravity can every muscle relax.

There are many definitions of good posture. The military definition orders "Chin in, chest out, shoulders back." Your mother's definition adds its own requirements. I like the definition which says that good posture is whatever upright alignment requires the least muscle work to maintain. This implies maximal muscle relaxation, saving energy and reducing stress throughout our entire musculoskeletal system. This concept is further explained in the tid bit box on page 204.

Figure 10 -3

In spite of no movement, muscles work continually to hold us up against the force of gravity.

Figure 10 - 4

Contraction of our skeletal muscles expresses how we feel emotionally.

5. Muscle contractions express our emotional feelings. On confident days, we dance with wide, open sweeps. On depressed days we droop. Anger juts us out in different directions (Fig. 10-4). In confusion, we get all pinched together or jerky. Any posture carries a certain emotional content and impact.

Changing our posture also can change how we feel. A droopy day feels brighter just by looking up and lifting our chest.

If you are curious about the emotion expressed by a certain posture, put yourself in that posture. Exaggerate it a little. What temperament does that put you in? Really get into it. Act it out. What role does it make you feel like playing? The emotional content of a specific posture is not exactly the same for every person, but that should give you some ideas.

6. We use skeletal muscle contraction to protect ourself against emotional and physical trauma. The pain of trauma and the fear of pain induce continuous, unconscious muscle contraction. When someone yells at us, we unconsciously contract muscles to brace ourselves against the abuse. Muscles surrounding an injured shoulder unconsciously increase their tone to splint against painful movement. During appendicitis, muscles of our anterior abdominal wall become hard, like a board, protecting our inflamed tissue from the questioning pokes of the doctor.

After a first brief attack, protecting muscles will slowly regain trust and relax. The more often we are threatened, the more continuously we assume our protective positions, until they can become our normal posture.

Eventually, the threat may be gone or the injury healed, but the muscles are still protecting. They don't know when to quit and we don't 'think' to tell them, since it all happened on an unconscious level. Thus, we may accumulate considerable unconscious muscle tone and contraction through past traumas not released from our bodies (Fig. 10-5). The sequel and chronic effect of this is developed further in the tid bit box on page 206.

7. Human beings are warm-blooded animals who require a constant, specific body temperature. Since the environment around us changes in temperature, and is frequently colder than our body, we need to have our own internal heat supply. Skeletal muscles are one important source of heat. As any cell releases the energy of glucose, about 25 percent of the energy is made available for cell functions. The other 75 percent escapes in the form of heat. Since skeletal muscles consume so much glucose, they provide the greatest amount of heat for our body.

A brisk walk or a little physical work will always warm us up on a cold day and our extra layers of clothing can be shed. Heating our home with a wood stove has at least one advantage. Splitting the wood for the day will already thaw us out on an icy morning, even if we have trouble starting the fire.

When we become deficient in heat and get too cold, what is one involuntary response? Shivering. When we shiver, we produce little contractions in our skeletal muscles. Not enough contraction for effective movement, but enough to force muscles to break apart glucose, this time for benefit of the escaping heat rather than any particular movement.

• These are the functions of skeletal muscle. Let us look closer at the whole musculoskeletal system to see how these functions are performed.

Figure 10 - 5

Through traumas not released form our bodies, we may accumulate considerable unconscious skeletal muscle contraction, holding us in ways we do not even realize.

Relaxed upright posture provides two challenges: **1.** a vertical alignment of body parts defying the force of gravity and **2.** using internal balance rather than muscle tension to hold that alignment.

When working with posture, we must always remember the tireless force of gravity, pulling everything down, toward the center of the earth. Whenever we hold any body part upward, we use muscular effort to resist gravity. Our muscles must pull up with a force equal to gravity pulling down.

The further anything protrudes out from its supporting pedestal, the more advantage or leverage gravity has. This means that the downward pull of gravity is accentuated so the thing is harder to hold up. In contrast, the more completely something is positioned over its supporting pedestal, the less advantage gravity has. The effect of gravity is reduced and the thing is easier to hold up. For example, stretch out your arm while holding this book in your hand. The weight of it is noticeable. Now support the book on your forearm, close to your elbow. Has the book lost weight? Your outstretched arm protrudes significantly from your supporting pedestal, giving the advantage to gravity, so you need to work harder to hold up the book.

The same principle is true for posture. In standing posture, our supporting pedestal is our feet. The more vertically we align our body over our feet, the more disadvantaged gravity becomes, and the less weight we feel (Fig. 10-6b). Very little effort overcomes the effect of gravity. We can save energy, stay relaxed, and reduce the pressure on our joints. While sitting, our pelvis supports us. If we keep our upper body over our pelvis, we stay comfortable and relaxed.

Stressful postures protrude some body part out from our pedestal. There are thousands of variations on this theme. In some people the knees are locked and the pelvis is brought forward (Fig. 10-6c), in others the head protrudes forward (Fig. 10-6a), in others the buttocks stick out backward, in yet others a shoulder lifts off to one side while the hip presses out to the other side, and so on. The body will balance one protrusion by throwing some other body part in the opposite direction. It is easy for gravity to latch its invisible hands onto all these outstretched structures and pull them down, so our muscles hang on, working hard just to hold us up.

The trick is to align the body parts so they stay in position by balance, with very little muscular effort. I received a great lesson in this while trying to wind surf. I was using every muscle in my body to fight against the wind and sail. What fun? After three days I had pain in the joints of my hands just from hanging on. But I was determined! Then a man in a row boat came up beside me and explained how I could just balance my body weight against the force of the wind in the sail. In balance, it should take less than the strength of one finger to hold the sail. It was marvelous.

This concept is effective for posture as well. So many attempts at good posture are like my first three days of wind surfing. Massive muscular effort is used to fight some body part into alignment. We cinch our shoulders back by contracting the muscles between our scapula. After holding and holding, we collapse in fatigue, have sore muscles, and give up hope.

Another way to have good posture is to listen to the man in the row boat. We can fiddle around inside our structure, seeking the delicate balance of each body part over the next. If we relax the muscles around our shoulders and feel where they hang, we can lean a little forward or backward so they fall into a more favorable position. When alignment is achieved by internal balance rather than antagonistic muscle tension, we feel relaxed. It takes only 'one finger' to hold us up.

After working at posture all our life, we may have a lot of chronically tight muscles. Releasing this tension is important so we can feel the weights of our body parts, to know how to achieve balanced alignment. Be patient, this takes time.

When we find a more restful, aligned posture, we may at first feel cockeyed or bent out of shape. Even though our old posture was leaned forward, our brain adjusted to call that posture straight up. A truly straight up posture then may temporarily confuse our brain so it feels that we are tipping backward. It takes time for our muscles and brain to understand, integrate, and incorporate a new position and have it feel normal.

A mirror may be helpful to give us visual feed back about our actual position while our muscles' sensory tools are readjusting.

On the other hand, when working with postural correction, mirrors sometimes should be avoided. Focus on the internal feeling of energy, relaxation, and freedom, rather than the actual appearance of structural change. The energy we feel and project influences how people see us as much as our precise structure. If you feel tall and erect, keep enhancing and projecting that feeling. Do not check in the mirror too often if you are discouraged to see there is still some hunch in your back. The hunched mental imprint does not need reinforcement. If you must look, look at the improvement.

Figure 10 - 6

The more vertically we align our body, the more disadvantaged gravity becomes, so we can relax.

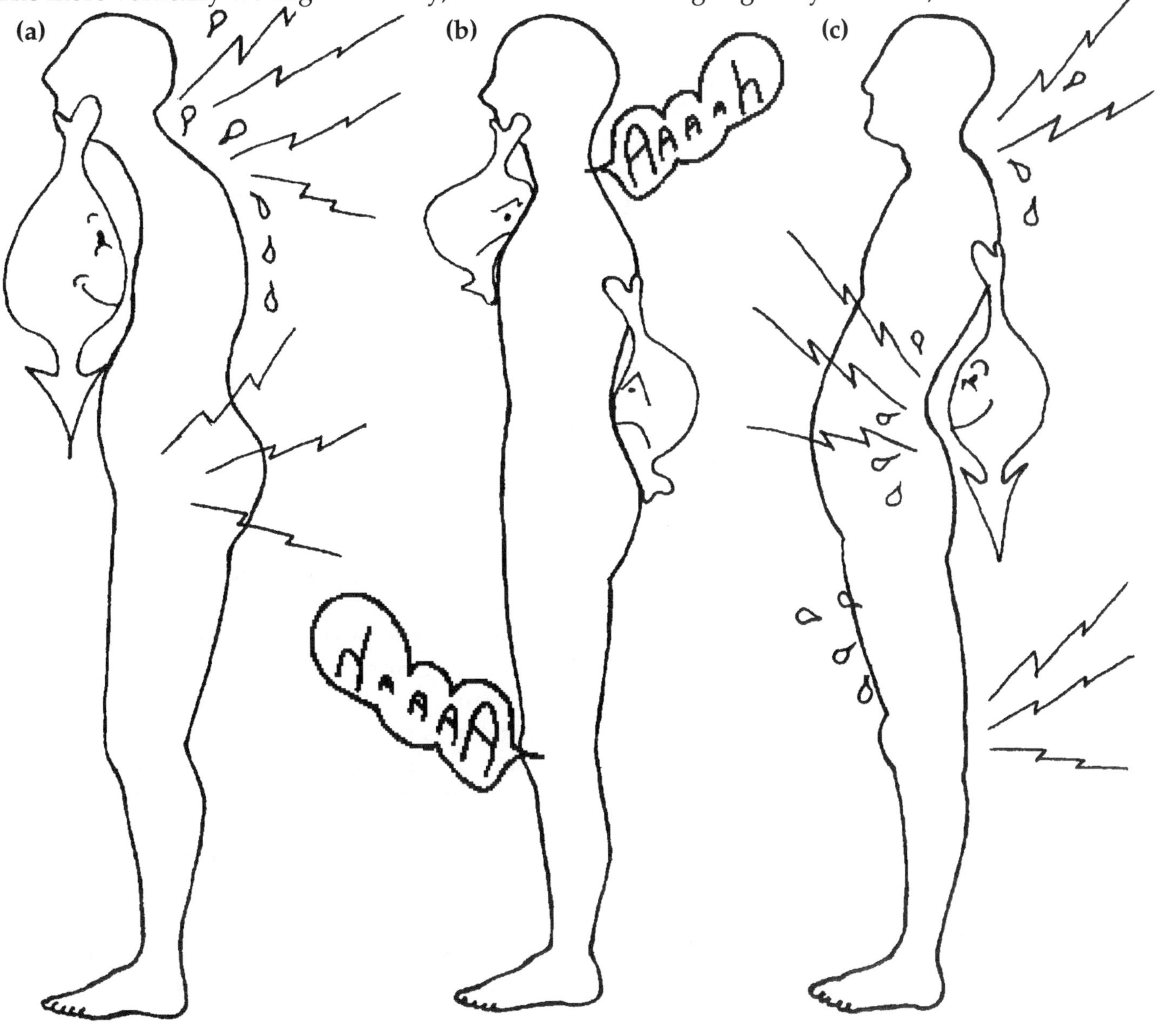

Three detrimental conditions evolve from any chronic, continuous muscle contraction, whether it holds a stressful posture, a protective stance, or excessive normal tone.

1. Since the muscular effort is a continuous contraction, without movement, the actual amount of work is disguised, but the task may be immense. Continuous contraction continuously consumes our energy, leaving us fatigued for no apparent reason.

When I am giving therapy sessions, my clients often question "Why am I sore there? I haven't done anything." Perhaps they have not done a vigorous, intentional activity, but those muscles continuously work overtime just protecting or holding a posture which gives too much advantage to gravity.

2. The continuous tug-of-war between gravity, traumas, and muscles is not only exhausting, but stressful on our joints. They are excessively compressed and distorted, taking weight off their optimum, central surface and bearing it on a thin edge. Ligaments are stretched, reducing joint stability. No wonder muscles go into spasm, backs ache, and joints crumble. This in itself is traumatic and painful and we are back at the beginning, increasing our protection.

3. What is even worse, the continuous contraction compresses nutritious blood flow through our muscle, so working conditions deteriorate. While muscles are prodded to keep working, they lack a good supply of nutrients and drown in their own wastes. The lunch truck visits less often and the garbage man skips. This disrupts homoestasis causing muscle damage and inflammation.

Inflammation itself produces pain. We are again back to the beginning of a vicious circle. Our unconscious response to pain is to increase muscle contraction. Thus, poor posture and excessive protection and tone become huge sources of chronic musculoskeletal problems as summarized below.

It is just this vicious circle which many therapies strive to break using emotional release or suppression, pain killers, muscle relaxation, circulatory stimulation, anti-inflammatory drugs, and so on. The therapist is limited and guided by what their license permits. Hands-on therapies work on many levels within this circuit, releasing emotions, reducing pain, relaxing muscles, stimulating circulation, balancing energy, and so forth.

Sequel of Continuous Muscle Contraction

emotional or physical trauma
↓
PAIN   holding stressful postures   excessive normal tone
↓   ↓   ↓
continuous muscle contraction
↓
reduced blood flow
↓
reduced nutrients and excess wastes
↓
muscle cell injury
↓
inflammation
↓
PAIN (→ back to continuous muscle contraction)

STRUCTURAL ORGANIZATION

Muscle Fibers

In the embryo, as skeletal muscle is first forming, primitive muscle cells reproduce at the site of a muscle-to-be (Fig. 10-7a). A large collection of cells is made. Then, in a linear orientation, adjacent cells fuse to form long slender structures called **muscle fibers** (Fig. 10-7b). Only the touching cell membranes disappear in the fusion. All the other cell structures remain, so a single muscle fiber has many nuclei and all the usual cell contents.

In chapter 4, on Muscle Tissue, all the characteristics attributed to skeletal muscle cells actually belong to muscle fibers.

Connective Tissue Wrappings

As muscle fibers form, they are individually wrapped in a thin, transparent sheet

Figure 10 - 7

The embryonic development of a muscle. **(a)** Primitive muscle cells reproduce at the sight of a muscle to be. **(b)** In a linear orientation, adjacent muscle cells fuse to form long, slender muscle fibers. **(c)** Connective tissue wraps each muscle fiber, extending beyond both ends to attach to the bones.

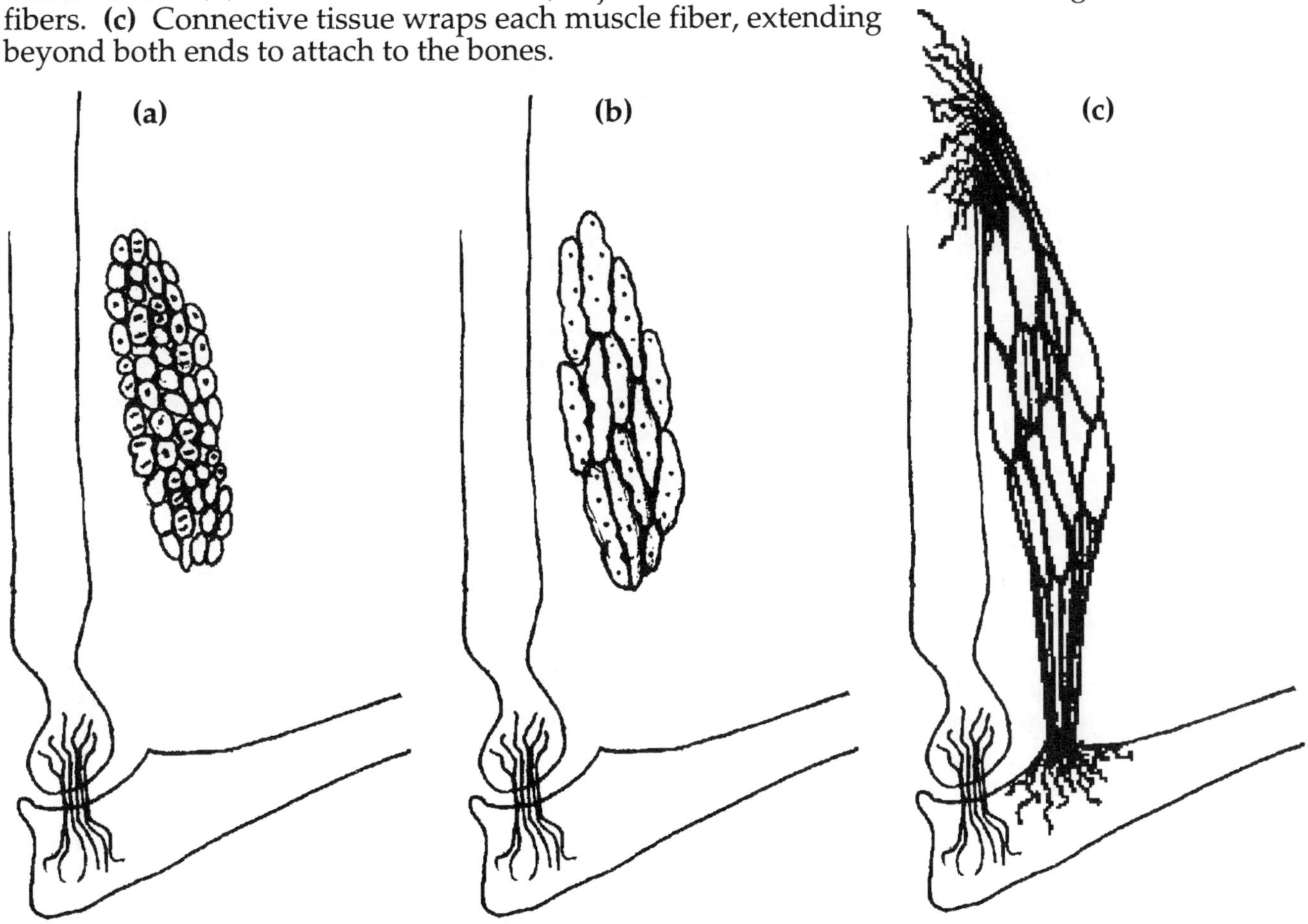

(d) A cross-section through a muscle, showing its fibers and three layers of connective tissue wrappings.

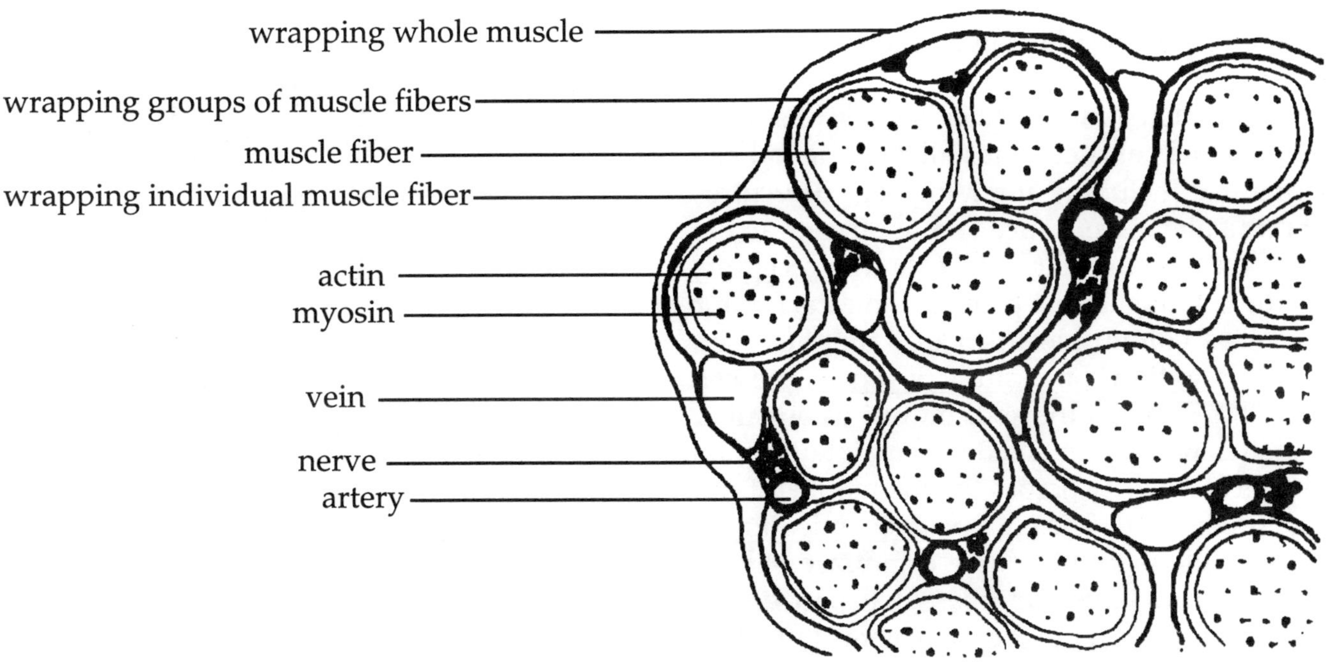

of connective tissue, like cellophane. A small bundle of these wrapped fibers is surrounded again by a second layer of connective tissue, and finally the whole collection is surrounded and supported by connective tissue (Fig. 10-7d). All these layers of connective tissue wrappings extend beyond both ends of our muscle fibers to weave into our periosteum and attach to our bones. This is how muscle tissue is attached to bone (Fig. 10-7c).

This conglomerate of muscle tissue and connective tissue wrappings is what we call **a muscle.** The area which contains muscle fiber and connective tissue is called the **belly** of our muscle. The connective tissue wrappings within the belly are called **fascia.** As the wrappings extend beyond the muscle fibers they are called the **tendon** of our muscle (Fig. 10-7d). 'Tendon' means stretching and refers to location rather than quality. A tendon is a dense, almost avascular cord of strong, non-stretchy collagen which stretches from muscle to bone, conveying the force of the muscle.

Only the muscle fibers of a muscle can contract, so it is only the belly of our muscle which shortens and bulges. Its pulling force passes through our tendon into our bone to create movement. Our tendon itself cannot create a force nor change its length, its only job is to hang on.

Contractile Elements

So what is the magic inside a muscle fiber which allows it to contract and relax? We have said that a muscle fiber is made of several merged cells and contains all the original cell structures including many nuclei. Structures of particular significance to muscle contraction are the following (Fig. 10-8): **1.** The receptors on the cell membrane provide a means for our nerve to communicate to our muscle fiber. **2.** The sodium-potassium pump in the cell membrane keeps the sodium out of our fiber and the potassium inside it, maintaining an electrical polarity across the membrane of our resting fiber. **3.** The endoplasmic reticulum in muscle fibers is regularly organized, like a lattice throughout our fiber. No part of our fiber is far from it. Our endoplasmic reticulum

> The process of skeletal muscle cell reproduction and fusion is completed at least by the age of two years old. At this point, the number of muscle fibers is permanently established. Fibers may be killed off but new ones are not made. There is no replacement of destroyed muscle tissue as discussed in chapter 4 under Muscle Tissue Repair. Indeed, in skeletal muscle there are no cells left to reproduce. All of the structure has been converted into fibers.

Figure 10 - 8

A muscle fiber contains all the original structures of the merged cells.

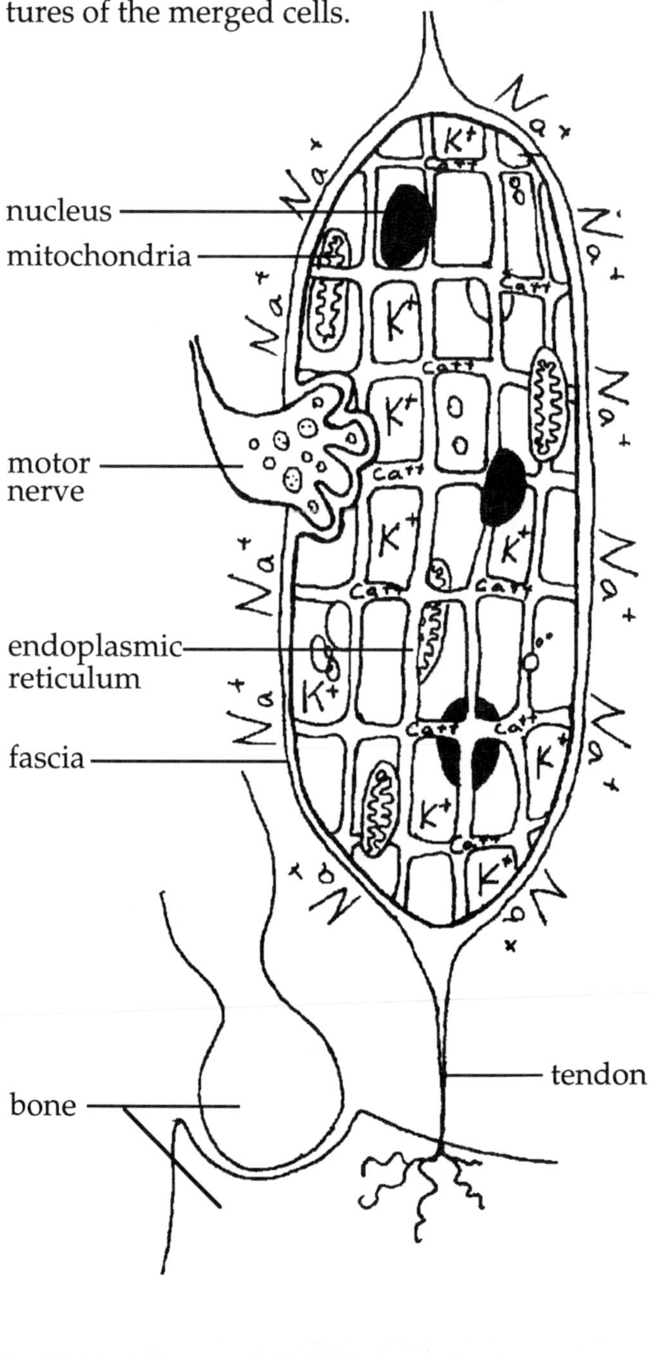

quickly communicates nerve instructions to every part of our fiber. In a resting muscle fiber, our endoplasmic reticulum also hordes most of our calcium inside its tubule, leaving only a skimpy share in the general fluid of our fiber. **4.** Large numbers of mitochondria are present to provide the large amounts of energy required to exert and release the force of contraction.

• An exclusive feature within our muscle fiber is a carefully organized **set of proteins,** strung amongst the other structures. The protein is what gives the striated appearance to skeletal muscle. When our muscle fiber is at rest, these proteins bond loosely with each other, leaving lots of space (Fig. 10-9a). Instructions from a motor nerve lead to a chemical sequence which attempts to tighten the protein bonds. Our proteins try to pull into each other until there is no empty space. This pulling creates the force of contraction and consumes energy. If our proteins are able to pull into each other, our fiber is shortened and movement occurs (Fig. 10-9b).

While the force of one protein gripping more closely to its neighbor is a very tiny force, when we add up the millions of proteins within a muscle, this summated force becomes significant and produces our muscular work.

The sequence leading to muscular work and back to rest is detailed in the tid bit box on page 210.

Figure 10 - 9

A longitudinal section through a muscle fiber and its insertion. **(a)** When the fiber is at rest, the proteins are loosely bonded to each other. **(b)** The muscle fiber contracts as the protein bonds pull into each other.

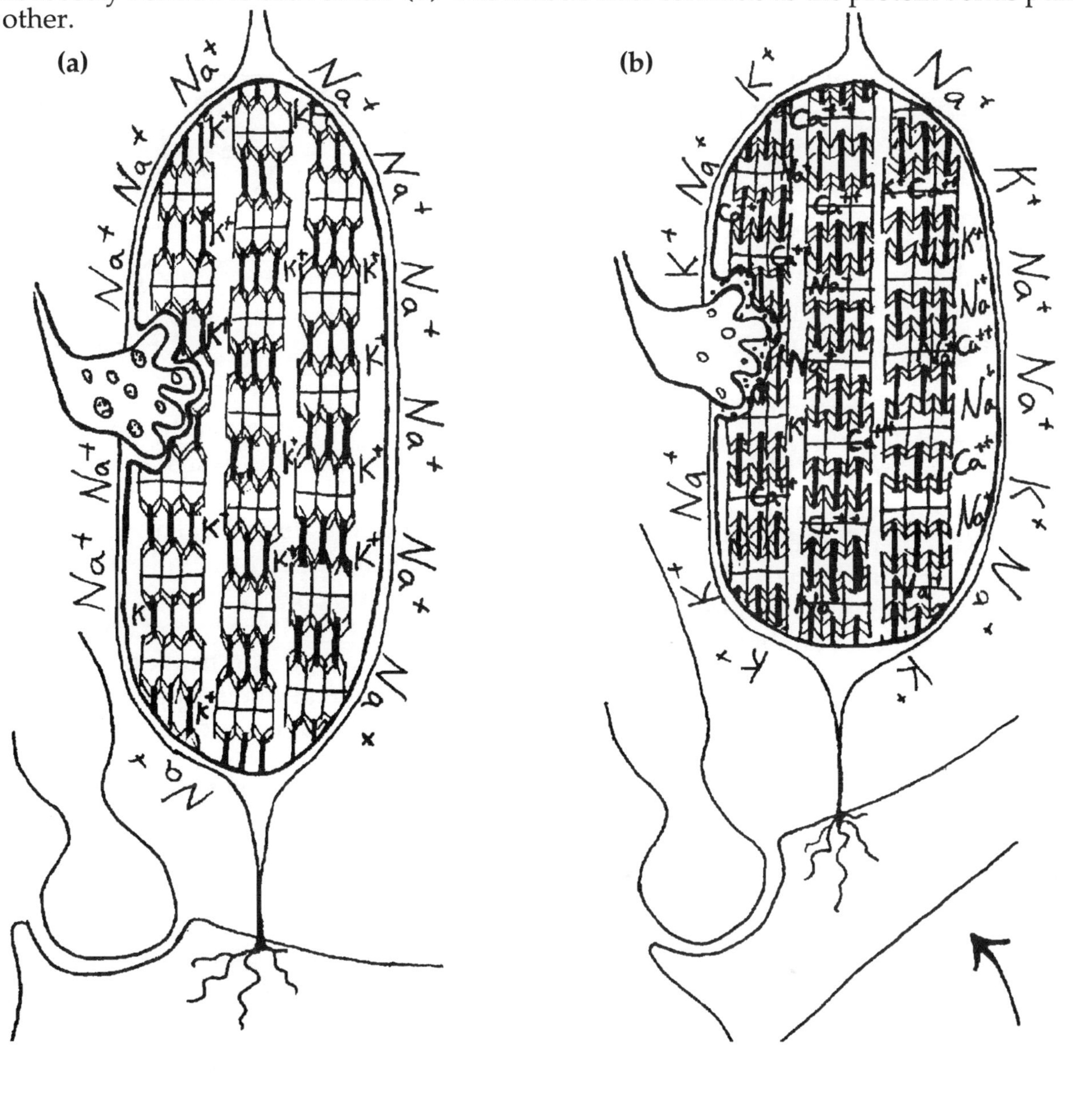

The appearance of calcium around our muscle proteins is essential for contraction, while the disappearance is essential for relaxation. Interestingly, the required increase to produce contraction is less than the required decrease to produce relaxation. For example, if it takes an increase of 50 calciums to contract our muscle, it takes a decrease of 75 calciums to relax it. This means that the first problem in calcium deficiency is being unable to relax, leaving our muscles stuck in contraction. There was enough calcium to contract, but not enough to relax. This is one of the causes of a muscle spasm. As calcium deficiency worsens, muscles may be unable to contract, leading to weakness or paralysis. Generally calcium concentration would exceed both minimums.

Our motor nerve passes its instructions to its muscle fiber in the form of chemicals released from the tip of the nerve cell tentacle. These chemicals touch special receptors on our muscle fiber's outer membrane (Fig. 10-10). This touching stops the action of the sodium-potassium pump, changing the electrolyte balance (Fig. 10-9b), and therefore, the electrical potential across our fiber's membrane. This change is conducted throughout the whole lattice of the endoplasmic reticulum so our whole fiber experiences it simultaneously. The electrical change brings the calcium out of our endoplasmic reticulum into the fluid surrounding our proteins (Fig. 10-9b). This increase in calcium, along with large amounts of energy, facilitate the pulling together of our proteins. The two proteins which pull together are called **actin** and **myosin** (Fig. 10-9b). There are several other proteins which assist the activity.

Our nerve chemicals only play tag. They touch our muscle receptor, say "You're it!", and then dash back into their nerve, or they are immediately broken down by enzymes. When our nerve chemicals no longer are touching our muscle receptor, our sodium-potassium pump turns back on to re-establish the resting electrolyte balance and electrical polarity. The calcium is sucked back into hiding. This drop in calcium and a small amount of energy breaks the tight bonding of our proteins and their force of contraction is gone (Fig. 10-9a).

Figure 10 - 10

Chemicals released from the tip of the nerve bond to the muscle fiber receptors.

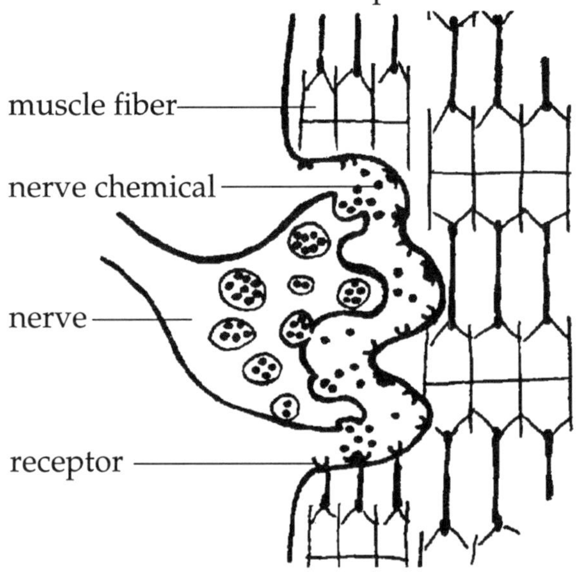

Origin versus Insertion

Of the two attachments of skeletal muscle, one always is designated as the **origin** and one as the **insertion.** There are rules which can help us decide which end of a specific muscle is which, but all these rules have exceptions.

In general, the **origin** of a muscle is its stabilized, stationary end. This usually is the inferior end in the trunk of our body and the proximal end in our appendages. We usually are supported on our feet or pelvis, moving the upper end of our body. The upper appendage usually is supported proximally with our hand thrust out to do something. But we can use muscles in either direction. We can hold onto a bar with our hand and chin up to it. In this case, our hand is stabilized and our proximal structures are moving toward it. Even when we use muscles in the opposite direction as

Figure 10 - 11

The origin **(1)** and the insertion **(2)** of two muscles, the pectoralis major **(A)** and the anterior deltiod **(B)**, indicating how the origins are more stationary, proximal, and broader than the insertion.

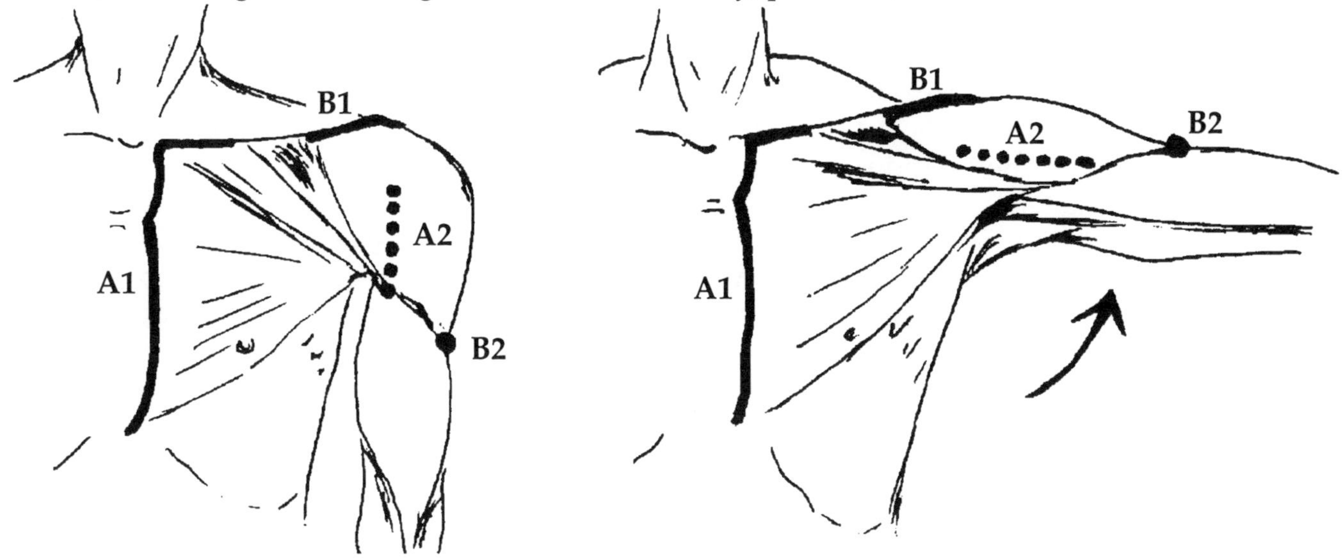

usual, its proximal end still is called its origin. The tendon of origin often is shorter and broader, dispersed over a greater area than the insertion tendon (Fig. 10-11).

Conversely, the **insertion** is the moving end of our muscle where the work is accomplished. The insertion is pulled toward the origin. The insertion usually is the superior end in the trunk of our body and the distal end in our appendages. Our tendons of insertion tend to be longer but more concentrated, attaching into a smaller area of bone, although it would contain the same amount of connective tissue as the origin.

Synergistic and Antagonistic Groupings

Most joints are crossed by a group of muscles which all do the same movement, although they each may have their best mechanical advantage at different degrees of the movement. All these muscles which can help each other are called **synergistic** muscles, 'syn' meaning together, and 'ergon' to work. Synergistic muscles usually cross over the same side of a joint (Fig. 10-12). For example, all the muscles which flex a joint would be on its ventral surface. At any point of the movement, the muscle with the best mechanical advantage is called the **prime mover.**

Joints also have a group of muscles which can undo the movement or create the

Figure 10 - 12

Muscles 1, 2 and 3 are synergistic to each other, all crossing ventral to the elbow. Muscles 4 and 5 also are synergistic to each other, both crossing dorsal to the elbow. The two groups of muscles, however, are antagonistic to each other, crossing over opposite sides of the elbow.

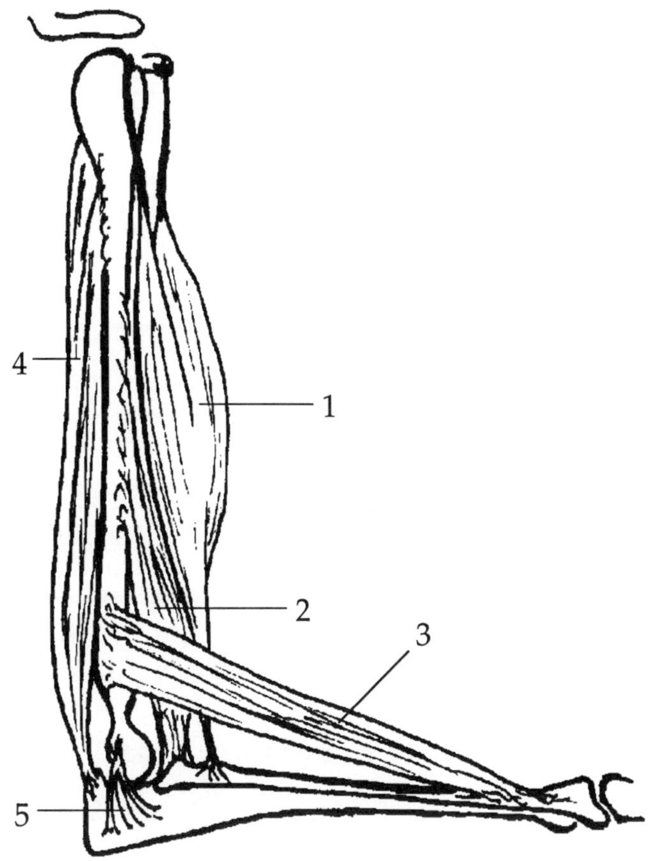

opposite movement, stretching the muscles which do the movement. These muscles are called **antagonistic** muscles, 'anti' meaning against, and 'ergon' to work. Antagonistic muscles usually cross over the opposite sides of a joint (Fig. 10-12). For example, muscles which flex a joint are on its ventral surface while their antagonistic extenders are on the joint's dorsal surface.

• When referring to muscles as synergistic or antagonistic, it is essential to say what joint and direction of movement are being considered. Two muscles may be synergistic in one movement but antagonistic in another. For example, the same muscles which help each other to flex the wrist, antagonize each other in adduction and abduction of the wrist (Fig. 10-13). The same muscles which help each other to extend the hip, antagonize each other in rotating the knee.

Figure 10 - 13

The same muscles which help each other to flex the wrist, oppose each other in adduction and abduction of the wrist.

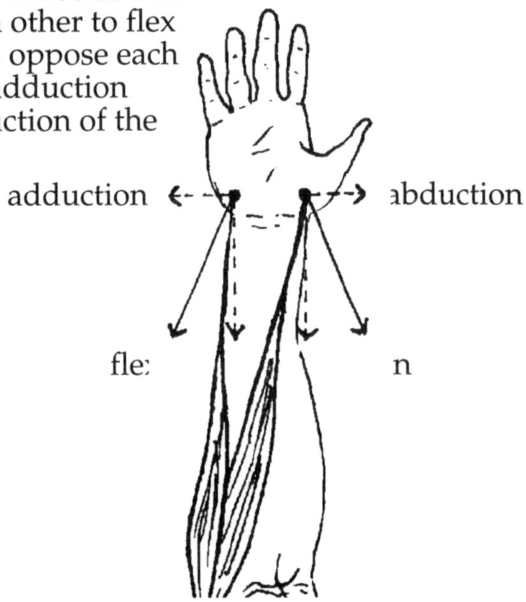

> The number of muscle fibers touched by one motor nerve cell varies from muscle to muscle, depending how finely or coarsely they contract. In delicate muscles, like those of the voice box, tongue, or hand, one motor nerve cell may touch as few as 15 muscle fibers. In course muscles, like those of the thigh, one nerve cell may touch as many as 300 muscle fibers.

NERVE INNERVATION

Motor Nerve

So far we have two bones, an intercepting, lubricated joint stabilized by ligaments, and skeletal muscles tied to our bones to produce the force of contraction. This still is a useless contraption. What else is essential before it can work? A **motor nerve.** Our skeletal muscles are dependent on our nervous system for instructions as to when and how much to contract. Nerves which relay these instructions are called motor nerves. Our muscles are our motors, they get us moving. Motor nerves enter muscles somewhere near the center of the muscle bellies (Fig. 10-14).

Skeletal muscles are called voluntary muscles in that the part of our brain that initiates movement is conscious and under our volition.

A motor nerve includes the tentacles of many nerve cells which can send instructions individually and independently of each other. Within our muscle, our nerve cell's tentacles spread out to touch and instruct individual muscle fibers (Fig. 10-14). The point of touching is called the **motor end plate** or the **neuromuscular junction** (Fig. 10-10).

Each motor nerve cell branches to touch many muscle fibers, but each muscle fiber is touched by only one nerve cell (Fig. 10-14). A single motor nerve cell with all the muscle fibers that it touches is called a **motor unit.**

> In a bodywork session, all our muscles may not need the same treatment. If one muscle is chronically tense and contracted, its antagonist will be chronically overstretched. While the first muscle needs to be relaxed, the antagonist needs to be stimulated. "Wake up! Remember, you can contract too!" In bodywork, this balance is seldom mentioned. Everyone talks about relaxing, lengthening, or sedating, but many bodyworks do not even have the capacity to stimulate, strengthen, contract, or tonify. In actuallity, they are both needed equally but in different places in our body.

Figure 10 - 14

A longitudinal section of a muscle showing its motor and sensory nerve connections.

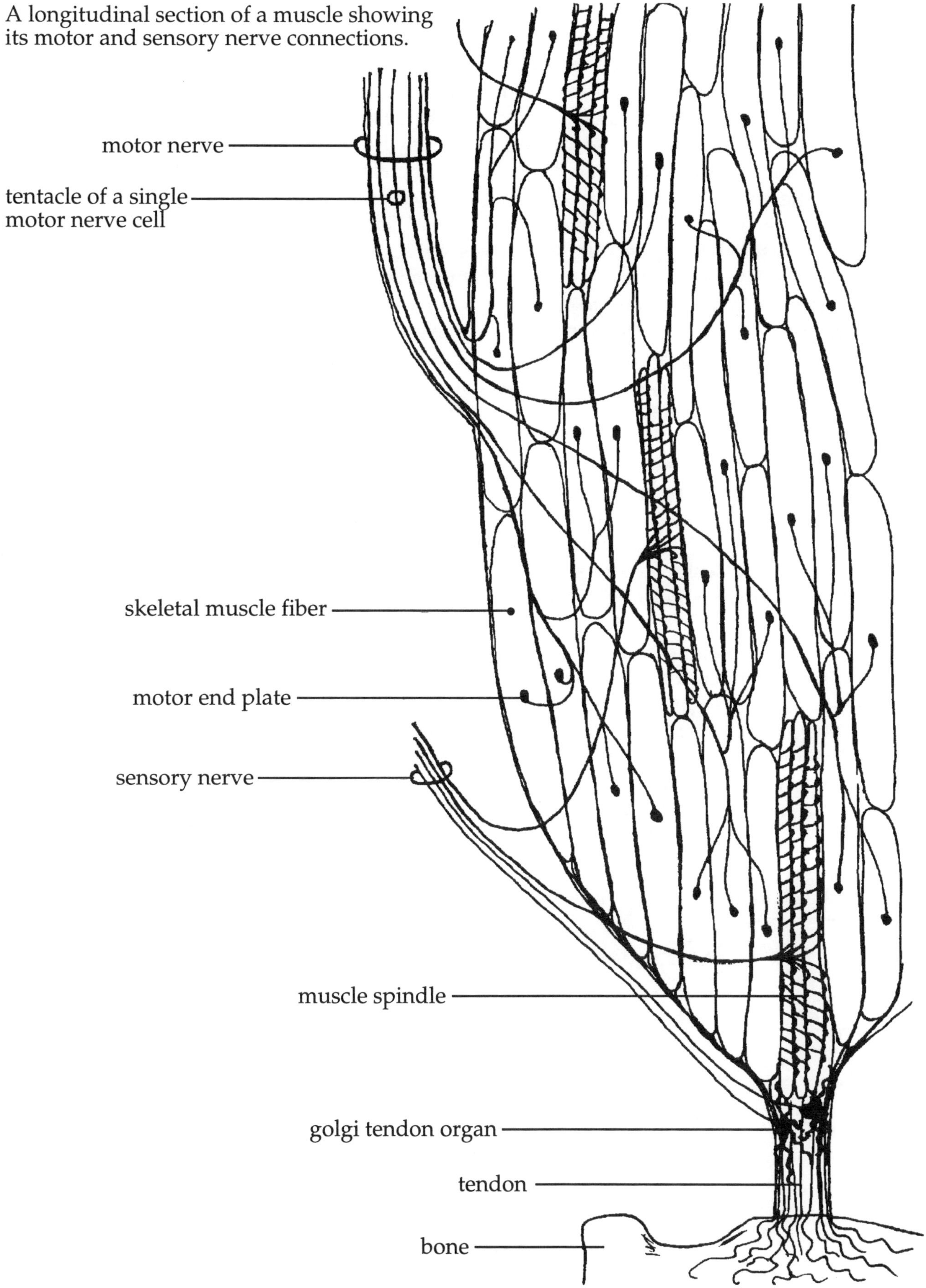

Sensory Nerves

At this point our muscular system still is useless. It can produce movements, but they are totally out of control. In order to be an effective instructor, our brain needs feedback from our muscles. Our brain needs to know if the number of motor units stimulated was adequate for the job. Our brain needs to know if some other process has stretched or compressed a muscle, changing our body's position. **Sensory nerves** monitor the activity of the muscle and keep the brain up to date on the latest changes. Sensory nerves measure changes in the force of contraction and changes in the length of the muscle belly as described in chapter 4, under Muscle Tissue.

• Our muscle's force of contraction is experienced as a pulling or tension in the tendon between our contracting muscle belly and our bone. Tension and tendon have the same Latin root meaning to stretch. Nerves which measure this pulling have sense receptors within our tendon. They are called our **golgi tendon organs** and look like octopuses with a central body and multiple radiating legs (Fig. 10-14). Increase or decrease in tension is registered by our golgi tendon organ, regardless of change in length or movement (Fig. 10-15).

The amount of tension which builds up in a tendon before movement occurs depends on the resistance of our bone to movement. In order to create movement, the tension must just overcome the resistance. While movement is in process, there is little change in tension.

The weight of a piece of paper provides minimal resistance to lifting. A few motor units create a small tension and the movement already is happening. Our golgi tendon organ will note this initial small change in tension and tell our brain.

The walls of most buildings have massive resistance to movement and usually human muscle cannot contract hard enough to overcome it. If our brain gives orders to try, our golgi tendon organ will register our muscles' massive attempt, even if there is no movement (Fig. 10-15). Our brain knows that our muscle is doing its best to follow orders.

• A muscle's change in length is experienced in the belly of the muscle. Nerves which measure the change in length begin with little contraptions called **muscle spindles.** They are like tiny springs coiled around specialized muscle fibers within the muscle belly (Fig. 10-14). The length of the spring changes when our muscle is stretched or compressed, both actively and passively, regardless of any force of contraction.

A muscle spindle does not note the absolute length of our muscle, but rather an interval of change in length. The interval in skeletal muscle is minute, so after the smallest movement, our brain gains knowledge of it.

Figure 10 - 15

A muscle may create a massive force but still effect no movement. The golgi tendon organ notifies the brain of the change in tension while the muscle spindle experiences no change.

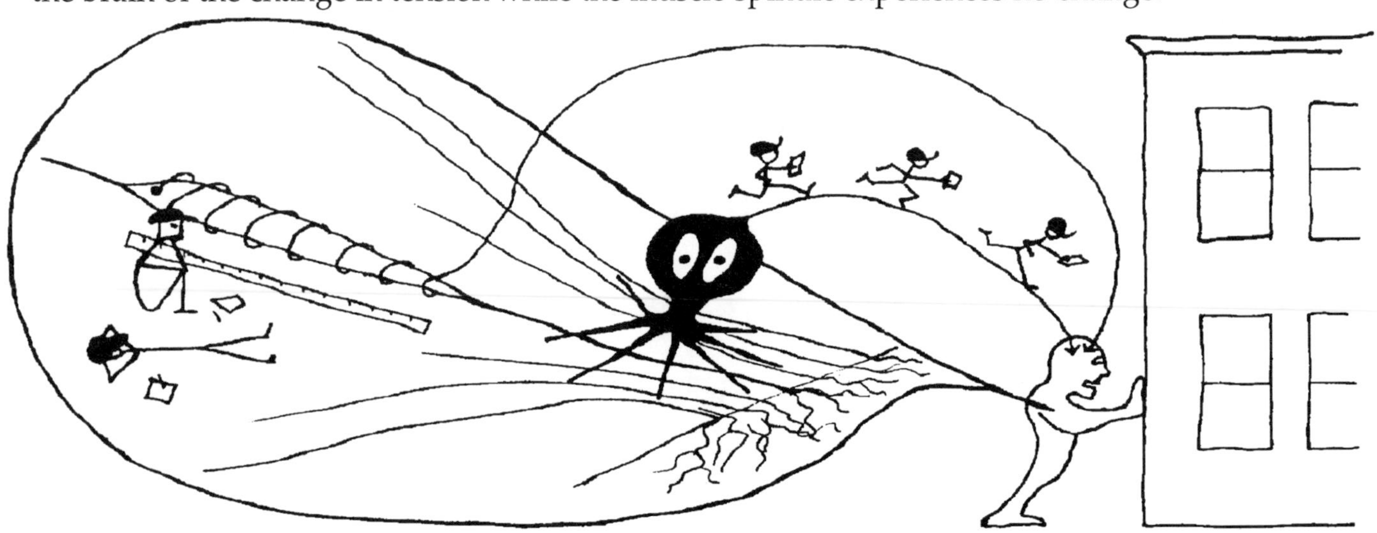

- Sensory information is essential for our brain to stimulate forces appropriate to a given task. When our brain has given orders to move something, it waits for feedback that its orders were followed, that the task was completed. If the sensory feedback does not match the given orders, our brain knows that the job was not completed.

Initially, when we see a rock that we want to lift, our brain has no idea how heavy it is, or how many motor units to stimulate to lift it. Our brain takes an experienced guess and waits for feedback. If the stimulation is enough, it will get the report of increasing tension and muscle shortening (Fig. 10-16b).

If the stimulation is not enough, the tension will increase so that our brain knows the muscle is trying, but no shortening occurs (Fig. 10-16a). Our brain knows that it needs to stimulate more motor units to get enough force to lift the rock. It will stimulate a few more, and wait for feedback on what happens.

Figure 10 - 16

(a) In this case, a motor message stimulates increased tension in the muscle but is not sufficient to produce movement. **(b)** An increased quantity of motor messages further increases the tension and is sufficient to effect movement.

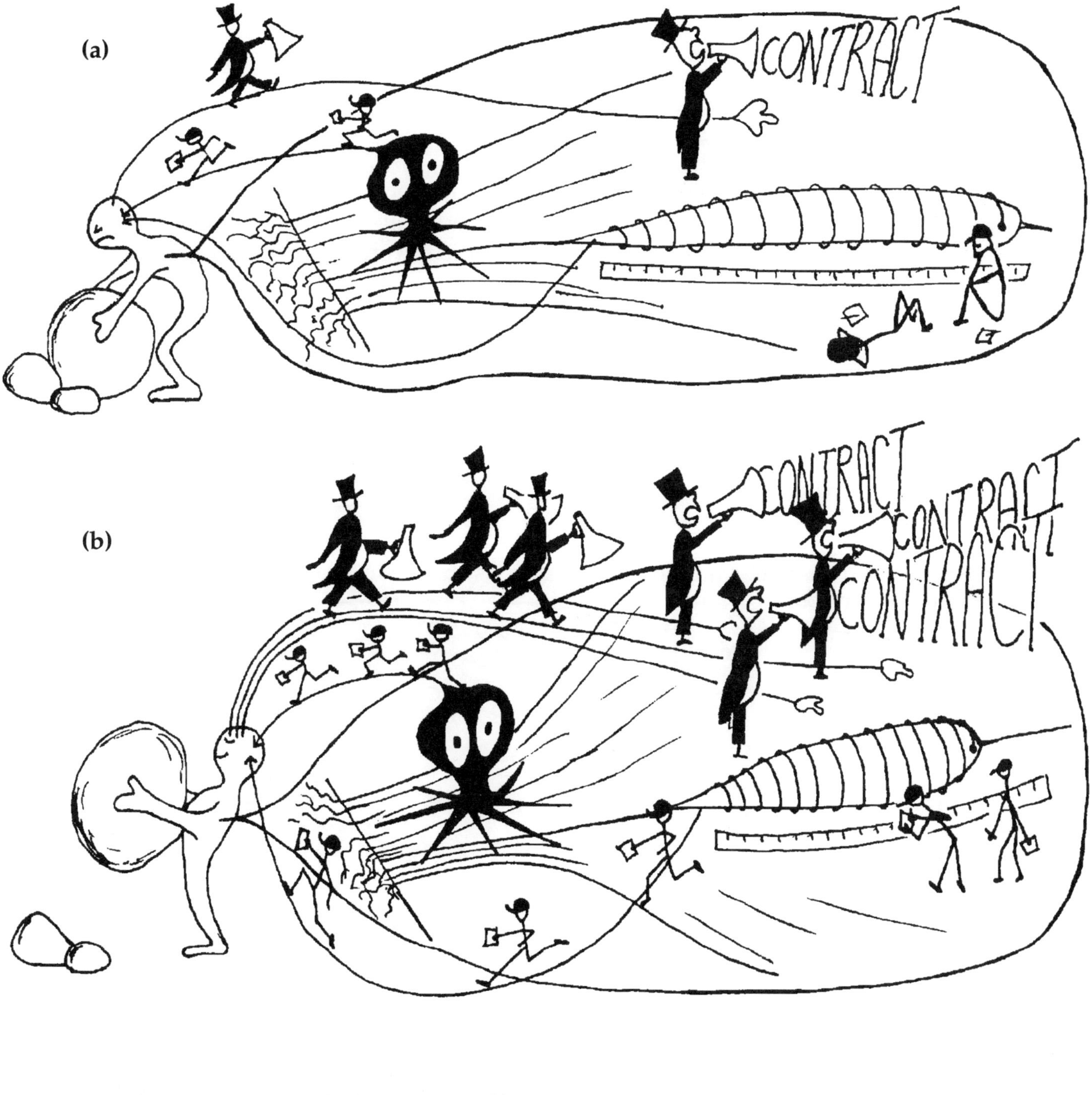

All this testing happens so quickly that it seems to us that we just grab the rock and lift it up. After years of experience in lifting rocks, our brain gets better and better at guessing how much force is required to lift any one rock and less testing is necessary.

We can be tricked when we are over-confident. For example, a piece of foam has been painted to look like a rock. As we go to lift it, our brain gives instructions as if it were as heavy as a rock that size. Of course, foam does not need that much force and we go flying up with the excessive effort. However, it should not throw us right over. As we fly up, our brain is surprised by the extent of movement from so little tension. Our brain sees its error and adjusts its instructions to fit the actual weight and move it more appropriately.

- When our body is moved passively, we feel it because of the sensory feedback to our brain. If something or someone else moves us, changing the length of muscles, our muscle spindles tattle (Fig. 10-17). Our brain knows immediately, on a conscious and unconscious level, that we have been moved.

Our brain may not like movement without its permission and may pull our body back to the original position. This explains why the leg jumps in response to the doctor's reflex hammer and the occasional jerk just as we fall asleep as discussed in chapter 4 under Muscle Tissue.

- Sensory feedback is essential for our brain to know how to hold our body in a specific position. When we decide to stand up, our brain instructs all our muscles how much to contract in order to stand us up and hold us there. If we stretch an arm to point at something, gravity pulls down on that arm and starts to tip us forward. As we tip forward, the muscles in the back of our calves are stretched. The muscle spindles immediately report this to our brain (Fig. 10-18a) so it knows how to readjust the muscles' instructions to accommodate for our pointing arm (Fig. 10-18b).
- The sensory feedback to our brain tells us where we are in space. Our brain translates from any specific change in muscle length into a new, specific body positions.

> Tricking our brain with unusual combinations of sensory feedback is the basis for some postural correction techniques such as proprioceptive neuromuscular facilitation or PNF stretches. Muscles are actively pulled into their longest position and then strongly contracted against a force that prevents movement. The abundant feedback from the golgi tendon organs tricks our brain into releasing underlying tensions which may have been limiting the length of the muscles. Releasing the chronic holding pattern allows the muscle to further elongate.

Figure 10 - 17

When we are moved passively, the length of muscles is changed without a motor message or a change in muscle tension.

Figure 10 - 18

(a) As gravity tips us from a given position, muscles are stretched without motor messages or a change in tension. **(b)** The brain changes its instructions to the muscles to bring us back to the original given position.

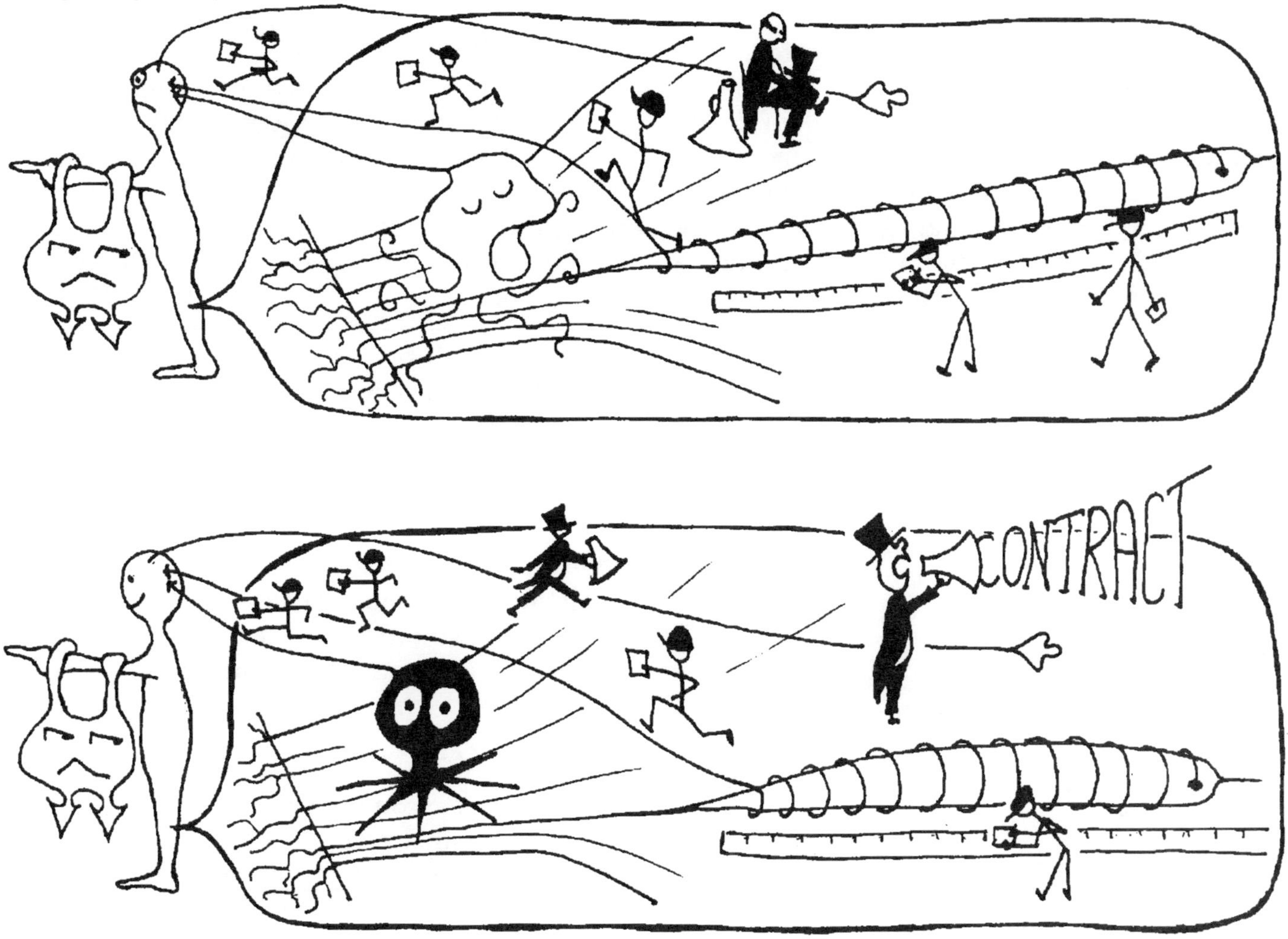

• Our brain's memory of sensory feedback is relatively short. If we have not changed the length or tension in a single muscle for 20 to 30 minutes, our brain loses the memory of the last position. After sitting very still or waking up without moving, we may have no sense of our position in space or where our arms are relative to our body. We may feel like we are floating or slowly tumbling through space. One small contraction, however, will reorient our brain so that it can tell us where we are. Not who we are but where.

PRINCIPLES OF MUSCLE ACTION

All-or-None Theory

The motor unit is the smallest functional unit in our muscle. When nerve instructions reach a motor unit, all its muscle fibers contract as much as possible or they do not contract at all. This formula is called the **all-or-none theory of muscle contraction,** however, it actually applies only to individual motor units. How does one muscle then, exert the different forces of contraction or different amounts of shortening, as we continually experience? How can we now lift a rock above our head, and later lift a paper to eye level?

Force of Contraction versus Distance Shortened

• When discussing the activity of a muscle it is necessary to distinguish between the force of contraction and the quantity of shortening. As the proteins of muscles pull into each other, they create a force of contraction.

They may (Fig. 10-16b) or may not (Fig. 10-16a) be successful in contracting or shortening to create movement. Shortening only occurs when there is a force of contraction, but a force of contraction does not always create shortening. It depends on the opposing forces or resistance to movement.

- Sometimes the intention of the force of contraction is not to produce movement, but only to create enough force to support or stabilize, to oppose or hold against movement (Fig. 10-3). For example, when we carry a bucket of water, gravity on the water exerts a downward pull on our arm. This could separate our glenohumeral joint or depress our scapula. Muscles contract to hold the bones together against the force of gravity. This muscular effort is just enough to overcome the force of gravity, without actually elevating the scapula, so it does not create movement.
- The activity or accomplishment of the muscle depends on: **1.** the number of motor units stimulated at any one instant, and **2.** their orientation to each other.

Individual muscle fibers and motor units do not go the entire length of our muscle belly (Fig. 10-14). Motor units are imprecisely organized in two directions. They lie in a series either side by side across the diameter of a muscle, or end to end along the length of a muscle. When motor units lying side by side are simultaneously stimulated, they summate the force of contraction. The more units stimulated, the stronger the force. It takes more to lift a rock, and less to lift a paper (Fig. 10-19a). When motor units lying end to end are simultaneously stimulated, they summate the amount of shortening. The more units stimulated, the greater the distance moved without changing the force (Fig. 10-19b). It takes more to lift over the head, and less to lift to eye level.

It follows that broad muscles have more motor units lying side by side and therefore have greater strength potential, while long muscles have more motor units lying end to end and therefore have more shortening potential, producing greater range of motion.

- When muscles are creating a force of contraction, they are working and consuming energy whether or not movement occurs (Fig. 10-3). A muscle can change its force of contraction or its amount of shortening independently or simultaneously with other muscles.

Fleeting versus Sustained Contraction

- A single nerve message to a motor unit causes only a fleeting force of contraction. As soon as the nerve stimulation stops, the force of contraction stops. Muscles do not work when their boss turns its back. The duration of the effect of one nerve message lasts an average of one tenth of a second (Fig. 10-20a).

> The all-or-none theory of muscle contraction actually should be called the all-or-none theory of motor unit contraction. It says that when nerve instructions to contract reach a motor unit, that whole motor unit contracts to its fullest potential or it does not contract at all. This is one motor unit out of a whole muscle full of motor units. Each motor unit acts independently according to its own nerve message. Some may contract while others stay relaxed.

> There is one major exception to the all-or-none theory of muscle contraction. When a single motor unit is restimulated while still in the process of releasing a contraction, before it is completely relaxed, the motor unit will contract further with each stimulation -- to a point -- then its real full potential is reached (Fig. 10-20). This progressive potential is called **treppe.**
>
> Treppe is part of the purpose of the warmup before an athletic event or the benefit of constant movement through an event like boxing. It brings or keeps the muscles closer to their absolute fullest potential.

> The fullest potential of a motor unit or muscle contraction depends on many variables such as their acid-alkaline balance, their calcium, sodium, or potassium concentration, and their energy availability.

Figure 10 - 19

(a) As more motor units lying end to end are stimulated, the distance moved increases.
(b) As more motor units lying side by side are stimulated, the force of contraction increases.

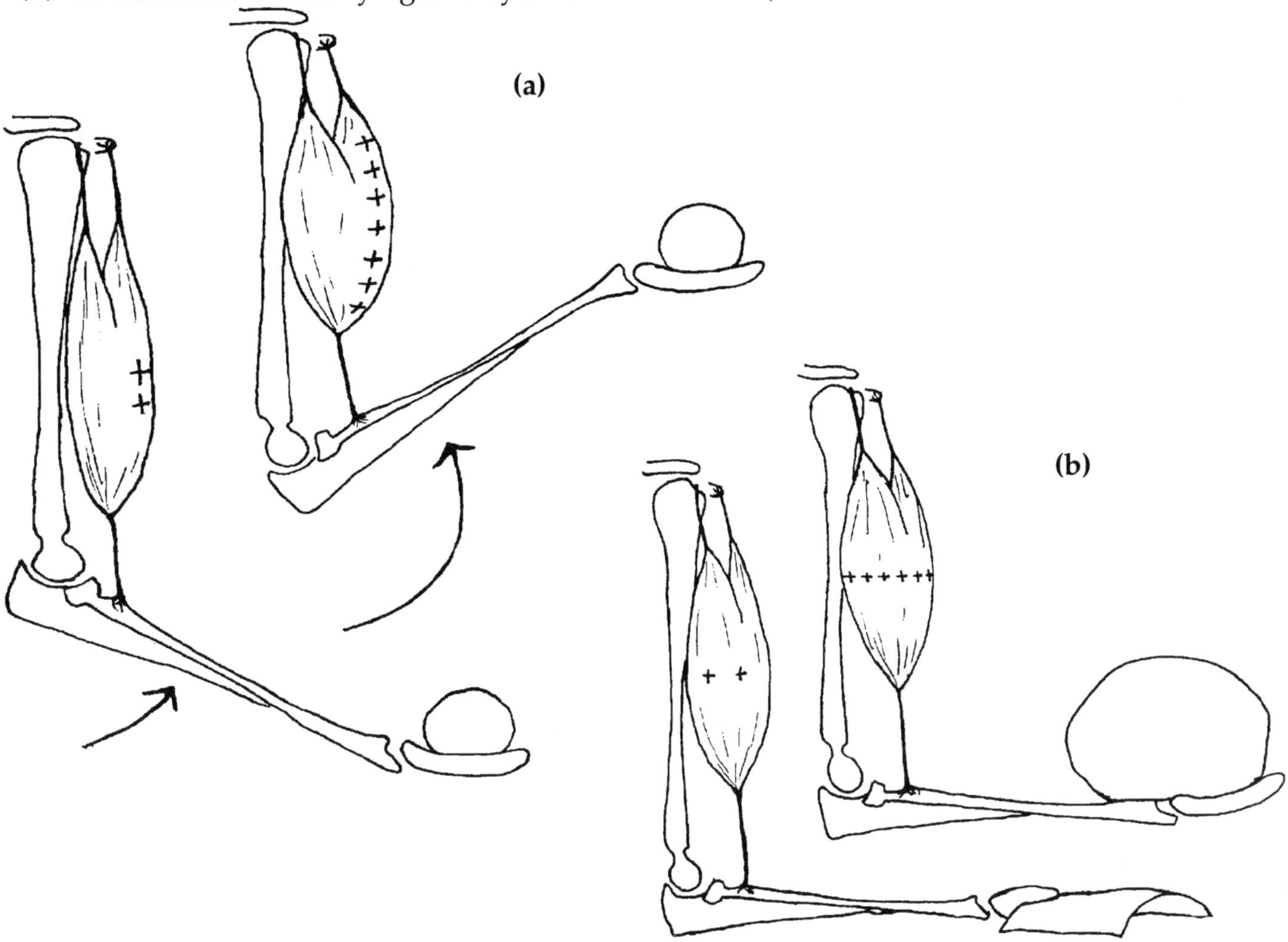

Figure 10 - 20

The arrows along the horizontal axis indicate the event of a nerve stimulation. **(a)** The height of the arch indicates the fullest potential of a motor unit contraction with a single nerve stimulation. **(b)** The upward scoops indicate the fullest potential of a motor unit contraction after a series of rapid nerve stimulations. This step-up in contraction potential is called Treppe.

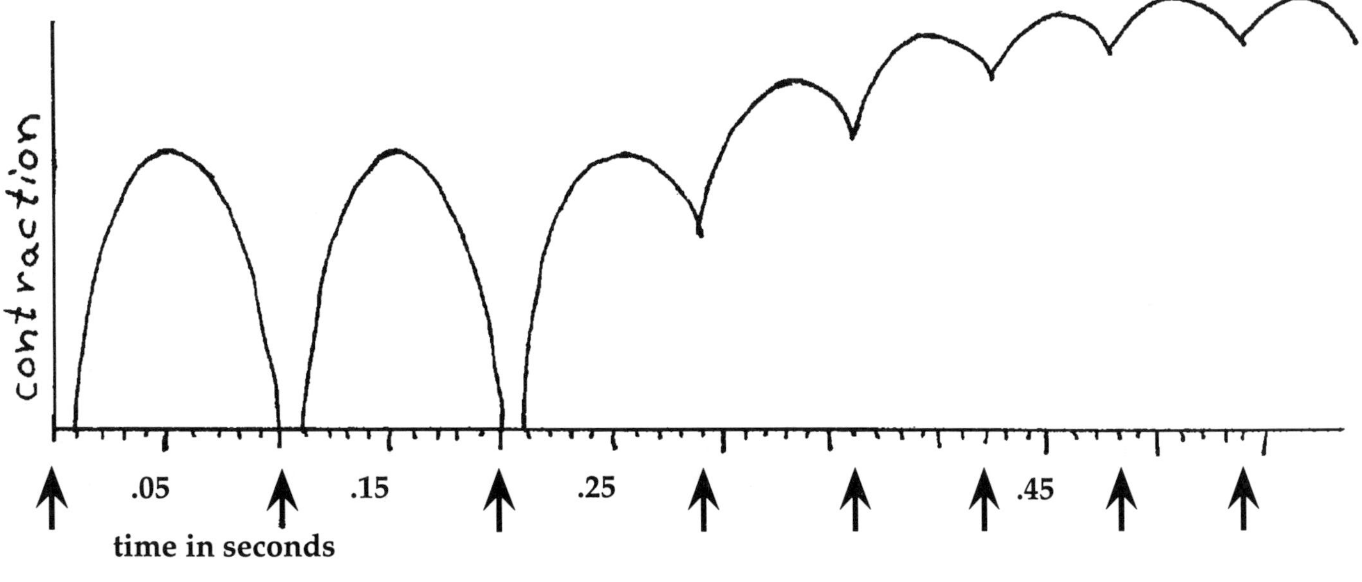

To keep a muscle working, such as through a sweep of movement or holding up against the continuous force of gravity, requires a continuous stream of instructions less than one tenth of a second apart.

Generally our motor units take turns. Our nerve messages rotate, so that every one-tenth of a second a different group of motor units does the work. One motor unit never is left to do all the work. This reduces fatigue since some motor units are resting even while the work continues. The lighter the object, the fewer motor units work at once, and the more time for rest between turns. We can hold up a piece of paper all day. We may be fatigued by boredom but not by the weight of the paper. Heavier objects use more motor units so there is less time for rest. Fatigue comes quicker. We may be able to carry a 90-pound bag of cement, but it better not have to go far.

Extent of Contraction

There is a limit to how close muscle proteins can pull into each other. When all the space is filled, it is the end. This sets the limit on how far a muscle can contract and how much it can bend its joint. A fully contracted muscle is roughly two-thirds of its resting length.

Normal Tone

- When a single motor unit is resting, there is no nerve message instructing it at that moment. When a whole muscle is resting, no matter how relaxed our muscle is, it always has a little trickle of nerve messages keeping one or two motor units contracted. This means that our muscles always have a little tone or tension, keeping them ready for action. This tension is present even in a state of deep meditation or sleep, as long as our neuromuscular system is intact and healthy.

This constant underlying tension is called **normal tone.** Normal tone lies somewhere on a long continuum from a completely tense or **spastic** muscle to a **flaccid** muscle with absolutely no tension. Both ends of this continuum are pathological. A healthy state is somewhere in between, more toward the relaxed end.

- Normal tone is controlled by unconscious parts of our nervous system in response to a host of memories, emotions, judgements, and sensory inputs. The effect is different from person to person and fluctuates from time to time, depending on the conditions.

After five cups of coffee to get our day going or a chaotic morning at work, our nervous system is over stimulated, conducting too many messages. Pain, stress, anxiety, noise, caffeine, and nicotine are common factors which increase the excitability of our nervous system and the number of messages sent to our muscle. These factors shift normal tone toward the spastic end. Our muscles generally become tighter and we feel more rigid, easily startled, jumpy, strained, and fatigued. Increased normal tone consumes muscle energy, compresses their blood supply, and accumulates wastes in our muscles. We are tired even when we have apparently done nothing.

After a quiet walk or meditation our nervous system becomes less excitable. Peace, quiet, comfort, and support stabilize our nervous system. Fewer messages are sent to our muscles so their tension is reduced. We are more relaxed, softer, and open. More blood flows to nourish all our tissues and organs.

- Normal tone, like muscle work, is rotated among all the motor units of our muscles. Each motor unit takes its turn and then rests while others produce the ongoing tension.

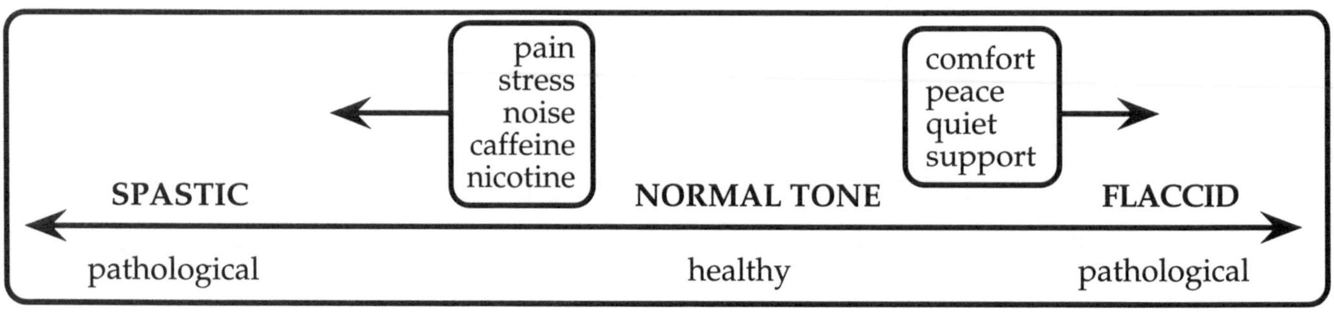

Limits of Movement

While the force of muscular contraction initiates movement, three things may limit, decelerate, and stop movement. **1.** The shape of our bones allows only a particular direction and degree of movement. **2.** The placement and length of ligaments limit and stop our movement. And **3.** the contraction of antagonistic muscles limits, decelerates, and stops movement. As muscles are being stretched, they receive increased motor nerve messages to increase their tone and resist being overstretched. This tone slows down our movement. When the force in our stretching muscles equals the force in our contracting muscles, our movement stops.

Muscles with increased normal tone will stop our movement more quickly, while reducing our normal tone may significantly improve our range of motion.

Isotonic and Isometric Movement

Many exercises are described as being either isotonic or isometric.

- 'Iso' means no change. 'Tonic' refers to the tone or tension created by a force of contraction. So an **isotonic** exercise produces no change in the force of contraction during the exercise. A 3-pound weight is always pulled down by gravity with a force of 3 pounds. To lift the weight always requires a 3-pound force of contraction against gravity. Whether we lift the weight higher or lower, it always requires 3 pounds. To move the weight, the force of contraction in a muscle does not have to change, only the amount of shortening.

In theory this is possible. In practice, as we move, the mechanical advantage of our structure is changing. Our force of contraction would have to change to accommodate the changing mechanical advantage of different positions. So, strictly speaking, if movement happens, there is no pure isotonic exercise.

> The design of some exercise equipment incorporates how movement changes the mechanical advantage. It incorporates an automatic change in the resistance to movement, so the force required from the muscle is as close to isotonic as possible.

In true isotonic exercises, our golgi tendon organs are not stimulated since there is no change in the tension on our tendon.

- 'Metric' comes from meter which is a measure of length. An **isometric** exercise, then, produces no change in length. When pressing our hands with equal force against each other, we can press harder and harder, but as long as the forces are equal, there is no movement. Also, when we push against something that resists movement with a greater force than we can produce, we can push harder and harder, but produce no movement. This is an isometric exercise. If we decide to push over a sturdy building with our bare hands and brute strength, it probably will be an isometric exercise.

In isometric exercises, our muscle spindles are not stimulated, but any change in tension is recorded by our golgi tendon organ (Fig. 10-15).

Unconscious Control

We have said that the main control of skeletal muscle is located in a conscious part of our brain and that skeletal muscles only work when we make a decision to work them. Our first intention to move is conscious, but many unconscious or crudely conscious centers of our brain help to produce the graceful movements that we wish to perform. We decide to point to the picture, but this simple task solicits participation from a vast collection of centers and circuits in our central nervous system.

Our unconscious or crudely conscious centers coordinate our force of contraction between our synergistic and antagonistic muscles, so movements happen smoothly. These centers adjust the tension in muscle all over our body to keep our body balanced during movement (Fig. 10-18b). They maintain normal tone. They hold assumed positions until orders for change are given. This includes the host of protective stances we find ourselves in. These centers may take over habitual activities like breathing or walking so we can do other things like chew gum or engage in a complex

Figure 10 - 21
(a) When muscles are physically challenged, they add more protein to the breadth of their existing fibers. **(b)** When muscles have nothing to do, they lose protein from the breadth of their existing fibers.

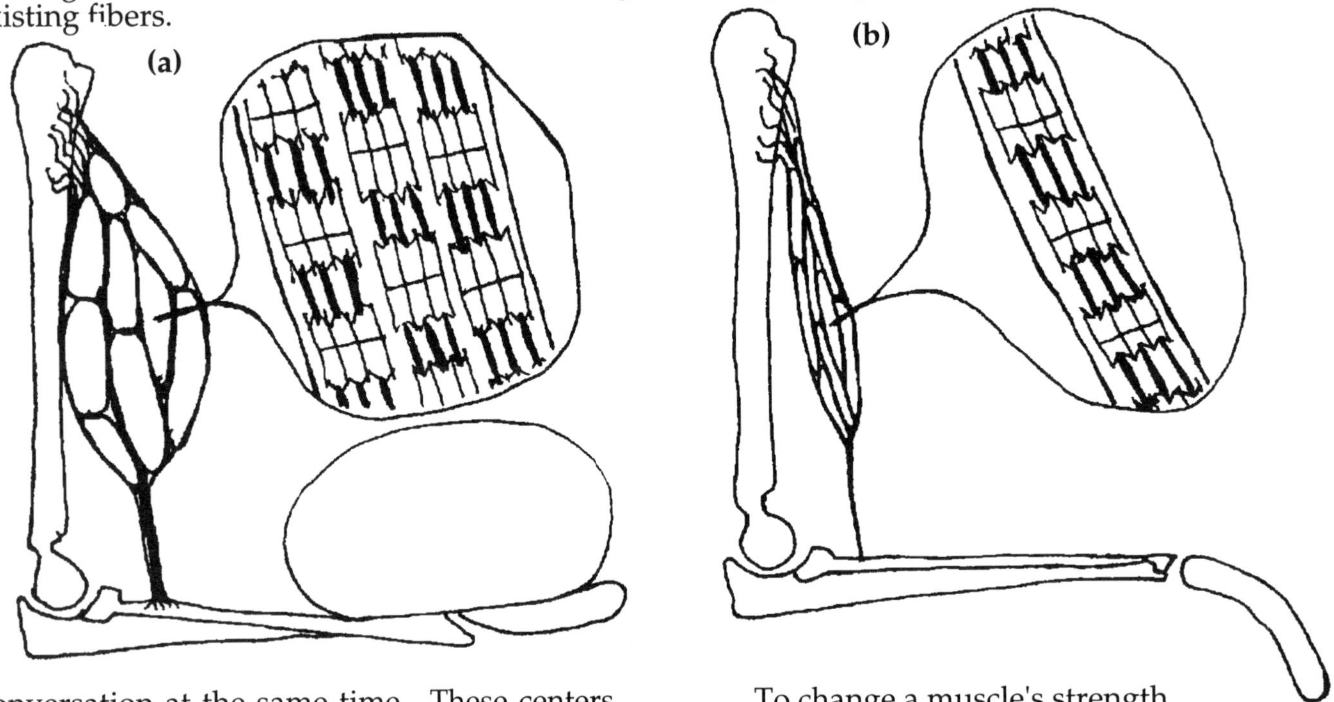

conversation at the same time. These centers work behind the scene, doing all the things required to support an intentional movement, things that we do not consciously consider.

While much of the sensory feedback from our muscles reaches conscious parts of our brain so that we are aware of what is happening, even more sensory feedback goes to our unconscious or crudely conscious centers so that they can adjust everything precisely.

QUALITIES OF MUSCLE ACTION

Muscles have three qualities: **1.** the strength of their force of contraction, **2.** their endurance to keep contracting over a period of time, and **3.** their flexibility, allowing change in their length. We will discuss each of these qualities and the different aspects of our muscle which contribute to them.

Strength

The strength of a muscle is how hard it can pull at any moment in time. What produces the force of contraction in a muscle? The pulling together of the proteins inside its muscle fibers. The sum of all the proteins pulling together is the strength capacity of that muscle.

To change a muscle's strength capacity requires a change in the quantity of protein. More protein in a muscle means that there are more protein bonds and therefore, more pulling force. Less protein means less bonds and less pulling force.

To add protein, each muscle fiber adds to what it already has. This is called muscular **hypertrophy.** Protein is added to the breadth of the fiber, making it more bulgy, but not longer (Fig. 10-21a). The strength is increased, but the range of motion is not changed.

Similarly, as muscles lose protein, they do not lose fibers but rather diminish the amount of protein in each fiber. This is called muscular **atrophy.** The muscle becomes less bulgy and less strong (Fig. 10-21b).

• Three factors influence the amount of protein inside a muscle fiber: physical activity, hormones, and diet. These three factors are the same factors which influence the amount of protein in connective tissue. They affect muscle the same as they affect connective tissue.

When muscles are worked heavily, they whimper "This is too much work! We need help!" The heavy work stimulates them to hypertrophy. This is what happens when we work out and get bigger muscles.

If muscles are not challenged with work, they will diminish their protein (Fig. 10-22). Muscles are not pack rats, carrying around everything they once used, in spite of no present use. Muscles quickly shrug off, "Who needs all this stuff? It hasn't been used in ages." Just like connective tissue, our muscles atrophy to the point of use. We do not save our body by resting it. We keep it by working it regularly.

Three hormones affect the quantity of protein in muscle fibers. Growth hormone and sex hormones stimulate hypertrophy, building more protein in our muscle. Cortisone stimulates atrophy, breaking down the protein out of our muscles.

In order to build protein into our muscles it is necessary to take adequate protein in our diet. Body builders need more protein to build and maintain a larger muscle mass. People deficient in protein or carbohydrate may break down their own muscle protein in order to maintain other body requirements.

Figure 10 - 22

We do not save the body by resting it. When muscles are not challenged, they will diminish their protein.

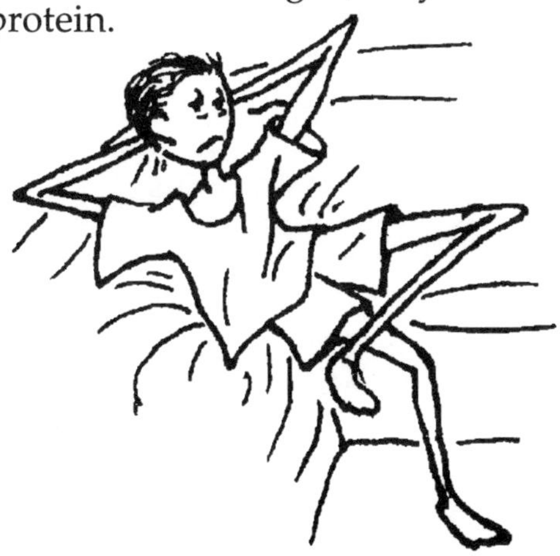

The **anabolic steroids** sometimes used by athletes are similar to sex hormones (many are derived from the male sex hormone, testosterone), stimulating muscles to build protein, thus increasing their size and strength. When taking any substance, however, it is important to remember that while it has one desired effect, it also has other effects which may not all be favorable in our body. Taking anabolic steroids may have many side effects including those of taking The Pill for contraception.

Eating increased protein does not, in itself, build bigger muscles. Our muscles still must be exercised to feel their need for more strength. Body builders cannot just gorge themselves with protein. They must be faithful to their rigorous workout routine.

Endurance

The endurance of a muscle determines how long the muscle can continue to perform. Long endurance requires an ongoing energy supply for continued pulling of the proteins in our muscle fiber. Our most efficient energy is suppled by using oxygen to break apart glucose, releasing its chemical energy. The enzymes for this process are contained in our mitochondria. Muscle fibers can increase their endurance by increasing any or all the elements that contribute to energy availability. This includes oxygen and glucose storage, the number of mitochondria, and the efficiency of the mitochondrial enzymes. It also includes the muscle's blood supply to replace consumed products and eliminate wastes.

The elements of endurance take up some space but not much, so the overall size of our muscle may not change significantly as its endurance changes. Continuous, repetitive demands on a muscle will stimulate it to increase endurance. When demands on a muscle are reduced, it will reduce its endurance.

Flexibility

To be flexible is to be pliable and yielding. A flexible muscle allows its length to be changed by different forces. It yields to the stretching of antagonistic muscles or other forces. It yields to the compression of its own contraction or the passive compression of synergistic muscles or other forces. Muscle flexibility influences the range of motion of the joints it crosses. The more flexible a muscle is, the further the joint can move.

The flexibility of a muscle depends on its normal tone and the quality of its fascia.

- Increased normal tone resists passive

stretching, reducing our muscle's flexibility.

• Since fascia shapes a muscle, the quality of our fascia influences the pliability of muscle shape. When individual wrappings are soft, long, and separate from their neighbors, their shape is more easily changed (Fig. 10-23b). If individual wrappings become hard, short, and stuck together, they become stiff and unyielding. They resist stretching and compression, thus immobilizing our joints (Fig. 10-23a).

Fascia becomes stiff and unyielding through repeated injury and inflammation. Fascia is injured by: **1.** working it too hard before it is warmed, **2.** lifting weights which exceed its strength, **3.** fast, jerking movements, **4.** leaving wastes in our muscles through inadequate cool down after working, and **5.** reduced flexibility. A vicious circle is created. Reduced flexibility potentiates injury and injury leads to reduced flexibility. Injury is reduced by avoiding the five activities listed above and maintaining the strength and length of our connective tissue through regular work.

When injury occurs, scar tissue replaces the injured connective tissue. Scar tissue has more collagen, shrinks as it matures, and tends to weave adjacent wrappings into each other. This produces the hard, short, and stuck-together condition which reduces flexibility.

Persons with reduced flexibility are sometimes called **muscle bound.** The condition of their own muscles limits their mobility.

MECHANICAL ADVANTAGE

Any muscle has certain **mechanical limitations** determined by the location of its attachments, its strength, its flexibility, the shortening potential of its belly, and the movement potential of the joint it crosses.

When we look at the **mechanical**

> A **Pulled muscle** is a frequent, but imprecisely used term. It implies damage to our muscle's connective tissue, either its fascia or tendon. It may be microscopic, tearing relatively few proteins, it may damage a large percentage of a muscle's protein, or a tendon may completely tear off its boney attachment. The damage may be scattered throughout the muscle or limited to a localized area. Pulled muscles are caused by anything in the fascia injury list described above.
>
> **Shin splints** are microsopic damage of tendons attaching into the tibia. **Tennis elbow** is damage to tendons originating on the epicondyles of the humerus.

Figure 10 - 23

(a) When muscle fascia is short and adhered, the range of motion is limited. **(b)** When muscle fascia is long and separate, the muscles change their shape easily, yeilding a good range of motion.

(a) (b)

Figure 10 - 24

When a wagon is pulled along a horizontal surface, a shorter person has a better mechanical advantage than a taller person. The applied force of the shorter person is more in line with the work to be done.

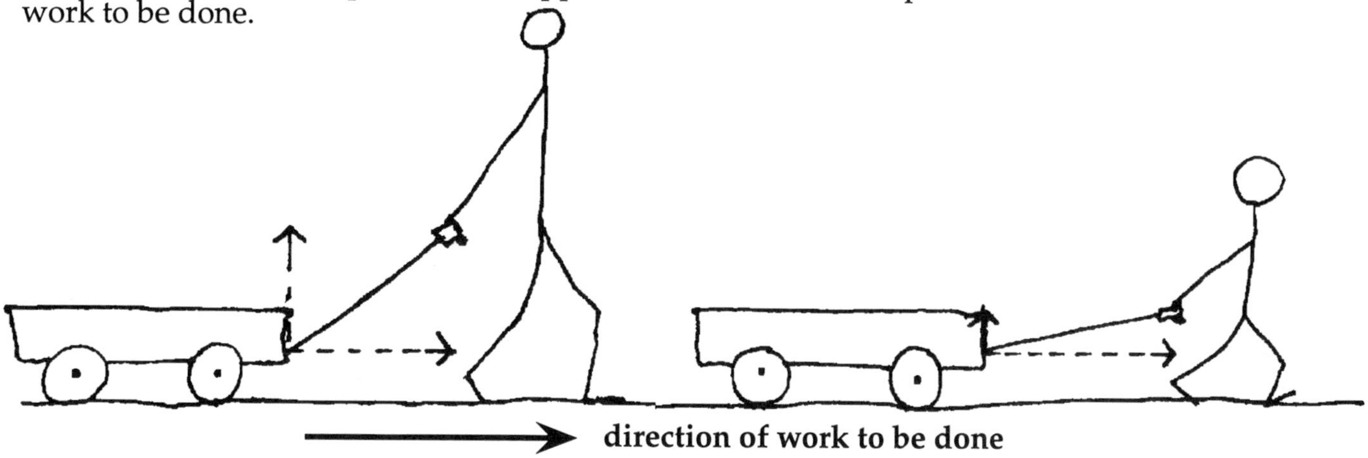

advantage of a muscle to do a certain job we ask "Given the requirements of this job and the limitations of this muscle, will this muscle's effort be effective, or is a lot of its effort inappropriate?" When a large proportion of the muscle's force is effective, it has good mechanical advantage. When a large proportion of the muscle's force is inappropriate, it has poor mechanical advantage and might be used only if an all out effort is required.

We will look at a muscle's mechanical advantage in two ways: **1.** its capable direction of force, and **2.** the distance from the joint that the muscle attaches relative to its strength and shortening potential.

Direction of Force

- When pulling a wagon, the direction of the handle implies the direction that the force will be applied to the wagon. The handle lies on some diagonal between horizontal and vertical, being more horizontal when pulled by a shorter person and more vertical when pulled by a taller person (Fig. 10-24). If we want to pull the wagon along a level sidewalk, the desired direction of work is horizontal, so a shorter person would have a better mechanical advantage. The direction that they pull is more horizontal, which is the direction we want to go. A taller person pulls more upward on the wagon, which is not the direction we want to go. That is why children pull their toys in wagons, while adults drive cars.

Figure 10 - 25

The direction in which a muscle can exert its force is limited to a straight line between its two bony attachments, indicated by line **A**. This force can be divided into its vertical aspect, line **B**, and its horizontal aspect, line **C**, to evaluate how the muscle participates in a specific direction of movement.

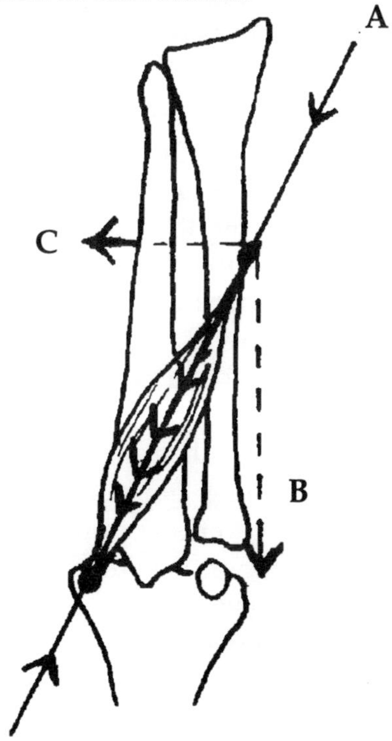

- The direction in which a muscle can exert its force is limited to a straight line between its origin and insertion. This is its only movement potential. Whenever it contracts, its two ends pull together along this line (Fig. 10-25).

In any given work situation, the useful force of a muscle is only as much as the line of the muscular force parallels the desired direction of work. If the desired direction of work is exactly in line with the muscle, then 100 percent of that muscle's force is useful. If the desired direction of work deviates from the line of the muscle, the degree of deviation is the amount of inappropriate effort by the muscle.

The closer that a muscle lines up with the desired direction of work, the better is its mechanical advantage for that work situation, the more advantageous to use it.

- Say that the straight line of a muscle lies diagonally between ankle extension and foot inversion (Fig. 10-26). If that is the exact movement we want, bravo, that muscle could do it without assistance or wasted effort. But let us say that we want to purely extend our ankle with no degree of inversion. Our muscle is stumped. Every time it tries to extend our ankle, inversion automatically happens. The location of its attachments demands this. That is the straight line between its origin and insertion so that is its only possible movement.

Now inversion is inappropriate for the desired direction. It does not help, in fact, it creates extra work. We have to exert an equal eversion force in some other muscle to antagonize the inversion, since our first muscle has no choice but to do it.

Thank God (or whomever you credit for your design) there is another muscle crossing our ankle. The straight line of this second muscle bends the ankle diagonally between extension and eversion. What luck! What great planning! The two muscles can work together to extend our ankle as desired, while their opposing actions pull against each other and cancel each other out (Fig. 10-26).

- Our muscles' effort to cancel out the inappropriate force is, in a sense, wasted, but with the muscles we have, this waste is necessary to accomplish the exact moves we want. Actually, the antagonizing forces are helpful to stabilize or support our joint while it moves. However, in excess, the antagonizing forces may create excessive wear on our joints.

In our bodies, there are very few muscles that pull in exactly the direction we want, especially in joints which move in many directions such as our hip. Most movements involve **1.** contracting muscles which have some advantage in the desired direction of work, and **2.** antagonizing all the unwanted directions they also pull. Most movements of our body are this kind of team effort. Rarely does our body perform single handed plays.

Figure 10 - 26

Muscle **A** extends the ankle and inverts the foot. Muscle **B** is synergistic with **A** in extending the ankle, but antagonistic in everting the foot.

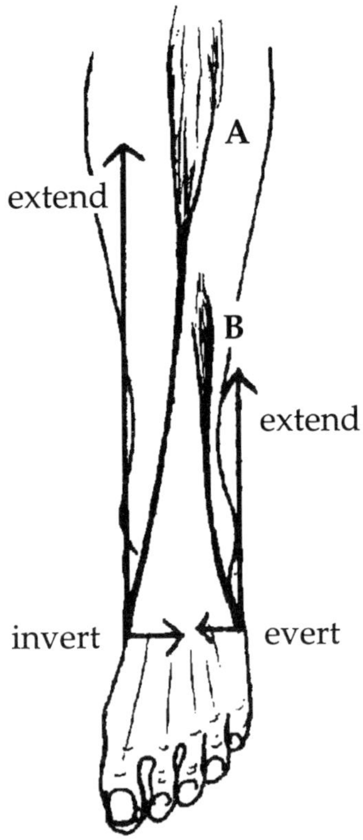

Work versus Working Force

The second concept of mechanical advantage addresses the distance that a muscle attaches from its joint relative to the strength and shortening potential of the muscle. This concept applies all the basic rules of levers.

A lever is a rigid bar which can pivot around a fixed point or fulcrum to accomplish work. How the work occurs depends on four factors: **1.** how much the work resists movement, **2.** where the work is placed on the lever, **3.** how strong the working force is, and **4.** where the working force is placed on the lever. There is always an exchange. When work is placed further away from the fulcrum, it will have greater resistance to movement. It

Figure 10 - 27

(a) When a weight is placed further away from the fulcrum, it feels heavier (bulging muscle) but can be moved a greater distance. **(b)** When the same weight is placed closer to the fulcrum, it feels lighter but cannot be moved as far (solid arrows) with the same amount of muscle shortening (dotted arrows).

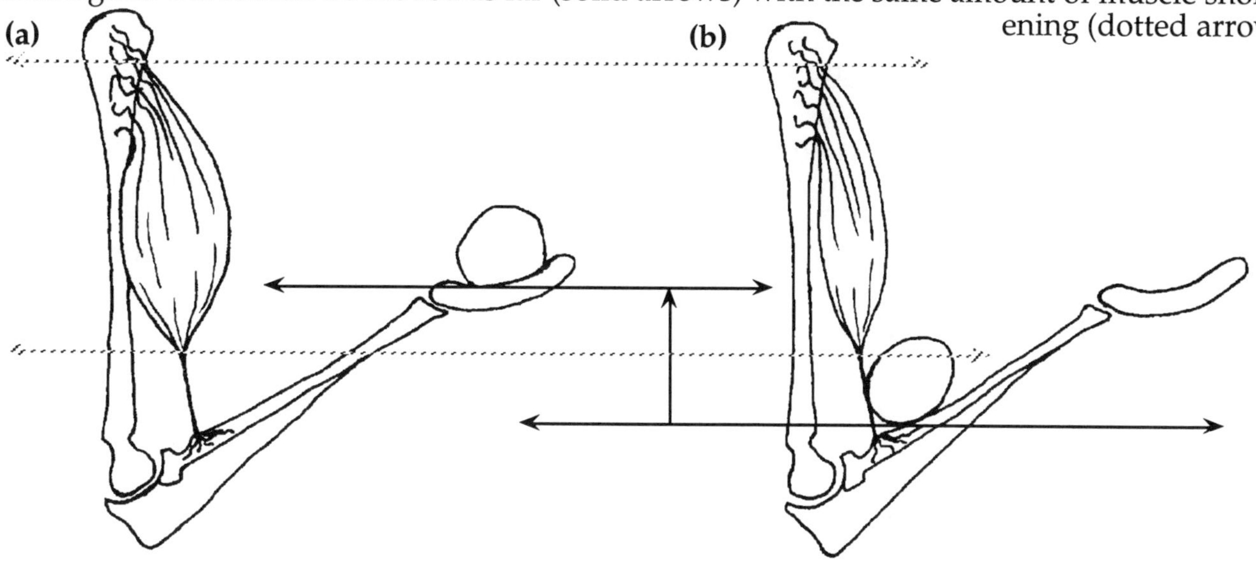

Figure 10 - 28

(a) When muscles attach further away from joints, they can lift heavier weights a short distance. **(b)** When muscles attach closer to joints, they cannot handle such heavy weights, but the same amount of muscle shortening (dotted arrows) moves the weight further (solid arrows).

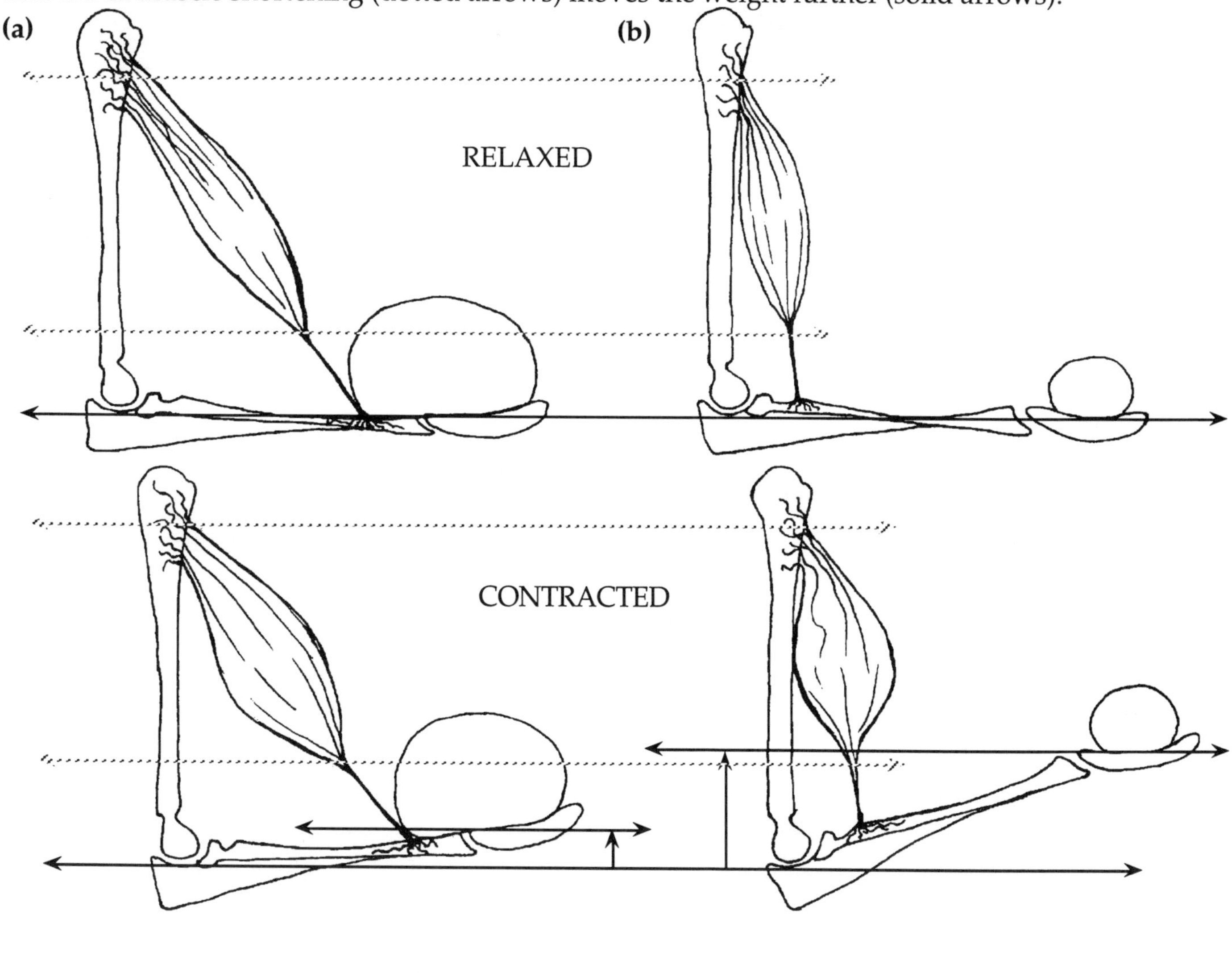

feels heavier, but it can be moved a greater distance (Fig. 10-27a). When the work is placed closer to the fulcrum, it will have less resistance to movement. It feels lighter, but it cannot be moved as far (Fig. 10-27b).

In contrast, when the working force is placed further away from the fulcrum, it can more easily overcome greater resistance to movement, but the distance of movement is small (Fig. 10-28a). When the working force is placed closer to the fulcrum, its effect will seem weaker but it can create much more movement (Fig. 10-28b). How the levers are arranged depends on the work one wants to accomplish.

When a mother balances her 7-year-old child on a teeter-totter, it can be a good combination. The mother may not like the idea of balancing way up in the air on a wobbly, little board. She does not have to. She can sit close to the axle where the movement is considerably less and use her extra weight to soar the child into his/her imagination.

Figure 10 - 29

(a) When a mom's and child's weights are equidistant from a fulcrum, the mom cannot get off the ground and the child cannot come down.
(b) When the mom moves closer to the fulcrum, the child's weight can balance with hers.

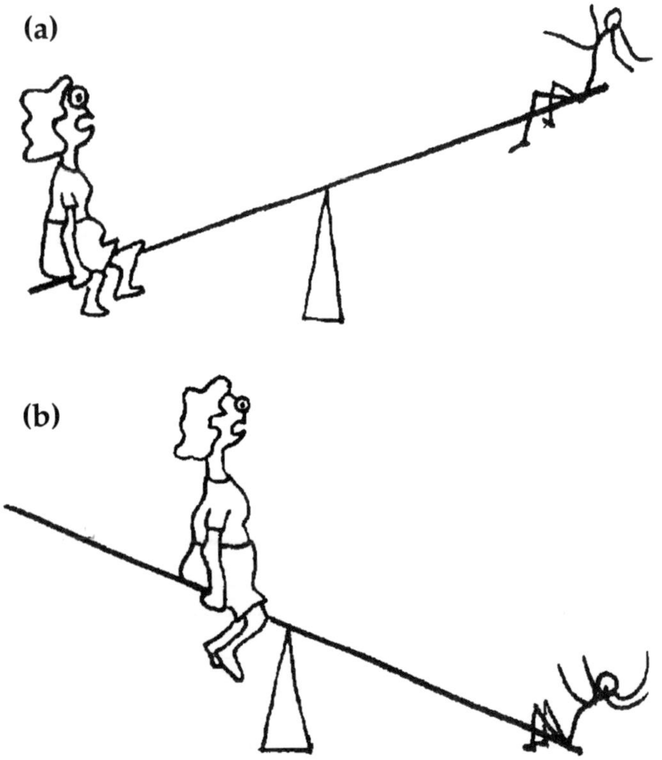

In fact, because the mother weighs more (more working force), she must weaken its effect by sitting closer to the axle if the child is to come down at all (Fig. 10-29). If the mother wants to balance way up in the air, she will have to play with someone her own size and find a board to match.

• All the above principles translate directly into our bodies. The levers are our bones, the fulcrums are our joints, the work to be done is the movement and support of our structure, and the working force is the force of contraction of our skeletal muscles.

In our bodies, the placement of the working force is fixed by the location of our muscle attachments. Our movement potential (range of motion) also is limited by the shortening potential of each muscle. With time, we can change the quantity of force available as we hypertrophy or atrophy our muscles. The most immediate aspect that we can change is the quantity and placement of the work to be done and which muscles we use to do it.

Muscles which insert closer to their fulcrum produce a greater range of motion at the end of their bone (Fig. 10-28b), but they must create a greater force to do it. They have to work hard, but a short contraction on their part will produce a wide sweep of movement at the end of their lever. Many muscles in the body are arranged in this manner, since their shortening potential is limited to one third of their resting length.

Muscles which insert further away from their joint, will more easily move a certain resistance, but cannot move it as far. These muscles are handy when the resistance to movement is enormous (Fig. 10-28a).

Which muscle has the best advantage in a given work situation depends on the quantity of weight to be moved and the distance it needs to be moved. For example, how heavy is a box of books and how high does it have to be lifted. Say that muscle A inserts close to our joint and that muscle B inserts further away from our joint. If the box is light and the shelf is high, use muscle A. It could probably produce enough force and it can go the greater distance. If the box is heavy, perhaps muscle B could manage the weight easier, but its distance is limited. We would have to make room on a lower shelf. If the high shelf is the only possible place, we may need to use muscle A to achieve the distance, but will have to take the books out of the box and move them one by one. Or exercise for a few weeks to strengthen muscle A before attempting to heave the whole box up on the high shelf.

- As we do our daily work, it is useful to keep the principles of levers in mind so that we perform actions to our best advantage, with the least strain on our structure. Lifting a paper in our hand is easy. Its weight is so small. When carrying a heavier object, we may not hold it in our hands so far from our supporting joints. We may hold it across our forearms much closer to our elbow, or even rest it on our shoulder.
- These same principles of mechanics are what make it easier to support upright posture when our body is positioned over our supporting pedestal rather than protruding outward on long levers.
- In our body we have many synergistic muscles with all different types of mechanical advantages, giving us the enormous range and subtle diversity of movement and strength that we all demonstrate constantly in our activities.

DEDUCING A MUSCLE'S ACTION

As discussed under mechanical advantage, each muscle pulls a joint in a certain direction depending on the location of its attachments. <u>We can know the direction of action of a specific muscle by drawing a straight line between its two bony attachments</u> (Fig. 10-25). All that muscle can do is attempt to bring its two bones together along that line. Understanding the possible actions for any specific muscle is that simple.

When we discussed the movement potentials of specific joints, we separated them into distinct directions based on the three dimensions of space: vertically forward or backward, vertically sideways, and horizontally. The actions of muscles are categorized in the same way. The direction of their actual force can be broken down into the porportions which fall horizontally or vertically relative to the basic directions.

The movements we perform, however, are not so regimented. We move in diagonals between the directions, using combinations of muscles. When we lift our arm at a diagonal between flexion and abduction, we combine the efforts of muscles that flex and abduct. Most movements are done by collections of muscles which add force to the directions we want and cancel out each other's forces in the direction we do not want.

If we want to see which muscles would do a certain movement, it is easiest to examine the movement in terms of which of the directions it combines. Does it combine vertically forward and horizontal, or vertically forward and vertically sideways, and so forth. Then we can deduce what muscles could be combined for that movement.

Joints that move in only one or two directions only need a few different muscle to create those movements.

Joints that move in many directions need many different muscles to create their great variety of movement. Our glenohumeral joint is one of the most moveable joints in our body. Muscles which move it cover our chest, shoulder, and most of our back, wrapping all the way around the joint. Since it moves in every direction, it needs a muscle line radiating out from it in every direction.

MEANING OF MUSCLE NAMES

The names of muscles may seem long and difficult, but they are often made of word parts which give us useful information about the muscle.

The name may tell us any of the following: **1.** the muscle's action (flexor, adductor, tensor), **2.** its fiber direction (oblique, transverse, rectus), **3.** its location (intercostal, latissimus dorsi, lumborum), **4.** its number of heads or origins (biceps, triceps), **5,** its shape (deltoid, quadratus, teres), **6.** its size (maximus, vastus, brevis), or **7.** its bones of attachment (sternocleidomastoid, brachioradialis).

When considering which muscles would perform a certain movement, we must always consider how gravity is affecting the situation.

As we squat down, our hips and knees are flexing. We may use flexing muscles to begin the movement, but once our bony alignment is broken, gravity gains advantage and pulls down vigorously, producing more than enough force of flexion. In order to squat comfortably, without a crash, we need to use extending muscles to oppose the force of gravity (Fig. 10-30).

When we set down a glass of water, gravity is all too anxious to help. Even while the glass is moving down, we must use lifting up muscles to prevent it from smacking on the counter.

Figure 10 - 30

In order to squat slowly, the extending muscles contract to oppose the flexing force of gravity.

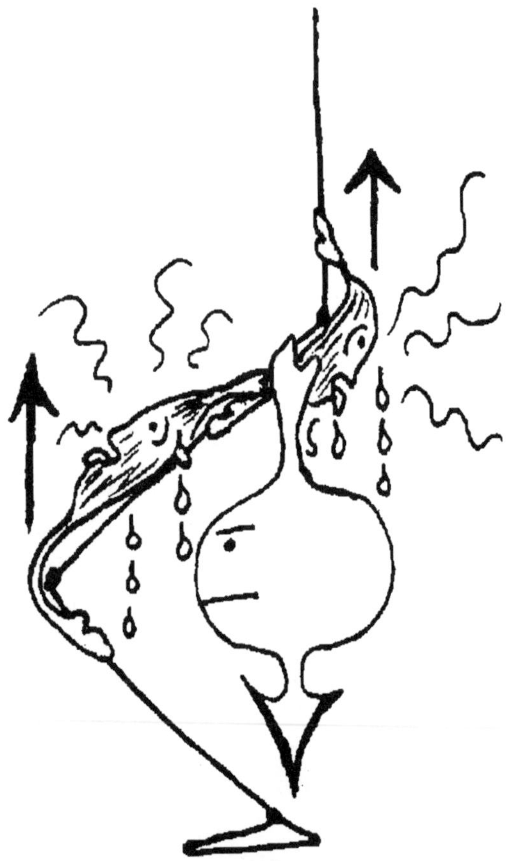

The very first step of releasing energy from glucose does not use oxygen. It is called **anaerobic glycolysis** (Fig. 10-31). Muscles which have run out of oxygen can still get the energy to continue working by this process.

The end product of anaerobic glycolysis is **lactic acid.** Muscles which are forced to work past their oxygen resources may accumulate a lot of lactic acid. This may throw off the acid-base balance of our muscle fiber, interfering with its basic metabolism. This may cause fiber injury, inflammation and sore muscles.

Lactic acid is called an **oxygen debt.** It is the product of lack of oxygen and we can only get rid of it by adding oxygen to it. With oxygen, the rest of the energy in lactic acid can be released by breaking it down into CO_2 and H_2O (Fig. 10-31) using the same process which breaks down glucose.

Any muscle can slowly metabolize lactic acid, paying its own oxygen debt as it recuperates its oxygen supply, but it may be injured in the meantime. This is one time when each cell of our body works for the whole. Every cell in our whole body is willing to help pay our muscle's debt. Lactic acid will diffuse out of our muscle fiber into our blood. All our cells, especially those of our liver and other muscles, will take in the lactic acid and add oxygen to it. The debt will be paid more rapidly, before injury happens.

In order for our whole body to help with the oxygen debt, blood needs to circulate through the indebted muscle to move the lactic acid out of that muscle and make it available to the rest of our body. Proper cool down after working should circulate the lactic acid and avoid a lot of soreness and inflammatory damage in our muscles.

• Now we are ready to examine the individual muscles of our body. The following three chapters discuss each muscle in terms of the location of its origin and insertion, the

Figure 10 - 31

Glucose which is broken down without oxygen yields lactic acid and a small amount of energy. When oxygen is added to the lactic acid it will further breakdown to carbon dioxide, water, and more energy.

$$\text{Glucose} \xrightarrow{\text{anaerobic glycolysis}} \text{lactic acid} + \text{Energy}$$

$$\text{lactic acid} + O_2 \downarrow$$

$$CO_2 + H_2O + \text{Energy}$$

It is important to warm up before performing rigorous activity. Warm-up puts our tissues into a soft, pliable consistency, increasing flexibility and reducing risk of injury. Almost anything can serve as a warm-up if it makes us sweat a little, speeds up our heart and breathing, and warms us up enough to almost take off our sweat shirt. Examples are to flip around like a puppet on a string, chase a butterfly, romp like a puppy, or run on the spot and swing our arms. Since our tissues are not yet warm, a warm-up should not be forceful or press the range of motion to its edges.

I often see stretching used as a warm-up. People come into a cold room, position themselves on a cold floor and lean for 10 minutes in one direction, stretching something. The tissues are all cold and stiff so the stretch is slow and arduous, and actually risks causing injury. There is not enough movement to warm up anything. Stretching is important, but warm up first and the stretch will be twice as effective in half the time.

How about a hot shower for a warm up? My answer to that -- there are two ways to cook a potato. Put it in a pot of boiling water for 25 minutes and the heat will slowly penetrate to the center of the potato, or put it in a microwave. With all the water molecules inside the potato jumping around, it is cooked in 5 minutes. I am not suggesting we do microwave warm-ups, but moving our own muscles warms us up from the inside like the microwave moving the water inside the potato. When standing in a hot shower, the heat has to penetrate slowly through the layers, but it can warm us up.

The purpose of the cool-down after rigorous activity is to replenish consumed nutrients and carry away lingering waste products. Movements which produce light, alternate contraction and lengthening of a muscle are the best way to circulate blood and re-establish homeostasis. Cool-down movements should sweep a full range of motion as effortlessly as possible, shimmering the tissues so the bones rattle loosely in the flesh.

Stretching at the end of a workout can leave one relaxed, open, and loose, but it does not stimulate circulation and so should not be the only cool-down process.

joints it can move, the directions it can move those joints, and how to palpate the muscle.

It is up to you to find your muscles in your own body. The instructions will help you locate their bellies and tendons. Move while you are exploring your muscles. Feel them contract. Feel them firm up, bulge, and gather toward their origins as they contract. Feel the tendons become taut. Feel which muscles do a specific movement. Then relax in your body and feel your muscle bellies soften and their tendons loosen.

If you are having trouble finding a muscle, then perform its movement. It is your muscle's contraction that produces the movement, so you can see and feel it shorten and bulge out toward you as it contracts. Like a waving hand in a crowd, it calls your attention.

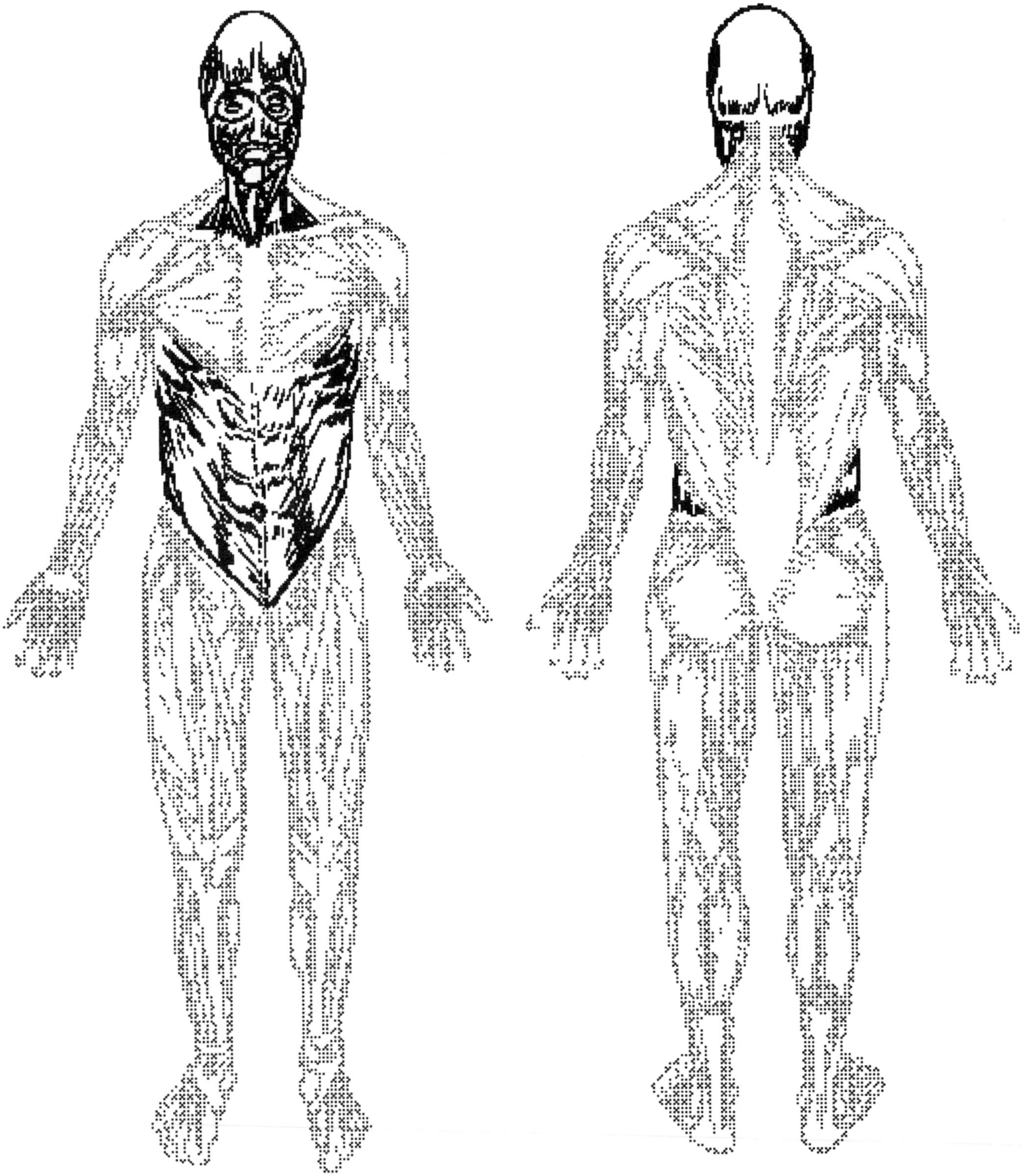

chapter 11: **Axial Skeleton Muscles**

• Thinking back to our axial skeleton, what kinds of joints are the most common? The slightly-moveable, fibrous and cartilaginous joints. The design of our axial skeleton is for increased stability, sacrificing mobility.

Do we need a big bunch of muscles to support and move joints which already are stable and only slightly-moveable? No, it takes only a few. Relative to the number of bones and joints in our axial skeleton, the muscle count is small.

Even with 24 sequential joints in our vertebral column, we cannot rotate our vertebrae 360° or curl them into one complete circle, so the muscles which move them are relatively few and simple. Imagine the spirals we could make with 24 movable joints in each arm, but don't turn into an octopus imagining the muscles it would take to use them.

Many axial skeleton joints have no muscles crossing them and are moved only by indirect stresses or pressures from adjacent tissues. For example, most of the sutures between our skull bones have no muscles oriented to move them, but they may be moved by changes in the internal or external skull pressure.

Many of the muscles of our axial skeleton are deep, covered by bones, cartilage, internal organs, or other muscles. We will discuss only the basic organization of many of these muscles and may not even mention their names. The emphasis is on muscles that are palpable or important for some special function.

• The distinction between origin and insertion in our axial skeleton muscles is not always clear and the written literature does not always agree. In many cases we can easily stabilize either end of the muscle and frequently use it in both directions, so the distinction is not so important. I may avoid taking sides by just referring to the attachments of the muscles without distinguishing origin from insertion.

• We will begin with the muscles in the skull.

MASTICATION MUSCLES

Remember the one palpable synovial joint in our skull? Our temporomandibular joint. Movement of our temporomandibular joint is used in eating and oral expression such as talking, singing, yawning, shouting, and so forth. The most power is required in bringing our teeth together, closing or flexing the joint when we bite or chew. To chew is to masticate, so muscles that move our temporomandibular joint are called muscles of mastication.

There are two pairs of powerful muscles which flex our temporomandibular joint. They are both oriented vertically to maximize their mechanical advantage in this movement.

Masseter

Our **masseter muscle** originates on the inferior edge of our zygoma. It inserts into the external surface of the vertical strut of our mandible (Fig. 11-1). As our masseter contracts, it carries our mandible up toward our zygoma, closing or flexing our temporomandibular joint.

Lay your fingers over the flesh between your zygoma and the angle of your mandible. Clench your teeth. The hard lump that gathers under your fingers is the contraction in the belly of your masseter muscle.

Figure 11 - 1
A lateral view of the skull showing the masseter and temporalis muscles.

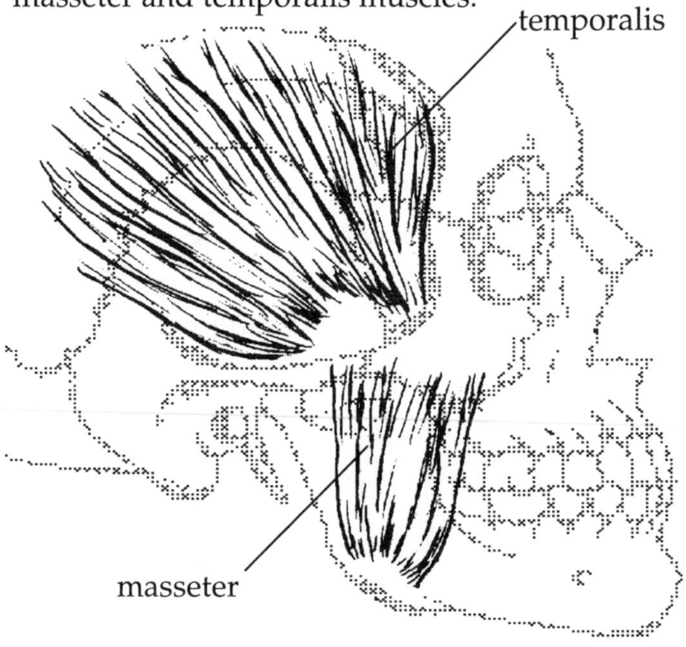

Temporalis

Our **temporalis muscle** originates on the external surface of our temporal bone, from immediately behind our eyes to above our ears. It passes under our zygomatic arch to insert into the top of our mandible, on the vertical point just in front of each condyle (Fig. 11-1). As our temporalis contracts, it carries our mandible up toward our zygoma, closing or flexing our temporomandibular joint.

Fan your fingers over the side of your head, from just behind your eye socket to behind your ear. Clench and release your teeth repeatedly. The movement under your fingers is the contraction and relaxation of your temporalis muscle.

Pterygoids

Two pairs of muscles assist the manipulation of our mandible, our **medial and lateral pterygoid muscles.** They are deep and palpable only from inside our mouth and throat, a dangerous place to go searching with too much diligence. These muscles help to close our temporomandibular joint, waggle it from side to side, pull it forward and backward, and open it.

In an upright position, gravity is always trying to pry our mouth open, so it takes our constant intention and muscle contraction just to keep it closed. (If you need an excuse for gabbing along, try that. The muscles that hold your mouth closed are taking a rest.) When lying on our side or standing on our head, gravity is not available to open our mouth, but we can still do it with the help of our lateral pterygoids.

MIDDLE EAR MUSCLES

We have one set of muscles which we never touch, not in our deepest work, even on ourselves. These muscles are inside our middle ears which are totally encased by our temporal bones. These muscles originate on the bony wall of our middle ear and insert into the tiny bones which vibrate to conduct the sound wave across our middle ears. When these muscles are startled by loud noises, they contract and pull on the vibrating bones. This dampens the vibration and reduces the amplitude of the sound in an effort to prevent tissue damage.

In the midst of a racket we may have to yell just to carry on a friendly conversation. Upon leaving the racket, the world may still sound distant and muffled. We still may be yelling over a friendly conversation because these muscles are too jangled to let go.

- Many muscles around our head and neck disregard one of the basic rules of skeletal muscles, both of their ends do not attach to bone. Their origin is stabilized on bone, but they insert into soft tissue. As the muscles contract, they move that tissue over our skeleton. These muscles include our muscles of facial expression, the external muscles of our eyeball, and the muscles of our tongue, pharynx, and larynx. These muscles have a variety of jobs, but they all make a unique and significant contribution to emotional expression.

Many animals have similar muscles going from bone to skin all over their body. For example, a cow may use these muscles to twitch the skin over her hind quarters, chasing off the flies. Her idle stare makes it look like a routine task, devoid of emotional content, but who knows.

FACIAL EXPRESSION MUSCLES

As we change from smiling to pouting to frowning, we slide our facial skin over the underlying bones, wrinkling, bulging, and tauting our skin in different places (Fig. 11-2). These movements are created by our **muscles of facial expression.** Joints are not involved except for the added gestures of our temporomandibular joint.

Our muscles of facial expression are organized around three **sphincter muscles,** one around each eye and one around our mouth. The other facial muscles originate on surrounding bones and insert into the sphincters (Fig. 11-2) or overlying skin. Insertions into our mouth sphincter are the most profuse.

Figure 11 -2

The muscles of facial expression originate on skull bones and insert into the skin or one of three sphincter muscles. As they contract, they pull the face into its many expressions by moving the soft tissues over the bones. Can you see which muscles are used in the following expressions?

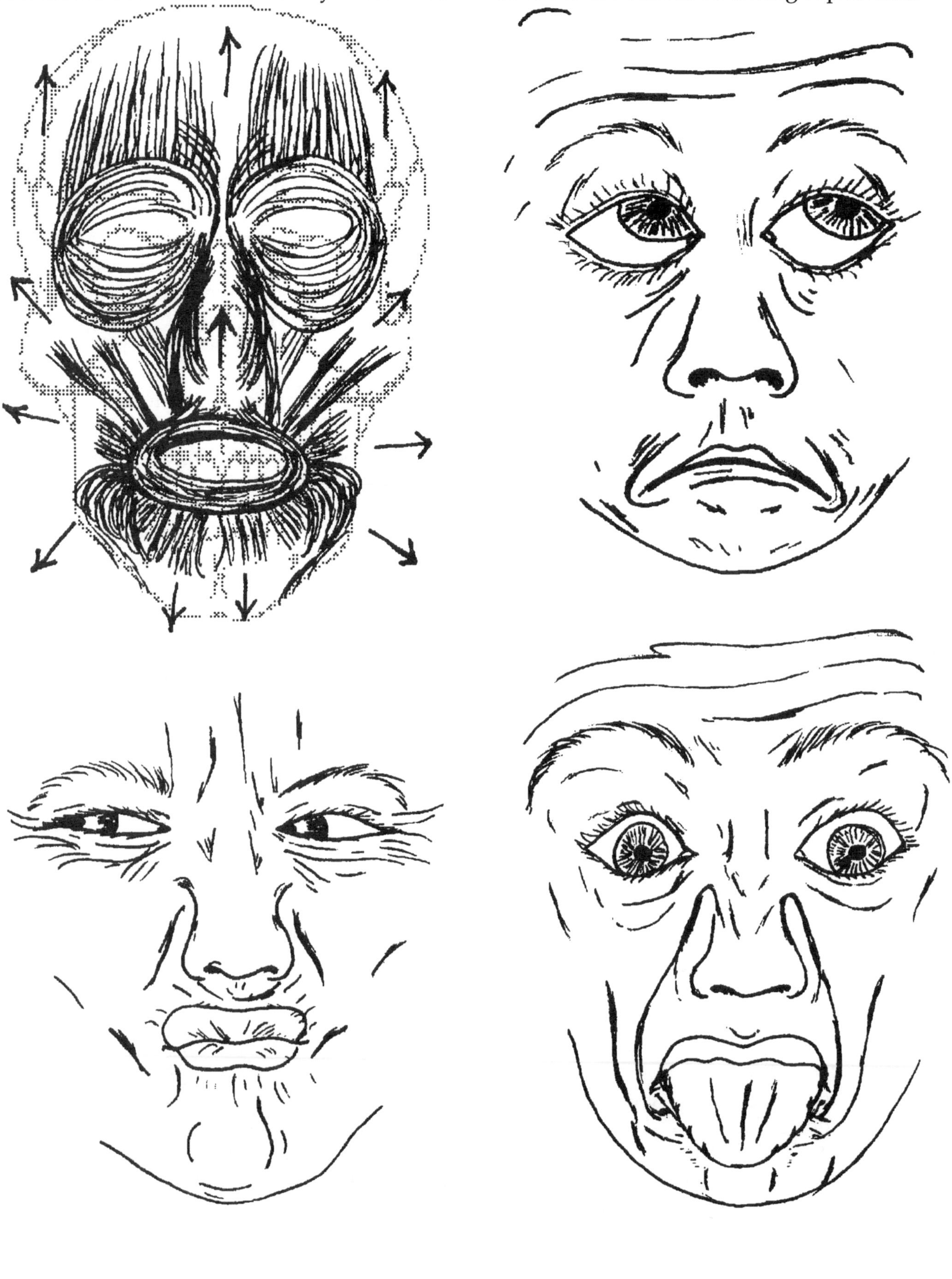

As the muscles contract, they pull the sphincter toward their own origin, distorting the sphincter and overlying skin in a certain direction. These distortions manifest our emotional expressions. Somehow we instinctually know what the different shapes mean. Even if we do not understand the language of a spoken message, we can know its emotional content just by watching the movements of the messenger's face.

> In a **sphincter muscle,** the muscle fibers form a ring or donut with no beginning, no end, and often no bony attachment. As the muscle contracts, the donut squeezes its hole closed. As the muscle relaxes, the donut hole opens wide.

TONGUE MUSCLES

Movement of our tongue is essential for speaking, manipulating food within our mouth, and swallowing. Muscles of our tongue originate on our hyoid, mandibular, and temporal bones, and the floor of our mouth. These origins form stabilizing roots from which our tongue can move and express freely. The structure of our tongue itself is the muscle bellies (Fig. 11-3). They weave into each other in such a way as to provide all the movements of our tongue: curling, pointing, folding, protruding, wagging, and so forth, a frequent volunteer in emotional expression.

EXTERNAL EYEBALL MUSCLES

Our field of vision is greatly enlarged by our ability to move our eyeballs within their sockets. This movement is produced by muscles which originate on our bony eye socket and insert into the outer layer of our eyeball (Fig. 11-4). The details of these muscles will be discussed when we discuss our eye.

Figure 11 - 4

A lateral view of the external muscles of the eye. The arrows indicate the potential movements.

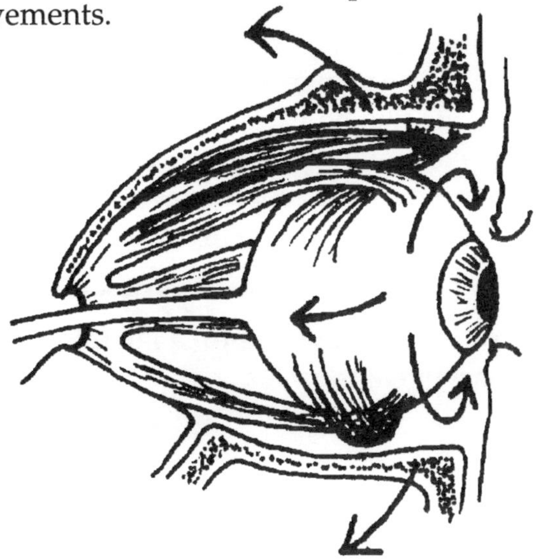

PHARYNGEAL MUSCLES

Our **pharynx** is the area behind our nose and mouth down to the top of our voice box and esophagus. It is a passageway for air, food, and water. As we continue with the muscles of our axial skeleton, we will see how

Figure 11 - 3

A lateral view of the muscles of the tongue. In **(a)** the tongue is sticking out. In **(b)** the tongue is tucked in behind the lower teeth.

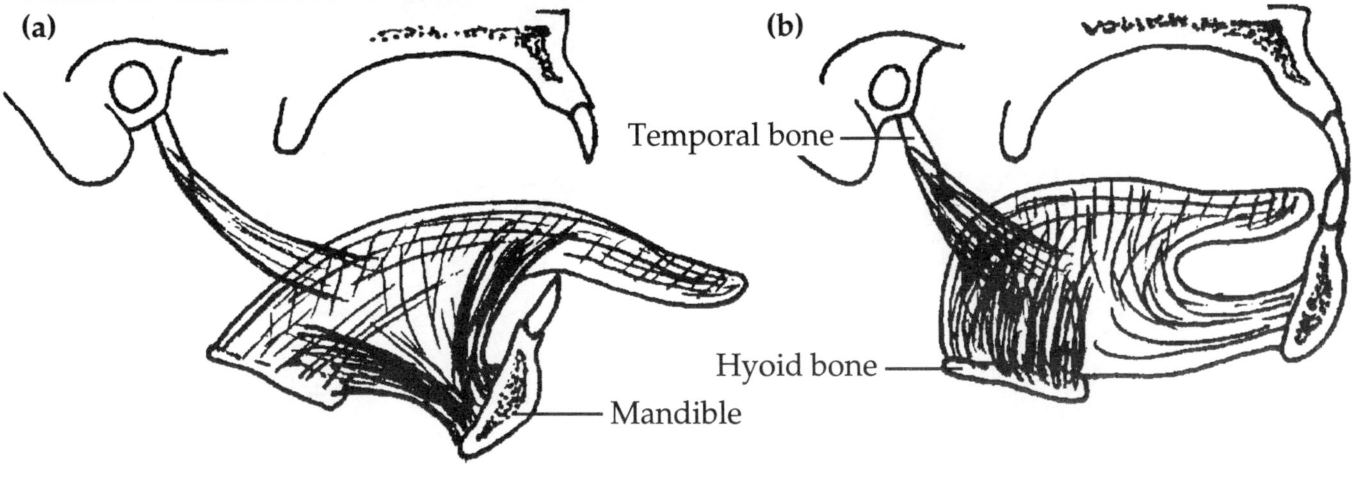

air is drawn into our chest by a vacuum created in the chest. Food and water, on the other hand, must be pushed through our pharynx.

The walls of our pharynx contain muscles which loop around the passageway. Our pharynx is like a sock suspended from its top on a single hook, like a Christmas stocking hanging from the fireplace mantel. The single hook for our pharyngeal sock is on the front of our occipital bone, just in front of our spinal cord. Our pharynx droops down and forward from this point. In the front it also is supported on our moveable mandible, hyoid bone, and larynx (Fig. 11-5).

Our pharyngeal sock is well worn. Its toe end has two holes, an anterior hole leading to our exclusive air passage and a posterior hole leading to our esophagus.

As our pharyngeal muscles contract, they squeeze into our pharynx. The contraction starts at the top and continues downward like a wave of constriction, pushing our food ahead of the contraction, lifting our hyoid bone and larynx in the process. Gently hold your hyoid bone and larynx. Feel them bob up as you swallow. This closes your airway so that curious or straying food has no place to go but into your esophagus.

LARYNGEAL MUSCLES

Our **larynx**, commonly known as our voice box, is a cartilaginous structure in the front of our neck. It is part of the tube connecting our nose to our lungs. Our larynx contains two membranes or vocal cords which may be placed into the flow of air from our lungs. The air vibrates the vocal cords, creat-

Figure 11 - 5

(a) A posterior view and **(b)** a lateral view of the pharyngeal muscles.

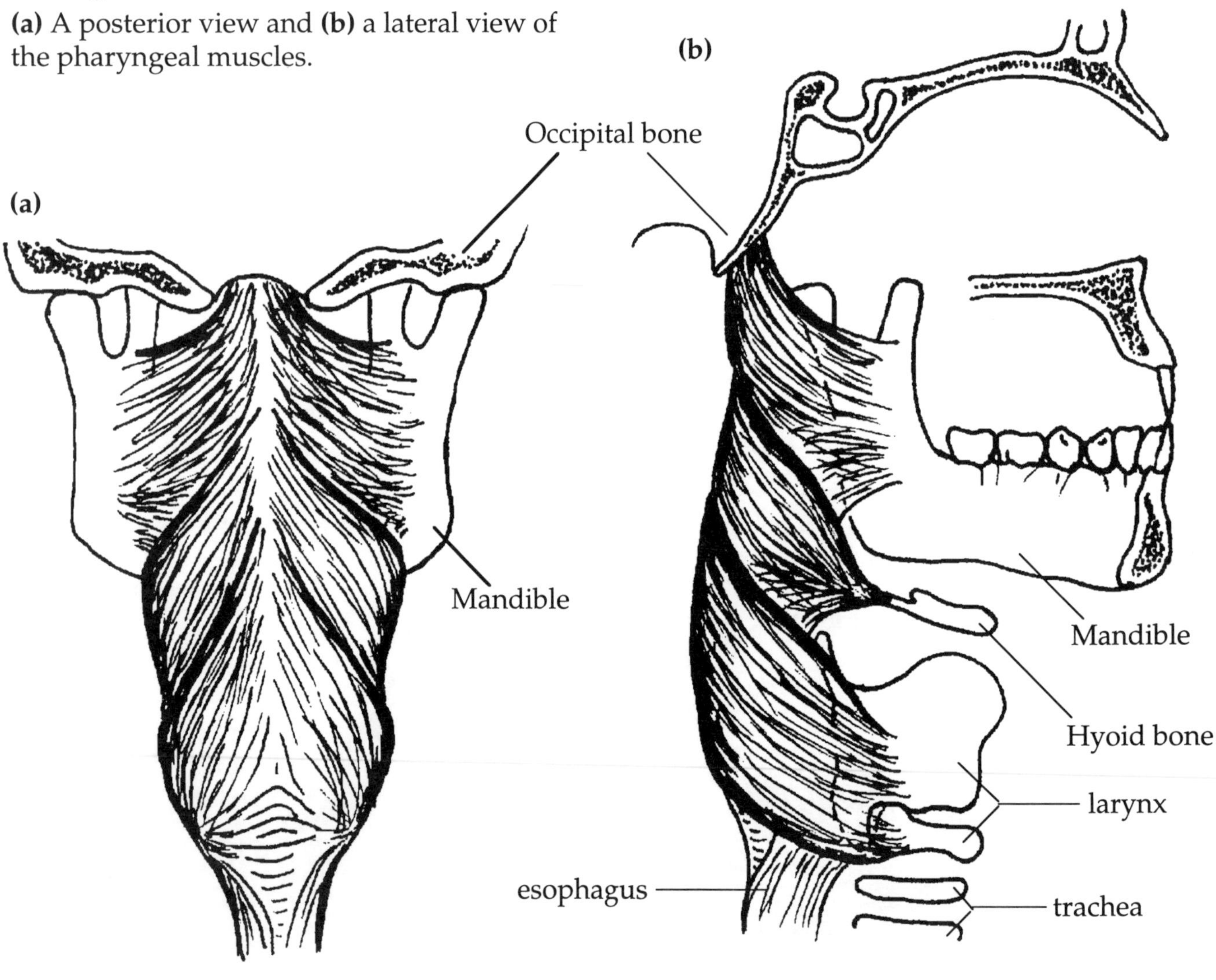

ing a sound wave. The position and tautness of the vocal cords determines what kind of sound we utter. Tiny, delicate muscles around and within our larynx (Fig. 11-6) change the position or tautness of our vocal cords to create and change our sounds. These muscles are not individually palpable, although their effects are audible and palpable in the vibrations of our larynx. Gently hold your larynx while saying your name. Feel the sound it makes.

Figure 11 - 6

A posterior-lateral view of the larynx.

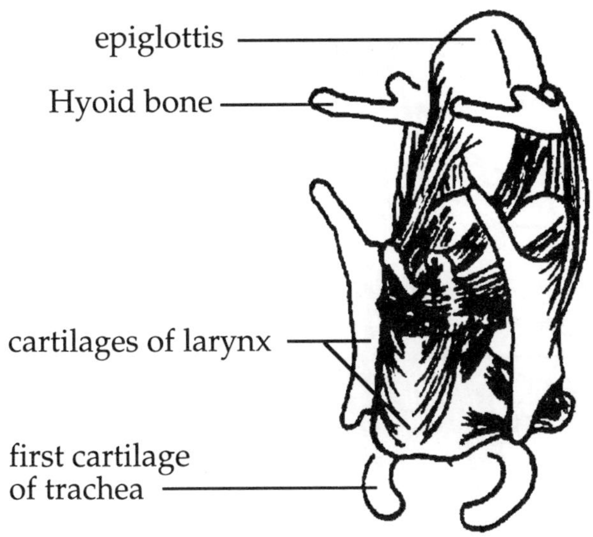

• What bones provide the structural support through the trunk of our body? Our back bone or column of vertebrae.

What are the movement potentials of our intervertebral joints? In the vertical direction, our joints flex, extend, and laterally bend. In the horizontal direction, they rotate.

Muscles anywhere in our torso may effect these movements, whether they are in the front of our neck and abdomen or in our back. The muscles do not need to originate or insert on the vertebrae, they only need to indirectly cross the joints in some manner. Many of these muscles have other direct, local functions besides their support and movement of our vertebral column.

Vertically oriented muscles would produce the vertical movements of our intervertebral joints. Any muscle which is ventral to the pivot point of joint movement would **flex** our vertebrae. Any muscle which is dorsal to the pivot point of the joint would **extend** it. Any muscle which is lateral to the pivot point of movement would **laterally bend** our vertebrae.

Horizontally oriented muscles would produce horizontal movement, **rotating** our intervertebral joints. The direction of rotation would vary with the specific orientation of the muscle. When a muscle on the left side of our body rotates our body toward the right, it is called **rotation to the opposite side.** When a muscle on the left side of our body rotates our body toward the left, it is called **rotation to the same side.**

ANTERIOR NECK MUSCLES

The muscles of our anterior neck include the muscles behind our chin, between the two sides of our mandible, and down the front and sides of our neck to our sternum, clavicle, and upper ribs. Collectively, these muscles have five jobs. **1.** They form and support the floor of our mouth. **2.** They anchor our tongue. **3.** They support and move our voice box. **4.** They support and move our cervical vertebrae. And **5.** They lift our ribs.

> Since the muscles of our anterior neck influence our tongue and larynx, they greatly affect our **speech.** Tightness or spasticity in these muscles may make speech difficult, while releasing these muscles may enhance verbal expression. thus becoming an important part of speech therapy.

We will discuss two of our anterior neck muscles, our **sternocleidomastoid** and our **scalenes.** The others are significant functionally, but small or not individually palpable.

Sternocleidomastoid

The name **sternocleidomastoid** incorporates the names of three bones, the sternum, clavicle, and mastoid process of the temporal bone. The first two bones tell us about the muscle's origin. A dense cord leaves from the top of our manubrium sternum and a wider

Figure 11 - 7

A lateral view of the neck: **(a)** is superficial with only the skin removed, and **(b)** has all the muscles removed except the sternocleidomastoid and the scalenes.

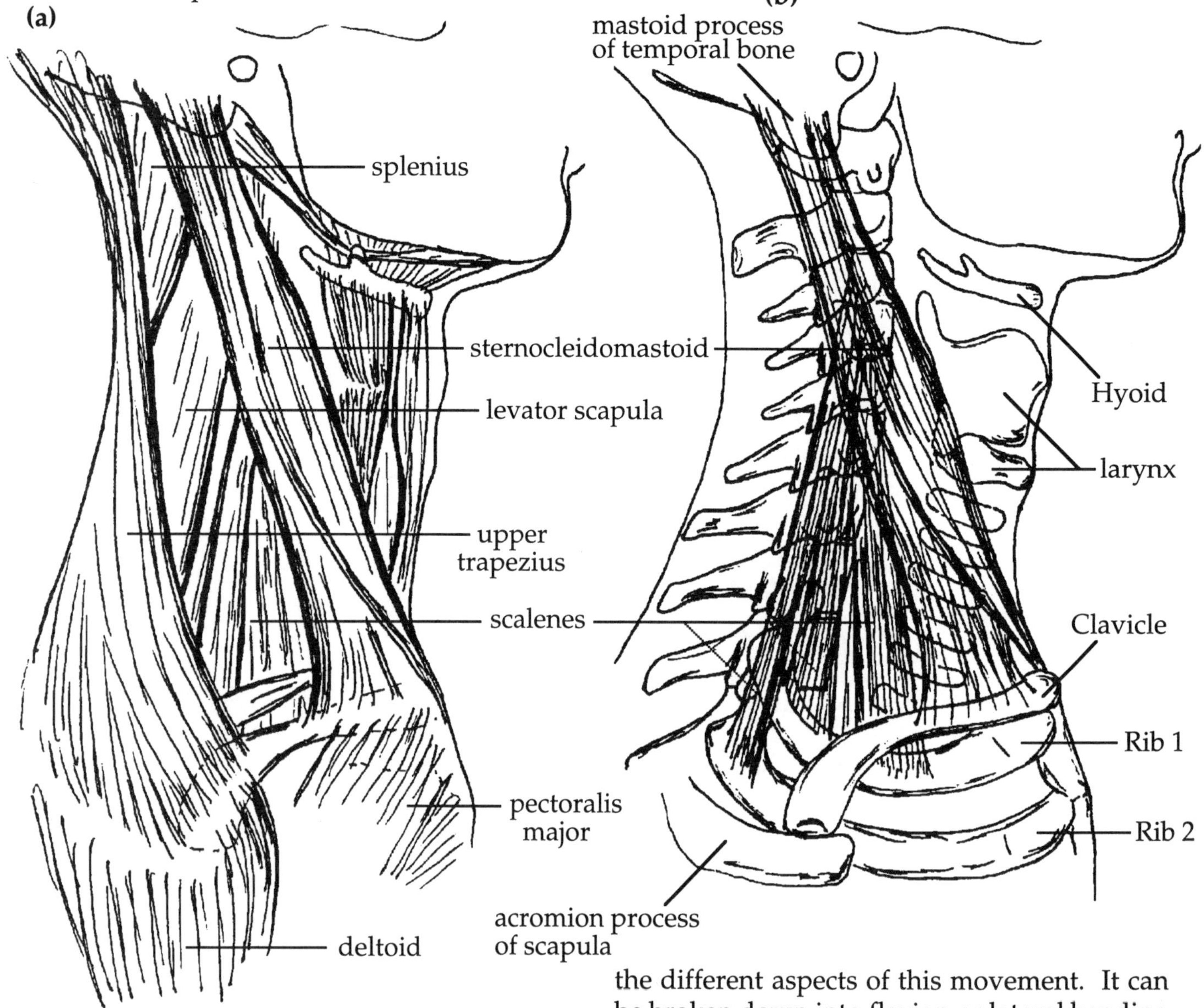

band from our medial clavicle. These two parts merge and run obliquely through our anterior neck to insert into our mastoid process (Fig. 11-7 and 11-9). All these bones, the tendons, and the muscle belly are easily palpable. Be sure you can find them.

With its oblique orientation, our sternocleidomastoid can help with vertical and horizontal movements (Fig. 11-9). When only one sternocleidomastoid muscle contracts, it would pull its mastoid process down and around toward our sternum. Perform this movement and feel your muscle as it stands out beside your voice box. Try to get a sense of the different aspects of this movement. It can be broken down into flexion or lateral bending of our cervical vertebrae and rotation to the opposite side.

When our left and right sternocleidomastoid muscles contract in unison, their opposing rotations cancel each other out so that only flexion of our cervical vertebrae occurs.

In a closer look at our sternocleidomastoid muscles, throughout most of their length they lie ventral to our cervical vertebrae and would, therefore, flex it. As the muscles reach for our mastoid processes, however, they fall dorsal to the pivot point of the movement of our occipital bone on our atlas (Fig. 11-7b). They would, therefore, extend this joint, tip-

Figure 11 - 8

A lateral view of the neck showing the complete action of the sterncliedomastoid, flexing the cervical vertebrae but extending the occiput on the atlas.

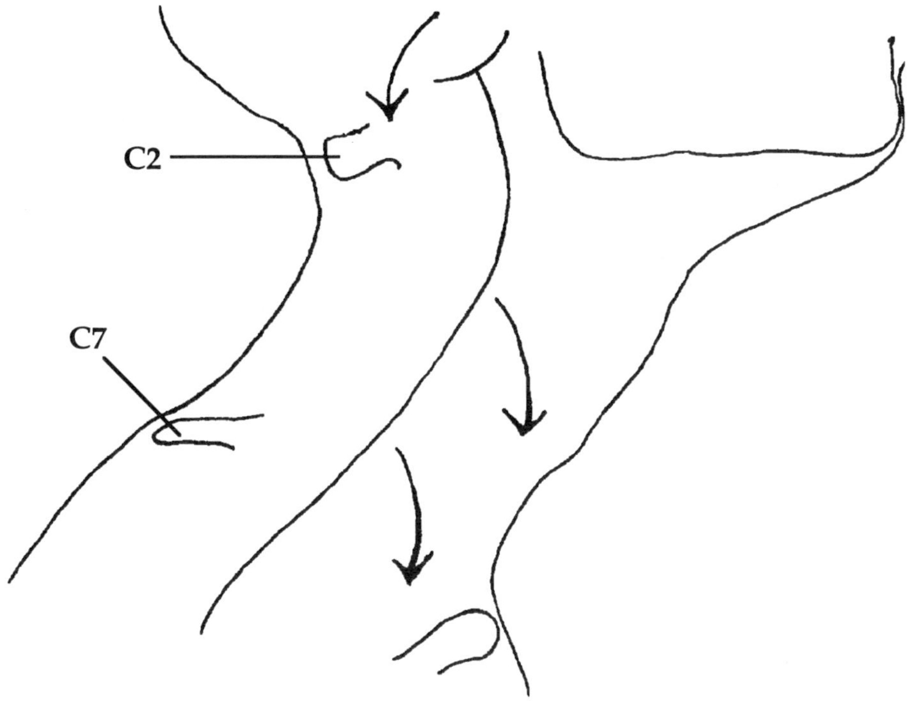

ping our head backward over our flexed neck. This action brings our head in front of our body, our chin out, and the back of our neck shortened and compressed (Fig. 11-8). This is a common posture in our society and may cause significant stress on the cervical vertebrae.

Scalenes

Our **scalenes** are thin bands of muscle which originate on the transverse processes of our second through sixth cervical vertebrae, posterior to our sternocleidomastoid. They run forward and downward, to insert into our first two ribs (Fig. 11-7 and 11-9). The name of these muscles describes their shape, a scalene triangle having no equal sides or angles.

To palpate the borders of your scalene triangle follow the posterior edge of your sternocleidomastoid, the anterior edge of your upper trapezius, and the superior edge of your middle clavicle. Your scalenes form a thin sheet within this triangle.

As our scalenes contract, they would pull our cervical transverse processes and ribs closer together. This movement can happen in either direction. If our ribs are stabilized, then our cervical vertebrae will flex or laterally bend. If our vertebrae are stabilized, then our ribs will be lifted. The latter is most common.

Lifting our ribs expands our chest, pulling air into our lungs. In restful breathing these muscles contribute very little, but in respiratory distress they actively participate and may stand out like straps on the side of our neck as they contract. Because of this behavior they are often called the strap muscles.

For most people it is difficult to isolate these muscles to demonstrate their contraction.

OTHER VITAL STRUCTURES

Besides muscles, there are many vital, sensitive structures located in our anterior neck (Fig. 11-9c). Our neck is traversed by so many systems, and yet stays so narrow that vital structures cannot be hidden, cushioned, or shielded like they are in many other places of our body. Most of these structures will be discussed in their appropriate system but they will be mentioned now, so we are aware of them as we first explore our anterior neck.

It is not that these structures should not be touched, only that when we do touch them,

Figure 11 - 9

Anterior view of the right side of the neck. In **(a)** only the skin has been removed, the arrows show the action of the sterno-cleidomastoid. In **(b)** the tapezius, sternocleidomastoid and sternohyoid muscles, and the distal clavicle have also been removed, revealing deeper muscles. In **(c)** all the muscles have been revoved, leaving the bones and other sensitive structures.

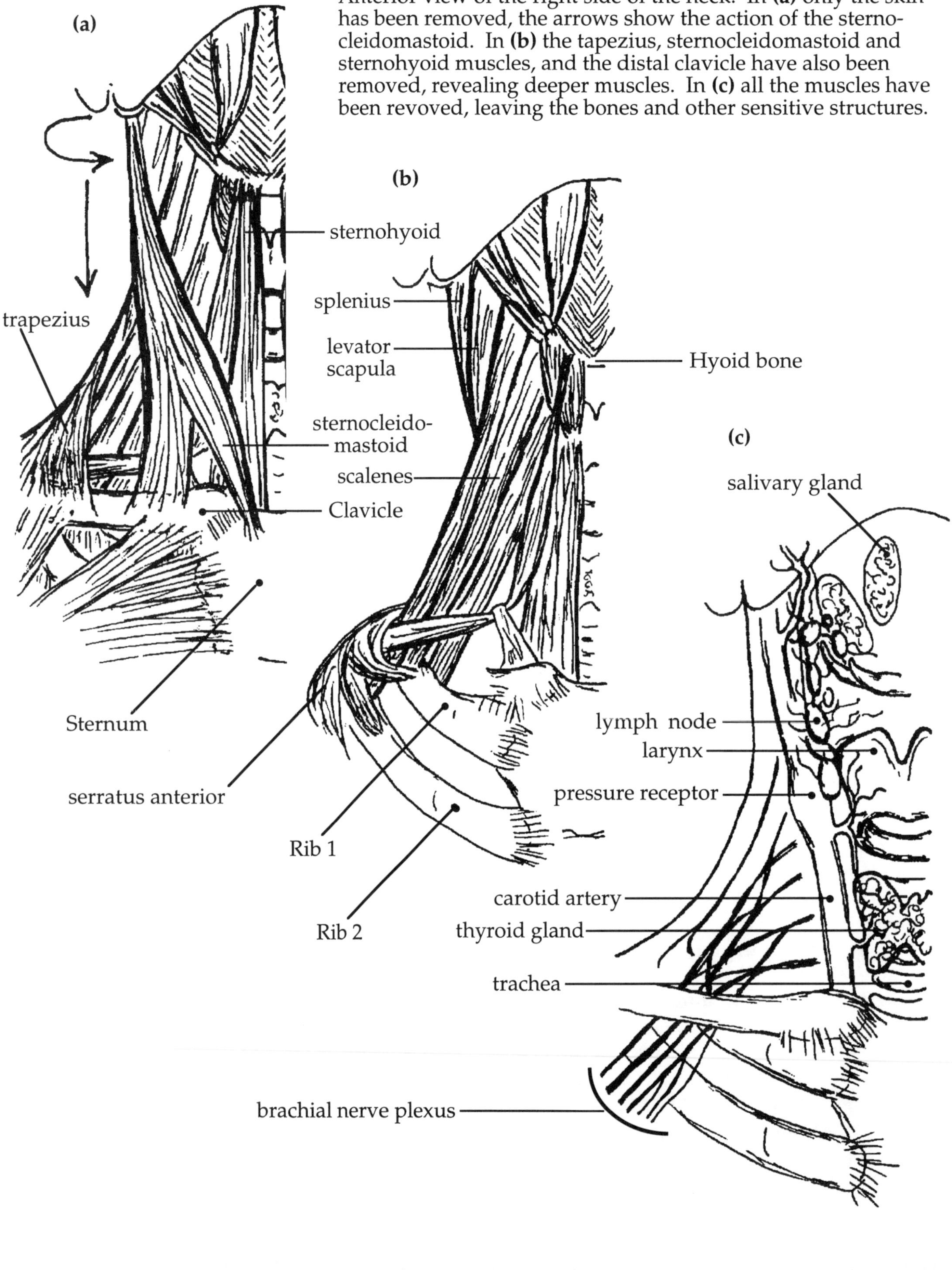

we know what they are and how their palpation may affect our body.

Two pairs of **salivary glands** sit behind our chin, between the two sides of our mandible. Manipulating our salivary glands may squeeze their saliva into our mouth and cause a swallow reflex. Try it on yourself right now.

Many **lymph nodes** and **vessels** sit under our mandible and along the anterior edge of our sternocleidomastoid. Generally they are not sensitive to palpation, but may be enlarged and tender during a sore throat or other infections. Sometimes lymphatic drainage is recommended, sometimes it can dangerously spread an infection. Until you know when it is appropriate, do not make assumptions.

The airway from our nose to our lungs lies in the midline in the front of our neck. It includes our **larynx** and **trachea.** Our larynx, commonly known as the voice box or Adam's apple, is located just inferior to our hyoid bone. We already have felt it vibrate and bobble as we talk and swallow. Our trachea continues downward from our larynx, slipping behind our sternum. Both are supported by cartilage giving them a firm but pliable nature. Most people feel choked by manipulation of the airway, but in some cases, working it may feel good and be exactly what our client needs.

Just above our sternum in front of our trachea is our **thyroid gland,** one of our important endocrine glands. Depending on its activity, manipulation may be helpful or harmful, so it is best left alone until you understand the effects of its function to make the appropriate judgement.

On each side of your larynx, feel the large, bounding pulse of your **carotid arteries.** Our left and right arteries together supply about two-thirds of the blood to our brain, so heavy, occluding pressure should be avoided.

Pressure receptors reside in the walls of our carotid arteries. These receptors measure our blood pressure. Our brain uses this information to regulate the activity of our heart and blood vessels. For better or worse, these receptors cannot tell the difference between internal blood pressure and external pressure on the vessel. They interpret all pressures as blood pressure and our brain adjusts our heart rate and vessel diameter accordingly.

If our receptors are compressed externally, our brain interprets this as high blood pressure and will slow down our heart and dilate our vessels to bring our blood pressure down. If, in fact, our blood pressure was normal, it may drop enough to cause collapse and loss of consciousness. In some cases of violence or abuse, this may be the desired affect, but it is not usually what clients come to us for.

In some cases of elevated blood pressure, brief, light stroking over the pressure receptors may help to reduce the blood pressure; however, this stroking needs to be done with mindfulness of the possible affects.

The last sensitive structure in our anterior neck is our **brachial nerve plexus,** communicating between our brain and our arm. These nerves leave our vertebrae through intervertebral foramen C5 through T1. They travel through the area of our scalene muscles, over our first ribs, and under our mid clavicle. Nerves are fussy about pressure. In our neck, where the padding is minimal, strong or continuous pressure may irritate our nerves causing inflammation or shooting pains down our arms, similar to hitting our funny bone.

CHEST WALL MUSCLES

Intercostals

The organs in our chest are supported and protected by our ribs. Tiny muscles fill the spaces between adjacent ribs (Fig. 11-10) and so acquire the name **intercostal muscles.** There are three layers of intercostal muscles, oriented in the same manner as our lateral abdominal wall. That discussion follows on page 247 and will not be repeated here.

Our **external** intercostal muscles lift the anterior end of our ribs to expand the anterior-posterior depth of our chest and reduce our thoracic pressure. The most commonly noted result is to draw air into our lungs or cause inhalation. Other effects are discussed under the diaphragm. To palpate your external intercostals, press your finger tips into the space between your ribs, under you arms.

Figure 11 - 10
An anterior view of the rib cage showing the two outer layers of the intercostal muscles.

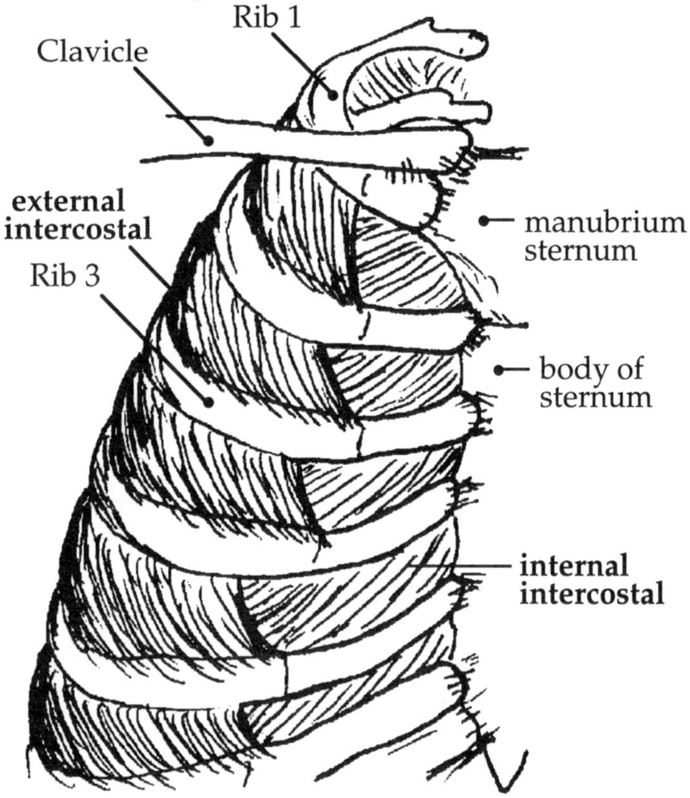

Figure 11 - 11
The abdominal egg with its bony support.

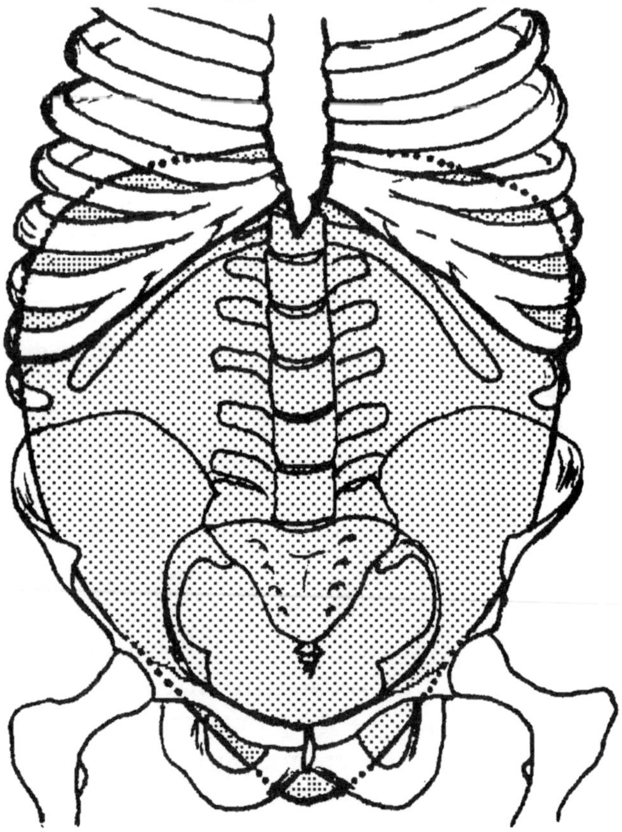

Take a few quick breaths and feel your external intercostals flickering.

Our **internal** intercostal muscles pull our ribs down, decreasing the space in our chest and increasing the pressure. This pushes the air out of our lungs or causes exhalation. These muscles are deep and not palpable

ABDOMINAL MUSCLES

• Our **abdominal cavity** is shaped like a hen's egg (Fig. 11-11). The broad end of the egg is at the top, surrounded by the xiphoid process of our sternum, our six lower ribs, and our lower thoracic vertebrae. The narrow end of the egg is at the bottom, surrounded by our pelvic bones and sacrum.

These surrounding bones provide protection and support for our abdominal organs, but also are an encumbrance, limiting the expansibility of our organs as well as the movement of our vertebrae. This encumbrance is offset by the mobility of the ribs and two large gaps in the bone. These gaps provide our abdomen with two pliable soft spots, one in the front where we can feel our belly bulge and recede, and one at the top where there is no bony partition between our chest and abdomen..

• Within its bony protection, our abdominal cavity is entirely wrapped in muscle and fascia. It has anterior, lateral, and posterior walls, a ceiling, and a floor. The soft, muscular

> The functions of the organs in our abdominal cavity are similar to those of an egg, nourishing and disposing of wastes. In women, abdominal organs also may hold a fetus and support it metabolically.

> The bottom end of our abdominal cavity often is referred to as our **pelvic cavity.** There is no structural barrier which separates our pelvic and abdominal cavities. Our pelvic cavity is simply the lower end of our abdominal cavity, and the floor of our pelvic cavity also is the floor of our abdominal cavity.

Figure 11 - 12
The muscular walls crossing the bony gaps of the abdominal cavity may **(a)** stretch and bulge under increased pressure or **(b)** contract to push in and support. The anterior wall usually is the opposite of the roof and floor.

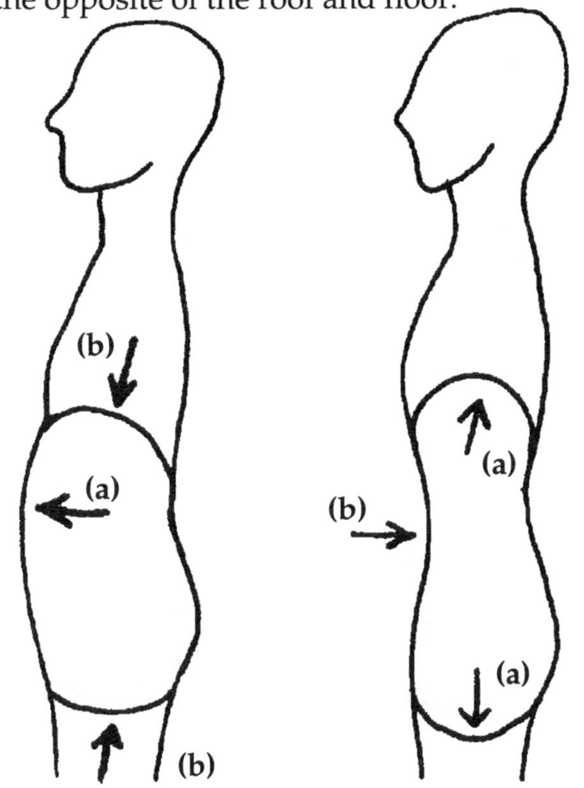

expanses support across the bony gaps, but allow stretch and recoil in response to changes in abdominal pressure (Fig. 11-12). It is important to have this flexibility since the abdomen contains organs which oscillate somewhere between bursting full and empty, according to our indulgences. Our stomach and urinary bladder are two such organs.

While muscular walls allow stretching and bulging under increased pressure, they may themselves contract to push back and hold in or up, protecting and supporting our abdominal organs in the absence of bone. Too much bulging or sagging may indicate inadequate tone in these muscle and may be reduced by exercise.

• The muscles of our anterior, lateral and posterior abdominal walls are all supported by one sheet of connective tissue which wraps around the entire cavity like a built-in kidney belt. This sheet is very thin, but tough, wide, and layered (Fig. 11-13 and 11-14), like several sheets of cellophane wrapped around our abdomen. This simple idea can be confused by the many names it is given such as the lumbar fascia, the lumbodorsal fascia, the thoraco-lumbar fascia, and the aponeurosis. In this book we will call this belt of connective tissue our **lumbar fascia.**

In our back, our lumbar fascia attaches into the periosteum of our lumbar spinous processes. In the front midline, the left and right fascia weave into each other with no bony interlude (Fig. 11-14a). At the top, our lumbar fascia attaches into our ribs, costal cartilages, and sternum. At the bottom, it attaches into our sacrum, ilium and pubis (Fig. 11-13).

As our lumbar fascia spans from our anterior superior iliac spine to our pubic bone, its bottom edge curls to form a dense cord called the **inguinal ligament** (Fig. 11-13).

As they pass from origin to insertion, all the muscles of our abdominal walls split between specific layers of our lumbar fascia (Fig. 11-14b). For our lateral abdominal muscles, our lumbar fascia forms much of their origin and insertion tendons.

Figure 11 - 13
A lateral view of the lumbar fascia showing the bony attachments.

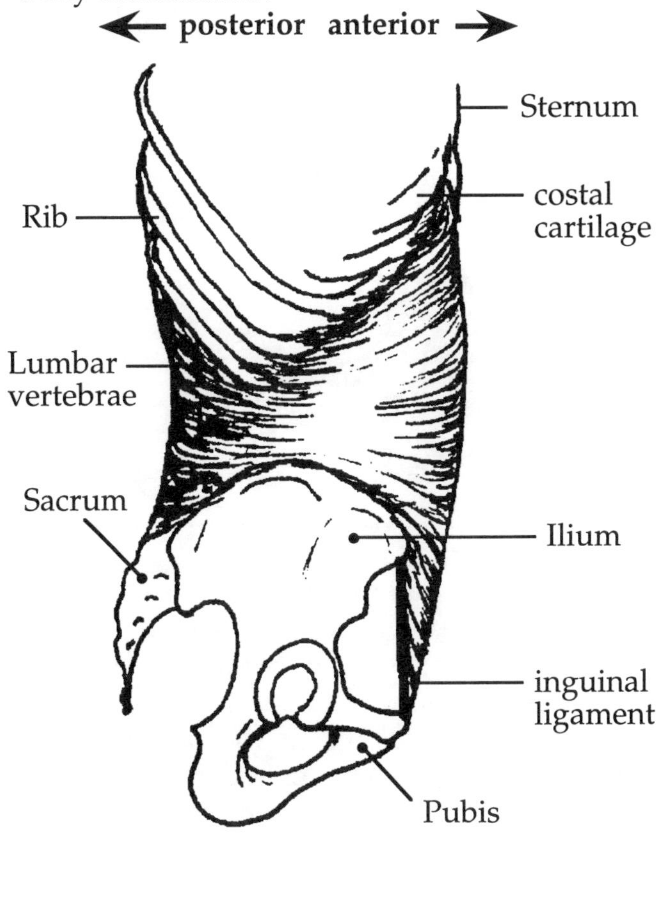

Figure 11 - 14

A cross section through the mid abdominal area showing **(a)** lumbar fascia with its posterior vertebral attachment and **(b)** the muscles of the abdominal wall. In butchered animals, the lateral wall is sliced into **(c)** bacon.

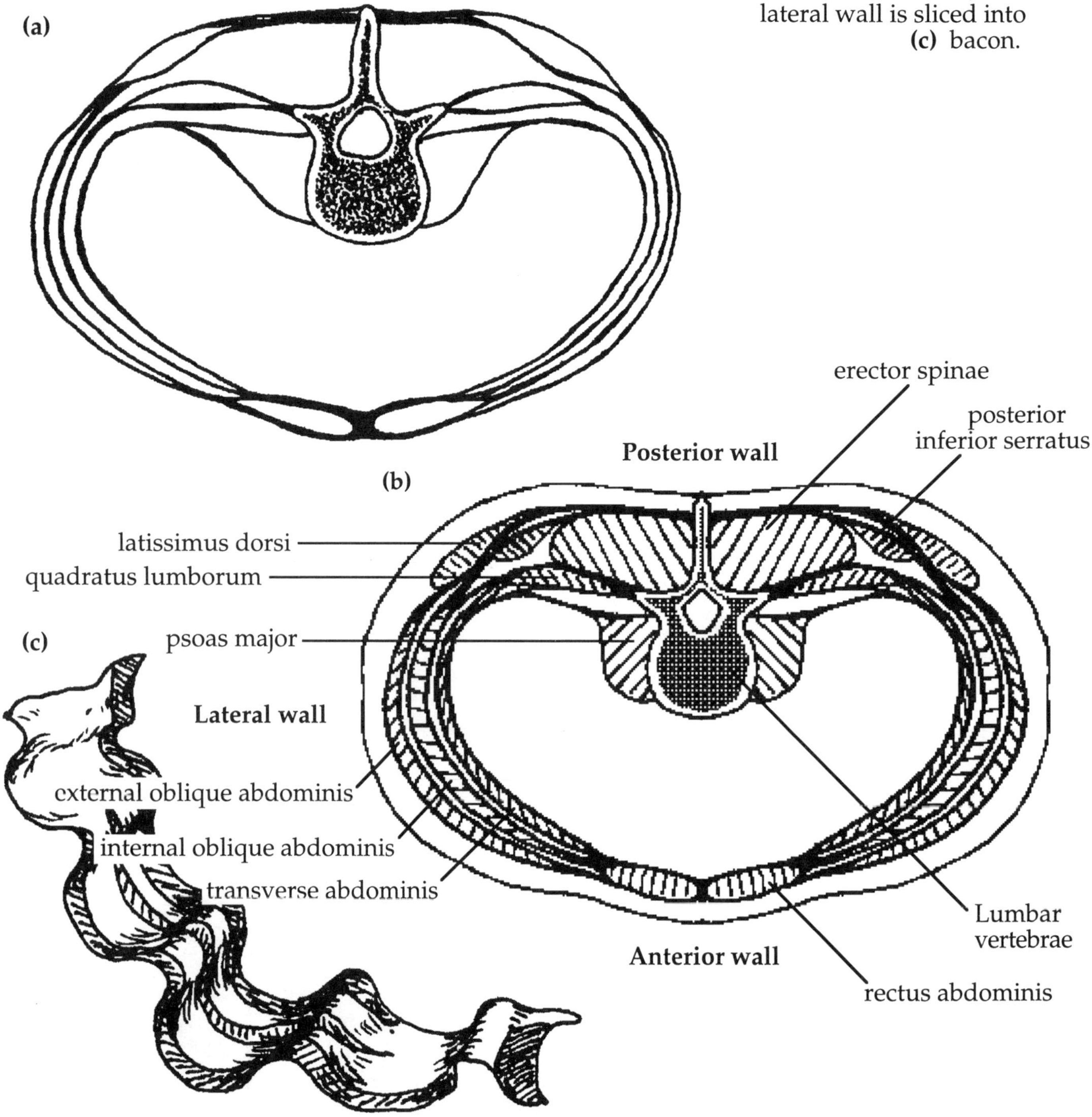

To portray how thin the three muscles of our lateral abdominal wall are, we can go to the butcher shop. The meat cut which is made into bacon is the lateral abdominal wall of a fat pig. Most of the size of the bacon strip is the fat which has accumulated in the fascia of the muscle. The tiny veins of red meat are the three layered muscles. The red tuft at the end is a snip from the rectus abdominis (Fig. 11-14). Is ignorance bliss? Enjoy your next bacon and egg breakfast, or will it be eggs and toast?

We can get a lot of information about our own muscles at the butcher shop since all mammals follow a similar pattern.

ANTERIOR ABDOMINAL WALL

Rectus Abdominis

To be 'erect' means to stand vertical and muscles which have 'erect' or 'rectus' as part of their name are vertical in their orientation. Our **rectus abdominis,** then, is a vertical muscle in the anterior wall of our abdomen. It lies just lateral to the midline and spans from our pubic bone to the costal cartilages of our fifth, sixth, and seventh ribs and xiphoid process (Fig. 11-14 and 11-15). The fascia of the muscle is continuous over this distance but the muscle fibers are collected into four shorter segments, one above the other, producing the rippled appearance of a well-conditioned belly.

Our rectus abdominis has three functions. **1.** It protects and supports our central abdominal organs. **2.** When we take in a big breath, our abdominal organs are pushed down and out against our rectus abdominis. It stretches to show a protruding belly (Fig. 11-23a). To breathe out we may contract our rectus abdominis to push the organs back in and up, compressing the air out of our lungs (Fig. 11-23b). The rectus abdominis then, is a muscle of exhalation. And **3.** The skeletal joints crossed by the rectus abdominis are the lumbar intervertebral joints. Since the muscle is ventral to these joints, it would flex them.

Figure 11 - 15

An anterior view of the rectus abdominis. The arrows indicate its action, bringing the anteroir ribs and pelvis together as it flexes the lumbar vertebrae.

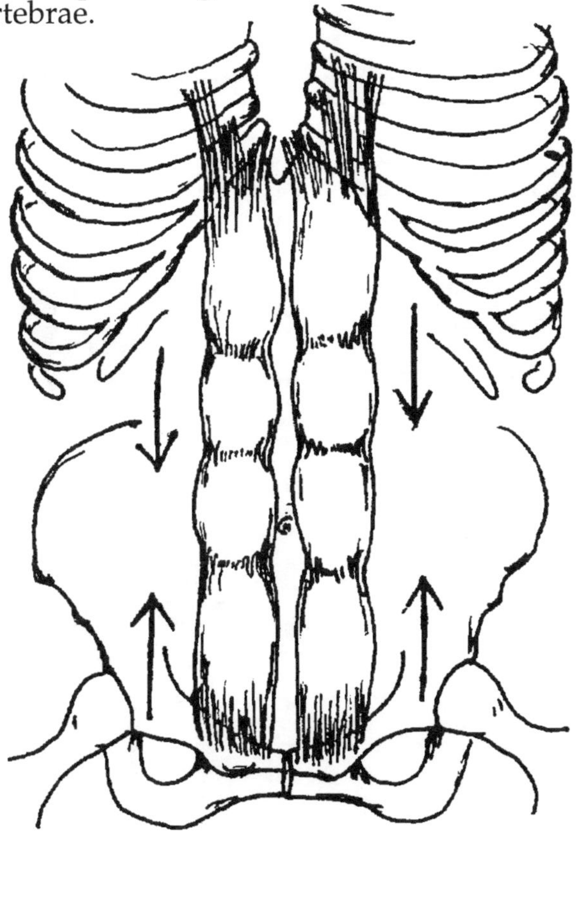

LATERAL ABDOMINAL WALL

The muscle bellies of our lateral abdominal wall begin just lateral to our rectus abdominis and wrap around to our side (Fig. 11-14b) between our ilium and ribs. Their origin and insertion tendons use the lumbar fascia and so extend beyond these limits.

Three thin muscles are layered over each other in this area. All three muscles together would be less than one-half inch thick. The fibers of each muscle are oriented in a different direction just like the thin sheets of wood glued together to make a single sheet of plywood or the layers in a rubber tire. The advantage in our abdominal wall is the same as in plywood and tires. A multi-directional support provides significant, multi-directional strength with minimal bulk. Our intercostal muscles forming our chest wall follow the same pattern as our lateral abdominal wall.

External Oblique Abdominis

The fibers of our outer muscle layer run obliquely in the same direction our fingers point when we slip our hands into our front pants pockets. This muscle originates from our last eight ribs and passes downward and medially. Its fascia and tendon is part of our lumbar fascia, inserting into our midline, pubic bone, inguinal ligament and anterior iliac crest. This muscle is called our **external oblique abdominis** (Fig. 11-16a).

Internal Oblique Abdominis

The fibers of our middle muscle layer run obliquely but perpendicular to our outer layer. They originate from our inquinal ligament, iliac crest, and lumbar fascia, pass upward and medially to insert into our anterior midline and lower four ribs This muscle is our **internal oblique abdominis** (Fig. 11-16b).

Figure 11 - 16
An anterior view of the three muscle layers in the lateral abdominal wall.
(a) external oblique **(b)** internal oblique **(c)** transverse abdominis

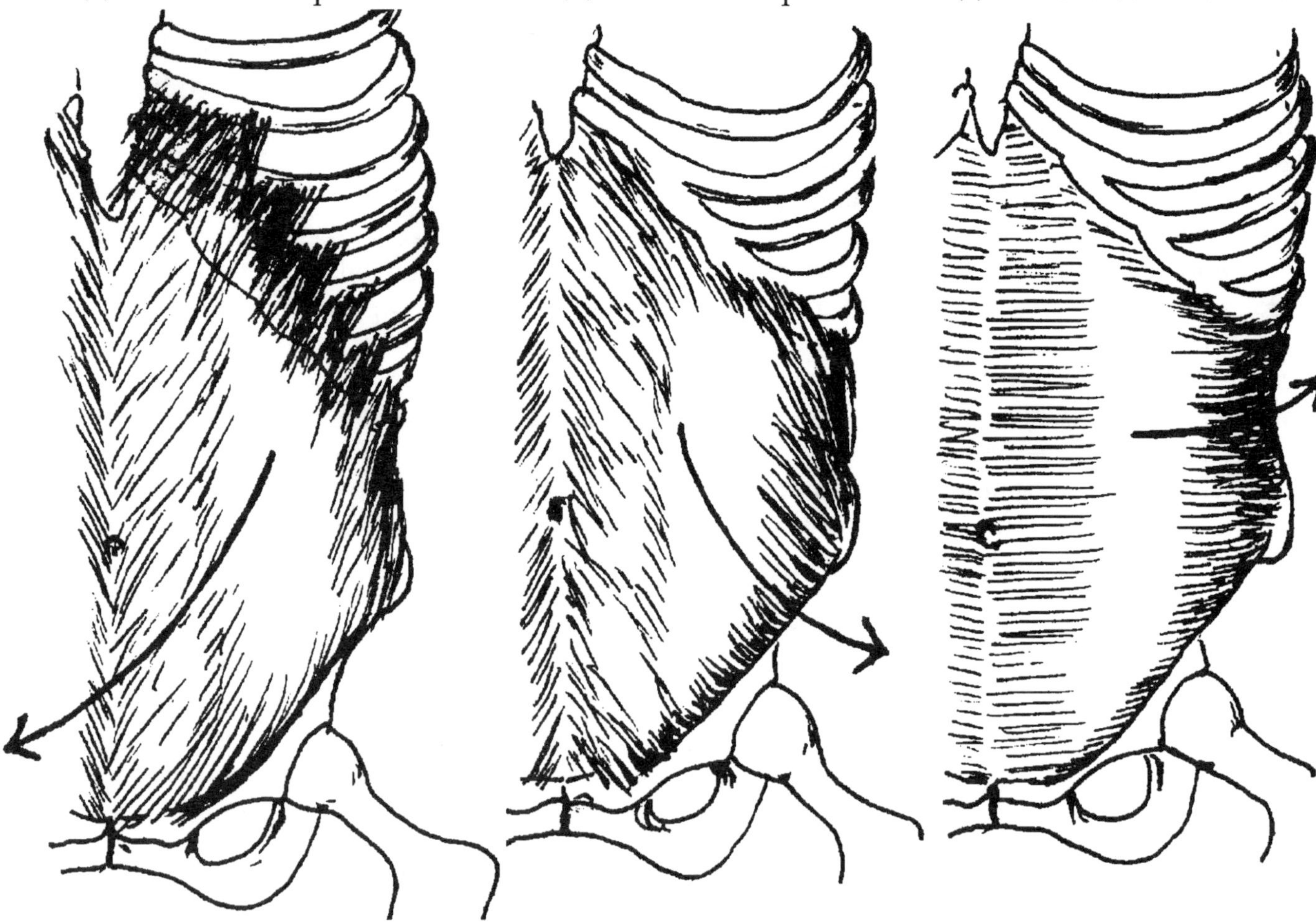

Transverse Abdominis

The fibers of our inner muscle layer lie transversely or horizontally. They originate from our lower ribs, lumbar fascia, and iliac crest, inserting into each other at our anterior midline. This muscle is called our **transverse abdominis** (Fig. 11-16c).

• The functions of our lateral abdominal muscles are similar to our rectus abdominis. **1.** They protect and support our abdominal organs. **2.** They push the abdominal organs inward and upward to compress the air out of our lungs, producing exhalation. Our transverse abdominis is especially good at this (Fig. 11-17). And **3.** These muscles also cross ventral to our lumbar intervertebral joints. Oblique muscles combine vertical and horizontal directions and so function to flex and rotate our lumbar vertebrae. Our external obliques rotate to the opposite side while our

Figure 11 - 17
Since the transverse abdominis wraps horizontally around the abdomen, its contraction will pull the soft anterior abdominal wall back toward the bones.

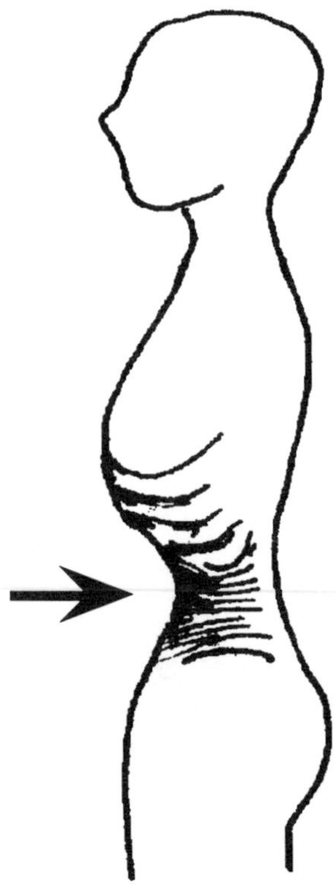

internal obliques rotate to the same side. Our transverse abdominis does not directly affect movements of our lumbar vertebrae.

POSTERIOR ABDOMINAL WALL

Three muscles support the posterior wall of our abdomen: our **psoas major, psoas minor,** and **quadratus lumborum.**

Psoas Major

'Psoas' refers to the loin, which means the long part and relates to our lumbar or waist area. In so many people this area is actually compressed or collapsed. This name then describes the challenge of what this area might become, rather than a picture of what it is.

Our **psoas major** has its origin on our anterior vertebral bodies from T12 to L5. Bundles of muscle from each vertebra join together to form a vertical column beside our vertebral bodies. They cross our sacroiliac joint, descend over our pelvis in front of our pubis, and insert into the lesser trochanter of our femur (Fig. 11-18a). Along this path our psoas major crosses eight joints: the six intervertebral joints from T12 to our sacrum, our sacroiliac joint, and our hip joint. Normally, our psoas major lies ventral to all these joints (Fig. 11-19a), and so would flex them.

The middle part of our psoas major is palpable. Lie on your back. Pull your knees up so your feet are flat on the floor. Relax your abdomen. Find the hollow area about one-third of the way from your pubic bone to your anterior superior iliac spine. This is where the psoas tendon crosses over your pubic bone. From this point, move upward into your abdomen. On deep palpation, your psoas may be felt as a cord running parallel to your midline, about one hand width lateral to it. If you cannot find your psoas major, make it contract by flexing your hip, lifting your foot off the floor. It will pop up at your fingers as it contracts.

Psoas Minor

Our **psoas minor** originates with our psoas major on the bodies of T12 and L1. It runs down with our psoas major, crosses our sacroiliac joint, and inserts into the junction of our ilium and pubis (Fig. 11-18b). Its main function is to flex our lumbar vertebrae, tilting our pelvis anteriorly.

Some studies show that the psoas minor is found in only 60 percent of the population.

Figure 11 - 18
An anterior view of the muscles of the posterior abdominal wall.
(a) psoas major **(b)** psoas minor **(c)** quadratus lumborum

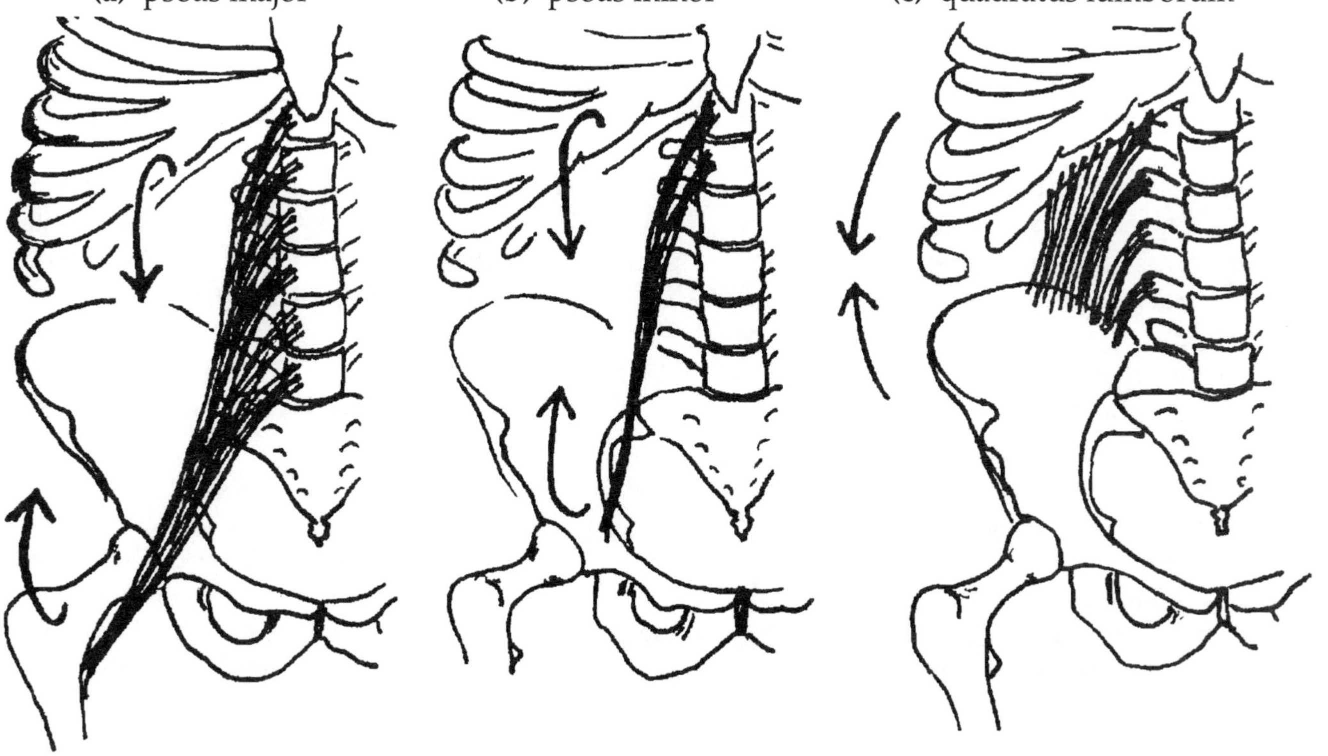

Figure 11 - 19

A lateral view of the psoas major. **(a)** shows a **normal** lumbar curve, the psoas falls ventral to the pivot point of the intervertebral joints, thus flexing them. **(b)** shows an **excessive** lumbar curve and posteriorly tilted pelvis. This position pulls the upper psoas dorsal to the pivot point of the intervertebral joints, thus extending them.

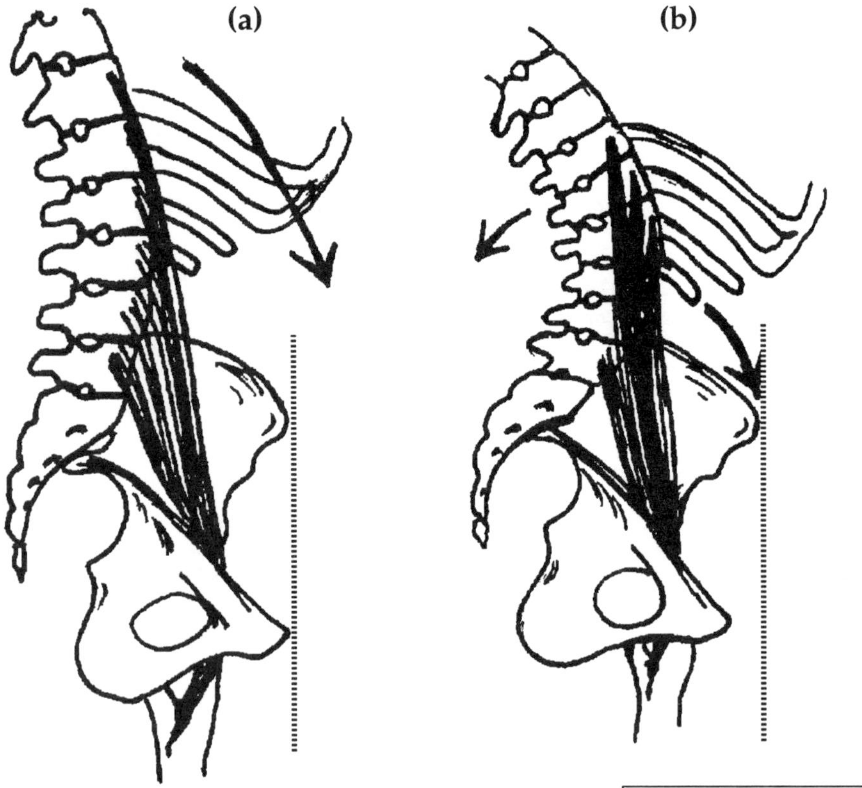

When our secondary lumbar curve is exaggerated and excessively extended, tilting our pelvis posteriorly, our psoas major may begin to fall dorsal to the pivot point of our lumbar intervertebral joints (Fig. 11-19b). As the orientation of the muscle changes, so does its function. Muscles which lie dorsal to joints produce extension, so our misplaced psoas will reinforce the extension in our lumbar curve rather than flexing it. Release of this condition then needs to include relaxation of our psoas major muscle.

Quadratus Lumborum

The name **quadratus lumborum** tells us that this muscle is a flat, four-sided quadrangle located in the lumbar area. It lies immediately posterior to our psoas, originating from our posterior iliac crest and inserting into our twelfth rib and upper lumbar transverse processes (Fig. 11-18c). As it contracts, it pulls our ilium and ribs closer together, producing lateral bending of our lumbar vertebrae.

While the ilium is called the origin, either end of the muscle can be stabilized. If we stabilize our pelvis, our upper body bends over toward the contracting side. If our rib cage is stabilized, our pelvis on the contracting side is lifted, taking the whole leg and foot up with it. This may be useful in walking for anyone who has a stiff or painful leg.

Walking forward requires two basic steps: **1.** get our foot up off the ground, and **2.** plunk it down somewhere in front of its original spot. In normal walking we get our foot up off the ground with a complex coordination of flexing our hip and knee, and extending our ankle and toes. This requires mobility in four major joints. If one of these joints is stuck, the whole process is limited. But we have just discovered another way to get our foot up off the ground. Stabilize our rib cage and contract our quadratus lumborum on our affected side to lift our pelvis, leg, and foot. Our whole leg can be stiff. It can be a prosthetic appendage. Nothing is required of it except to tag along. By medial rotation of our unaffected hip, we can move our lifted foot forward and plunk it down. All we need from our affected leg is to hold us up while we shift the position of the unaffected leg. This is not the easiest way to walk, but it moves us along.

ABDOMINAL ROOF - THORACIC FLOOR

Diaphragm

Our **diaphragm** is one of the most important skeletal muscles in our body, not so much for a particular movement of our skeleton but for its profound effect on many of our internal organs.

'Diaphragm' means partition, and our diaphragm is a thin, horizontal muscle which divides our torso into two cavities. It forms the floor of our thoracic cavity or chest, a chest being a rigid box which stores treasures. It forms the roof or ceiling of our abdominal cavity or belly, a belly being a leather bag, or a soft, floppy container. There are many images for the shape of our diaphragm It is like a dome, an upside down bowl, an umbrella, or a wilted daisy (Fig. 11-22).

The origin of our diaphragm forms a complete ring all the way around our torso. Small bundles of muscle fibers, referred to as **slips**, arise from different bones, cartilage or fascia. These include the anterior bodies of our upper lumbar vertebrae, the transverse processes of L1, the fascia of our psoas and quadratus lumborum, and the internal surface of our floating ribs, costal margin, and xiphoid process of our sternum (Fig. 11-21). This origin is the stabilized end of our diaphragm.

From their originating circumference, the muscle slips of our diaphragm point upward and inward, toward the center of our body. The fasciae of all the muscle slips weave into each other at the center to form a **central tendon** (Fig. 11-21). This is the insertion of our diaphragm. It has no bony insertion, but simply meets itself. The central tendon is the highest part of our diaphragm with the muscle slips bending down toward their origin, like the drooping petals of a wilted daisy.

At rest, the central tendon of our diaphragm floats above its origin. As our diaphragm contracts, its central tendon is pulled down toward its stabilized origin, flattening the dome (Fig. 11-20). The floor of our chest falls, creating more vertical space within our chest and reducing thoracic pressure to form a partial vacuum. This extra space is stolen from our abdomen. The domed ceiling of our abdomen comes down, increasing abdominal pressure and compressing abdominal organs.

As our diaphragm relaxes, the compressed abdominal organs push their ceiling back up. The abdomen reclaims its space and reduces its pressure. The chest is now compressed, but only for an instant. Our diaphragm contracts again and the process repeats itself.

This rhythmic pressure fluctuation produces many physiological effects in our abdomen and thorax. In our abdominal cavity, it **1.** stimulates the movement of food through our digestive canal, especially through our transverse colon, enhancing digestion and elimination, **2.** helps to pump blood through all our abdominal organs, especially our spleen, pancreas, liver, adrenals, and kidneys, and **3.** promotes the movement of venous blood and lymph from our legs and abdomen toward our chest. No wonder so many organs nestle up against our diaphragm to be

Figure 11 - 20

An anterior schematic of the torso, showing how the domed diaphragm divider moves during contraction.

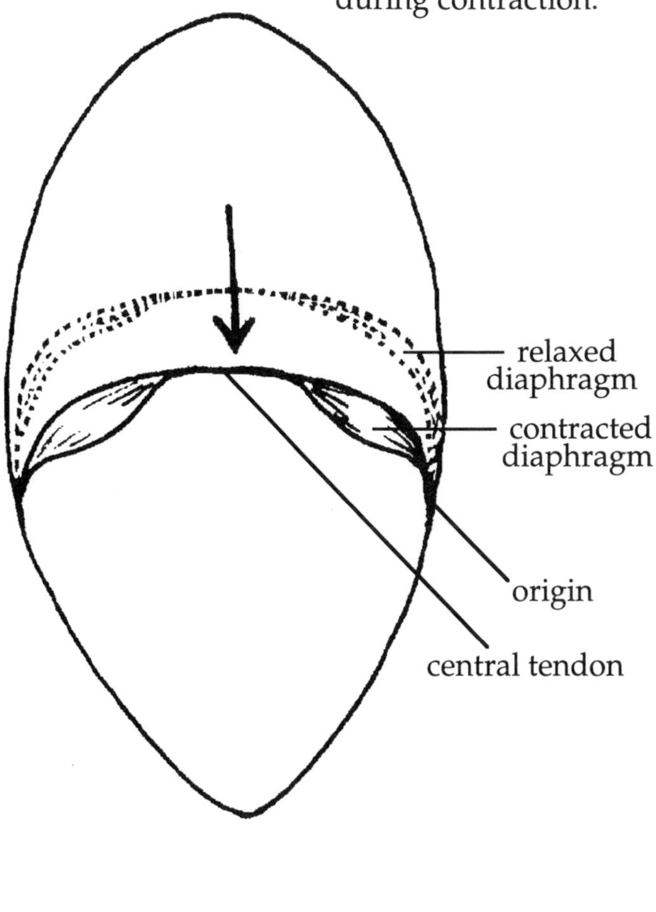

Figure 11 - 21

An anterior view of the diaphragm with the anterior left and central portion of the diaphragm and rib cage cut away.

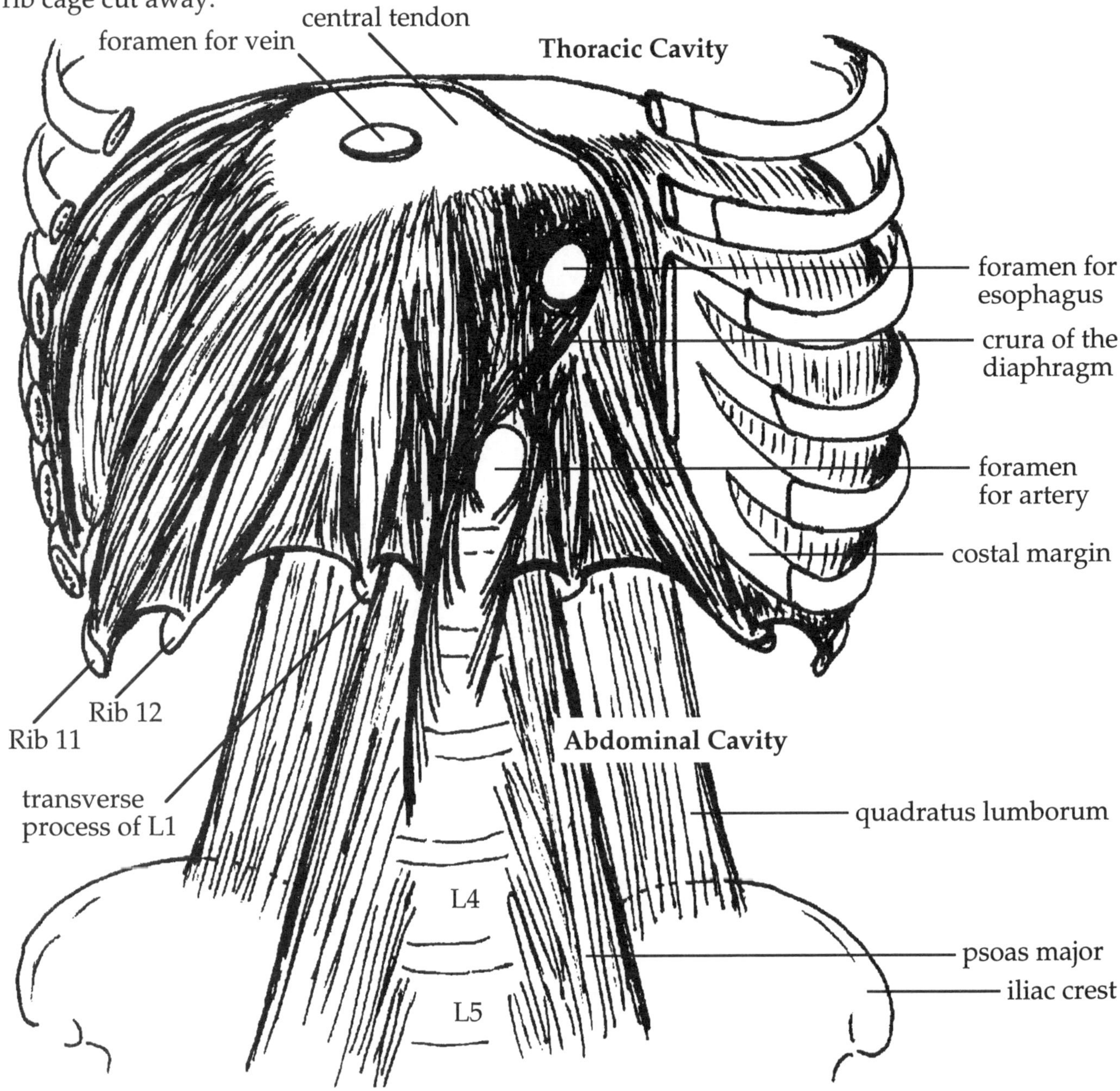

massaged by it.

When our diaphragm contracts, the partial vacuum in our chest sucks into itself, pulling **1.** air into our lungs and **2.** venous blood and lymph toward our heart. As our diaphragm relaxes, air is pushed out of our lungs and the flow of blood and lymph momentarily is reduced.

One effect of our diaphragm is problematic. Our esophagus spans through our chest and connects to our stomach immediately inferior to our diaphragm. As our diaphragm contracts, the pressure on our stomach and the vacuum around our esophagus tend to pull stomach contents back into our esophagus. This backward flow is called **gastric reflux.** If reflux happens, our esophagus is burned by the acid and enzymes from our stomach. Special muscle slips of our diaphragm, called the **crura,** encircle our esophagus like a noose (Fig. 11-21). In an effort to prevent gastric reflux, our crura cinch closed as our diaphragm contracts and our abdominal pressure increases.

Figure 11 - 22

The umbrella and wilted daisy are familiar images which give us a sense of the shape of the diaphragm.

Figure 11 - 23

A lateral view through the body showing **(a)** the opening and **(b)** the closing of the torso. The H indicates high pressure and the **L** indicates low.

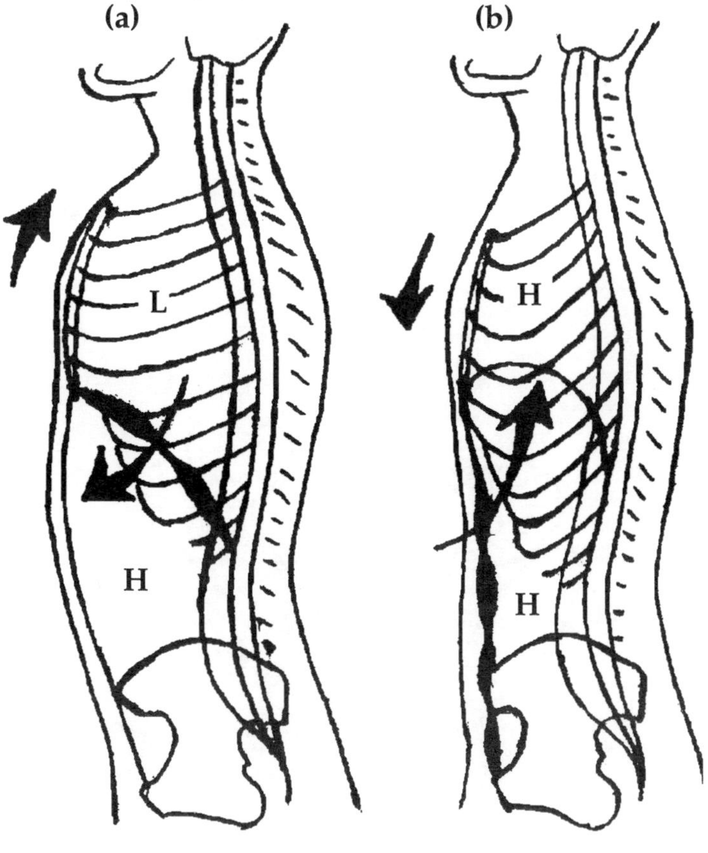

The overall size of our torso is in constant flux. Like the daily tides of the ocean, our body rises and recedes 15 to 20 times every moment (Fig. 11-23). This rhythmic, alternating expansion and contraction produces an overall fluctuation in our torso, open and closed, open and closed. When our torso opens, our thoracic pressure is low, while our abdominal pressure is high. When our torso closes, our thoracic and abdominal pressures are both high. This leads to profound physiological events in all our internal organs. The most significant events are discussed as effects of our diaphragm beginning on page 251, but our whole torso participates, including all our muscles of inspiration and expiration, and the movement of our ribs.

The two muscular expanses which most clearly reciprocate and work antagonistically to accomplish these events are our diaphragm and our anterior-lateral abdominal wall (Fig. 11-23), referred to as the two soft spots of our abdominal cavity. Our diaphragm is hidden inside our torso, but our anterior-lateral abdomen is exposed and vulnerable like the soft spots on a baby's head. Our two soft spots contract and relax like two alternating pistons. As our diaphragm contracts, it comes down and pushes downward on our internal organs. Our organs have to go some-where, so they push outward against our abdominal wall. A relaxed anterior-lateral wall will surrender to be stretched outward. As our anterior-lateral abdominal wall contracts, its inward pressure reflects upward through our organs, stretching our relaxed diaphragm upward.

Depending on the conditions of breathing, more or less of our muscles of inspiration may contract as our diaphragm lowers, helping to open our chest. Similarly, more or less of our muscles of expiration may contract with our abdominal wall, helping to compress our chest. The larger, more forced, or more difficult our breath, the more muscles work to accomplish it.

We have said that our diaphragm is a skeletal muscle and that skeletal muscles are voluntary, contracting only by conscious decision. But who thinks about every breath they take? The dolphins and whales do, but not people. Just like so many other muscle responses, unconscious reflexes from our brain cause our diaphragm to contract when we forget about it, which is most of the time. We can, however, interrupt and consciously contract our diaphragm any time we want.

ABDOMINAL FLOOR

The floor of our abdominal or pelvic cavity is supported by a small bowl composed of many muscles. They originate from the diamond-shaped inferior brim of our pelvis, including the body of our pubic bone, our inferior pubic ramus, our ischial tuberosity, and a ligament which goes from our ischial tuberosity to our coccyx (Fig. 11-26). A few slips of muscle attach to higher points on our pelvis. All these muscles are sometimes called our pelvic diaphragm. They function like a mirror image of our diaphragm, lifting upward into our abdominal cavity (Fig. 11-24), compressing it, and supporting our lower pelvic organs against gravity. The supported organs include our rectum, anus, male or female reproductive organs, and urinary bladder.

Figure 11 - 24

An anterior schematic of the torso, showing the mirrored movement of the diaphragm and the abdominal floor. The position of the muscles in relaxation is shown in dotted lines, while their contracted position is shown in solid lines.

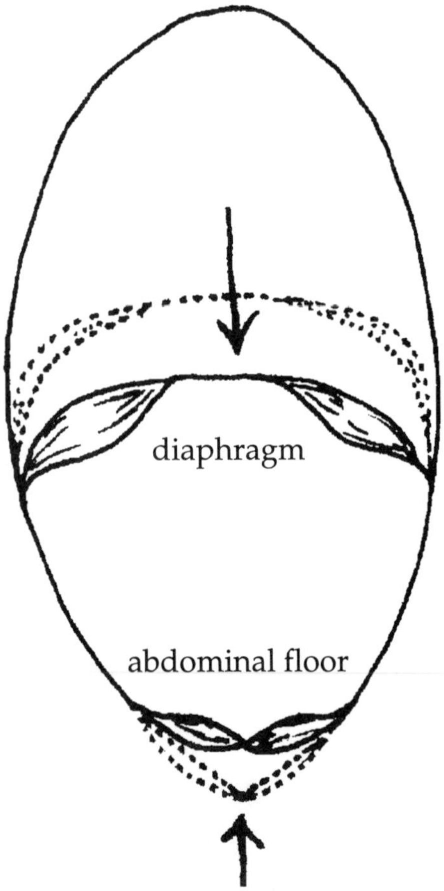

Figure 11 - 25

A lateral view of a midline cut through the lower abdominal organs. **(a)** shows their normal location. **(b)** shows a dropped abdominal floor and uterus, setting the urinary bladder up over its urethra.

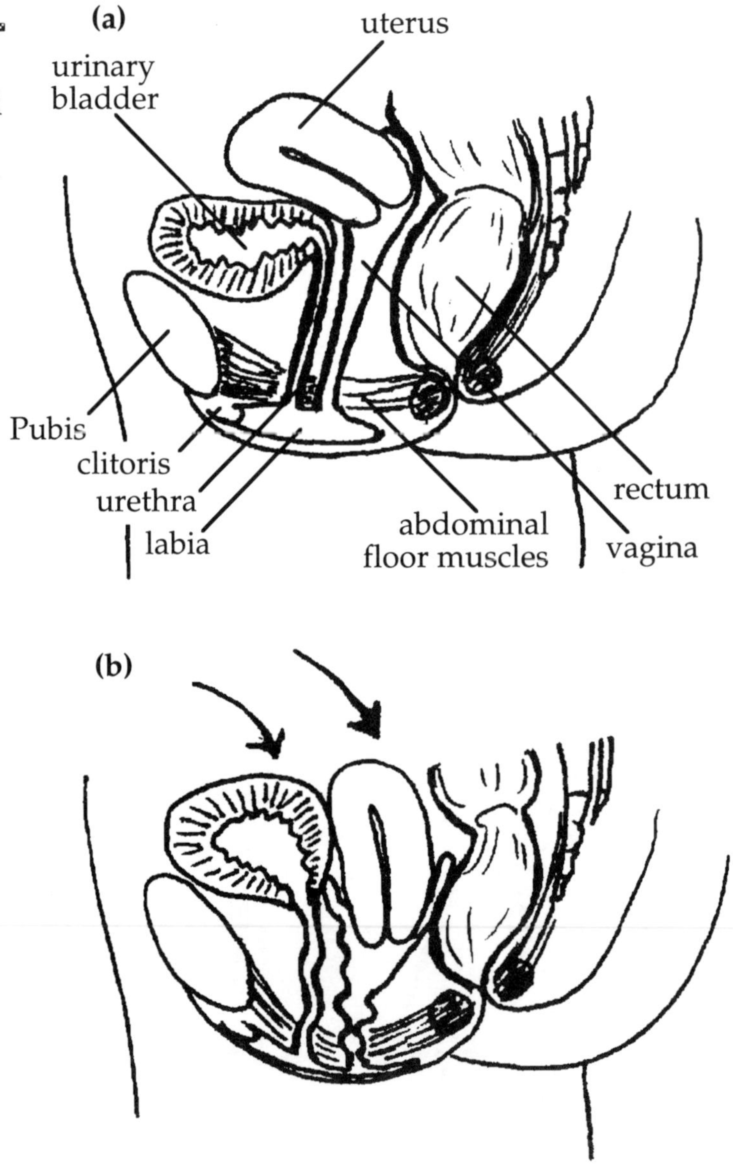

Figure 11 - 26

An inferior view of the muscles of the abdominal floor, differentiating female and male

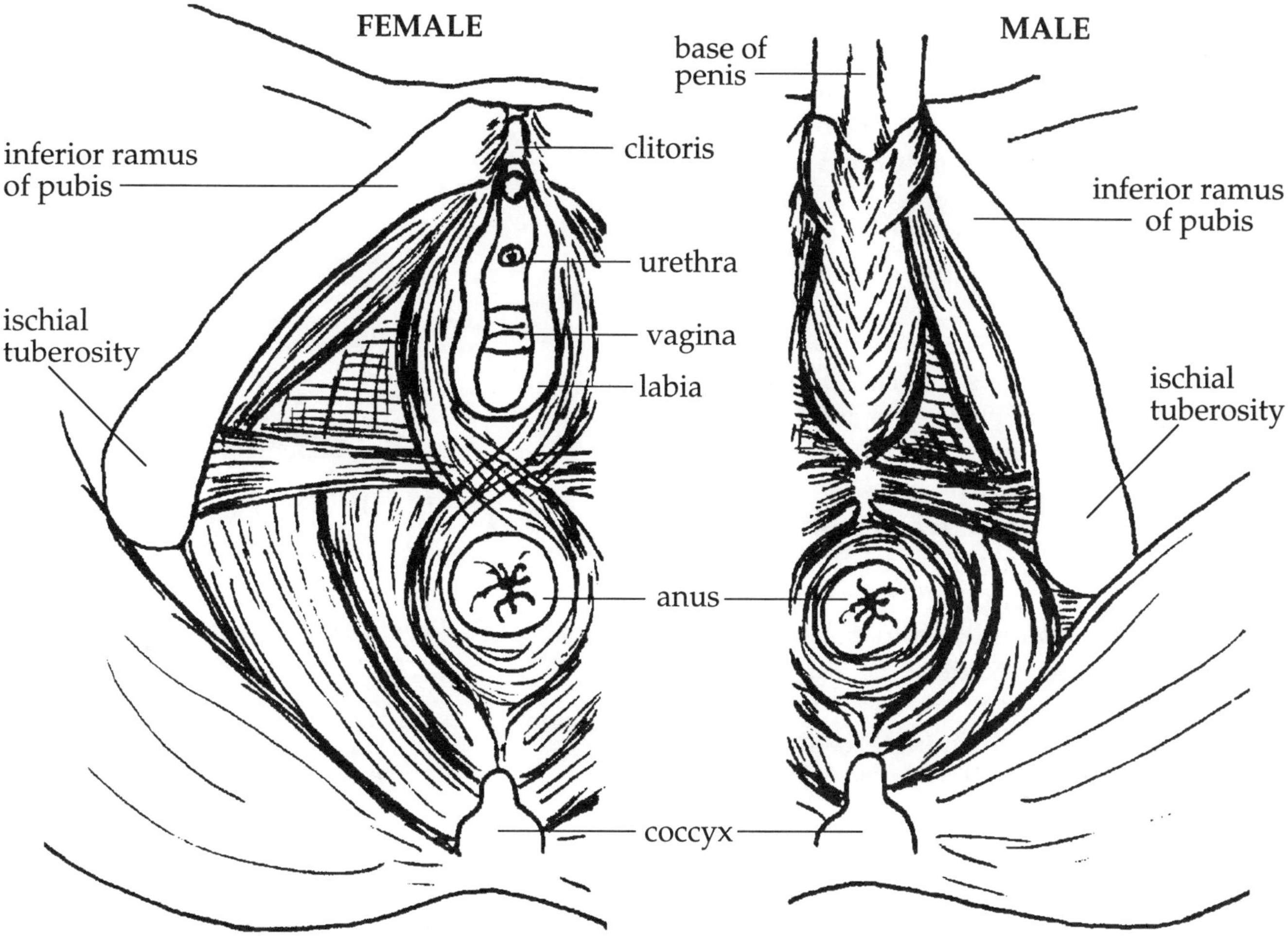

<u>Adequate tone in our pelvic floor muscles keeps our urinary bladder lifted up and tipped forward in its relaxed, filling position</u>. Weak or atrophied muscles let our bladder drop down so that it sits like a funnel over its urethra or exit tube (Fig. 11-25). This is its emptying or urinating position. An unsupported bladder sits ready to go all the time. It then takes the smallest increase in abdominal pressure to overcome our bladder sphincters and cause unplanned loss of urine. This is called **stress incontinence.** The stress is any increase in abdominal pressure such as coughing, sneezing, laughing, or picking up something heavy. Women have stress incontinence more often then men because: **1.** a woman's pelvic floor is broader and more difficult to support (Fig. 11-26), **2.** her urethra is shorter and has only one sphincter, and **3.** all their supporting tissues may be distorted during sexual intercourse and childbearing.

<u>Exercising our pelvic floor muscles increases their tone</u> so they can tip our bladder forward, out of its urinating posture and reduce stress incontinence. How do we exercise these muscles? Most people have no sense of what it would take to contract them. Awareness can be developed by pinching off our flow of urine while urinating. This automatically contracts our pelvic floor muscles and at the same time, demonstrates their ability to reduce urinary loss. When we have the idea of how to contract these muscles, they can be exercised any time and any place. Since it produces no skeletal movement, no one else can see that we are doing our exercises while we wait in the checkout line or for the bus.

In men, increased tone and support from the pelvic floor muscles may reduce premature ejaculation.

DEEP BACK MUSCLES

Our **deep back muscles** are our deepest muscles posterior to our vertebrae, covered by our upper appendage muscles. All our deep back muscles can be placed in one of three categories: **transversospinalis, suboccipitals,** and **erector spinae.** These names give important information about the orientation and therefore, the function of these groups.

Transversospinalis

Our **transversospinalis** include five different muscles which are the deepest of our deep back muscles and are not easily palpable. They are short, delicate muscles which span a distance of one to six vertebrae. Many of the transversospinalis incorporate some transverse orientation, lying obliquely (Fig. 11-28). They cross, for example, from the transverse processes of one vertebra to the spinous process of the vertebra above. An oblique orientation divides the strength of the muscle between vertical and horizontal movements (Fig. 11-28c). Since our transversospinalis are dorsal and lateral to our vertebrae, their vertical potential extends or laterally bends our vertebrae. Their horizontal potential rotates our vertebrae. Our transversospinalis then antagonize our oblique abdominals in vertebral flexion and extension, but are synergistic with them in rotation and lateral bending.

Suboccipitals

The most specialized joints of our vertebrae are between our occiput, atlas, and axis. To actualize their movement potential, we have special **suboccipital muscles** traversing these joints (Fig. 11-27). As their name states,

Figure 11 - 27

A posterior view of the suboccipital muscles.

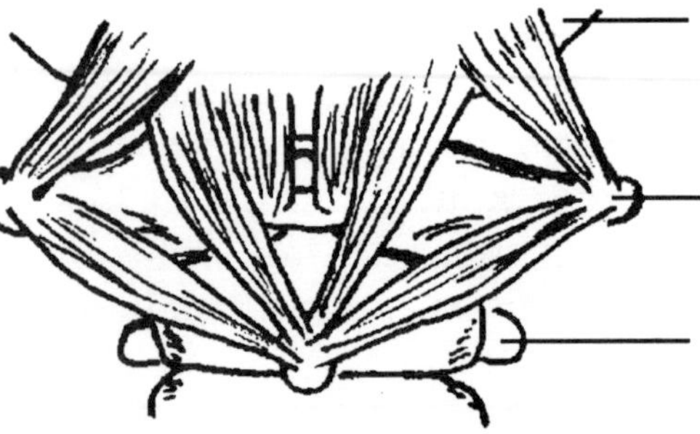

Figure 11 - 28

A posterior view of the thoracic vertebrae showing segments of specific transversospinalis muscles. For the most part, these muscles would continue up and down the entire length of the vertebral column.

(a)

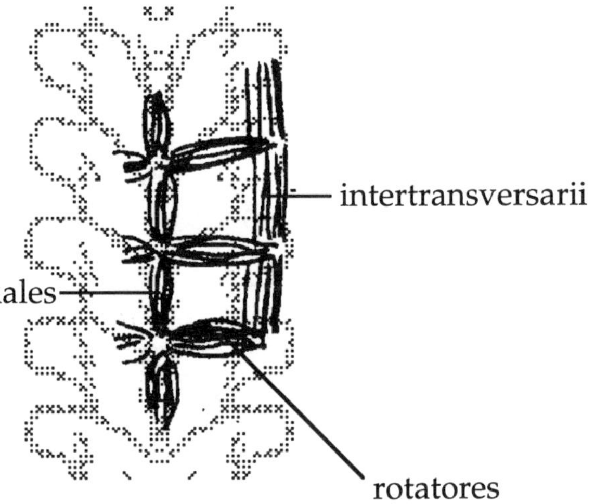

(b) the semispinalis and multifidus overlying each other.

(c) The oblique nature of these muscles divides their strength between vertical and horizontal movements.

they sit immediately under our occiput and so assist in extending our head over our neck and rotating our atlas over our axis.

Erector Spinae

Our **erector spinae** muscles are erect or vertical, and dorsal in their orientation so they keep us erect, extending our vertebral column.

Our erector spinae muscles sandwich between our transversospinalis and upper appendage muscles, filling the depression between the spinous process of our vertebra and angle of our rib. While they are entirely covered by muscles of our upper appendage, their bulk and fibrous direction often is still palpable through the thin overlying muscles.

Our erector spinae are long, interwoven muscles passing up to twelve vertebrae. Thin bundles of muscle cross over each other, originating from and inserting into progressively higher vertebrae. They are divided into three columns (Fig. 11-29). The medial column is called our **spinalis,** originating from the spinous processes of lower vertebrae and

Figure 11 - 29

A posterior view of the three columns of the erector spinae.

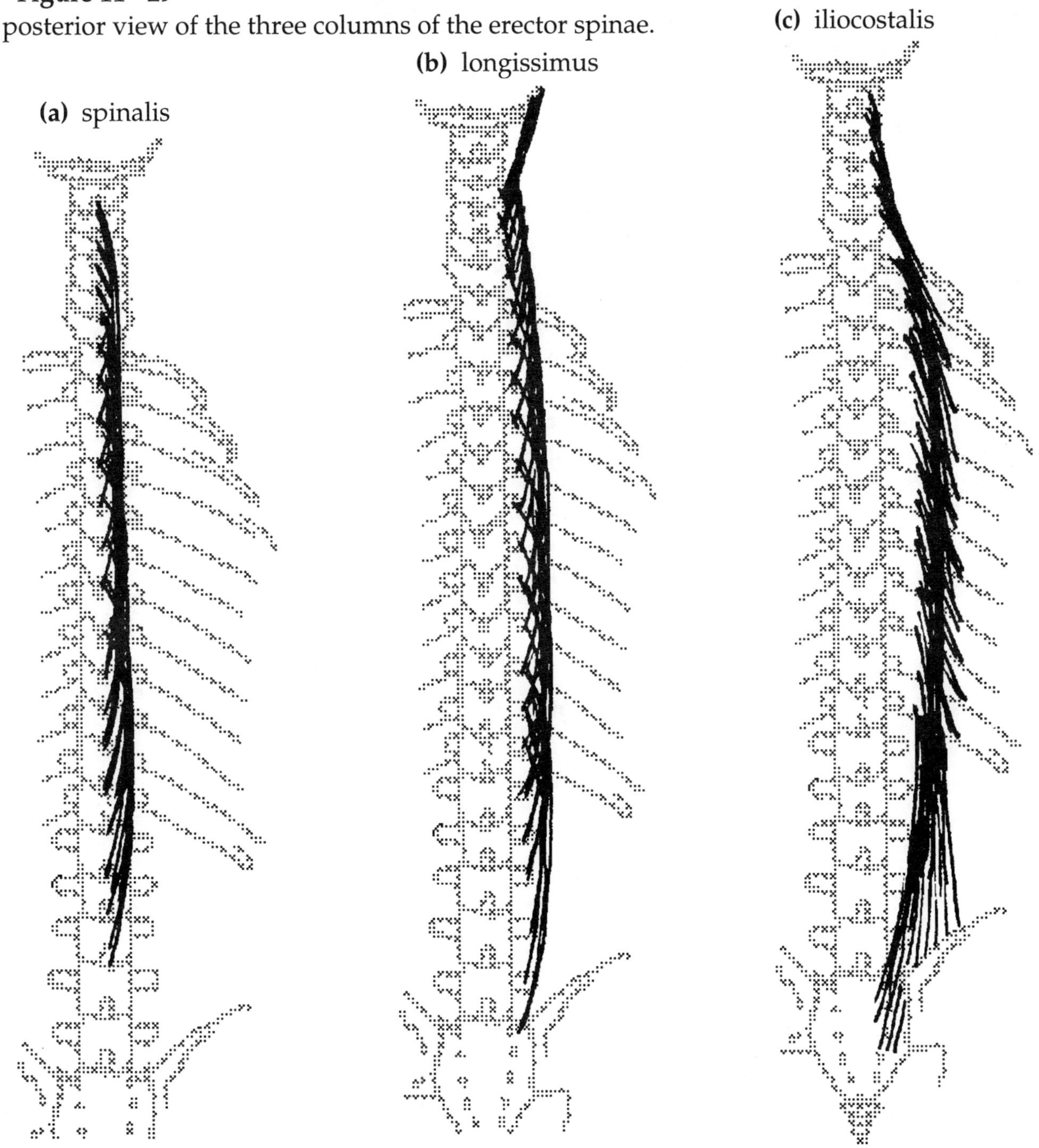

inserting into the spinous processes of vertebrae some distance above. The middle column is called our **longissimus,** originating from the transverse processes of lower vertebrae and inserting into the transverse processes of vertebrae some distance above and the mastoid process of our temporal bone. The lateral column is called our **iliocostalis,** originating from our iliac crest and ribs and inserting into ribs and transverse processes of cervical vertebrae.

• Three small muscles lie between the deep muscles of our back and the muscles of our upper appendage. They are the **splenius,** the **posterior superior serratus,** and the **posterior inferior serratus.** They are all difficult to palpate.

Splenius

A 'splenium' is a bandage. Our **splenius** muscle wraps around the back of our neck much like a bandage over our deep muscles in that area (Fig. 11-30). Our **lower** splenius fibers originate on the spinous processes of T3 through T6 and insert into the transverse processes of C1 through C3. Our **higher** splenius fibers originate from the lower nuchal ligament and the spinous processes of C7 through T3 and insert into the lateral base of our occiput and our temporal mastoid process. All the muscle fibers lie obliquely so their

Figure 11 - 30

A posterior view of the splenius.

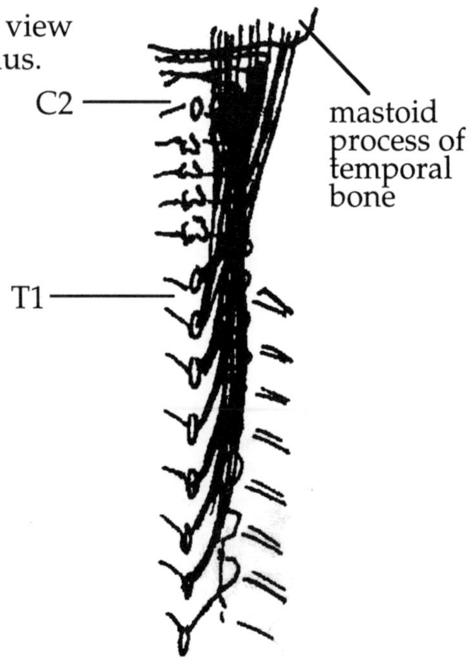

Figure 11 - 31

A posterior view of **(a)** the posterior superior serratus and **(b)** the posterior inferior serratur muscles on their boney attachments.

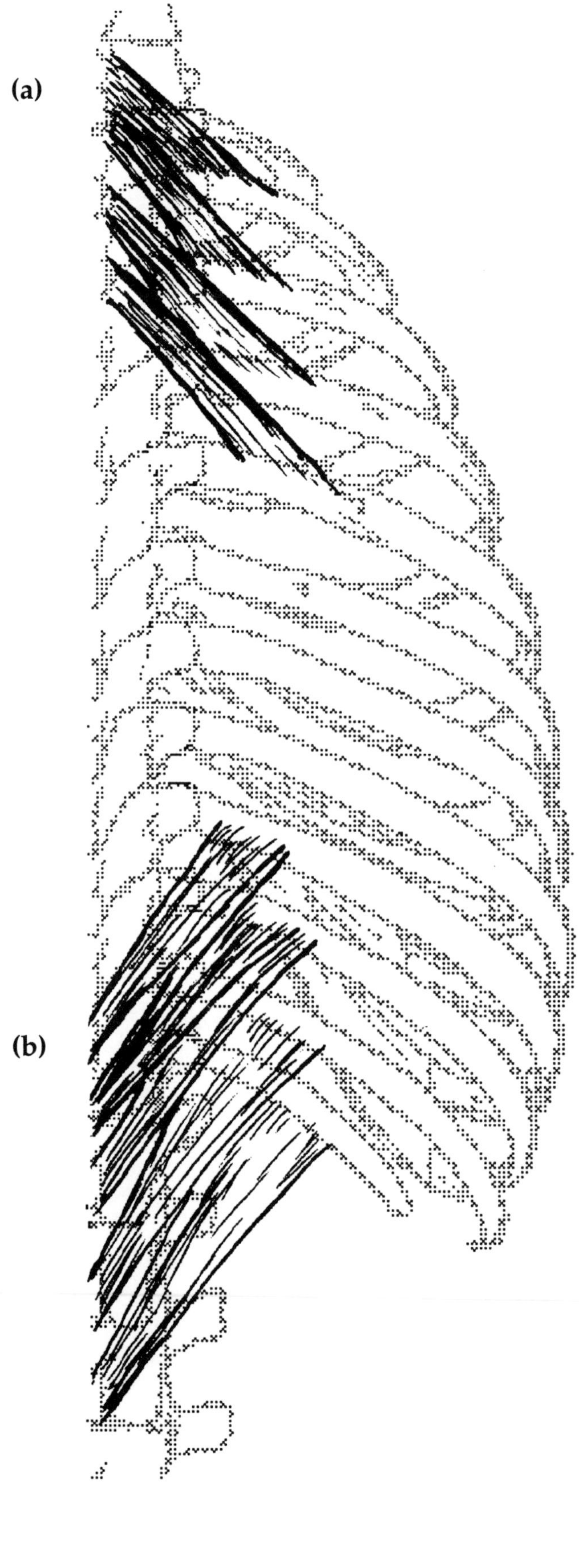

force is divided between vertical and horizontal movements. They lie dorsal to the pivot point of vertebral movement so that when both sides pull together, they extend our cervical vertebrae (or neck). When only one side contracts it produces lateral bending of our cervical vertebrae. Their horizontal effect produces rotation to the same side.

Posterior Superior Serratus

Our **posterior superior serratus** lies in our upper back. The first two words in its name give us this information. It originates from our lower nuchal ligament and vertebral spinous processes C7 through T2. From here it drops downward and laterally to insert into the lateral aspect of ribs two through five (Fig. 11-31a). As it contracts, it lifts these ribs up toward our vertebrae, increasing the space of our thoracic cavity, pulling air into our lungs.

When something is serrated, it has a scalloped or toothed edge like a saw. The insertion of this muscle forms a scalloped edge as it divides to attach to our individual ribs.

Posterior Inferior Serratus

As its name implies, our **posterior inferior serratus** also is located in our back, but lower down. It originates on our spinous processes T11 through L3. From here it runs upward and laterally to insert into the lateral aspect of ribs nine through twelve (Fig. 11-31b). As it contracts, it pulls these ribs down toward our vertebrae, decreasing our thoracic space, pushing air out of our lungs.

The insertion again forms a scalloped edge as the muscle fibers group into bundles to insert onto individual ribs.

Figure 11 - 32

Long-term habits which build muscles asymmetrically, such as always carrying a purse on the same side, can curve the vertebrae laterally, leading to scoliosis.

When muscles which laterally bend or extend our vertebrae are not balanced left to right, they may pull our vertebral column into a lateral curve. Any degree of lateral curve is abnormal and is called **scoliosis.**

Imbalances can develop from subtle, long-term habits which stress and build muscles asymmetrically and bend our vertebrae laterally. One example is counterbalancing a weight such as a purse (Fig. 11-32) or baby or books always carried on one side of our body. Another is to accommodate an asymmetrical body position such as always leaning on the same elbow while eating.

With the passage of time, muscles on the one side of our vertebrae develop more while the other side is chronically stretched and hypotonic. The normal tone of our stronger muscles will pull our body in their direction, curving our vertebral column.

Once a curve begins, muscles fall into abnormal positions around our vertebrae, altering their action. This may exacerbate the abnormal curvature, starting a vicious cycle which becomes more and more difficult to change.

Scoliosis may gain some relief from exercises which strengthen and increase the tone of our over-stretched muscles while stretching out and relaxing our tight muscles.

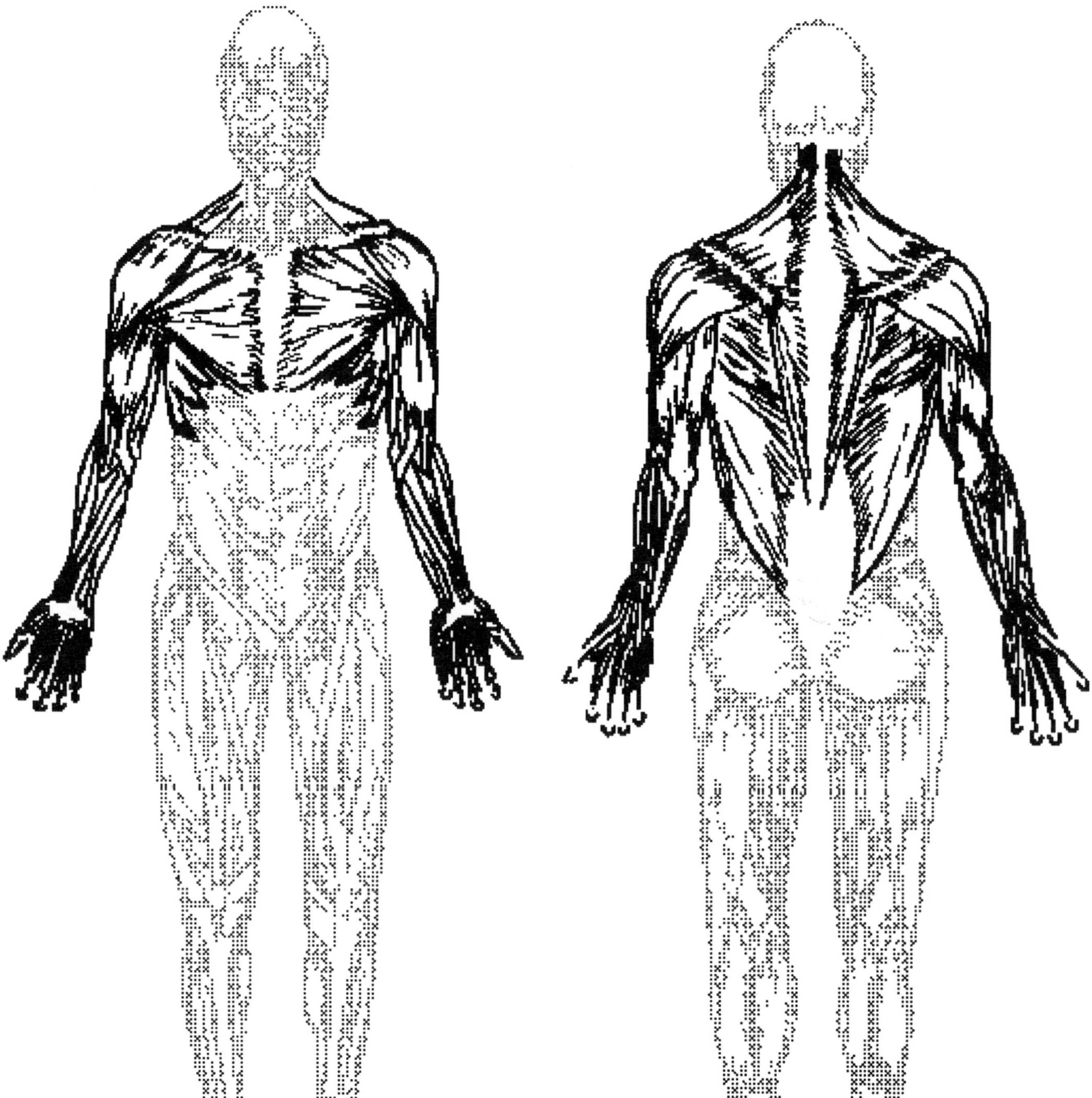

chapter 12: **Upper Appendage Muscles**

In general, the job of our arms is to be out there doing things for us. The skeleton of our upper appendage is designed to enhance mobility, and thus sacrifice stability. We compensate for our skeleton's unstable nature with muscles which support across our joints to maintain the alignment of our skeleton. This has the great advantage of incorporating movement into our supporting system.

SCAPULAR STABILIZATION

- Our upper appendage attaches to our axial skeleton at our sternoclavicular joint. With our clavicle suspended outward horizontally and the pectoral girdle deficit between our scapulae, the mechanics of this joint are very unstable. If all we had were bones and ligaments, our arms would dangle straight down from our sternum in front of our chests (Fig. 12-1). It is muscles that support our pectoral girdle in its more favorable position.

Since these muscles are stabilizing our pectoral girdle onto our axial skeleton, we can assume that their origin or stabilized end will be somewhere on our axial skeleton while their insertion is somewhere on our pectoral girdle. Most of these muscles attach to our scapula and so are called the muscles of scapular stabilization. Any scapular stability or movement always does the same for our clavicle, so it really affects the whole pectoral girdle.

- In our discussion of skeletal muscle we said that a muscle can create movement or hold against an opposite movement. The muscles of scapular stabilization are the epitome of this. Any one muscle can move our scapula in a specific direction or prevent movement in the opposite direction. We can enhance any arm movement by adding the distance that our scapula moves in that direction or we can make our scapula into a stable platform from which our arm can do work and receive external forces without strain or damage. We stabilize our scapula by making a force of contraction which opposes the forces going though our arm. If the weight of a bag is pulling down through our arm, our muscles pull up on our scapula. This contraction only needs to equal the external force, to balance it without creating its own movement. Most of the time our amazing body is actually combining movement and stabilization.

Figure 12 - 1
Bones and ligaments alone cannot keep our arms in place. Gravity will overcone the skeletal instability, leaving our arm dangling straight down from our sternoclavicular joints.

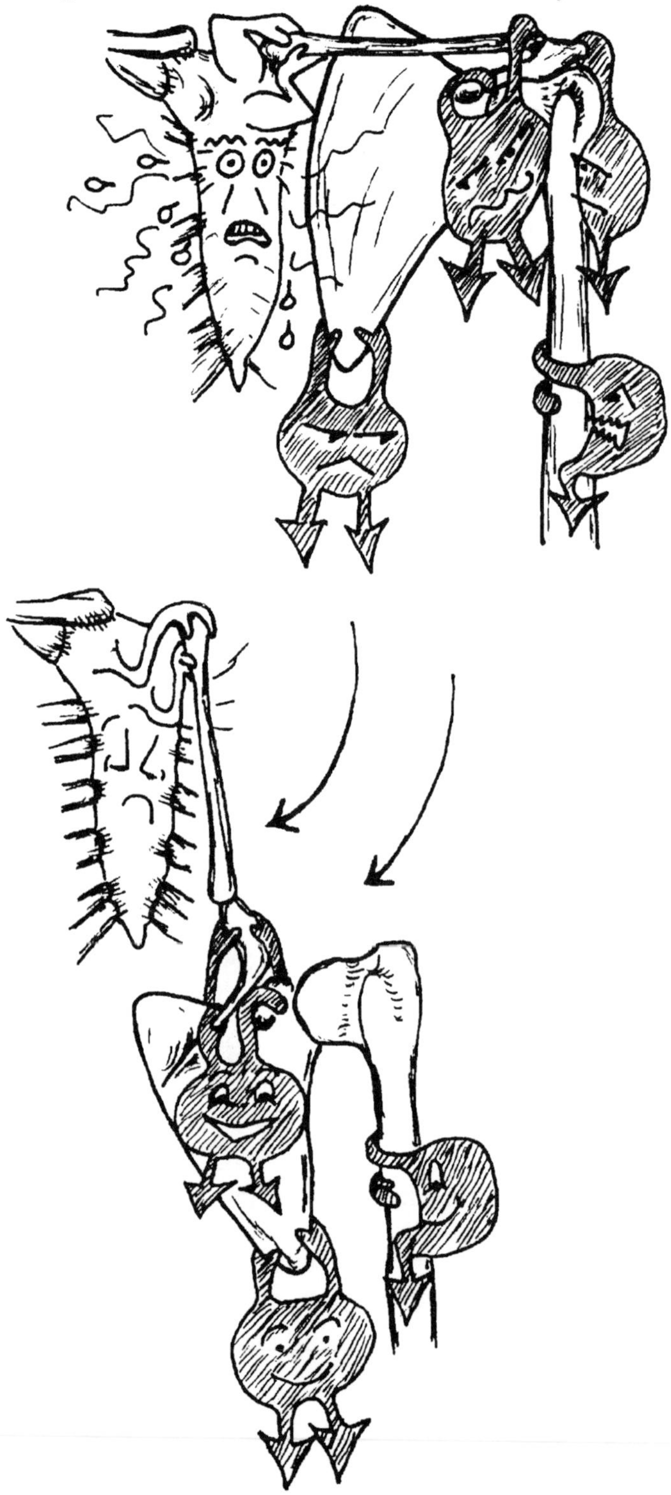

Trapezius

Our **trapezius** is a superficial muscle, lying immediately under our skin, nowhere covered by other muscles. The root of 'trapezius' means a four-sided, single-planed object. It is a broad, flat, triangular muscle, our left and right trapezii together making a flat diamond shape on our back (Fig. 12-2a).

Because of the expanse of our trapezius, it is divided into three parts which may work independently of each other and sometimes even antagonize each other. The three parts are upper, middle, and lower.

- Our **upper** trapezius originates from the base of our occiput and the spinous processes of our cervical vertebrae via our nuchal ligament. It inserts into our lateral clavicle and the acromion process of our scapula (Fig. 12-2b and 12-20).

Since our upper trapezius is above our scapula, it reaches down obliquely from the midline to pull up on the lateral tip of our scapula, retracting, laterally rotating, and elevating it (Fig. 12-2b). Anything we carry in our arms or hands exerts a force of medial rotation and depression on our scapula. Just the weight of our arm, constantly pulled down by the force of gravity, will do the same. Increased tone in our upper trapezius will stabilize our scapula against these forces.

Too often our upper trapezius overdoes its job, exerting so much tension that our pectoral girdle is excessively elevated, keeping our shoulders up by our ears and our clavicles V-necked (Fig. 8-2).

- Our **middle** trapezius originates from our lower cervical and upper thoracic spinous processes. It inserts into the spine of our scapula (Fig. 12-2b and 12-20). Its dominant action is to retract our pectoral girdle. Increased tone in our middle trapezius can stabilize our scapula against being pulled too far forward such as when something drags us forward (Fig. 12-3) or we drag something backward.

On the other hand, when our pectoral girdle is chronically protracted, our middle trapezius will be chronically overstretched. In such a case, it needs to be stimulated or awakened, rather that relaxed.

Figure 12 - 2

The trapezius muscle. **(a)** shows the diamond shape of the left and right trapezii together. **(b)** divides the trapezius into its parts: **A** upper, **B** middle, and **C** lower. The arrows show potemtial movement.

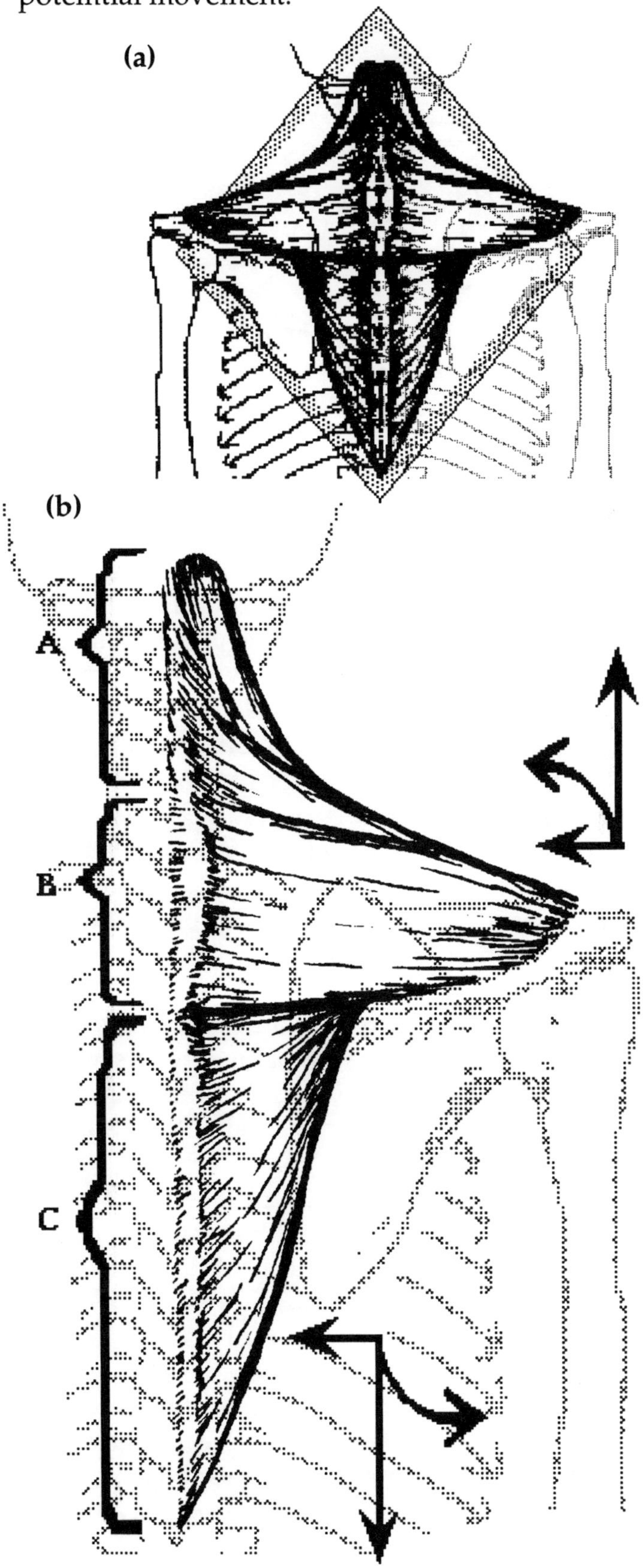

Figure 12 - 3

A force which drags us forward tends to protract our scapulae.

The retracting force of our trapezius can counteract this effect.

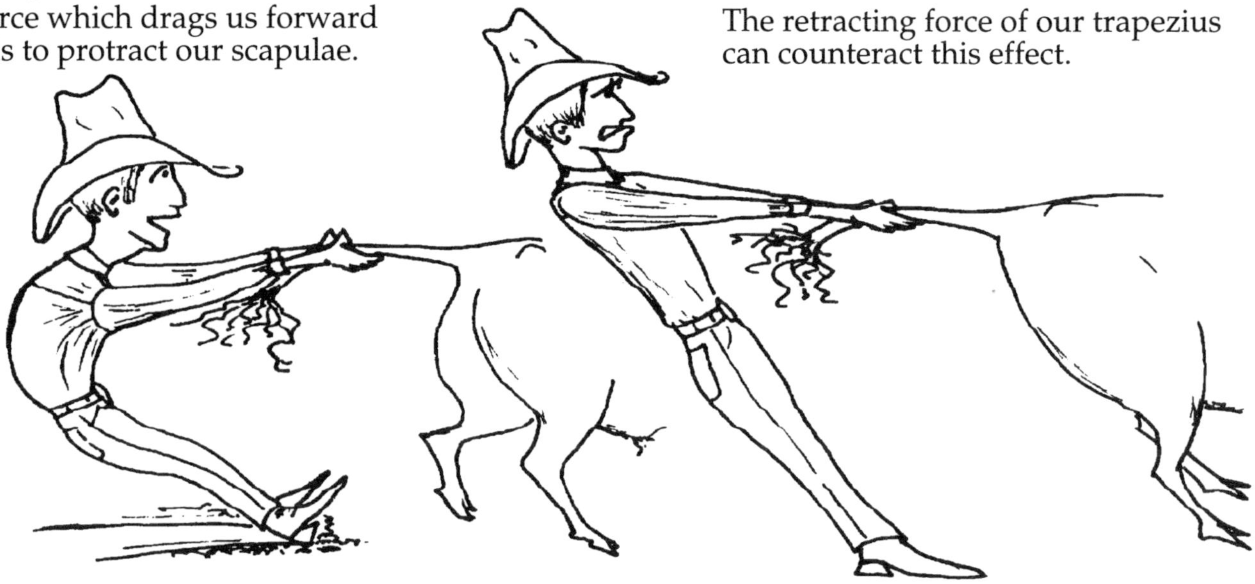

• Our **lower** trapezius originates from the spinous process of our thoracic vertebra from T 4 to T12. It inserts into the medial end of our scapular spine. Its force of contraction retracts, laterally rotates, and depresses our scapula (Fig. 12-2). When something is pulling upward on our scapula, as when we hang from our arms, our lower trapezius will support our scapula onto our axial skeleton, preventing excessive elevation. On the other hand, if our scapula is chronically elevated, our lower trapezius is chronically overstretched.

So while the overall effect of our trapezius is to retract and laterally rotate our scapula, it antagonizes itself in elevation and depression of our scapula (Fig. 12-2).

• All the bony attachments of our trapezius are easily palpable and since only skin covers its belly, it is palpable over its whole surface (Fig. 12-23). Remember that it is a thin flat muscle, so we are not looking for a bulging form.

Find the spinous processes of your upper cervical vertebrae. Slide laterally one to one and one-half inches to a subtle indentation posterior to your sternocleidomastoid. The hump that you slide over is the upper trapezius.

Go to the hollow above the middle of your clavicle. Slide upward and backward feeling for the anterior-superior edge of your trapezius (Fig. 12-25). Elevate your scapula to work the muscle and make this edge more prominent. Follow this edge from your lateral clavicle up to your occiput, passing through the indentation described above.

With your left thumb lying along that edge of your right trapezius, wrap your left fingers around to touch the spine of your scapula. Squeeze into the muscle you are surrounding. This is your upper and middle trapezius. Your middle trapezius continues medially to cover the area between your vertebra and upper scapula.

The inferior edge of the left and right trapezius form a V-shape on the middle back (Fig. 12-2). It is difficult to feel on oneself so call over a friend. Locate the spinous process of T12 and the medial end of your friend's scapular spine. Imagine a line joining these two points and slide perpendicular back and forth over it. This edge of the trapezius is thin but you may be able to sense it. Have your friend contract the muscle to make it easier to feel. What move would contract our trapezius? Pulling our shoulders back (retract) and down (depress) works our trapezius. Another movement which opens the space and increases visibility is to abduct our arm up to our ear. This laterally rotates our scapula, using our whole trapezius (Fig. 12-7), so we can see both its lower and upper edges. Have your friend raise (abduct) and lower (adduct) the arm to watch the trapezius work.

Levator Scapula

The name **levator scapula** tells us the action of this muscle. 'Levo' means to lift up, so the levator scapula lifts the scapula upward. Knowing how muscles work, where would we expect to find a muscle that elevates our scapula? Somewhere above our scapula.

Our levator scapula originates on the transverse processes of our cervical vertebrae, C1 through C4 (Fig. 12-4). To feel this origin, line up your four fingertips and press them into the vertical hollow just lateral to your upper trapezius, medial and posterior to your sternocleidomastoid, and just below your occiput (Fig. 12-23 and 12-25). The deep firmness under your pressure is the transverse processes of your cervical vertebrae (Fig. 7-15).

Our levator scapula inserts into the vertebral border of our scapula above the scapular spine (Fig. 12-4 and 12-20). The upper vertebral border is hard to palpate, but if we locate the medial end of our scapular spine, we know it is just above this point.

Contracting our levator scapula, then, brings the medial edge of our scapula up into our neck, causing elevation and medial rotation of our scapula (Fig. 12-4). If an external force is depressing or laterally rotating our scapula, increased tone in the levator scapula would hold the scapula against this force.

We already have palpated the origin and insertion of our levator scapula. Find its belly as a strap in the line between these two places. Near its insertion, feel your levator scapula slip under the upper edge of your trapezius. Your trapezius may be thin enough to feel your levator scapula through it.

Figure 12 - 4

A posteroir view of the levator scapula with arrows showing the direction it can move the scapula.

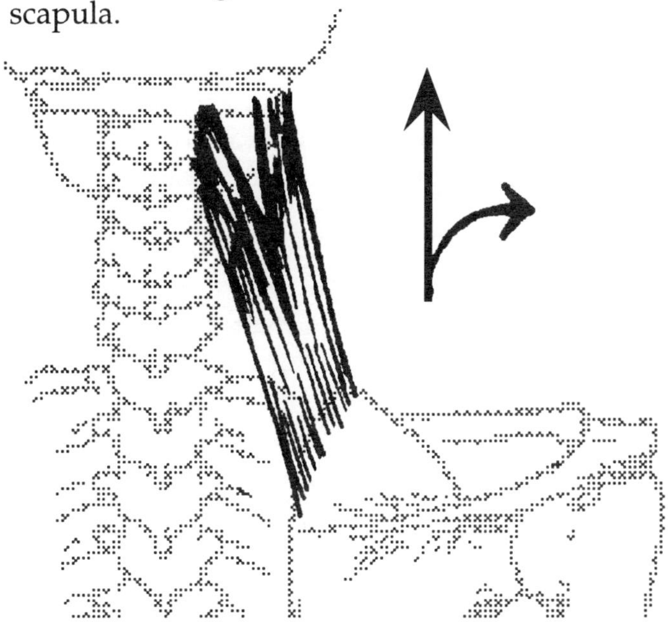

Our upper trapezius and levator scapula are a perfect example of muscles which are synergistic in one action and antagonistic in another. They both elevate our scapula, but one laterally rotates while the other medially rotates. If we want to elevate the scapula with no rotation, we would use them equally and their rotation forces would cancel each other out. If we want some degree of medial rotation, we would emphasize the contraction of our levator scapule. If we want some degree of lateral rotation, we would emphasize the contraction of our trapezius. Every movement in the body is done by a group of muscles which move, antagonize movement, and support in this way.

Rhomboids

Rhomboids is in plural because it is often divided into major and minor. We will examine it as one muscle.

Our left and right rhomboids together form an A-frame across our upper back (Fig. 12-6), their fibers running deep and perpendicular to our lower trapezius (Fig. 12-5). The name of this muscle tells us its shape. A 'rhombus' is a kite and refers to the classic shape that has four sides, with opposite sides parallel, but no right angles (Fig. 12-6).

Our rhomboids originate from the spinous processes of vertebrae C7 through T5. They insert on the vertebral border of our scapula below the scapular spine (Fig. 12-20). As our rhomboids contract, they elevate, retract, and medially rotate our scapula (Fig. 12-6). The direction of the fibers of our rhomboids is similar to that of our levator scapula so the actions of the two muscles also are similar.

Palpate your rhomboids most easily near the inferior angle of your scapula where they come out from under your trapezius to lie immediately under your skin (Fig. 12-23). Have a friend put an arm up beside the ear. This laterally rotates the scapula so that its inferior angle moves well away from the trapezius to expose more of the rhomboids (Fig. 12-7). Find the lower edge of your friend's rhomboids passing medially upward from the inferior angle of the scapula.

The rhomboid fibers are perpendicular to those of the lower trapezius and slip under the trapezius before reaching the vertebrae. Try to follow the lower edge and belly of your friend's rhomboids through the thin trapezius. Have your friend bring his/her arm down to the side and watch the rhomboids medially rotate the scapula. For increased visibility, hold against your friend's downward movement to make the muscle work harder.

To differentiate between the lower trapezius and rhomboids, have your friend elevate his/her scapula leaving the arm down by the side. Hold upward around your friend's upper arm and have him/her push down against you (Fig. 12-5). The lower trapezius will contract to try to pull down the scapula. The lower rhomboids will contract to try to medially rotate the scapula. You can see both of the muscles working at the same time.

Figure 12 - 5

When a person resists an upward force on the arm, both the lower trapezius and rhomboids come into action and can be visualized independently.

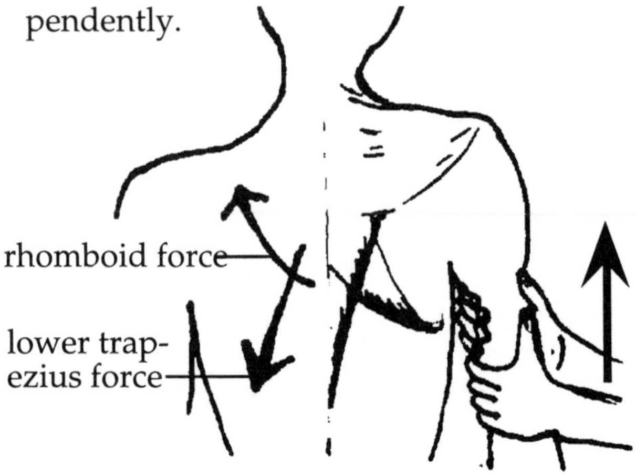

Figure 12 - 6

A posterior view of the rhomboids, showing its kite shape and the directions it moves the scapula.

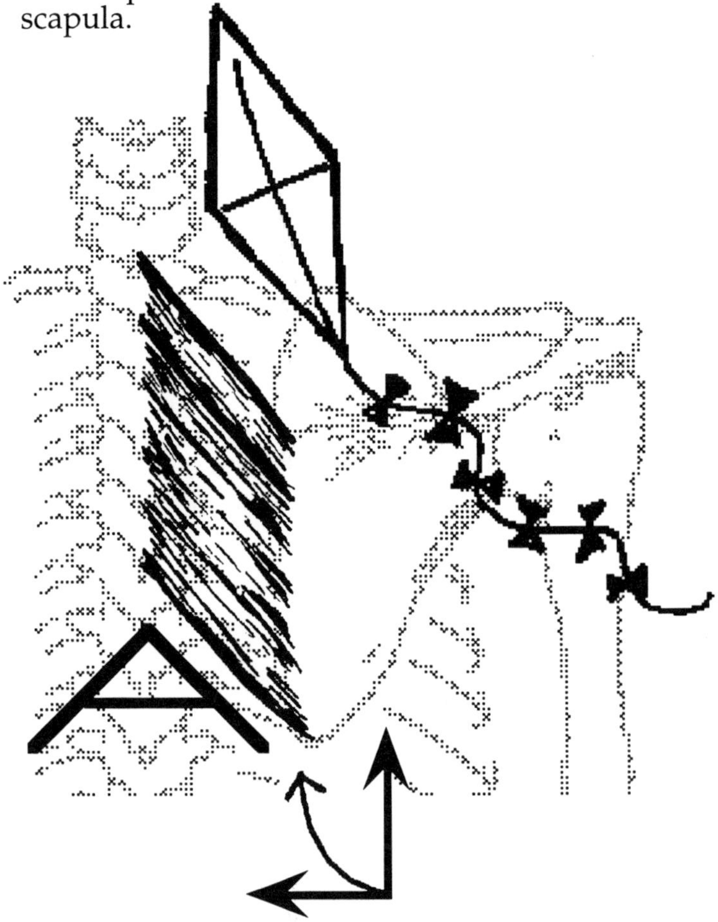

Figure 12 - 7

When the scapula is laterally rotated, more of the rhomboids are exposed lateral to the lower trapezius.

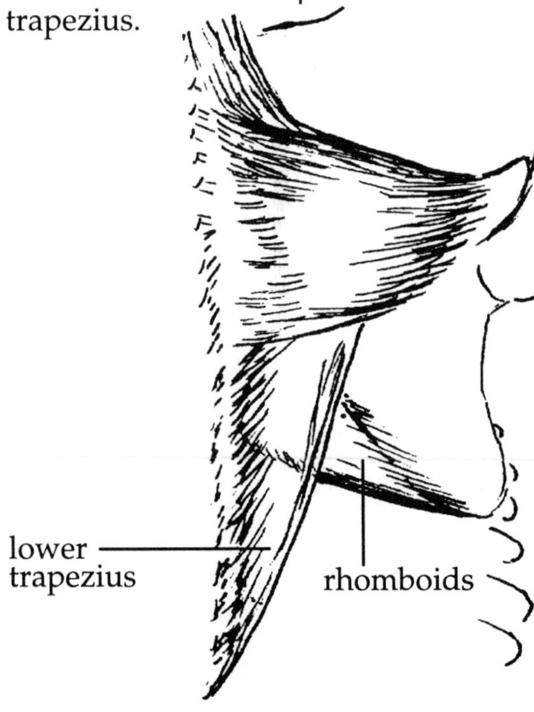

Serratus Anterior

The name **serratus anterior** tells us about the shape and location of this muscle. Its origin forms a serrated edge on the anterior surface of our body (Fig. 12-8).

Our serratus anterior originates in eight to ten bundles, one from each of our first eight to ten ribs. It originates on the external surface of the rib, at their anterior-lateral corner. In using the word corner, we must think of the cross-sectional shape of our chest as a rectangle rather than an oval (Fig. 12-24b).

From its rib origin, the muscle passes backward through our arm pit, between our ribs and scapula to insert into the anterior or internal surface of the vertebral border of our scapula. The concentration of the insertion is near the inferior angle of our scapula.

A nickname for our serratus anterior is the banana muscle because it looks like a bunch of bananas tucked under our arm (Fig. 12-8).

The rib attachment is the stabilized end of our serratus anterior, so contraction will protract our scapula (Fig. 12-8), pulling it forward around our ribcage. Conversely, when something is forcefully pushing backward on our arm or scapula, as when we are trying to push a heavy object forward in front of us, our serratus anterior will brace the scapula against excessive posterior movement. Since the insertion concentrates on the inferior angle of our scapula, the serratus anterior also will laterally rotate the scapula (Fig. 12-8).

With its insertion on the subscapular surface, our serratus anterior holds our scapula snugly against our ribs. Weakness in our serratus anterior allows the vertebral border of our scapula to pop away from our body wall, like wings sprouting on our back. The idea of wings might be fun but it is not mechanically advantageous for our pectoral girdle.

Most of our serratus anterior is covered by other muscles and bones. Its origin end is under our clavicle and pectoralis major. Its insertion end is under the posterior wall of our arm pit and scapula.

There is one place where we get a

Figure 12 - 8

An **(a)** anterior and **(b)** lateral view of the serratus anterior, with arrows showing how it can move the scapula.

(a)

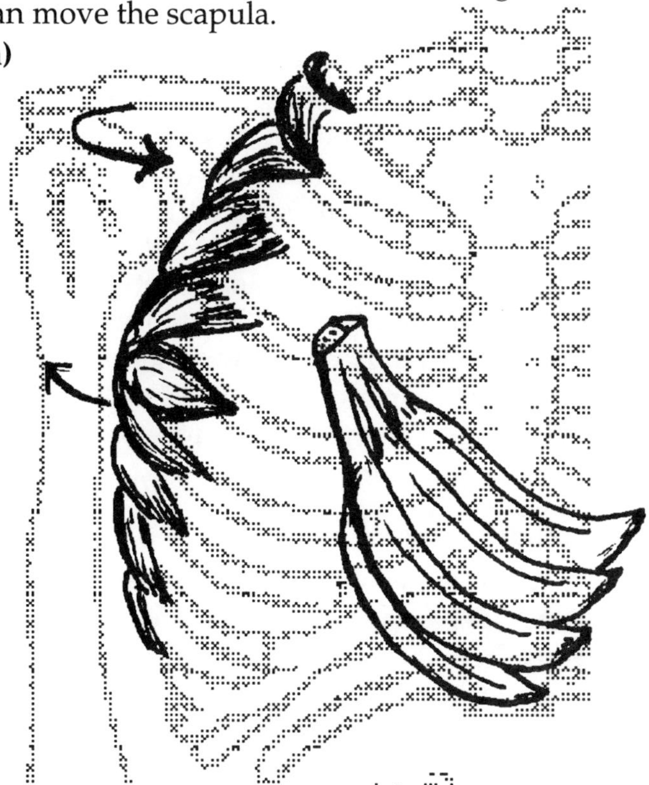

(b)

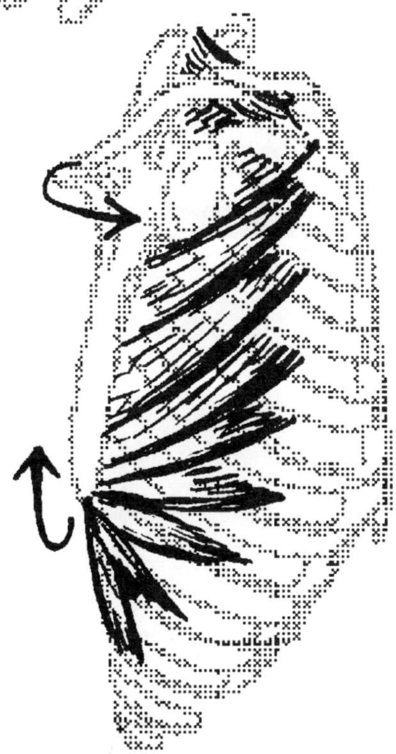

Our trapezius and serratus anterior form another synergistic-antagonistic couple. Both muscles work together to laterally rotate our scapula, but our trapezius retracts while our serratus anterior protracts.

glimpse of three or four segments of our serratus anterior's serrated origin. Lift your arm up against your ear. Look at your chest wall, under your arm, lateral and inferior to your breast. (Women may have to pull their breast forward.) The horizontal, linear mounds, like fingers under your skin, are a part of your serratus anterior (Fig. 12-23 and 12-25). Press your arm against the side of your head to see them contract as they further laterally rotate your scapula. Let your arm hang down beside you to see them relax.

The inferior edge of your serratus anterior may be seen as a band passing from the inferior angle of your scapula forward toward your costal margin (Fig. 12-23).

Pectoralis Minor

'Pectoralis' refers to the location of this muscle, 'pectoris' meaning the breast. 'Minor' tells us there is another, more important, or at least larger muscle in this area.

Figure 12 - 9

An **(a)** anterior and **(b)** posterior view of the pectoralis minor with arrows showing the direction it moves the scapula.

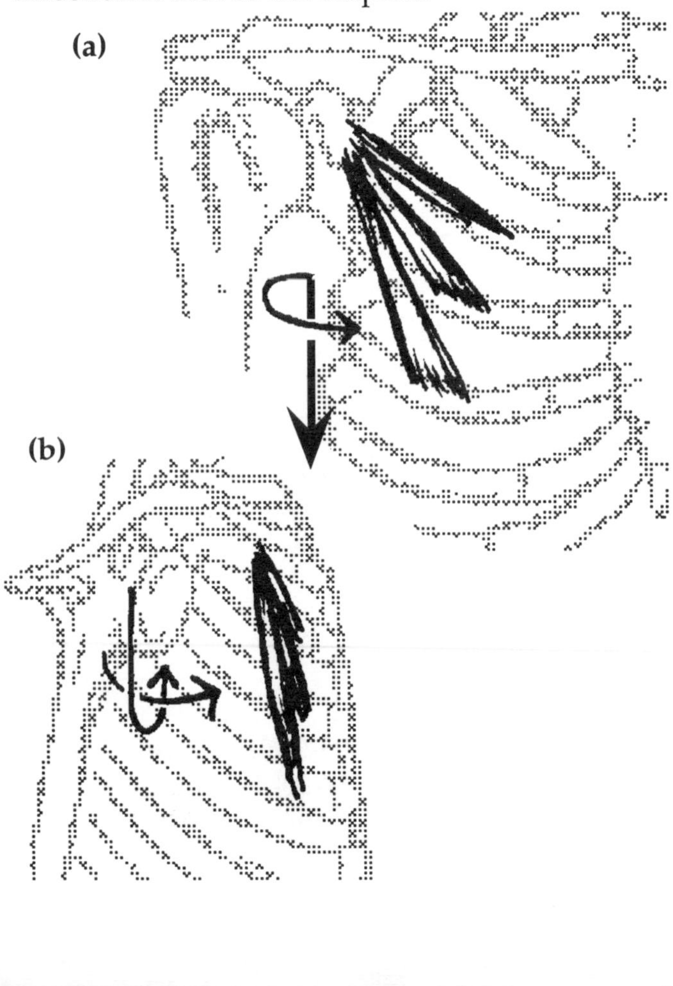

The origin of our **pectoralis minor** is under our breast, on the anterior surface of ribs two or three through five (Fig. 12-9). The muscle fibers pass vertically and laterally under our pectoralis major to insert into the coracoid process of our scapula (Fig. 12-21).

As our pectoralis minor contracts, it will pull the lateral corner of our scapula downward and around to the front of our body, producing depression, medial rotation, and protraction of our scapula (Fig. 12-9).

Our pectoralis minor is completely covered by our pectoralis major, but the lateral edge of our pectoralis major is easy to slip under. Place your fingertips in your arm pit. Press them medially under the edge of your pectoralis major, above your breast following your rib cage forward. Your fingers should bump into the vertical fibers of your pectoralis minor. If you cannot feel it, hold your fingers in place and protract your scapula. Your pectoralis minor will pop out against your waiting fingertips.

Subclavius

Subclavius tells us the location of this muscle. It is under our clavicle.

Our subclavius is a small muscle which originates on the anterior surface of our first rib and inserts into the inferior surface of the middle of our clavicle (Fig. 12-10 and 12-20). As our subclavius contracts, it pulls our clavicle down against our ribs, preventing excessive elevation or protraction of our pectoral girdle.

Because our subclavius is small and tucked away under our clavicle and pectoralis major, it is difficult to palpate. When it contracts, it pulls our clavicle over itself to secure its hiding place.

Figure 12 - 10

An anterior view of the subclavius.

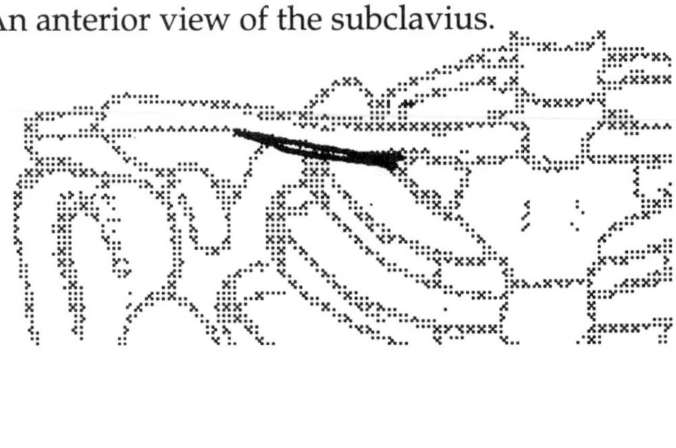

• We have found muscles radiating in every direction from our scapula - above, medially, below, laterally, and anteriorly (Fig. 12-11), so the pectoral girdle can move over the ribs in all these directions and be stabilized against forces from all these directions.

Figure 12 - 11
The arrows indicate all the potential movements of the muscles that support the scapula onto the axial skeleton.

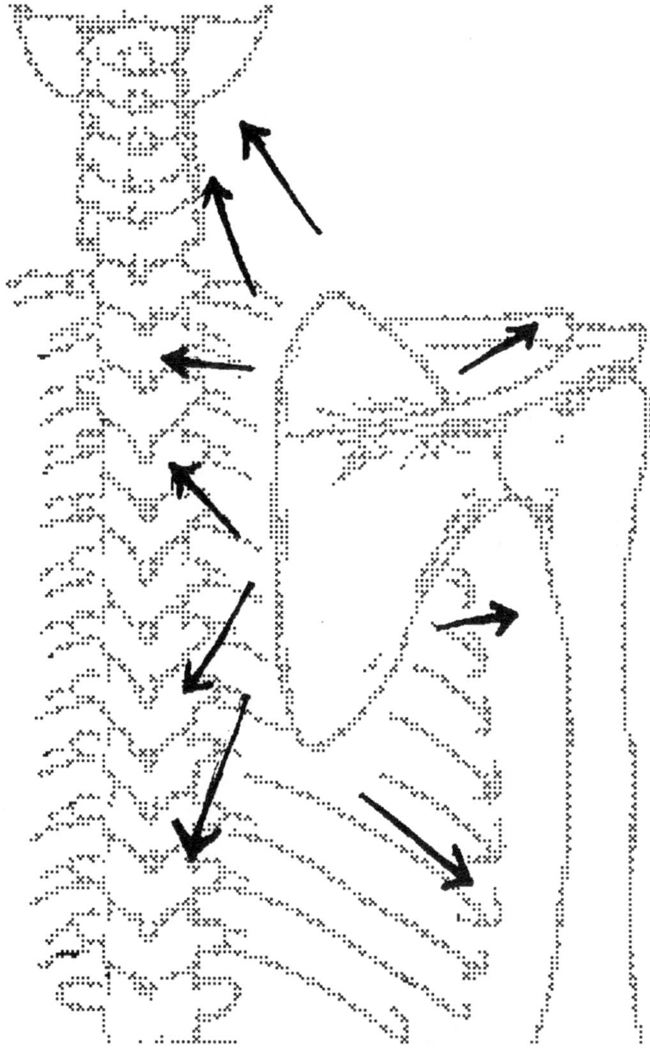

GLENOHUMERAL STABILIZATION

Having supported our pectoral girdle onto our axial skeleton, the next joint we come to is our glenohumeral joint. This joint is freely mobile but unstable and dependent on muscular support. Just the weight of our arm hanging down will dislocate our glenohumeral joint if all its supporting muscles go flaccid. Since our glenohumeral joint moves in every direction, it needs muscles radiating in every direction around it to accomplish the appropriate support and movement.

• All the muscles crossing our glenohumeral joint can lend a supporting hand, but four muscles are particularly designed to support. Collectively these muscles are referred to as our **rotator cuff.** A 'cuff' is a band that goes around the arm at the bottom of a sleeve. All these muscles together form a sleeve and cuff which go around our glenohumeral joint. The action of three of the muscles includes rotation, and so the name. All these muscle attachments are very close to the joint, which decreases their leverage for movement but increases their stability. Because of their shortness and propensity to stabilize, these muscles take the brunt of rapid or excessive arm movements, often being injured.

Our four rotator cuff muscles are the **subscapularis,** the **supraspinatus,** the **infraspinatus,** and the **teres minor.**

> When our glenohumeral joint dislocates, our humerus moves down into our arm pit. A hollowness forms between our humerus and our acromion process (Fig. 12-12), while the head of our humerus presses against our ribs and the axillary border of our scapula. The nerves, blood vessels, and lymphatics servicing our arm go through our arm pit and may become compressed or irritated by even mild dislocations.

Figure 12 - 12
A posterior view of **(a)** a properly located, and **(b)** a dislocated glenohumeral joint.

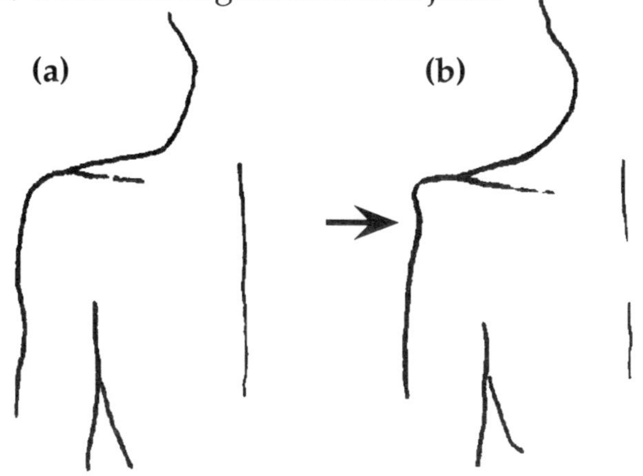

Subscapularis

Our **subscapularis** is named by the location of its origin and belly. We have already located our subscapular fossa on the anterior surface of our scapula, lying against our serratus anterior and ribs. Our subscapularis muscle originates from our subscapular fossa. Its muscle fibers pass horizontally in front of our glenohumeral joint and insert into the lesser tubercle of our humerus (Fig. 12-14, 12-17, and 12-21). Contraction of our subscapularis produces medial rotation at our glenohumeral joint and stabilizes this joint during other movements.

Palpation of the belly of your subscapularis is difficult, but you can gain a sense of it by pressing your fingers backward under the axillary border of your scapula, between your scapula and ribs. This separates the serratus anterior from the subscapularis (Fig. 12-24b). With your fingers in this position, medially rotate your humerus to feel the muscle working. You may find the tendon of your subscapularis high up across the anterior surface of the posterior wall of your arm pit.

Supraspinatus

Our **supraspinatus** is named by the location of its origin in our supraspinous fossa, the upper, external surface of our scapula (Fig. 12-13 and 12-20). From here this muscle runs laterally, deep to the acromion process of our scapula to insert into the anterior-superior aspect of the greater tubercle of our humerus (Fig. 12-15 and 12-17). As the supraspinatus contracts, it initiates abduction of the glenohumeral joint and helps to stabilize this joint during other movements.

The belly of our supraspinatus is covered by our upper trapezius, but can be palpated through it. With one arm dangling loosely by your side, press the fingers of your other arm into the supraspinous fossa, immediately above the spine of your scapula.

Figure 12 - 13

A posterior view of the supraspinatus, with an arrow showing how it would move the humerus.

Figure 12 - 14

An anterior view or the subscapularis.

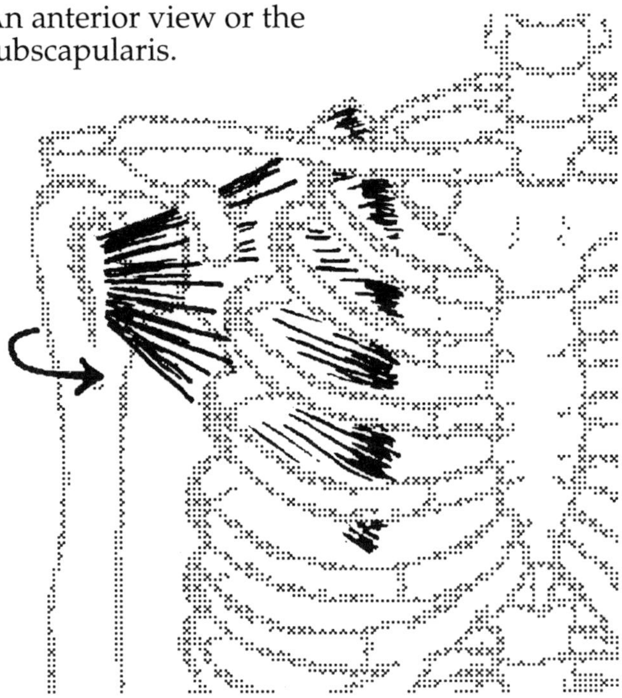

Figure 12 - 15

A lateral view of the shoulder area with the glenohumeral joint extended, making the insertion of the supraspinatus palpable.

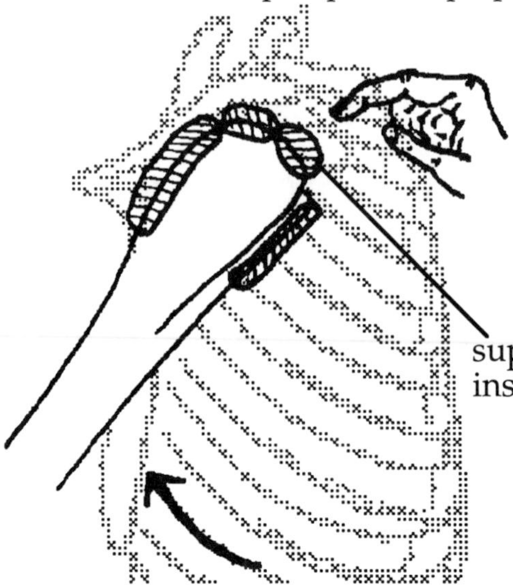

Lightly bounce your hanging arm away from your side and let it fall back. Feel your supraspinatus bulge and relax under your listening finger. If you lift your arm too high, your trapezius also will contract, making it difficult to feel through to your supraspinatus.

In anatomical position, the insertion tendon of our supraspinatus lies under the front of our acromion process. To feel it, press a finger downward just in front of your acromion process. Extend your glenohumeral joint, swinging the top of your humerus forward, out from under the acromion process (Fig. 12-15). You may feel a cord slip under your finger. This is the insertion tendon of your supraspinatus. You have to feel through your deltoid, so it is not easy.

Figure 12 - 16

(a) A posterior view of the infraspinatus with an arrow showing how it moves the humerus. **(b)** A lateral view of the shoulder exposing the insertion of the infraspinatus by flexing the glenohumeral joint.

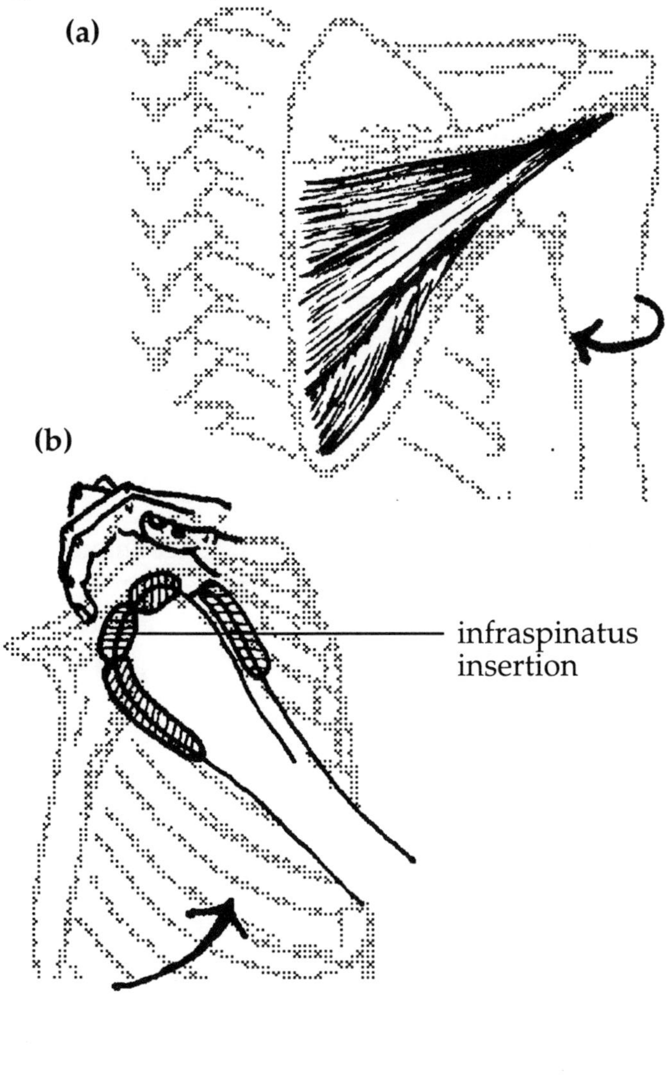

Infraspinatus

The origin of our **infraspinatus** is in our infraspinous fossa, the lower external surface of our scapula (Fig. 12-16a and 12-20). From here this muscle runs laterally and upward to insert into the posterior-superior aspect of the greater tubercle of our humerus (Fig. 12-16b and 12-17). As our infraspinatus contracts, its most significant action is to laterally rotate our glenohumeral joint and to stabilize this joint during movement. It also may assist in extension and adduction of our glenohumeral joint from a flexed or abducted position.

Our whole infraspinous fossa is covered by the belly of our infraspinatus. Near our inferior angle, this muscle is immediately under our skin and directly palpable (Fig. 12-23). Its upper part lies under our deltoid, but still can be felt when it is contracted. Place your fingertips just below the spine of your scapula. Laterally rotate your glenohumeral joint to feel your infraspinatus work.

In anatomical position, the insertion tendon of our infraspinatus lies under the back of our acromion process. To feel it, press a finger forward just behind your acromion process. Flex your glenohumeral joint, swinging the top of your humerus backward, out from under your acromion process (Fig. 12-16a). You may feel a cord slip under your finger. This is the insertion tendon of your infraspinatus. You have to feel through your deltoid, so it is not easy.

Figure 12 - 17

A lateral view of the humerus showing the insertions of the rotator cuff muscles.

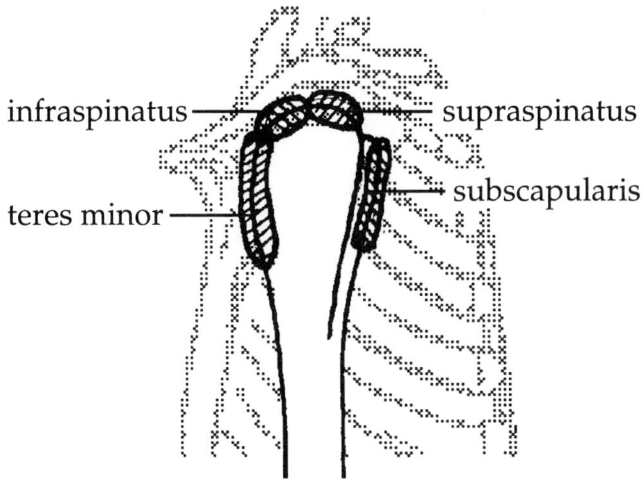

Teres Minor

'Teres' means round and describes the round cross section through our **teres minor.** It is called 'minor' because we have another muscle close by which is similar in shape but larger.

Our teres minor originates from the upper axillary border of our scapula (Fig. 12-18 and 12-20) and inserts into the posterior aspect of the greater tubercle of our humerus (Fig. 12-17). Its fibers lie parallel to those of our infraspinatus giving it very similar actions across our glenohumeral joint, laterally rotating and assisting in extension, adduction, and stabilization of that joint.

To palpate your teres minor, feel with your fingers along the axillary border of your scapula. Start at your inferior angle and move upward, lumping over a mound of muscle and dropping into a hole. Immediately above this hole is your teres minor (Fig. 12-23). To feel your teres minor contract, press your fingers into its belly and laterally rotate your outstretched arm.

A second way to palpate your teres minor is to begin from above. Bring your fingers back through your armpit to touch the spine of your scapula. Stretch your arm out to the side, abducting your glenohumeral joint. Slide your fingers down toward the axillary border of your scapula. The first tissue that rolls under your fingers is your posterior deltoid and infraspinatus. In the hollow immediately below them is your teres minor. With your arm still up, laterally rotate your glenohumeral joint to make the muscle bulge into your fingers.

Figure 12 - 18

A posterior view of the teres minor.

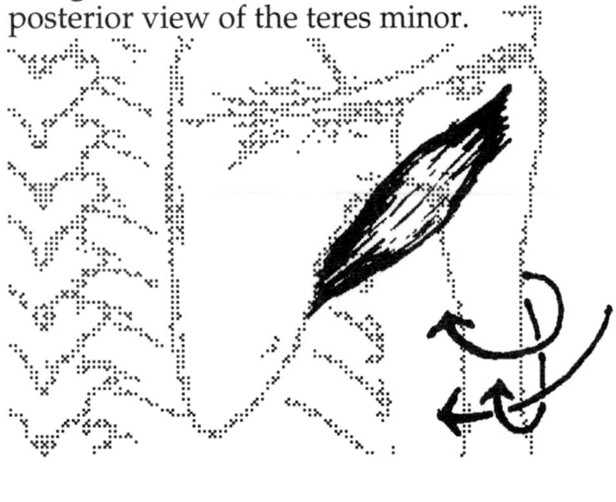

- See how the insertions of our rotator cuff muscles wrap around our glenohumeral joint, covering its anterior, superior, and posterior aspects (Fig. 12-17). Our rotator cuff has a hole inferiorly, and most dislocations of the glenohumeral joint go in that direction (Fig. 12-12).

GLENOHUMERAL MOVEMENT

Muscles that move our glenohumeral joint step further away from the joint to increase their leverage and make movement easier Both their origins and insertions are further from the joint.

Most often, when our glenohumeral joint is in action, our scapula is stabilized while our humerus moves, so in general, these muscles originate from our scapula or pectoral girdle and insert into our humerus. We will see that some of these muscles originate even more proximally from our axial skeleton.

Latissimus Dorsi

The name of our **latissimus dorsi** refers to its location. Do you see the two anatomical orientation words in its name? Lateral and dorsal. Our latissimus dorsi is a broad, flat muscle running laterally over the dorsal surface of our torso. It is narrow and close to the midline at our waist, spreading out to our upper arm (Fig. 12-19). In this way it accents a small waist and broad shoulders.

Our latissimus dorsi has an extensive origin which includes the inferior angle of our scapula, the spinous processes of vertebrae T7 through S5, and our posterior iliac crest (Fig. 12-19, 12-20, and 12-21). The muscle fibers run laterally and upwards, passing between our arm and torso to insert into the intertubercle groove on our proximal anterior humerus.

As our latissimus dorsi contracts it will extend, medially rotate, and adduct our glenohumeral joint. Its greatest effect in extension is from a flexed position down to anatomical position, as in bouncing a ball or doing a chin-up. Since our latissimus dorsi has its major origin on our axial skeleton, it also can indirectly medially rotate, depress, and retract the pectoral girdle as it contracts.

Our latissimus dorsi is superficial throughout most of its distribution, covered only by the inferior tip of our trapezius at the midline. Lift up your arm and reach under it, around the side of your body. Feel the lateral edge of your latissumus dorsi running vertically from your posterior iliac crest up to your arm (Fig. 12-23). Accent the edge by pulling your scapula down toward your waist while keeping your arm lifted. Put your thumb along this edge, letting your fingers wrap around your back. Your fingers now are covering a large part of the belly of your latissimus dorsi. Follow the muscle up beside the axillary border of your scapula, forming the edge of the posterior wall of your armpit. Hold this edge while pretending to bounce a ball and feel the muscle working. Follow it across your armpit anterior to the long head of your triceps and posterior to the short head of your biceps and coracobrachialis (these three muscles will be discussed later in this chapter) heading for your anterior humerus. Its insertion is too buried in muscle to feel.

Our latissimus dorsi twists 180 degrees between its origin and insertion, with the lowest part of its origin inserting into the top of our intertubercle groove and the highest part of its origin inserting into the bottom (Fig. 12-19). This twist gives the muscle better mechanical advantage in the vertical directions of movement.

For anyone who is into pyramids, you may be delighted to learn that we always carry a pyramid tucked up into each armpit (Fig. 12-25). Our ribs and serratus anterior form its medial wall. Our latissimus dorsi, teres major, and teres minor form its posterior wall. Our upper arm and its muscles, including the coracobrachialis, form the lateral wall of the pyramid. Our pectoralis major forms its anterior wall. The point of the pyramud is tickling high up in our armpit and the base is as far out as you care to imagine it.

Figure 12 - 19

A posterior view of the latissimus dorsi. The dotted insertion on the humerus indicates that it is anterior to the humerus and would really not be visible in this view. The arrows indicate how the latissimus dorsi would move the humerus.

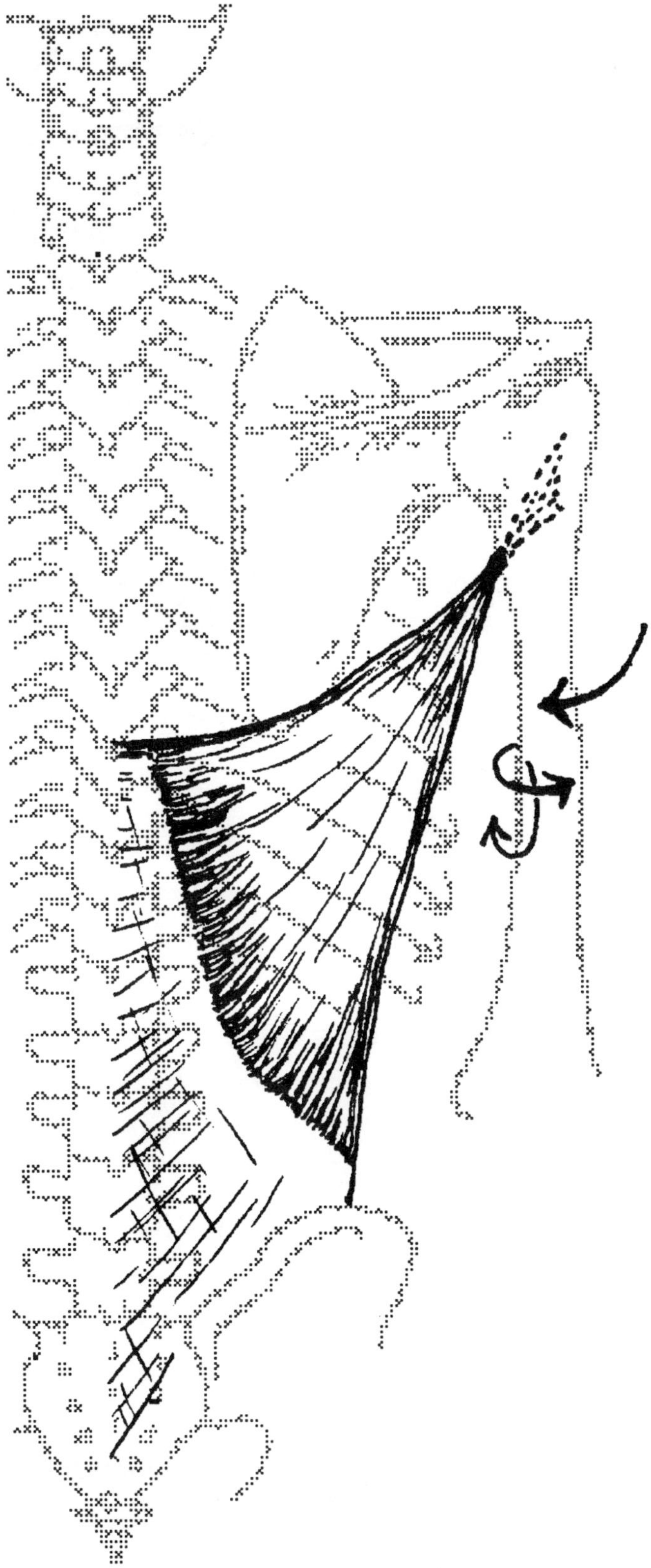

Figure 12- 20

A posterior view of the scapula and humerus showing the origins of muscles in diagonally-lined patches and the insertions of muscles in dotted patches.

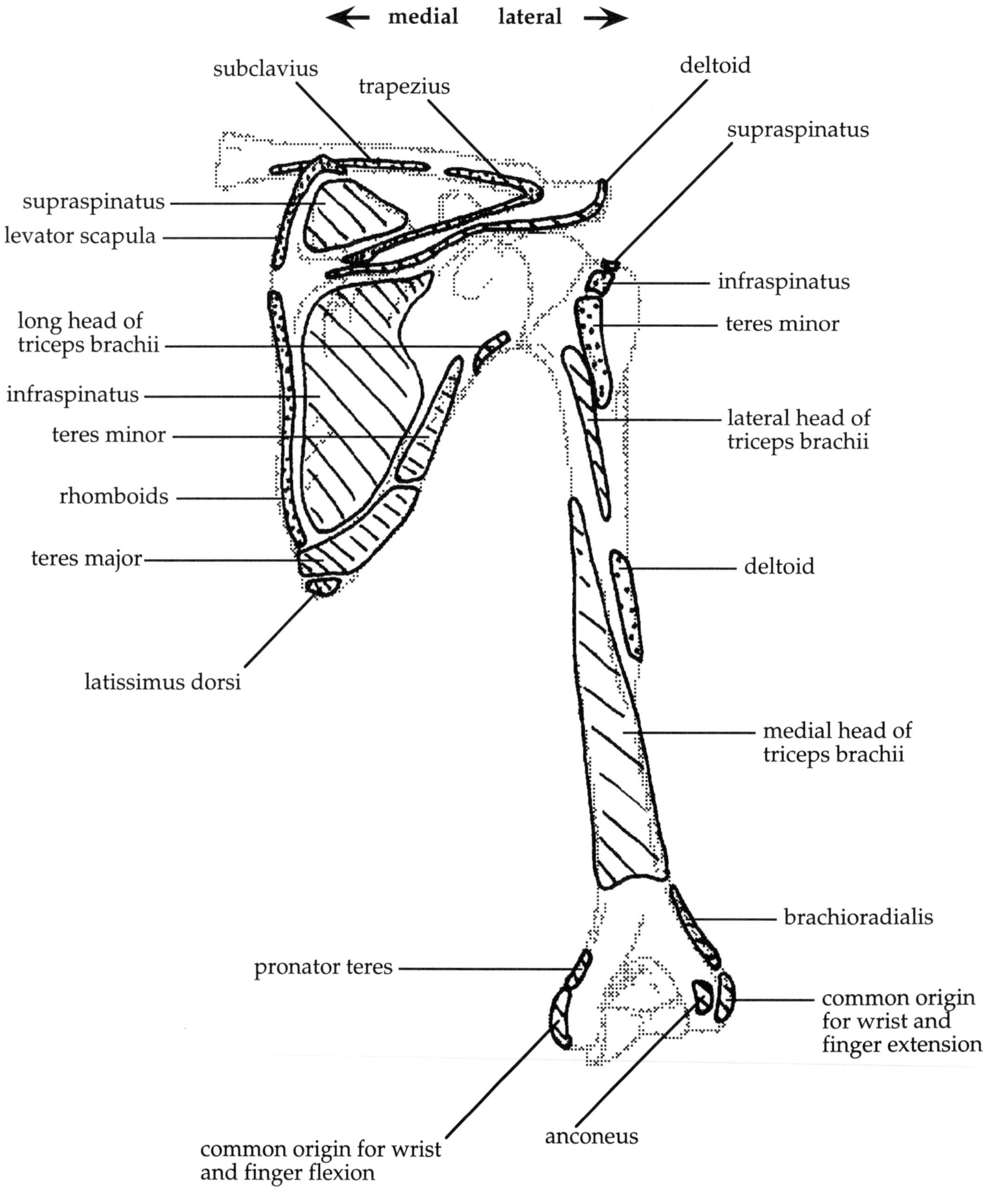

Figure 12- 21

An anterior view of the scapula and humerus showing the origins of muscles in diagonally-lined patches and the insertions of muscles in dotted patches.

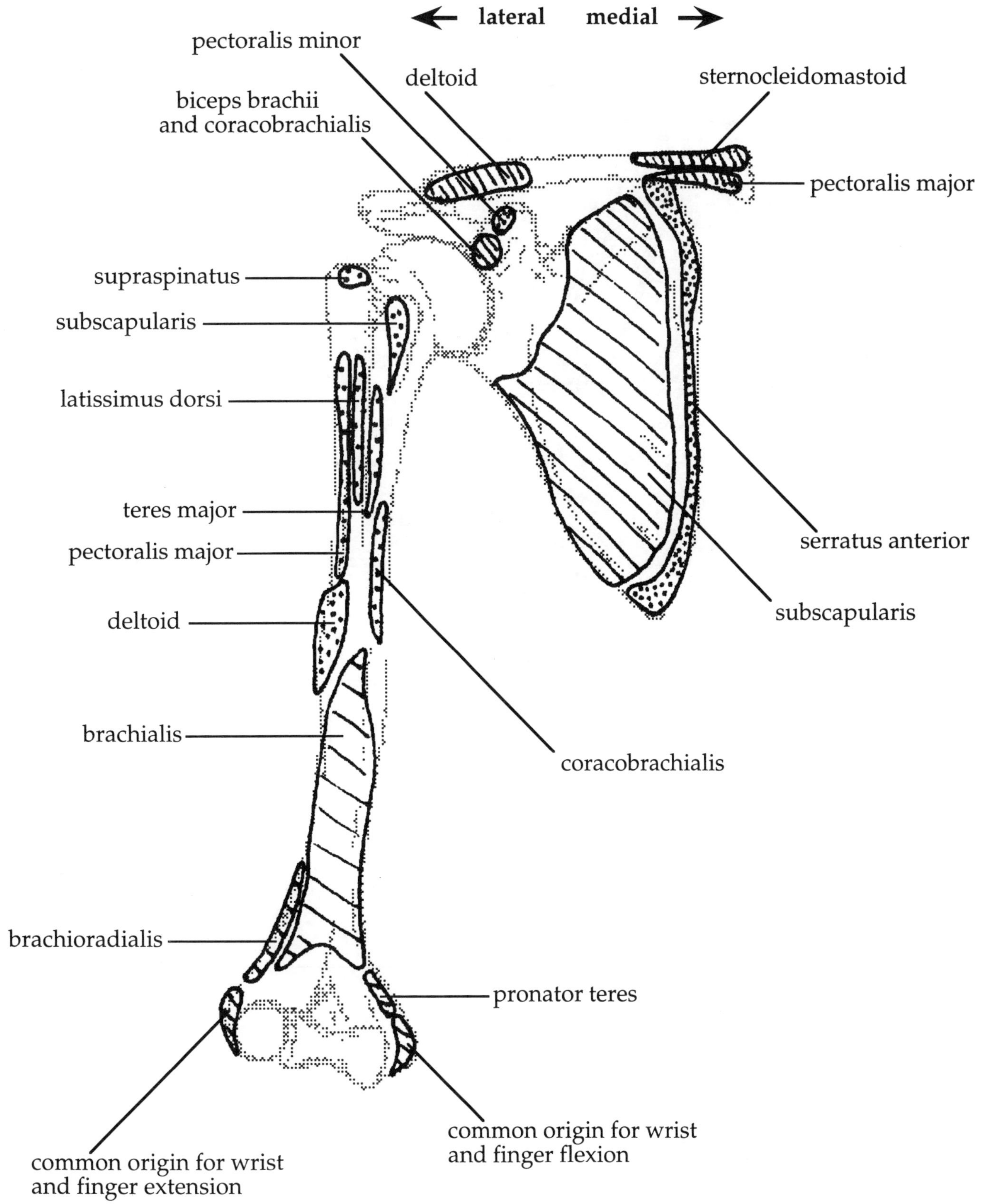

Figure 12- 22

A posterior view of the muscles of the shoulder area. The three most superficial muscles, the trapezius, latissumus dorsi, and deltoid, are only outlined with a dotted line. The deeper muscles are drawn more fully.

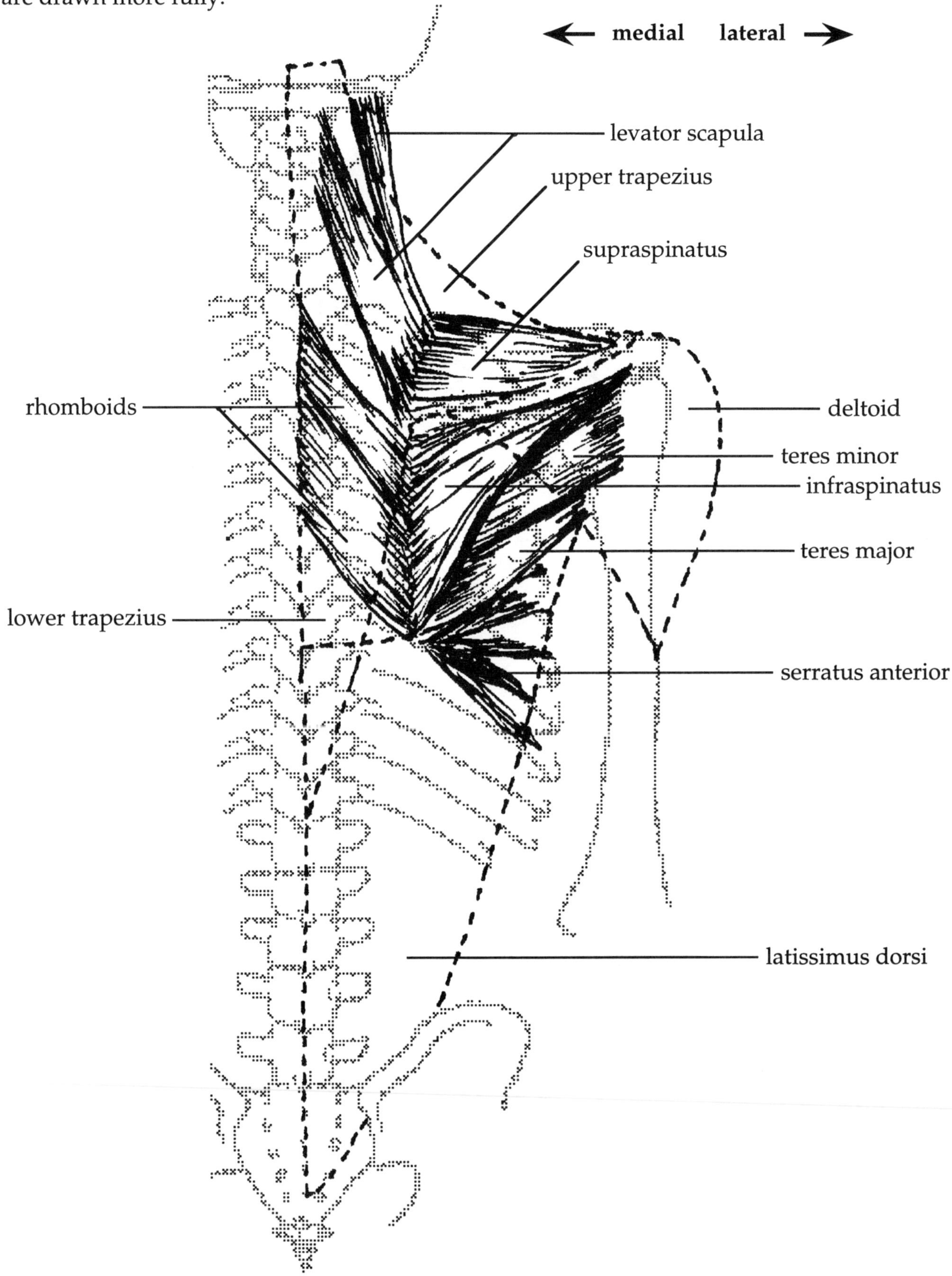

Figure 12- 23

A posterior surface view of the shoulder area indicating the superficial muscles.

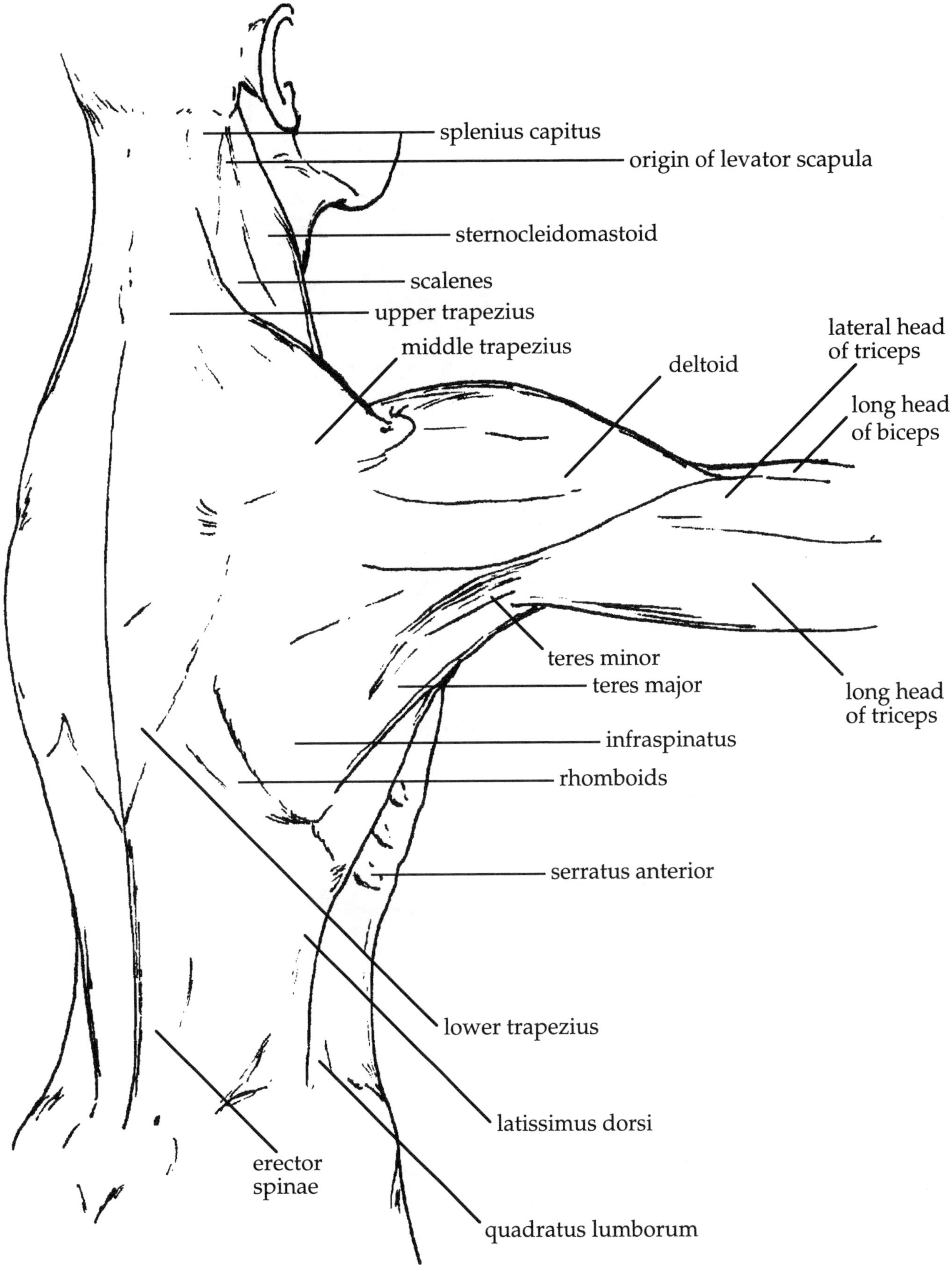

Figure 12 - 24

(a) An anterior view of the muscles of the shoulder area. The two most superficial muscles, the deltoid and pectoralis major, are outlined with a dotted line. The deeper muscles are drawn more fully.

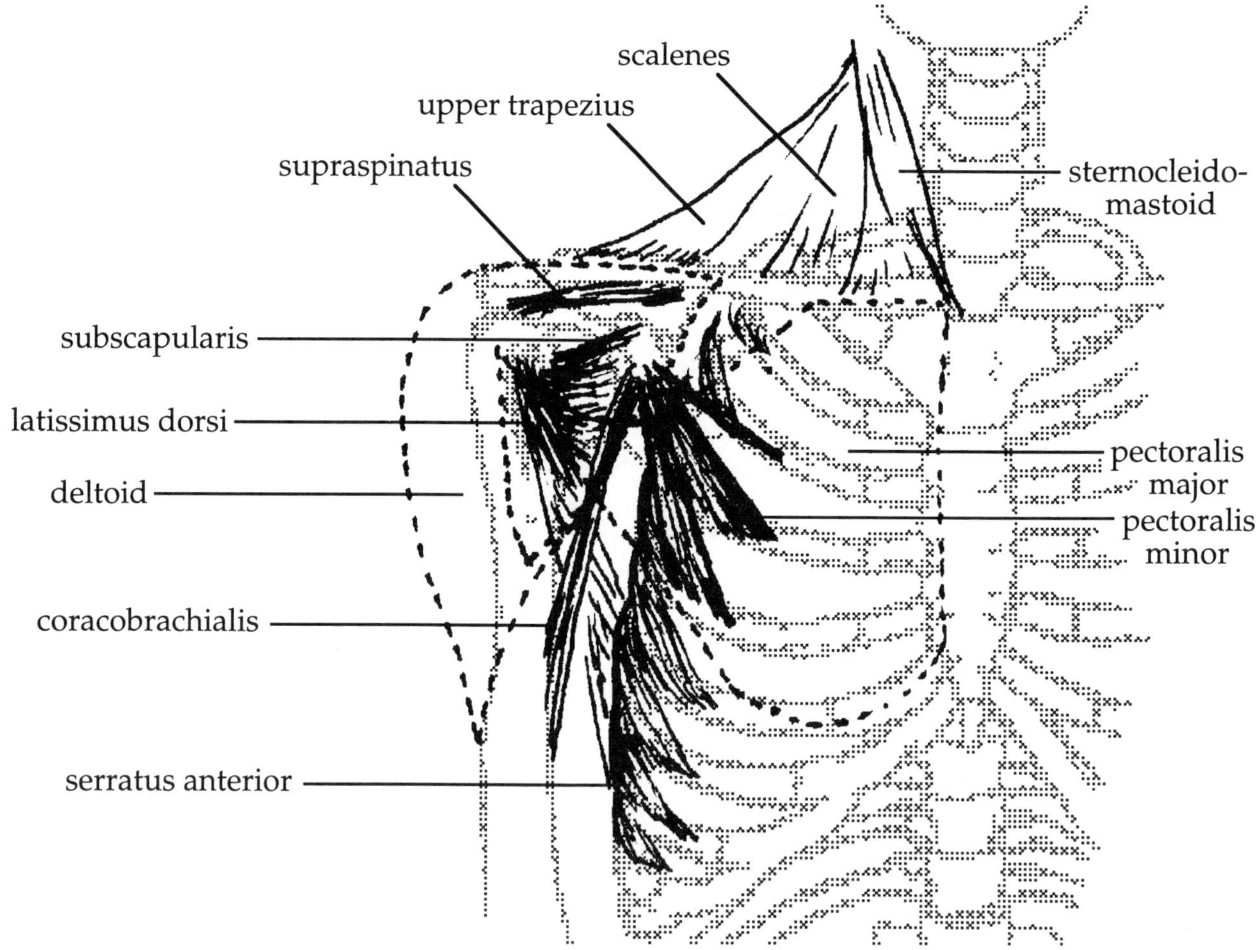

(b) A horizontal cross section through the shoulder area showing the orientation of several muscles. The rectangle projects where corners of the chest would be.

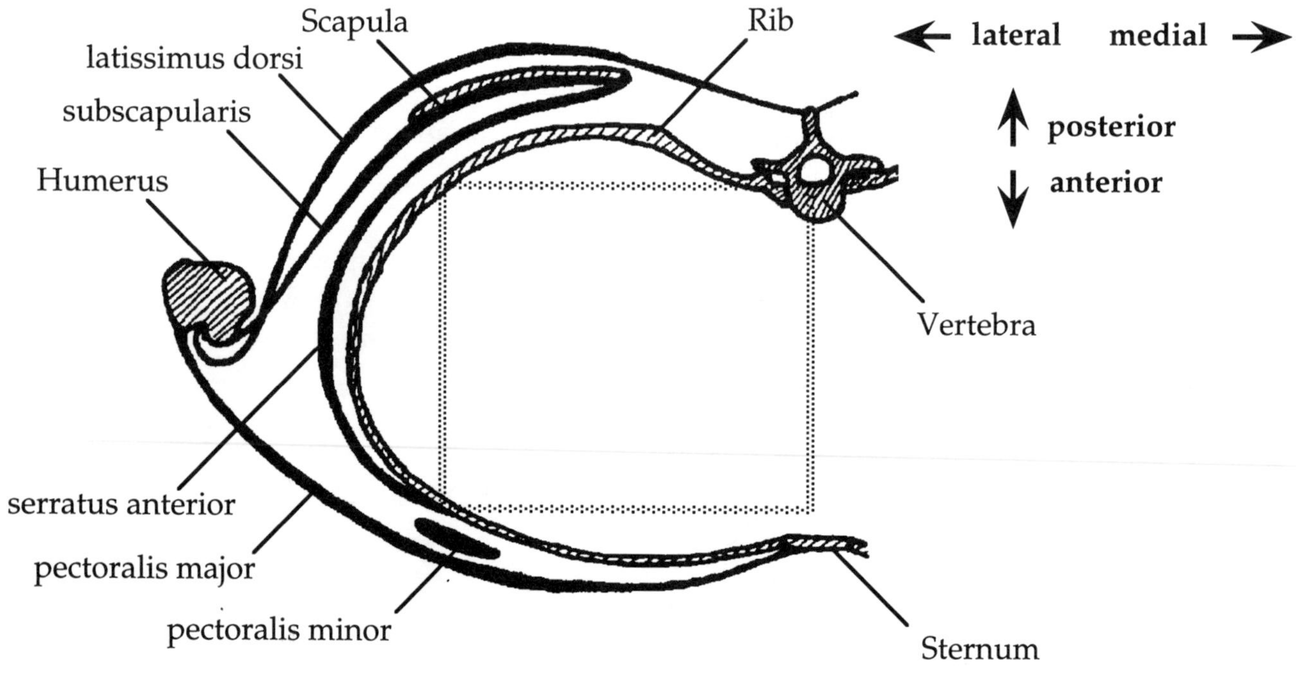

Figure 12 - 25

An anterior view through the shoulder area indicating superficial muscles.

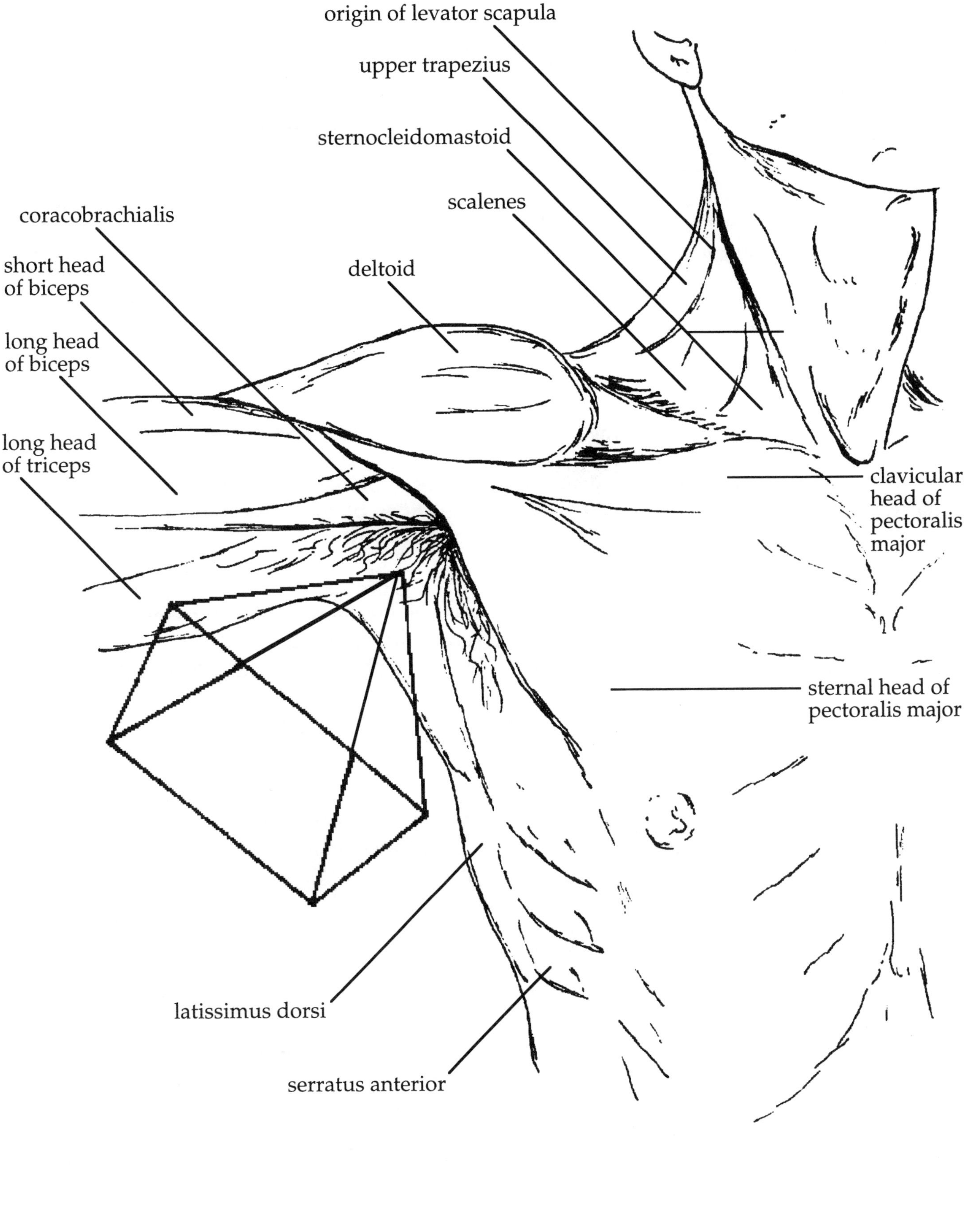

Teres Major

Our **teres major** is the larger of the two round, cylindrical muscles. It originates from the lower axillary border of our scapula near the inferior angle (Fig. 12-20 and 12-26). The muscle fibers pass between our arm and body to insert into the medial edge of the intertubercle groove of our humerus, just medial to our latissimus dorsi insertion.

The fibers of our teres major parallel those of our upper latissimus dorsi so the actions of the two muscles are similar, extending, medially rotating, and adducting our glenohumeral joint (Fig. 12-26).

Your teres major is palpable as part of the posterior wall of your arm pit (Fig. 12-23). Between your thumb and fingers, pick up the posterior wall of your armpit near the lower scapula. Find your latissimus dorsi at the edge of the wall, then slide toward the axillary border of your scapula. The tissue rolling under your fingers is the fibers of your teres major. In following the axillary border of your scapula to locate your teres minor, the first hump you went over was your teres major.

Figure 12 - 26

A posterior view of the teres major. The dotted insertion on the humerus indicates that it is anterior to the humerus and would not really be visible in this view. The arrows show how the teres major moves the humerus.

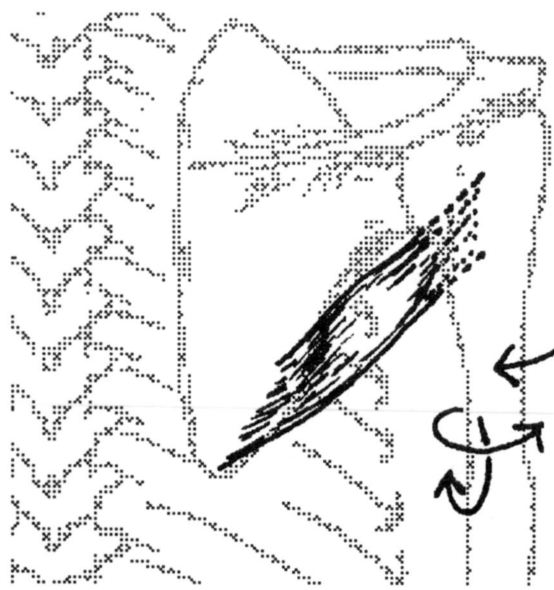

Deltoid

A 'delta' is a triangular Greek letter and the name for the triangular deposits at the mouth of a river. Our **deltoid** is a triangular muscle at the mouth of our arm, attaching it to our pectoral girdle.

The origin of our deltoid forms the base of the triangle, coming from our lateral clavicle and the acromion process and spine of our scapula (Fig. 12-20 and 12-27). As the muscle fibers descend from these bones toward their insertion, they surround our glenohumeral joint on three sides (Fig. 12-27c). On the basis of this orientation, our deltoid is divided into three parts, the **anterior** deltoid coming from our lateral clavicle, the **middle** deltoid from our acromion process, and the **posterior** deltoid from the spine of our scapula. All these muscle fibers insert into a small area on the middle of the lateral shaft of our humerus, forming the apex of the triangle. From anterior to posterior, the belly of our deltoid covers the insertions of the subscapularis, teres major, latissimus dorsi, pectoralis major, and teres minor (Fig. 12-22 and 12-24).

Since the main portion of our deltoid and its insertion are lateral to our glenohumeral joint, contraction of all parts of the muscle produce abduction of our glenohumeral joint (Fig. 12-27). When our arm is down by our side, the mechanical advantage of our deltoid is reduced, so our deltoid lets our supraspinatus initiate abduction, joining in to become the prime mover as the movement continues.

Since each part of our deltoid has a different orientation to our glenohumeral joint, each part has some different action potentials. Our anterior deltoid, lying anterior to our glenohumeral joint can flex, horizontally adduct, and medially rotate the joint. Our posterior deltoid, lying posterior to the joint, can extend, horizontally abduct, and laterally rotate the joint. So while our whole deltoid synergistically abducts our glenohumeral joint, its anterior and posterior portions antagonize each other in all their individual moves.

Our deltoid is superficial throughout its whole distribution and so is easy to palpate.

Figure 12 - 27

(a) An anterior view, **(b)** a posterior view, and **(c)** a superior view of the deltoid with arrows showing how it moves the humerus.

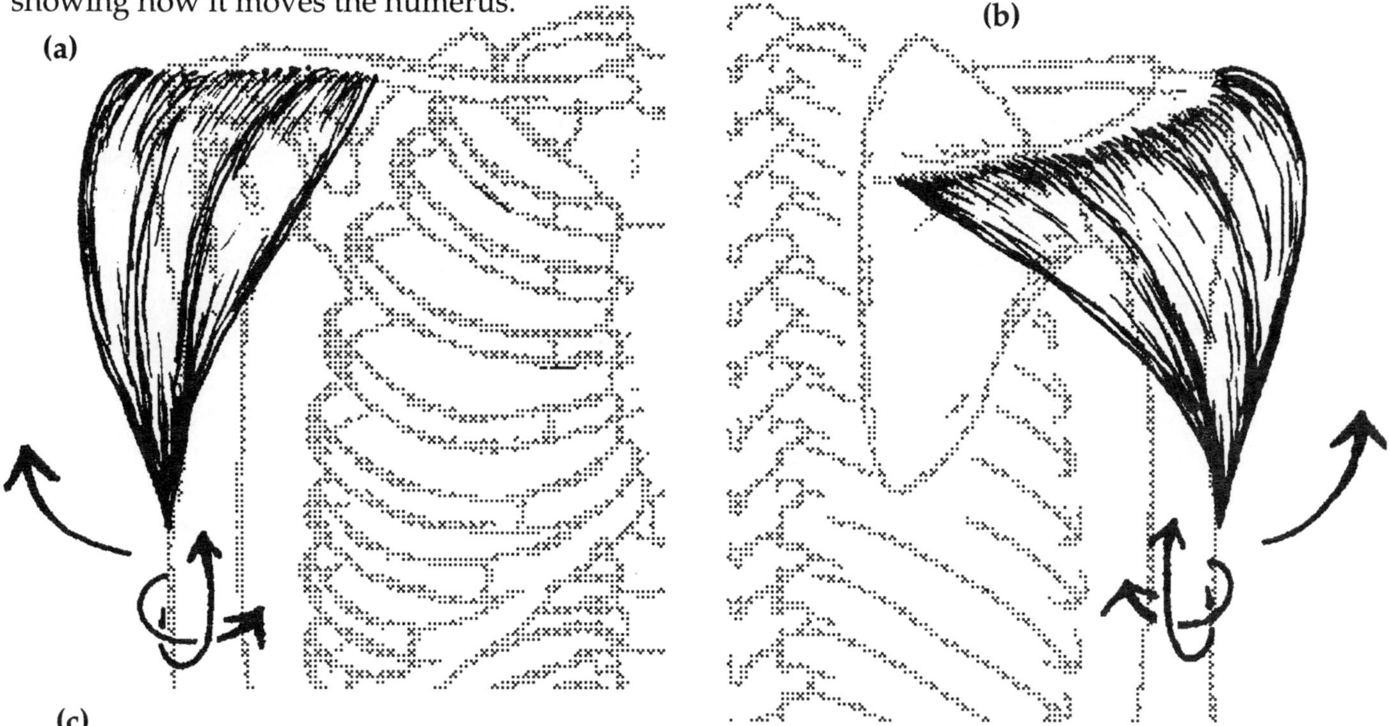

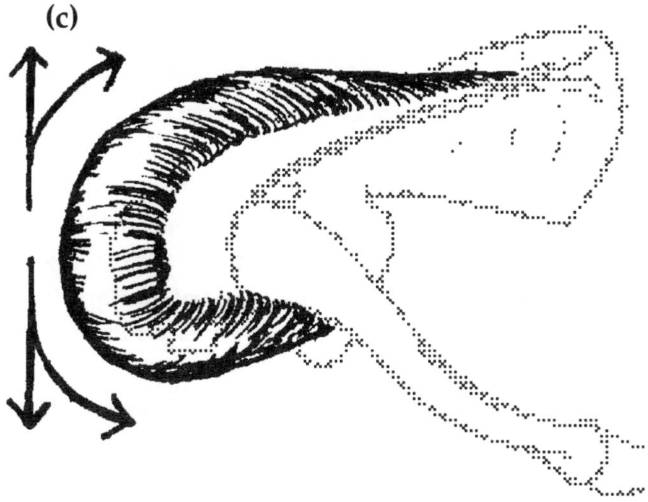

Follow its anterior edge from the lateral hollow below your clavicle, over the coracoid process of your scapula to its insertion. Flex and medially rotate your glenohumeral joint to separate your pectoralis major from your anterior deltoid (Fig. 12-25). With your thumb on this edge of your deltoid, cup your hand around your proximal humerus to cover the belly of your deltoid. Follow the posterior edge of your deltoid obliquely from your medial scapular spine to its insertion on your humeral shaft. To separate your posterior deltoid from other posterior muscles, horizontally abduct your arm.

Is the origin of the deltoid a familiar bony collection? Remember the insertion of the trapezius. Our deltoid begins where our trapezius leaves off (Fig. 12-22 and 12-24), continuing the broad triangle of muscle from our head and vertebrae out onto our arm.

Pectoralis Major

The name **pectoralis major** tells us that this is the larger of two muscles in the area of our breast. The upper part of the breast lies directly over it.

Our pectoralis major originates in two parts, a **clavicular head** from our medial clavicle and a **sternal head** from the entire length of our sternum and the first six costal cartilages (Fig. 12-28). The muscle fibers run laterally, slipping under our deltoid as they insert into the lateral edge of the intertubercle groove of our humerus, just beyond our latissimus dorsi insertion (Fig. 12-21). Our pectoralis major twists between its origin and insertion, the clavicular head attaching low in the groove and the sternal head attaching above. This potentiates the mechanical advantage of our pectoralis major for vertical movements.

As our whole pectoralis major contracts, our glenohumeral joint is adducted, horizontally adducted, and medially rotated. The clavicular head alone pulls our humerus up toward our clavicle, flexing our glenohumeral joint, while the sternal head pulls down, extending our glenohumeral joint (Fig. 12-28). So again, different parts of a muscle may be synergistic in some actions and antagonistic in others.

Our pectoralis major covers most of our anterior chest, beside and above our breast, and below our clavicle. Except for its insertion under the deltoid, it is not covered by any other muscle or bone. Only our breast covers its lower central portion (Fig. 12-25). Press the length of your fingers against your chest, parallel to your ribs, lateral to your sternum. Horizontally adduct your glenohumeral joint, bringing your arm straight out in front of you. Feel your pectoralis major working to pull your arm. As your arm moves forward, the insertion end of your pectoralis major webs forward, bringing your pectoralis away from your chest wall. Feel where it all gathers together toward its insertion.

Figure 12 - 28

An anterior view of the pectoralis major, with arrows showing how it moves the humerus.

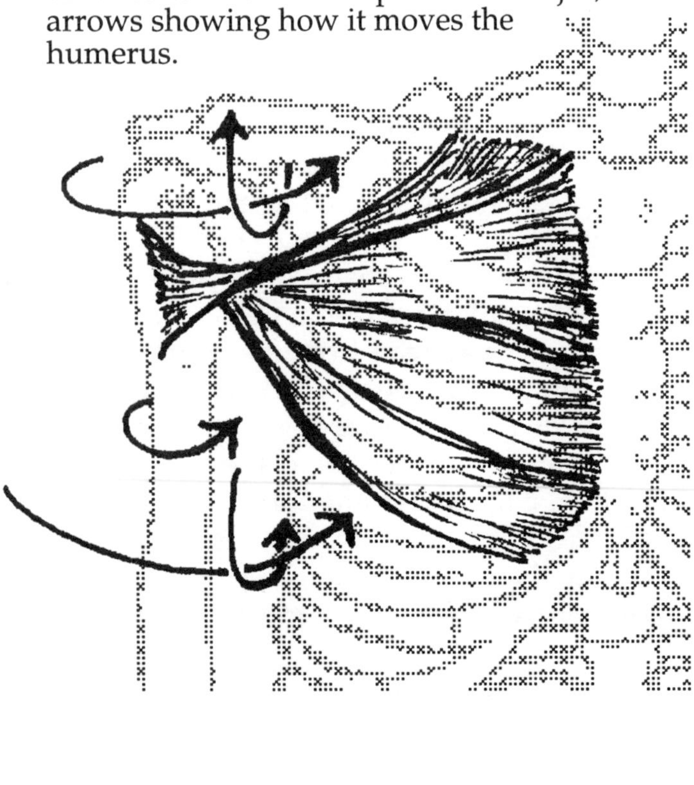

To isolate the clavicular head, perform its special action by flexing your glenohumeral joint. The muscle will pop out prominently from the medial clavicle, leaving a superior hollow between itself and your anterior deltoid, and an inferior hollow over the relaxed sternal head of your pectoralis major (Fig. 12-25). The clavicular head forms the upper part of the anterior wall of your armpit.

The sternal head forms the lower edge of your anterior armpit wall. Follow its web from the side of your breast to your arm. Feel it contract as you quickly extend your glenohumeral joint. To watch it contract, hold your other hand against the back of your upper arm. Press your arm down into your resistant hand. Your pectoralis major will bulge, pressing out your breast.

> There often is a distinct hollowness in our upper chest, immediately below our middle clavicle, accented by a protracted pectoral girdle. One reason for this hollowness is our clavicle sloping backward. Another is the absence of muscle descending from our middle clavicle (Fig. 12-24). Our pectoralis major originates medially and our deltoid laterally, but our middle clavicle is vacant. Modern medicine has given it a use, however, pacemakers are implanted here.

Coracobrachialis

The name **coracobrachialis** identifies this muscle's origin and insertion. What two bony references do you see? Our coracobrachialis originates from the coracoid process of our scapula and inserts into the mid-medial shaft of our humerus (Fig. 12-21 and 12-29), across from the insertion of our deltoid. As our coracobrachialis contracts, it flexes and adducts our glenohumeral joint.

Palpate your coracobrachialis in the medial, proximal aspect of your upper arm. It lies immediately behind the short head of your biceps and is hard to separate from it since they have the same origin and fiber direction. Flex your glenohumeral joint Find the band of muscle coming out of your armpit immediately inferior to the insertion of your pectoralis major, parallel to your humerus. This band of muscle includes the short head of your biceps and your coracobrachialis. The belly of the coracobrachialis is the more proximal and posterior part (Fig. 12-25). Still holding your arm up, straighten your elbow. Slide your fingers across the band of muscle, pressing toward the bone. When your fingers find a cleft, they will be separating the two muscles. Your coracobrachialis ends half way down your humerus, while your biceps continues into your forearm.

Figure 12 - 29
An anterior view of the coracobrachialis with arrows showing how it moves the humerus.

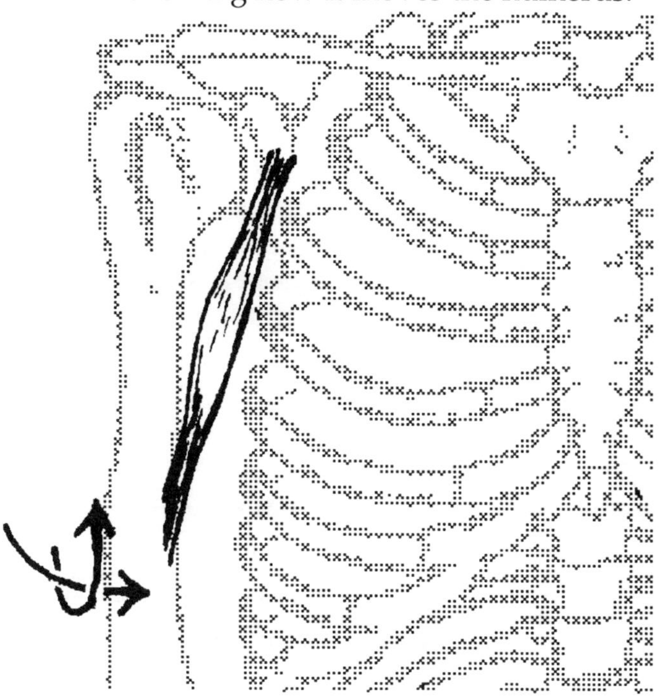

UPPER ARM MUSCLES

Upper arm muscles primarily act across our elbow. The principle movements of our elbow are flexion and extension. How should we organize our upper arm muscles to accomplish this? Such a simple movement pattern allows for a simple muscle orientation. We need one group of muscles ventral or anterior to our elbow to flex it and another group of muscles dorsal or posterior to our elbow to extend it. That's it. In our upper arm there are two muscle compartments, an anterior one and a posterior one.

- The **anterior muscle compartment** in our upper arm contains two muscles, our **biceps brachii,** and our **brachialis.**

Biceps Brachii

This muscle is on a first name basis with many people, simply being called the biceps, however, **biceps brachii** is its full name.

'Bi' means two. 'Ceps' comes from 'kephale' which means head. The 'head' of a muscle is its origin. 'Brachii' refers to the area of the humerus. So biceps brachii tells us that this muscle has two origins and lies in the area of the humerus.

The two bellies of our biceps brachii are actually called heads, one short and one long. The **short head** originates from the coracoid process of our scapula along with our coracobrachialis. The **long head** originates from the superior brim of the glenoid fossa of our scapula (Fig. 12-21 and 12-30). Its tendon passes through the synovial cavity of our glenohumeral joint, coming to lie in the intertubercle groove of our proximal humerus (Fig. 8-27). The groove secures the position of the tendon but the tendon is not attached, sliding freely through it. Within the groove, the tendon lies over our latissimus dorsi insertion and under our pectoralis major insertion (Fig. 12-31).

Both origins of the biceps brachii join together to insert into the radial tuberosity (Fig. 12-30 and 12-47).

Since our biceps brachii lies on the ventral surface of our elbow, as it contracts it will flex our elbow. But what other movement happens at our elbow? The radius spins to supinate and pronate. When our forearm is supinated, our radial tuberosity faces anterior. In pronation, our radial tuberosity faces more posterior. As our biceps contracts, it pulls the radial tuberosity forward to supinate our forearm.

Hold your elbow flexed at 90 degrees. Watch your biceps while moving back and forth between supination and pronation. See your biceps shorten and lengthen as it contracts and releases.

Figure 12- 30

(a) An anterior and **(b)** lateral view of the biceps brachii and **(c)** the arm position for maximin biceps contraction.

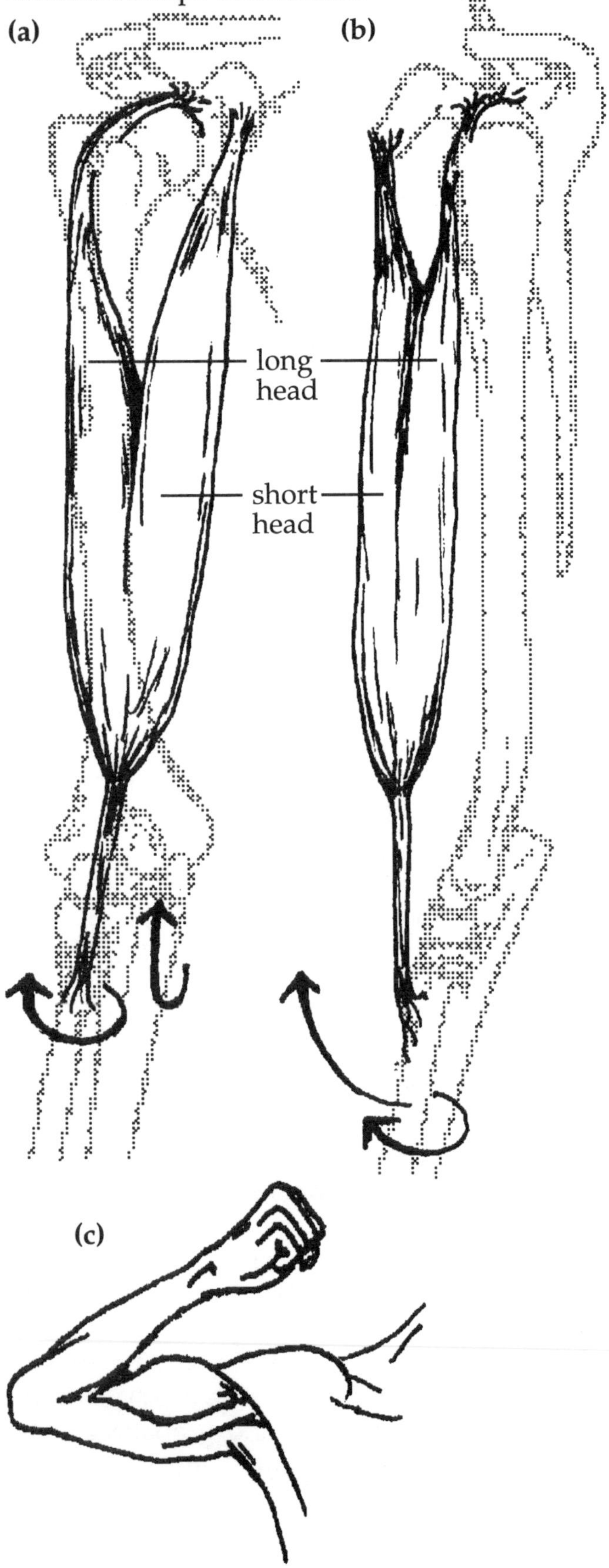

Recall that both heads of our biceps brachii originate from our scapula and therefore cross our glenohumeral joint. Both tendons are ventral to the glenohumeral joint and therefore assist in flexing it.

The classic position to show off our biceps combines all its actions to maximally contract it, flexing our glenohumeral joint and flexing and supinating our elbow (Fig. 12-30c).

Both origins of your biceps are covered by the deltoid but its belly is superficial. Pick up your biceps anterior to your humerus in the middle of your upper arm as it emerges from under your deltoid (Fig. 12-48). It is easiest to feel when your elbow is flexed to 90 degrees. The short head is smaller and lies medially, in front of your coracobrachialis. The long head is the larger, more lateral portion of the belly. To find the division between the two heads of your biceps brachii, press your fingers into your muscle just below your deltoid. Search from side to side across it. At some point, usually toward the medial side of the biceps, your fingers will fall through to the bone. This is the division between the two heads. It may be easier to feel when your biceps is stretched by extending your elbow and extending and laterally rotating you glenohumeral joint. Follow your biceps belly down until it tapers into its insertion tendon. Feel the tendon, like a thick cord crossing in front of your elbow.

Figure 12- 31

A cross section through the upper humerus showing the orientation of significant tendons.

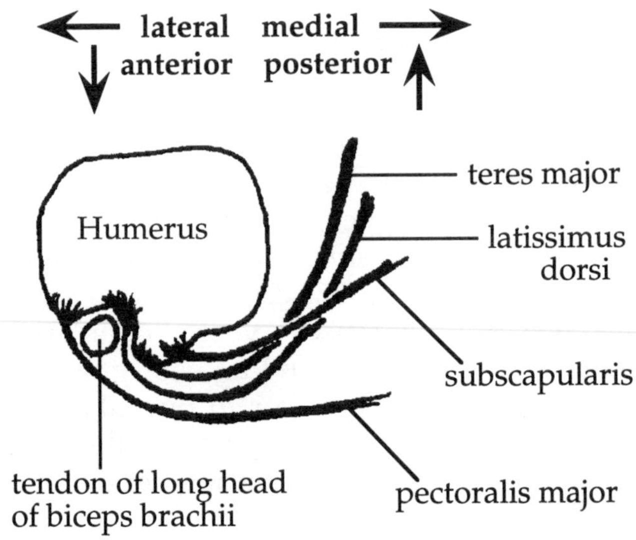

Brachialis

The name **brachialis** tells us that this muscle relates closely to the humerus. Our brachialis originates from the anterior distal surface of our humerus (Fig. 12-21 and 12-32) and inserts into our anterior proximal ulna (Fig. 12-47), crossing ventral to our elbow. As our brachialis contracts, it flexes our elbow. Will it supinate or pronate our elbow? Neither one. Since supination and pronation are movements of the radius, a muscle must insert into the radius to produce these movements.

Our brachialis is posterior to the lower part of our biceps brachii but anterior to our distal humerus (Fig. 12-47, 12-48, and 12-50). Pick up the distal belly and insertion tendon of your biceps. Slide back toward your humerus. All the soft tissue that you palpate, on the medial and lateral sides, between your biceps and humerus is your brachialis muscle.

- The **posterior muscle compartment** of our upper arm contains one muscle, our **triceps brachii.**

Triceps Brachii

'Tri' means three, 'ceps' is heads, and 'brachii' refers to the area of our humerus, so we know that our **triceps brachii** is a three-headed creature in the area of our humerus.

The three heads or origins of our triceps are distinguished as long, lateral, and medial. Its **long head** originates from the inferior brim of the glenoid fossa of our scapula (Fig. 12-20 and 12-33). The other two heads originate from our humerus and are therefore shorter, giving our long head its identity.

Since we have two shorter heads, they are actually called lateral and medial depending on whether they lie lateral or medial to a nerve that spirals down the back of our humerus. Our **lateral head** originates lateral to the nerve, quite proximal on our posterior humerus, deep to our deltoid (Fig. 12-20 and 12-33). Our **medial head** originates medial to the nerve from our distal posterior humerus.

All three heads of our triceps brachii merge into one short tendon, inserting into our ulna's olecranon process (Fig. 12-33 and 12-49).

Figure 12- 32
(a) An anterior and **(b)** lateral view of the brachialis with arrows showing how it moves the forearm.

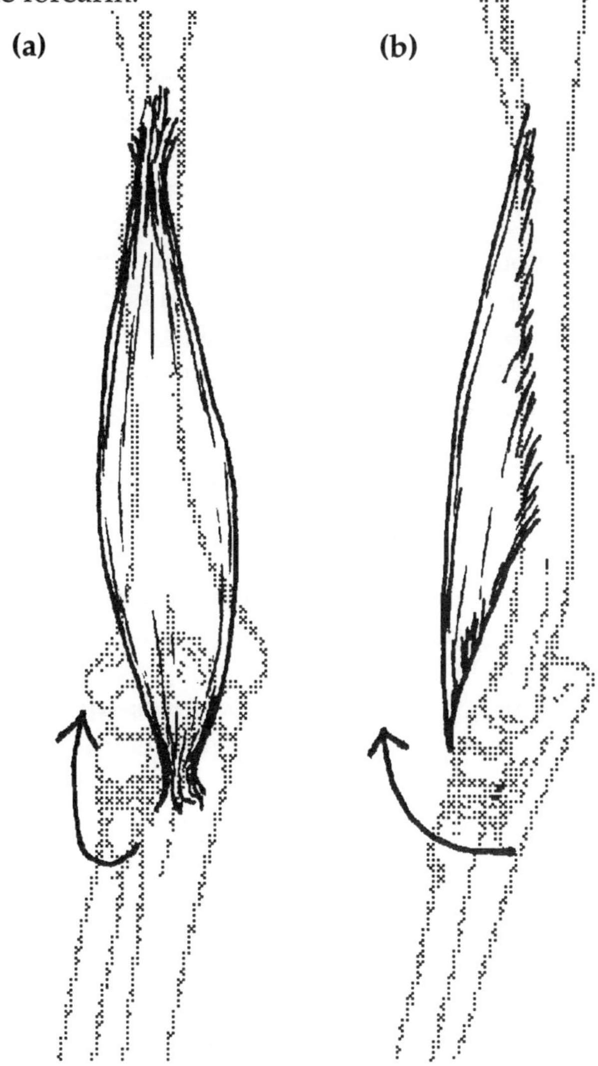

As our triceps brachii contracts, it extends our elbow. Most often we are working with our arms out in front of us and elbow extension is a downward movement. In this position gravity always is eager to extend our elbow so life for our triceps is easy. Our triceps becomes a slouch, letting gravity do all its work. When we use our arms above our head, gravity flexes our elbows. Our triceps is left on its own and any weakness is quickly exposed.

Since the long head of our triceps crosses our glenohumeral joint, it also can act across that joint, extending it.

Hold your arm straight out, palm up. The saggy, waggly part hanging down like a hammock from your humerus (Fig. 12-48 and 12-50) is the long head of your triceps brachii. Find its beginning in the posterior wall of your armpit, between your latissimus dorsi and posterior deltoid. Follow the long head down the medial-posterior aspect of your humerus. Its belly ends about two-thirds of the way down your humerus, with its insertion tendon lying over the other two triceps bellies.

Still holding your arm up, medially rotate your glenohumeral joint. As your elbow swings back and up, watch the pendulous nature of your triceps disappear as it begins to work to keep your elbow extended. In this position, gravity's effort will flex rather than extend your elbow.

To find the lateral head, contract your triceps but keep your elbow somewhat flexed. Feel the lateral head bulging over the lateral posterior humerus, from just behind the insertion of your deltoid down to the lateral epicondyle of your humerus (Fig. 12-50). Bend and straighten your elbow, back and forth. Watch your lateral head change length.

The medial head of your triceps lies deep to the insertion ends of the other two heads, adding to the distal mass of your triceps. It is easiest to palpate just proximal to the medial epicondyle of your humerus.

To palpate the insertion tendon of your triceps brachii, relax your arm and press two fingers into the hollow just above your olecranon process. Tense your triceps. The tendon will push out against your fingers.

Figure 12- 33

(a) A posterior and **(b)** lateral view of the triceps brachii.

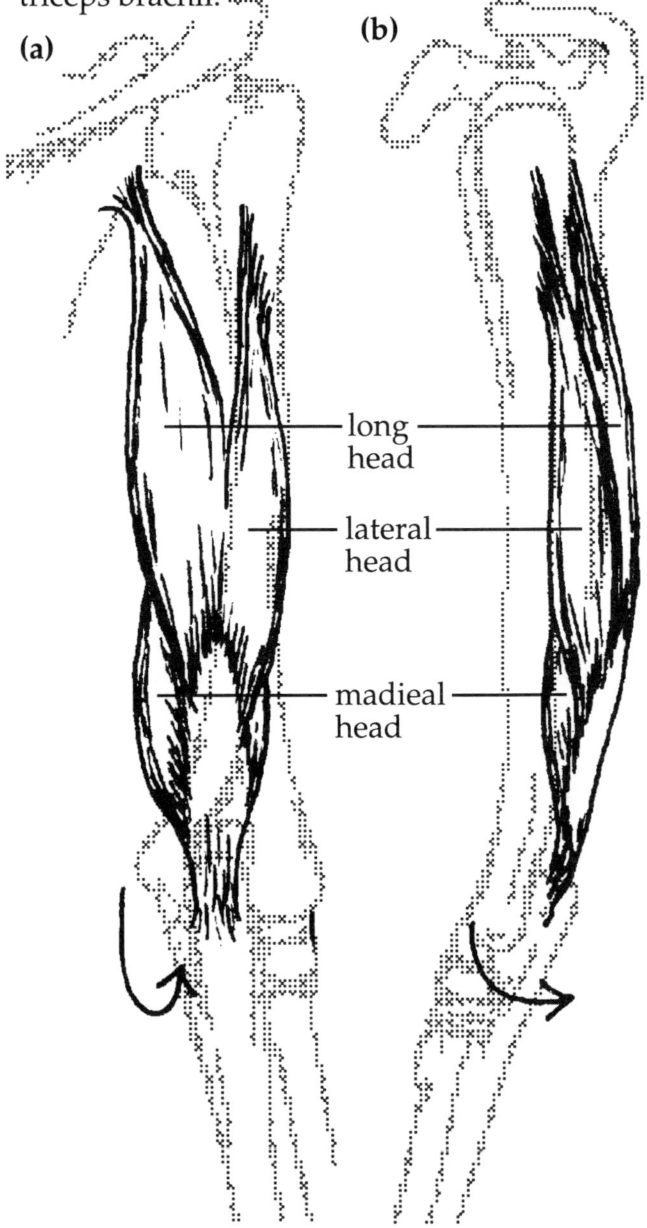

FOREARM MUSCLES

There are many little muscles all crowded together in our forearm. Most of these muscles have long, cumbersome names. To memorize these without understanding is an overwhelming task. On the other hand, if we learn a few simple rules and see what the muscle name tells us, everything falls into place quite neatly. First we will discuss these general rules, then each muscle individually.

- The muscles of our forearm act across our elbow, wrist, and finger joints. All these muscles can be divided into two compartments, an **anterior-medial compartment** which crosses the ventral surface of the joints, and a **posterior-lateral compartment** which crosses the dorsal surface of the joints. The landmarks which designate the borders of the compartments are the posterior shaft of our ulna, palpable the entire length of our forearm, and the lateral edge of one of our flexing muscles called the brachioradialis (Fig. 12-34). Since this muscle is an important landmark, we will discuss and find it before we continue with our general overview.

Figure 12- 34

A cross section through the forearm showing the flexor and extensor compartments and their boundaries.

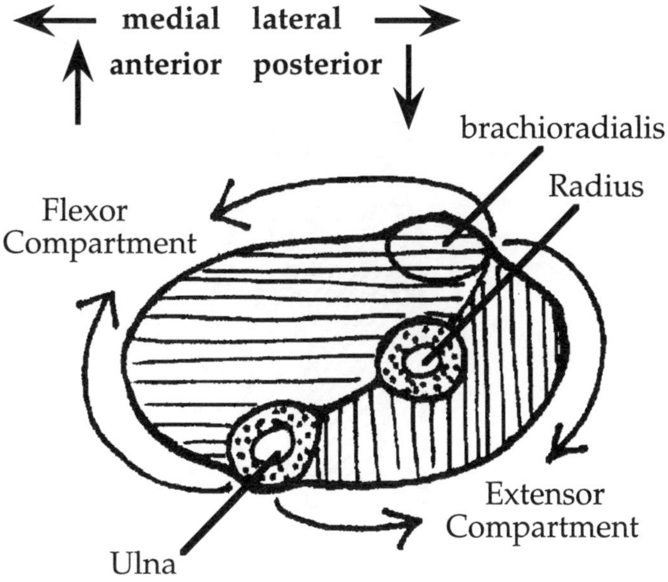

Brachioradialis

The name **brachioradialis** tells us about this muscle's bony attachments. 'Brachii' refers to the humerus and 'radialis' the radius. More precisely, our brachioradialis originates from the lateral edge of our distal humerus, about two inches above our lateral epicondyle (Fig. 12-21 and 12-35). It inserts into the styloid process of our radius (Fig. 12-47 and 12-49).

While the belly of our brachioradialis is in the forearm, the only joint it crosses is our elbow. In anatomical position, it lies ventral to the elbow and so will flex it.

Since our brachioradialis inserts into our radius, what other potential actions does it have? To supinate or pronate. The shortest position of our brachioradialis is neutral between supination and pronation. It can initiate both supination and pronation, bringing our forearm to the neutral position; however it cannot complete either move. From anatomical position, contraction of our brachioradialis will initiate pronation, but once we pass the halfway point to complete pronation, the muscle is being stretched. The same condition applies to supination.

Your brachioradialis forms the lateral edge of the anterior-medial compartment of your forearm. Flex your elbow to 90 degrees. Face the palm of your hand medially so your forearm is neutral between supination and pronation. Brace your arm so your elbow cannot flex, then try to flex it anyway. The muscle that pops upward is your brachioradialis (Fig. 12-48 and 12-50). It pops out further than its neighbors because it originates well above the lateral epicondyle of the humerus while the neighbors all originate on the epicondyle itself. Feel the insertion tendon of your biceps brachii immediately medial to your brachioradialis (Fig. 12-46). Follow your brachioradialis all the way from your humerus to the styloid process of your radius. Rotate your forearm into a supinated position to see clearly that the brachioradialis is ventral to your elbow.

Figure 12- 35

An anterior view of the brachioradialis with arrows showing how it moves the elbow.

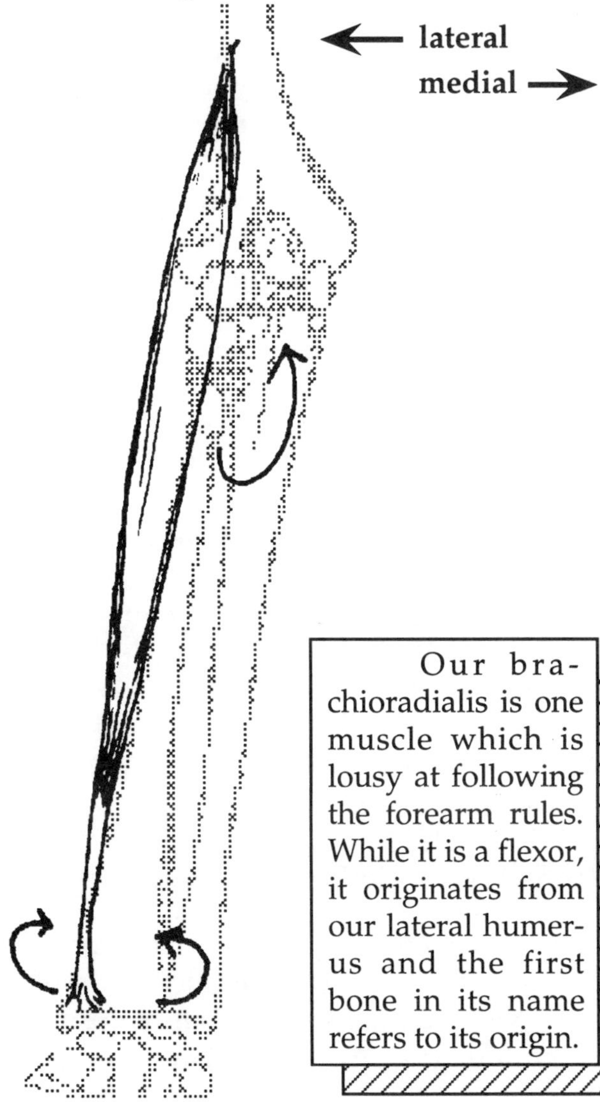

> Our brachioradialis is one muscle which is lousy at following the forearm rules. While it is a flexor, it originates from our lateral humerus and the first bone in its name refers to its origin.

Now that we have found our brachioradialis we can go back to the general rules of our forearm muscles. First, we will palpate the two compartments.

Place the tip of your left thumb just lateral to your right brachioradialis. Let your hand curl around the anterior-medial aspect of your forearm until your fingers touch your ulna. This is the **anterior-medial compartment** of the forearm. All these muscles cross ventral to the elbow, wrist, and/or fingers and can flex the joints they cross.

Now place your thumb along your ulna, pointing toward your hand. Let your hand curl around the posterior-lateral aspect of your forearm until your fingers touch your brachioradialis. This is the **posterior-lateral compartment** of the forearm. All these muscles cross dorsal to the elbow, wrist, and/or fingers and can extend the joints they cross.

The terms flexor or extensor appear frequently in the naming of the forearm muscles. These words tell us the general action of the muscle, and in this case, the compartment in which it is located.

The muscles in both compartments are layered over each other. Several consistent rules help us remember the origins and insertions of the layers.

With regard to origins: **A.** Superficial muscles originate from the humerus, while deep muscles originate more distally from the ulna and radius (Fig. 12-36). **B.** With the forearm in neutral position between supination and pronation, what part of the humerus pokes out, offering a handle for the flexing compartment? The medial epicondyle. Superficial muscles in the flexing compartment originate from the medial epicondyle of the humerus. With the forearm still in neutral position, what part of the humerus offers its support to the extending compartment? The lateral epicondyle. Superficial muscles in the extending compartment originate from the lateral epicondyle of the humerus. **C.** Even though many forearm muscles cross the elbow, only a few have significant action at the elbow.

With regard to insertions of the muscles of our forearm: **A.** All bony references in the

Figure 12 - 36

A lateral view schematic of the forearm and hand showing the pattern of origins and insertions of the muscle layers.

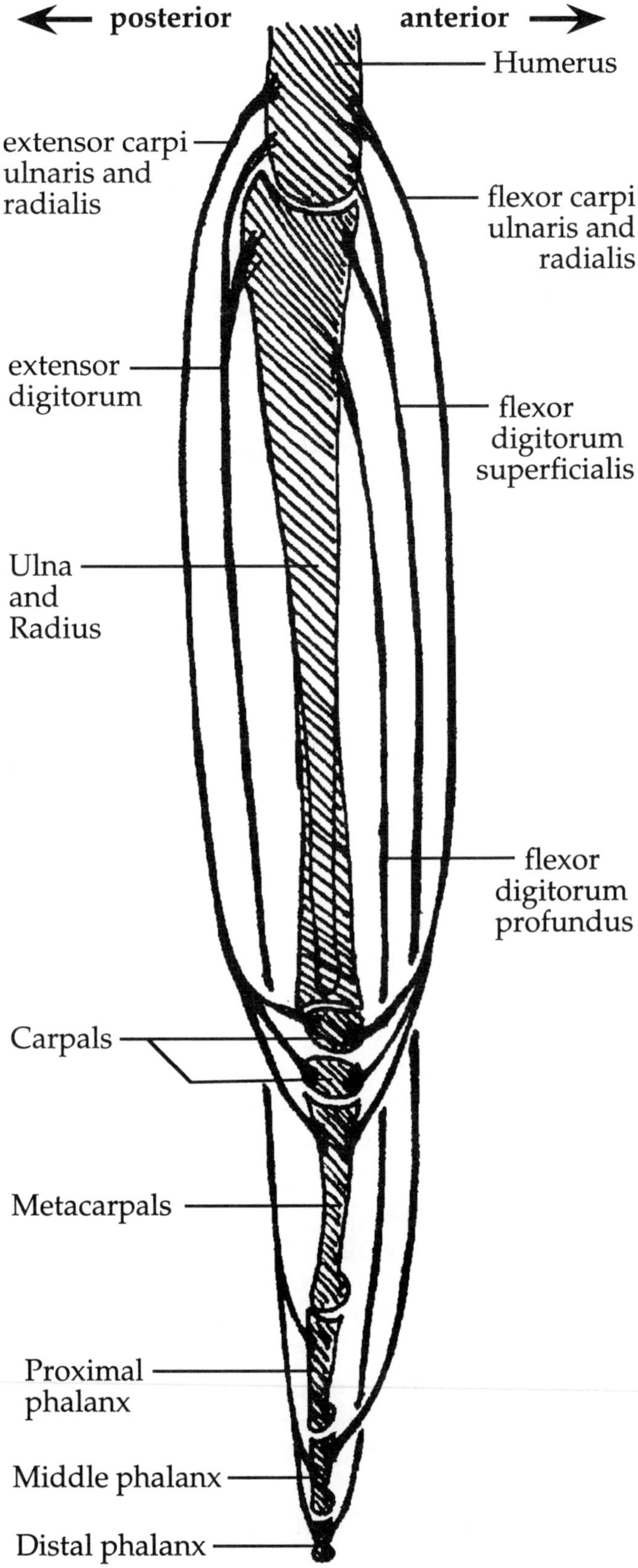

muscle's name tell us about the insertion of the muscle. For example, the name flexor carpi ulnaris tells us that this muscle inserts into the carpal bones on the ulnar side of the hand. **B.** Superficial muscles insert proximally in the hand, while deeper muscles insert distally in the hand (Fig. 12-36).

Are we ready? Let us begin. First we will discuss how each muscle fulfills the basic rules, then learn its specific details or exception.

ANTERIOR-MEDIAL COMPARTMENT

We will begin with the **anterior-medial compartment of the forearm.** It has four layers of muscles.

The **superficial layer** has four muscles, in addition to the brachioradialis. We will discuss them from medial to lateral.

Flexor Carpi Ulnaris

'Flexor' tells us the action of this muscle and that we are in the anterior-medial compartment. 'Carpi' tells us that the muscle inserts into the carpal bones. Carpal bones are proximal in the hand so we know by the rules that this muscle must be superficial. If we know that the muscle is a superficial flexor and follow the rules, where must it originate? The medial epicondyle of the humerus (Fig. 12-37 and 12-47). A portion of the **flexor carpi ulnaris** also originates from the proximal ulna. 'Ulnaris' tells us that the insertion is on the ulnar side of the carpal bones, more precisely the pisiform, hamate, and base of the fifth metacarpal.

As the flexor carpi ulnaris contracts, it pulls the pisiform bone toward the medial epicondyle of the humerus. This movement combines flexion and adduction of the wrist.

To feel the belly of your flexor carpi ulnaris, pick up the tissue medial to your ulna, just distal to the medial epicondyle of your humerus (Fig. 12-48). Perform the action of the muscle and feel it bulge. Follow the muscle distally along your ulna, moving from medial to anterior. On the medial edge of your anterior wrist, pick up its insertion tendon crossing toward your pisiform.

Figure 12 - 37

An anterior view of the flexor carpi ulnaris.

Palmaris Longus

Our **palmaris longus** lies just lateral to our flexor carpi ulnaris in the superficial layer of the flexing compartment of our forearm (Fig. 12-47 and 12-48). From this we can deduce that it originates on the medial epicondyle of our humerus (Fig. 12-38).

The name 'palmaris longus' tells us that it is a long muscle which inserts into the palm of our hand. It does not insert into bone, but into the dense connective tissue in our palm, beginning with our flexor retinaculum.

As our palmaris longus contracts, it assists in flexing our wrist.

In the bony alignment of certain animals, the palmaris longus locks the wrist joint so the animal can sleep in a standing position without collapsing. I don't know too many people who can stand on their hands when they are awake let alone when they are asleep. The main use of this muscle is in four-legged animals who balance easier. In humans, if tendon tissue is needed for some surgical repair or transplant, the tendon of the palmaris longus may be used.

The most palpable part of your palmaris longus is its insertion tendon. Flex your wrist. Two tendons pop out in the center on the ventral surface of your wrist (Fig. 12-48). The medial of the two is your palmaris longus. It is usually smaller, but stands out further since it inserts into the superficial connective tissue of your hand, rather than the deeper bones.

With your wrist pulled into flexion, follow the tense cord of your palmaris longus to the medial epicondyle of your humerus.

If you can only find one tendon crossing the center of your wrist, or there are a bunch of deep tendons but not one that particularly stands out, it is possible that you do not have a palmaris longus. It is missing in about 20 percent of arms. In any person, it may be present on one side, but not the other.

Flexor Carpi Radialis

The 'flexor' and 'carpi' of **flexor carpi radialis** tell us the same information as they did for the flexor carpi ulnaris. It is a superficial muscle in the anterior-medial compartment of our forearm. It originates from the medial epicondyle of our humerus and inserts proximally in our hand (Fig. 12-39 and 12-47).

'Radialis' tells us that the insertion is toward the radial side of our hand, specifically the base of our second and third metacarpals.

Crossing ventral to the wrist, our flexor carpi radialis will flex our wrist.

Its insertion tendon lies slightly lateral to the midline of our wrist, so it also has a mechanical advantage to abduct our wrist, but the potential degree of abduction beyond anatomical position is small.

Figure 12 - 38

An anterior view of the palmaris longus.

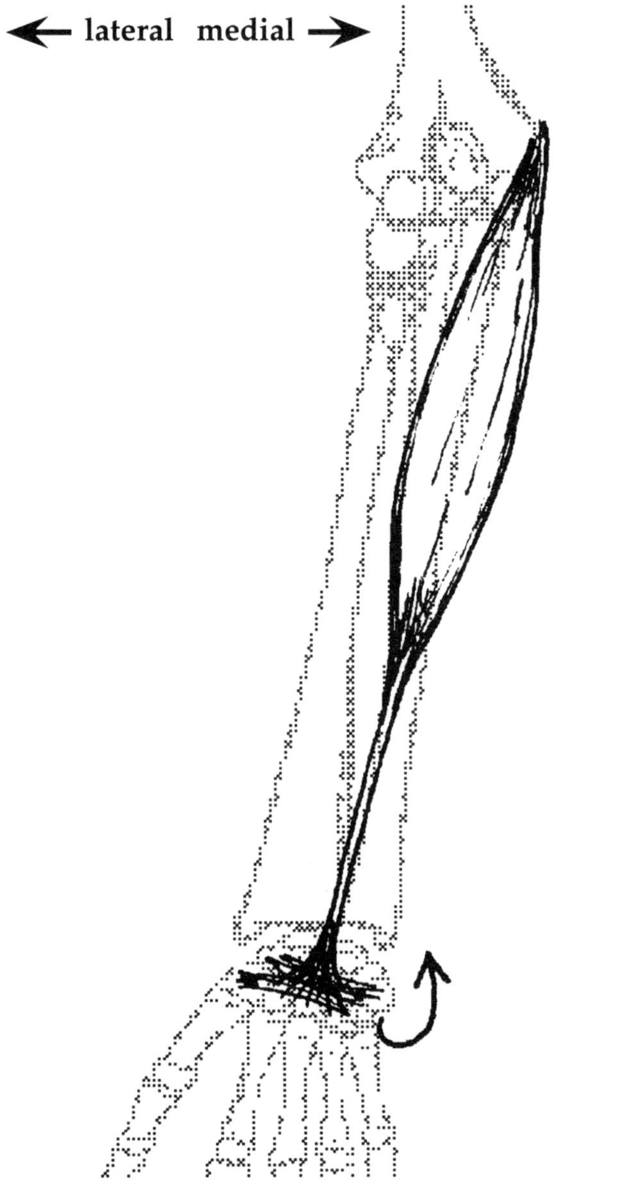

The most palpable part of your flexor carpi radialis is its insertion tendon. Flex your wrist slightly. Two tendons pop out in the center on the ventral surface of your wrist. The medial tendon is your palmaris longus, if you have one. The most lateral of the two is your flexor carpi radialis (Fig. 12-48). It usually is larger but less prominent as it inserts deeper on the bones. The belly of the flexor carpi radialis is in a straight line between its tendon and the medial epicondyle of your humerus, lateral to the palmaris longus,. It often is difficult to isolate clearly.

Figure 12 - 39

An anterior view of the flexor carpi radialis.

Figure 12 - 40

An anterior view of the pronator teres.

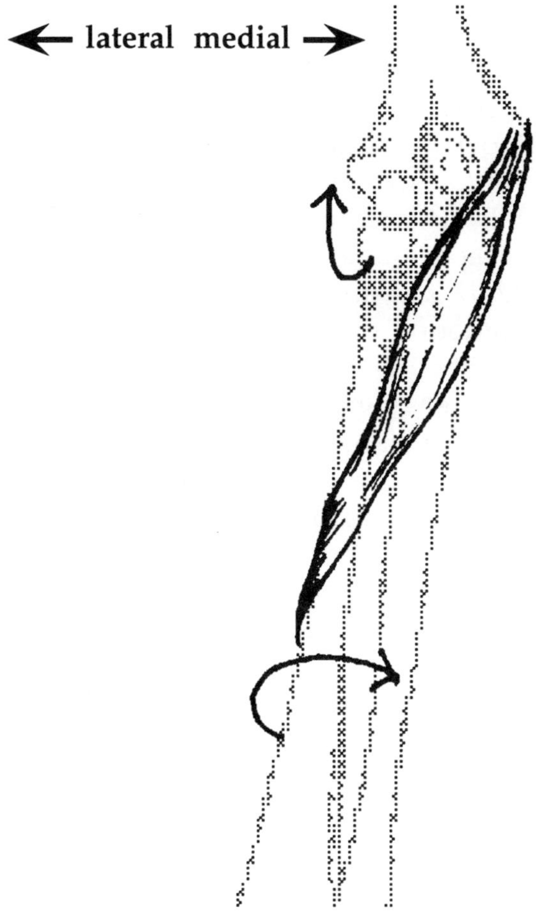

Pronator Teres

Our **pronator teres** lies lateral to our flexor carpi radialis and medial to our brachioradialis. It originates from the medial epicondyle of our humerus and our anterior proximal ulna (Fig. 12-40 and 12-47). It passes diagonally across our forearm to insert into the mid-lateral shaft of our radius.

Like our brachioradialis, our pronator teres only crosses our elbow joint, even though its belly is mainly in our forearm. Since it crosses ventral to our elbow, it will flex our elbow. The name of this muscle tells us of another action. Since it crosses diagonally from the medial side to insert into our radius, it can spin our radius to cross over our ulna into pronation.

'Teres' tells us that this muscle is round in cross section. It is easiest to palpate when your elbow is flexed to 90 degrees and your forearm is pronated (Fig. 12-46 and 12-48). In this position, find a deep indentation in the center in front of your elbow. Your biceps brachii insertion tendon plunges into this indentation. The brachioradialis forms its lateral wall, lateral to the biceps tendon. Press your thumb into the dent and pull medially. The muscle you come up against is your pronator teres.

Slide along your pronator teres from the medial epicondyle of your humerus to your radius. Near your radius, your pronator teres slips under your brachioradialis. Move back and forth between pronation and supination to feel your pronator teres contract and relax.

Flexor Digitorum Superficialis

'Flexor' tells us that this muscle is in the flexing compartment in our anterior-medial forearm. The **flexor digitorum superficialis** forms the second layer of this compartment.

'Superficialis' tells us that, while this is the second layer of muscle, it still is relatively superficial. Therefore, it still originates from the medial epicondyle of our humerus (Fig. 12-41 and 12-47). It also takes part of its origin from the anterior ulna and radius.

'Digitorum' tells us that the muscle inserts into our fingers. In our distal forearm, our flexor digitorum superficialis divides into four tendons. These tendons pass deep to our flexor retinaculum to insert one tendon into the middle phalanx of each of our four fingers. As each tendon approaches its insertion, it forks, so that each tendon attaches to both sides of its phalanx (Fig. 12-41). As our flexor digitorum superficialis contracts, it flexes our wrist and fingers at the metacarpal phalangeal (MP) and proximal interphalangeal (PIP) joints.

Wiggle your fingers forward and backward. Watch the ripple in your anterior-medial compartment. Press with your finger tips into the tissue between the tendons of your flexor carpi ulnaris and palmaris longus near your wrist (Fig. 12-47 and 12-48). Close and open a fist. Feel the tendon of your flexor digitorum superficialis sliding back and forth as the muscle contracts and is stretched.

Figure 12 - 41

An anterior view of the flexor digitorum superficialis.

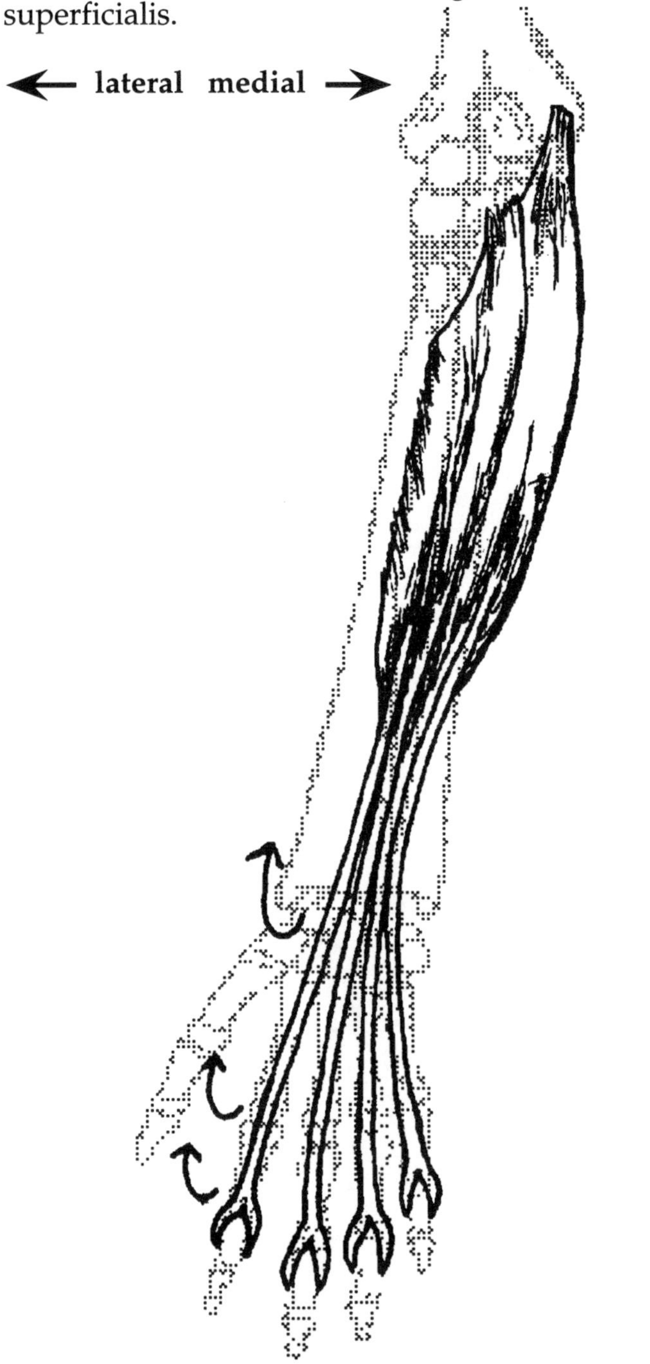

Flexor Digitorum Profundus

'Flexor digitorum' tells us that this muscle is in the anterior-medial compartment and flexes our fingers. 'Profundus' tells us that the muscle is deep, just as profound thoughts are deep. Our **flexor digitorum profundus** is in the third layer of this compartment.

In accordance with our rule that deeper muscles originate more distally, our flexor digitorum profundus originates from our anterior ulna and interosseous membrane (Fig. 12-42 and 12-47a). Near the wrist it divides into four tendons which travel through the hand deep to the tendons of the flexor digitorum superficialis. As the tendons of the flexor digitorum superficialis fork to insert, the tendon of the flexor digitorum profundus emerges through the fork (Fig. 12-47b) to insert into the ventral surface of the distal phalanx of our four fingers. Our flexor digitorum profundus is the only muscle which can flex our distal interphalangeal (DIP) joints and it assists in flexion of our MP and PIP joints.

Our flexor digitorum profundus is difficult to isolate from the superficial muscles

Figure 12 - 42

An anterior view of the flexor digitorum profundus.

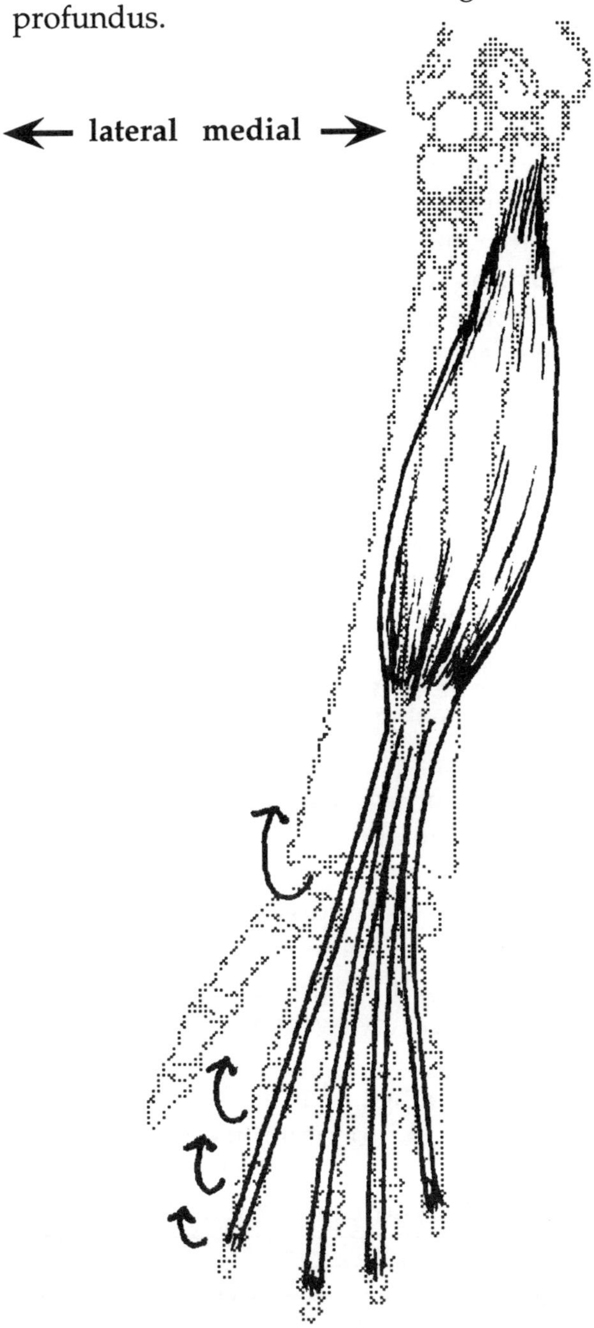

Both flexor muscles of the fingers illustrate a feature of any muscle which crosses more than one joint. If the muscle is asked to work at several joints at once, its effect is distributed and therefore less at any one joint. On the other hand, if the muscle is stretched at one joint, its effect is enhanced and concentrated at the other joints.

As our wrist is flexed, our finger flexors are already shortened. Their ability to work our fingers is weakened. As our wrist is extended, our finger flexors are stretched. Their ability to work our fingers is strengthened. Test it. Extend your wrist and flex your fingers into a strong fist. Keeping the fist, flex your wrist. What happens? The further you flex your wrist, the more the fist releases in spite of your efforts.

since they usually work together. It is part of the ripple when the fingers are working. This muscle is superficial in one place, as across the middle phalanx of our four fingers, after it emerges from under our flexor digitorum superficialis (Fig. 12-72). Hold one of your middle phalanges so that you press into its anterior surface. Curl and release the held finger. Feel the tendon tense and release with the movement.

Flexor Pollicis Longus

'Flexor' tells us that this muscle is in our anterior-medial compartment. 'Pollicis' is the word for thumb. 'Longus' tells us that it is longer than another similar muscle. Our **flexor pollicis longus** is in the third layer, deep to our flexor digitorum superficialis and lateral to our flexor digitorum profundus (Fig. 12-47).

Our flexor pollicis longus originates from the mid-ventral surface of the radius and interosseous membrane (Fig. 12-43 and 12-47). It inserts into the ventral surface of the distal phalanx of the thumb. The flexor pollicis longus contracts to flex the thumb, curling its tip into the palm. Since the flexor pollicis longus crosses ventral to the wrist, it can assist in wrist flexion.

To palpate the belly of your flexor pollicis longus, press into the flesh anterior to the middle of your radius, between your brachioradialis and flexor carpi radialis. Repeat flexion and extension of your thumb to feel the contracting muscle bulge under your fingers. As you move your thumb, watch the area near your wrist, just lateral to the tendon of your flexor carpi radialis (Fig. 12-48). See how it hollows as your thumb flexes. As the muscle shortens, it pulls toward its origin, vacating

Figure 12 - 43

An anterior view of the flexor pollicis longus.

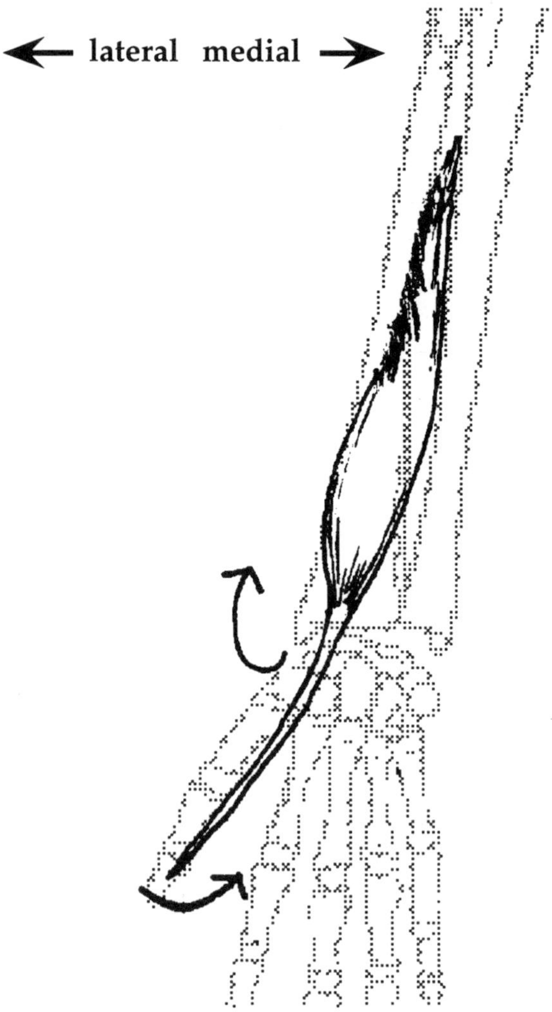

As muscles contract, they span the shortest and most direct route possible between their origin and insertion. In that case, as our fingers bend forward, flexors of the fingers would spring away from our arm and hand bones to go directly to our fingers, turning our hand into a webbed contraption (Fig. 8-24). This would drastically interfere with the efficient use of our hand.

Our finger flexors have a strong tendency to spring away, but their tendons are held down against the bones in order to maintain mechanical advantage for their actions and give the hand a cupped place to work. The **flexor retinaculum** is the beginning of this process (Fig. 8-24). Throughout the whole palm the connective tissue of the skin is adherent to the periosteum of the underlying bones to continue constraining the tendons.

Since our superficial muscles insert proximally, they also hold the ongoing muscles against our bones. Each insertion of our flexor digitorum superficialis forks to insert on the sides of our middle phalanx (Fig. 12-41 and 12-47). The tendon of our flexor digitorum profundus emerges in the center of the fork and is held in place by it.

Flexion of our fingers causes the flexor tendons to beat away at the connective tissue constraint. As time passes, this may cause enough damage, scar tissue, and compression in the hand to inhibit its function.

This is discussed further under Carpal Tunnel Syndrome in chapter 8.

this place, leaving a depression. Palpate the tendon of your flexor pollicis longus on the ventral surface of your thumb's proximal phalanx. Partially flex your thumb to make it easy.

Pronator Quadratus

Our **pronator quadratus** forms the fourth and deepest layer of our anterior-medial compartment. It originates from our distal anterior ulna (Fig. 12-44 and 12-47). Its muscle fibers pass transversely to insert into our distal anterior radius. As our pronator quadratus contracts, it pulls our radius to cross over our ulna, pronating our forearm. Our pronator quadratus is not palpable since it is covered by all the insertion tendons of the flexors of the wrist and fingers.

Figure 12 - 44

An anterior view of the pronator quadratus.

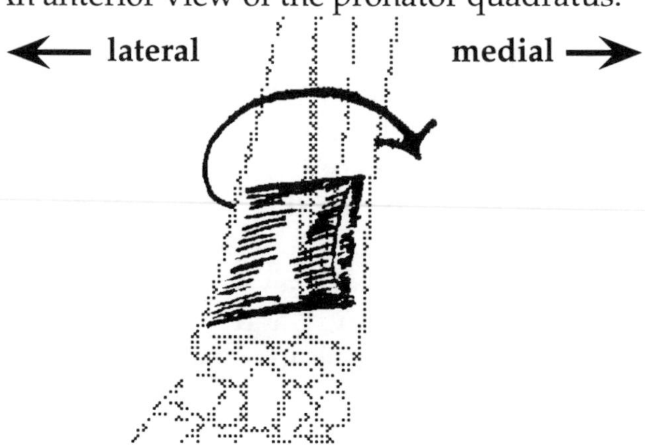

POSTERIOR-LATERAL COMPARTMENT

The muscles in the **posterior-lateral compartment of our forearm** follow a very similar pattern to those of the anterior-medial compartment. There are a few variations, but mostly we just have to replace 'flexor' with 'extensor', and 'anterior/ventral' with 'posterior/dorsal'. As you examine the back of your forearm and hand, remember that when you turn over your hand to look at the dorsal surface, medial still is to the ulnar or little finger side and lateral is to the radial or thumb side.

Our posterior-lateral compartment has two layers of muscles. The **superficial layer** includes muscles that extend our elbow, wrist, and fingers. We will discuss them in that order, which is different from the order of their location. There are six muscles in all.

Anconeus

Our **anconeus** is the shortest and most proximal muscle in the superficial layer of the posterior-lateral compartment of our forearm. An 'ancon' is a cantilever or stone projection. That is what elbows can feel like in a jostling throng or crowded dance floor. Our anconeus is a small muscle which wraps around our elbow, using it as a cantilever to gain mechanical advantage.

Our anconeus originates from the lateral epicondyle of our humerus (Fig. 12-45 and 12-49). It crosses dorsal to our elbow to insert into the lateral side of the olecranon process of our ulna. Since the anconeus passes dorsal to our elbow, it extends our elbow as it contracts.

Figure 12 - 45

A posterior view of the anconeus.

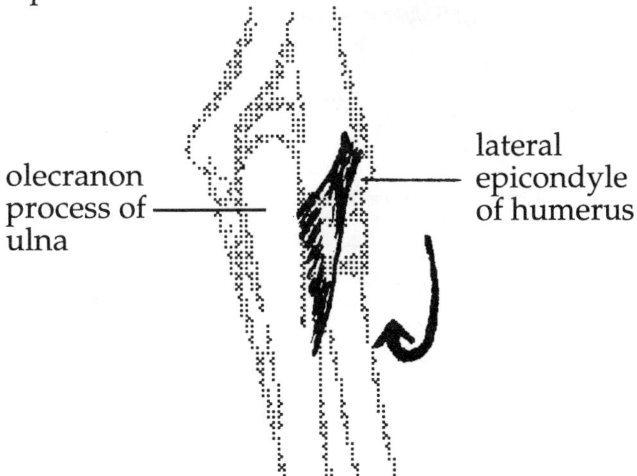

Figure 12 - 46

An anterior view of a pronated forearm.

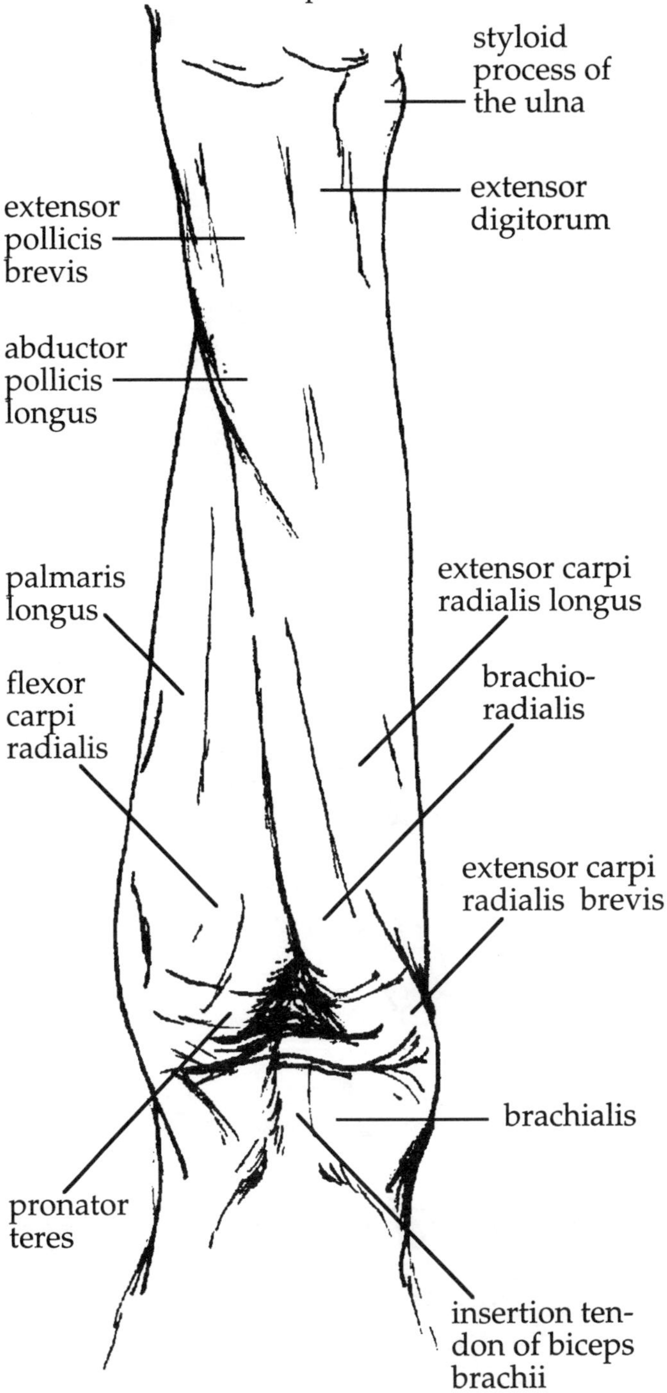

To palpate your anconeus, press into the tissue just lateral to your ulna, from the point of your olecranon process for a distance of about three inches down your arm (Fig. 12-50). Flex your elbow to 90 degrees and brace your forearm against something so your elbow cannot extend. Push against the bracing object, trying to extend your elbow, and feel the contraction as your anconeus works.

Figure 12 - 47

An anterior view of **(a)** the ulna and radius showing the origins of muscles in diagonally-lined patches and the insertions of muscles in dotted patches, and **(b)** the superficial muscles of the forearm with the flexor digitorum superficialis contrasted in black.

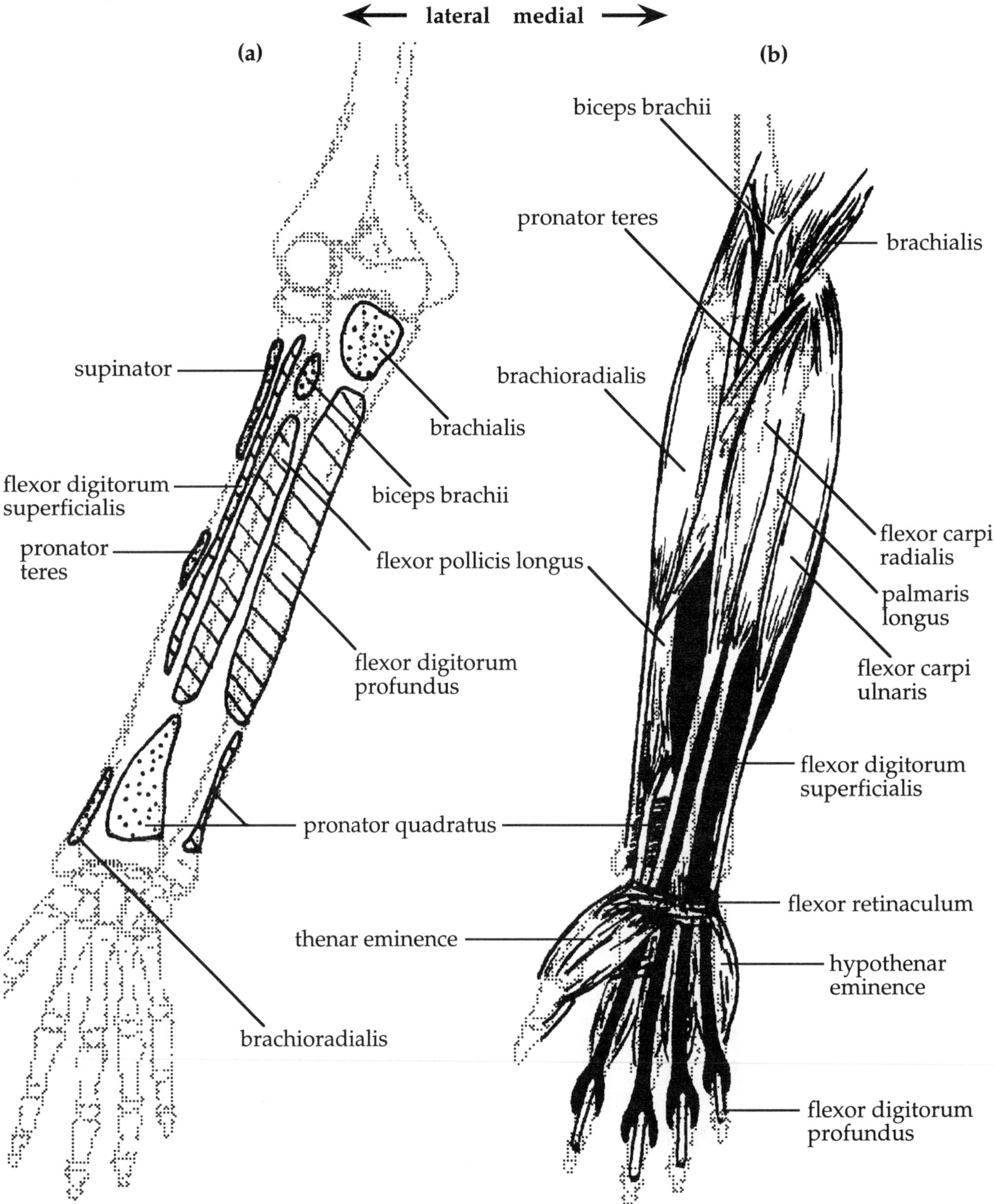

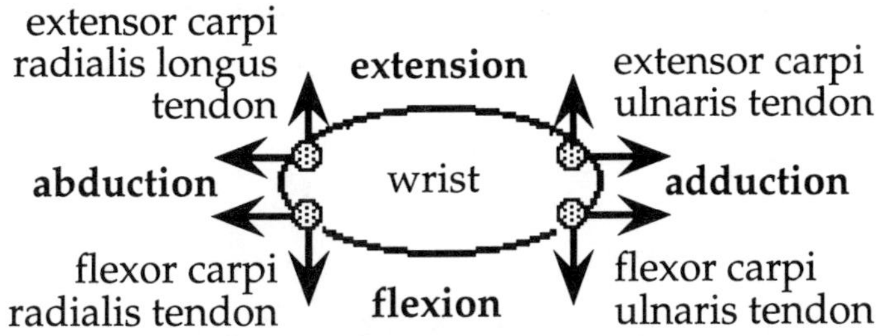

A cross section through the wrist showing how the four named muscles are synergistic in one direction and antagonistic in another.

Figure 12 - 48

A surface view of the medial upper arm and the anterior forearm, labeling superficial muscles.

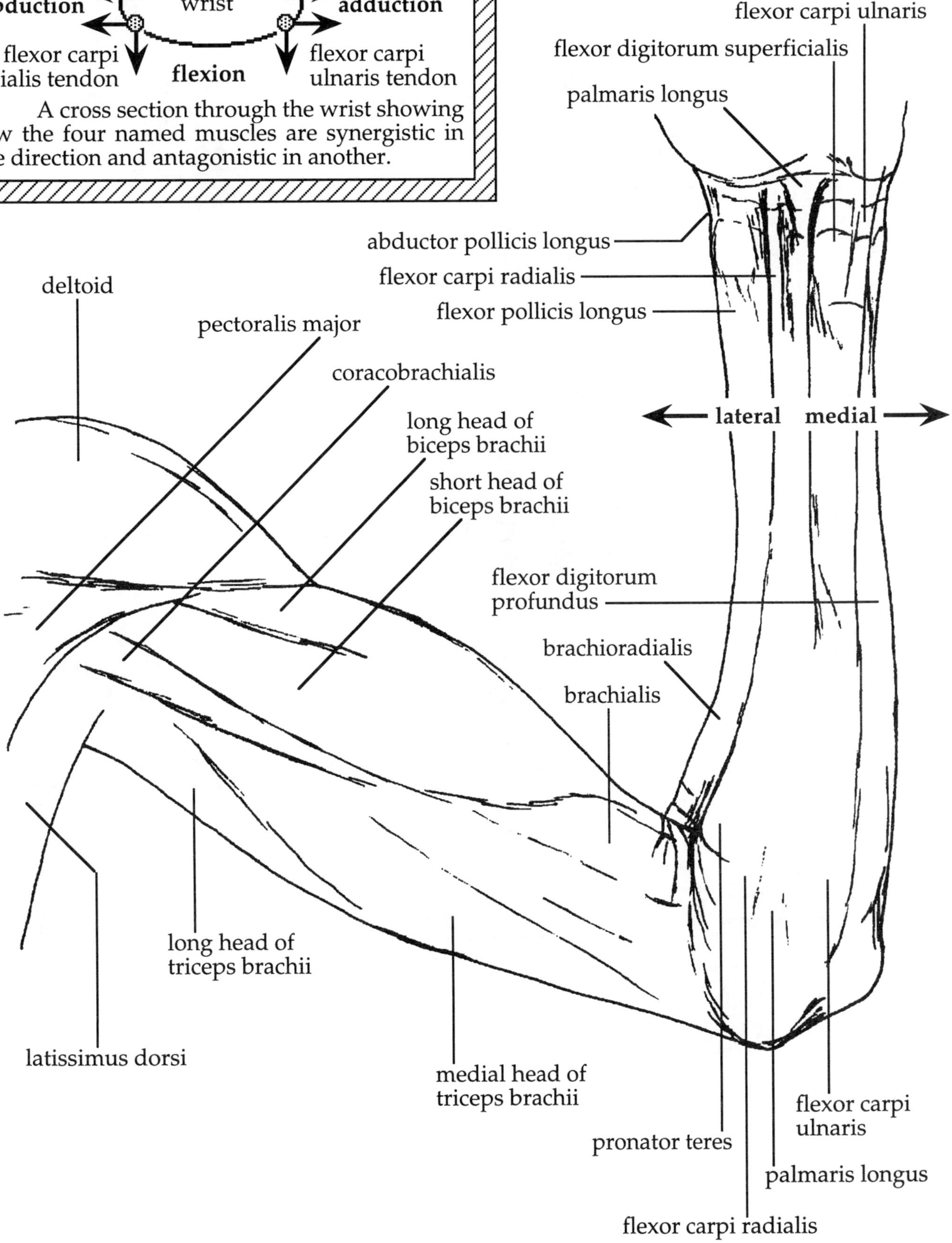

Figure 12 - 49

A posterior view of **(a)** the ulna and radius showing the origins of muscles in diagonally-lined patches and the insertions of muscles in dotted patches, and **(b)** the superficial and **(c)** the deep muscles of the forearm.

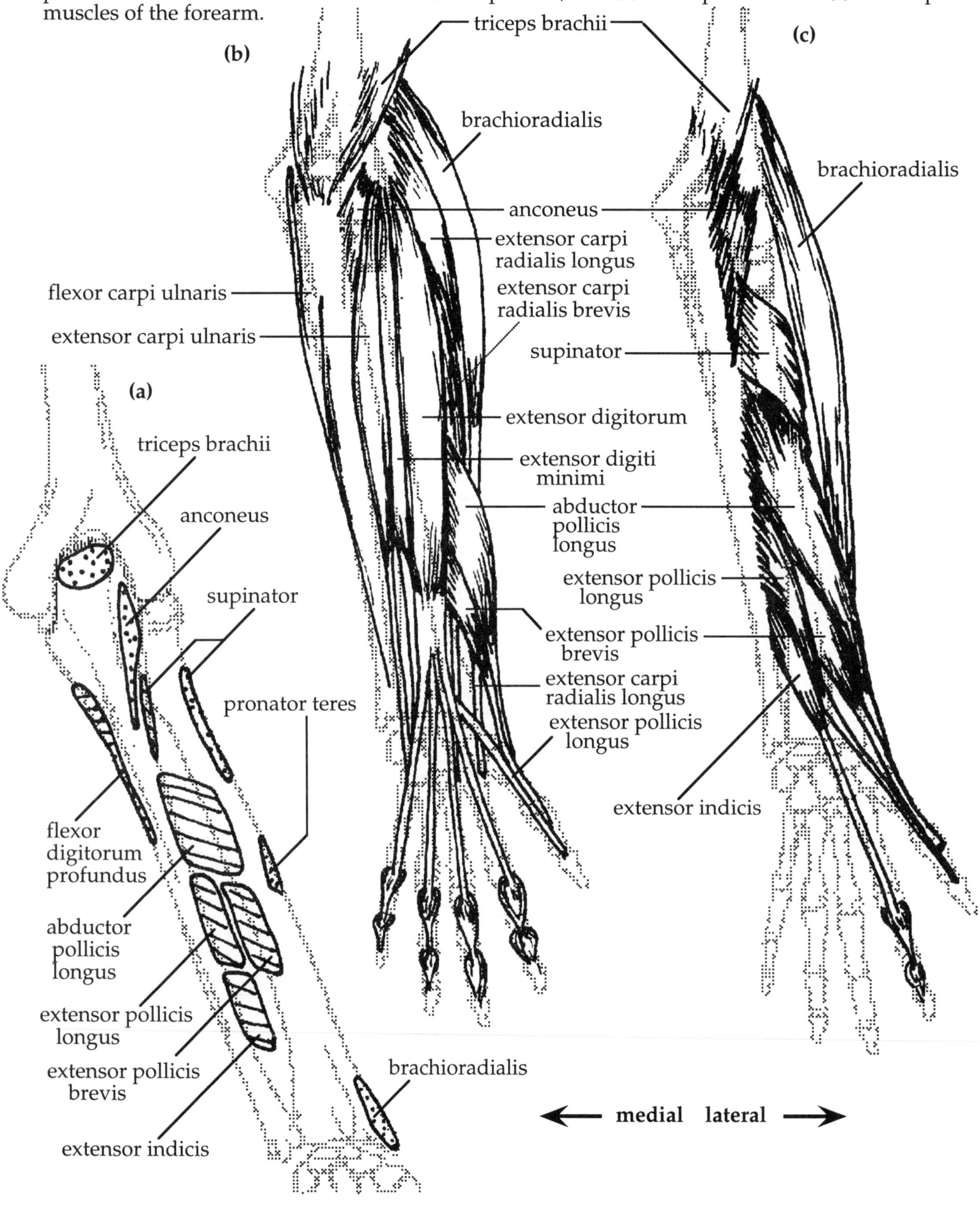

Figure 12 - 50

A surface view of the lateral upper arm and posterior forearm, labeling the superficial muscles.

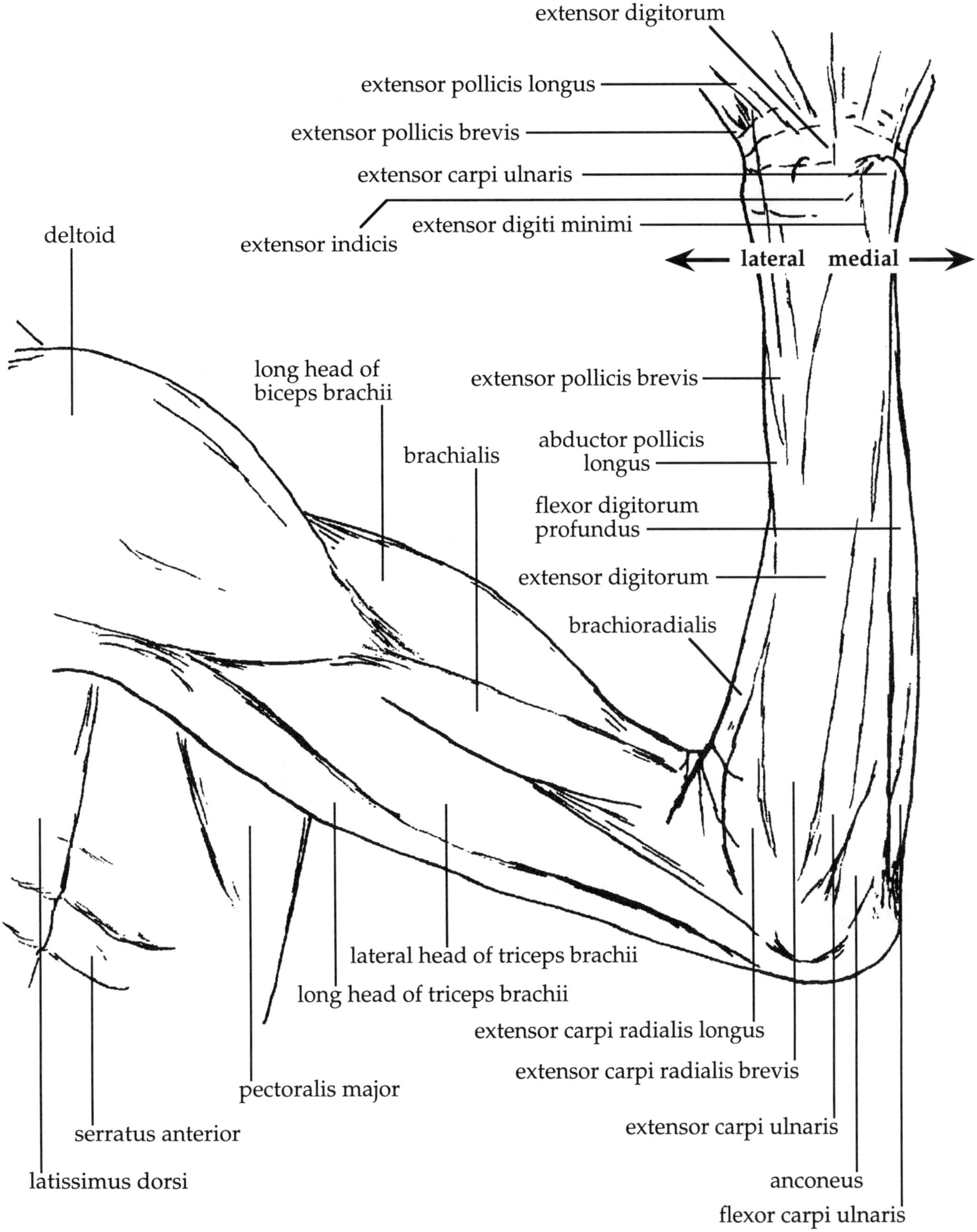

Extensor Carpi Ulnaris

Our **extensor carpi ulnaris** is similar to the flexor carpi ulnaris but lies on the opposite side of the ulna. 'Extensor' tells us the action of the muscle and implies its location in the posterior-lateral compartment. 'Carpi' tells us that the muscle inserts near our carpal bones. Carpal bones are proximal in the hand, so we know by the rules that this muscle must be superficial. Knowing this much about the muscle and following the rules, where must it originate? On the lateral epicondyle of our humerus (Fig. 12-49 and 12-51).

'Ulnaris' tells us that the insertion is on the ulnar side of our hand, more precisely the dorsal base of our fifth metacarpal.

As our extensor carpi ulnaris contracts, it extends and adducts our wrist. While our flexor carpi ulnaris and extensor carpi ulnaris antagonize each other in flexion and extension of our wrist, they work together to adduct it.

Our extensor carpi ulnaris lies immediately distal and lateral to the anconeus in our posterior forearm (Fig. 12-49 and 12-50). After the end of the anconeus, our extensor carpi ulnaris is directly lateral to our ulna. Follow your extensor carpi ulnaris all the way from the lateral epicondyle of your humerus, past the lateral side of the styloid process of your ulna to its insertion. Extend and adduct your wrist to feel its belly bulge and tendon become taut. When your forearm is supinated, the extensor carpi ulnaris is most clearly posterior across the wrist and lateral to the ulna. Pronate your forearm and feel how it almost seems to fall medial to the ulna.

Figure 12 - 51

A posterior view of the extensor carpi ulnaris with arrows showing extension and adduction of the wrist.

Extensor Carpi Radialis Brevis and Extensor Carpi Radialis Longus

Our flexor carpi radialis has two opposing partners on the extensor side, a short one and a long one. 'Extensor' tells us that the muscle is in the posterior-lateral compartment.

'Carpi' tells us that the insertions are near our carpal bones which are proximal in our hand, so these muscles are superficial and originate from the lateral epicondyle of our humerus (Fig. 12-49 and 12-52). The origin of the long one also reaches up the lateral edge of our humerus toward the origin of our brachioradialis.

At their insertions these muscles separate. While our flexor carpi ulnaris inserts into the bases of both our second and third metacarpal, our **extensor carpi radialis brevis** inserts into the base of our third metacarpal only, and our **extensor carpi radialis longus** inserts into the base of our second metacarpal.

Both our extensor carpi radialis brevis and longus antagonize our flexor carpi radialis

Figure 12 - 52
A posterior view of the extensor carpi radialis longus and brevis.

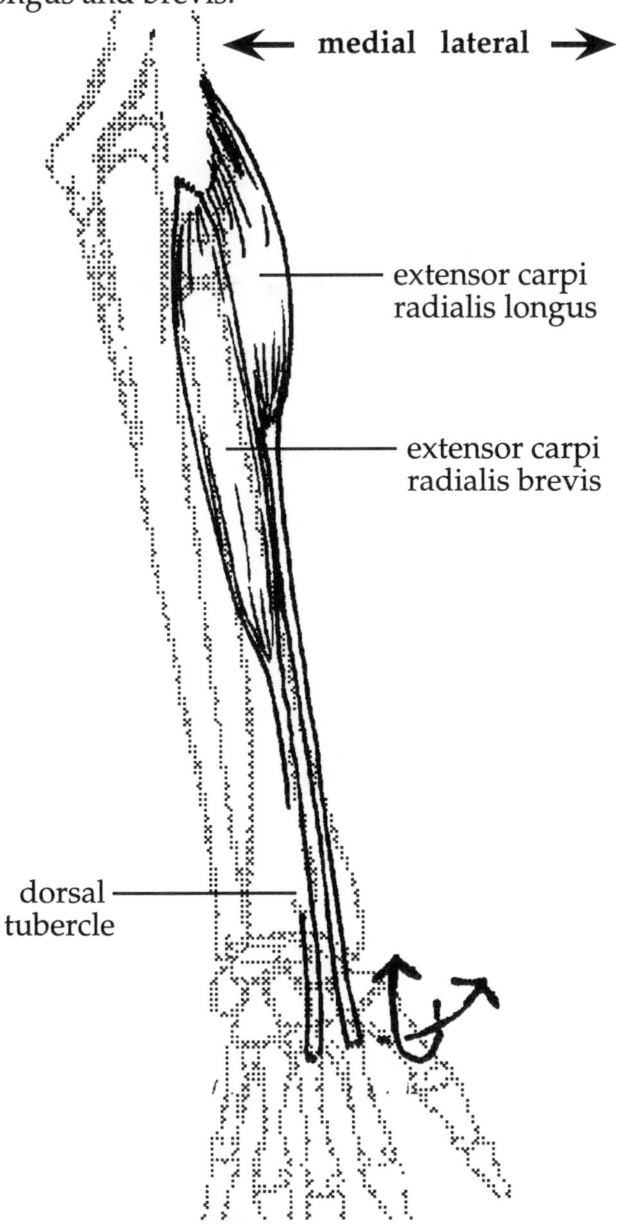

to extend our wrist. Since they insert lateral to the midline of our hand, they also can abduct our wrist, assisting our flexor carpi radialis in this action. The extensor carpi radialis longus is the most lateral and therefore the strongest in abduction.

Your extensor carpi radialis longus lies immediately lateral and posterior to your brachioradialis (Fig. 12-49 and 12-50), interfacing with the flexor compartment. Flex your elbow to 90 degrees. Locate the origin of your brachioradialis. While extending your wrist, watch or feel in your proximal forearm, just lateral and posterior to your brachioradialis. Two little bulges develop.

One bulge is proximal and quite short, between the origin of your brachioradialis and the lateral epicondyle of your humerus. Follow its tendon distally beside your brachioradialis to its insertion on the thumb side of the base of your second metacarpal. To palpate the tendon as it crosses over the carpal bones, curl your fingers and thumb. Feel over your carpal bones at the base of your second metacarpal while extending and adducting your wrist. The tendon will press out against your finger. Ironically this is the extensor carpi radialis longus. Its belly bulge is short, but its overall length is longer.

The second bulge is just distal to the first. Follow the muscle belly from the lateral epicondyle of your humerus, passing your extensor carpi radialis longus bulge and continuing distally to its tendon. Palpate the tendon just lateral (on the thumb side) to the dorsal tubercle of your radius (Fig. 12-52). Follow the tendon across the carpal bones to the base of your third metacarpal. While the belly is longer, the whole muscle is shorter, so this is the extensor carpi radialis brevis.

Just proximal to your wrist, the tendons of both these muscles are covered by muscle of your thumb (Fig. 12-49b).

Extensor Digitorum (Communis)

The name **extensor digitorum** tells us that this muscle extends our four fingers. This muscle is part of the superficial layer in the posterior-lateral compartment of our forearm, exposed between the extensors of our wrist, so it originates from the lateral epicondyle of our humerus (Fig. 12-49 and 12-53). As its tendon crosses our distal forearm, wrist, and hand, it splits to insert into the dorsal expansion of our four fingers.

As our extensor digitorum contracts, it extends our MP joints and assists in extension of both interphalangeal joints of our fingers.

Relocate your extensor carpi radialis brevis. Your extensor digitorum lies medial or on the ulnar side of it (Fig. 12-49b and 12-50). Alternately flex and extend your MP joints to

Figure 12 - 53
A posterior view of the extensor digitorum.

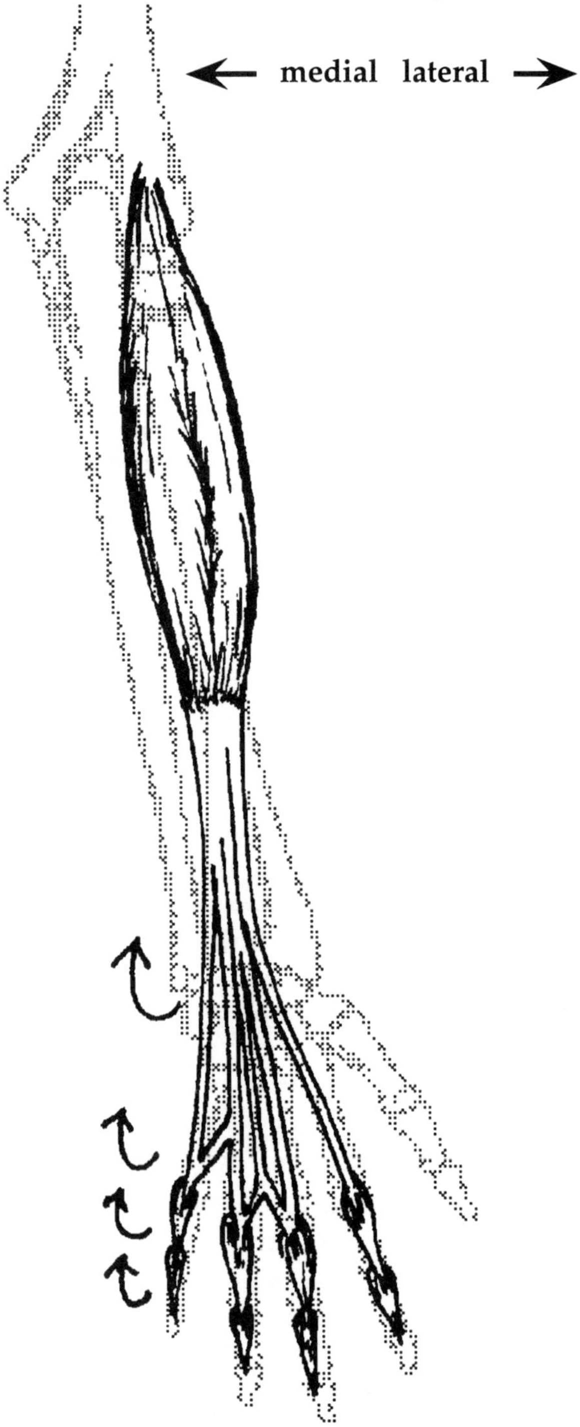

The insertion tendon of your extensor digitorum splits into at least four parts. Watch the back of your hand as you flutter your MP joints between flexion and extension. In between the four bold tendons which clearly go to a specific finger, there may be smaller ones which split near your knuckles to go to two adjacent fingers (Fig. 12-53). Each person has his/her own pattern of splitting the tendons of the extensor digitorum.

When one muscle works all four fingers, it is difficult to move any one finger independently of the others. This applies to flexion and extension. Practice and concentration will increase the independence, but there is always some simultaneous movement. Try to extend just your ring finger, leaving your other fingers curled. Do the other fingers creep along no matter what you do to stop them?

The fingers which are most independent are our index finger and little finger. Each of these fingers have their own individual muscles as well as the common ones.

A dense connective tissue covers the dorsal surface of each finger and attaches into all three phalanges (Fig. 12-53). It is called the **dorsal** or **extensor expansion.** Many of the muscles that insert into the dorsal aspect of the fingers insert into the dorsal expansion, and therefore affect both interphalangeal joints.

feel it working. Follow it from the lateral epicondyle of your humerus toward the center of your dorsal wrist. By the time it reaches your wrist it has divided into at least four tendons. Feel these prominent tendons fanning across the back of your metacarpals. Notice that the tendons disappear over your proximal phalanx, as they insert into your dorsal expansion.

Extensor Digiti Minimi

Our 'digiti minimi' is our little finger and this muscle extends it. Our **extensor digiti minimi** originates from the lateral epicondyle of our humerus (Fig. 12-54) and inserts into the dorsal expansion of our little finger. As it contracts, it will extend the MP and IP joints of our little finger.

Figure 12 - 54
A posterior view of the extensor digiti minimi.

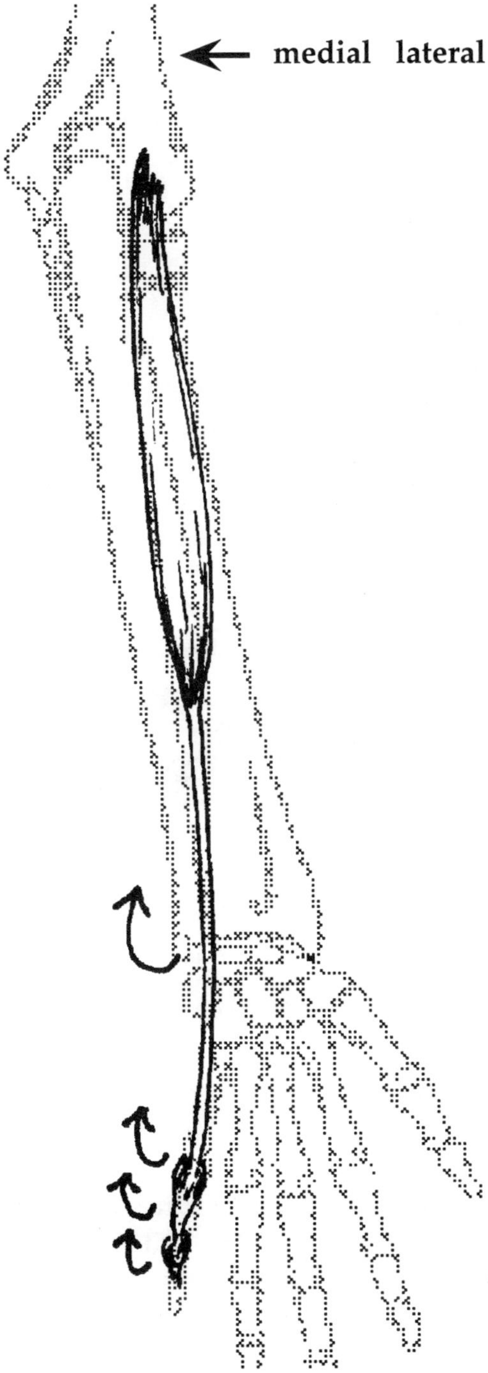

Figure 12 - 55
A posterior view of the extensor indicis.

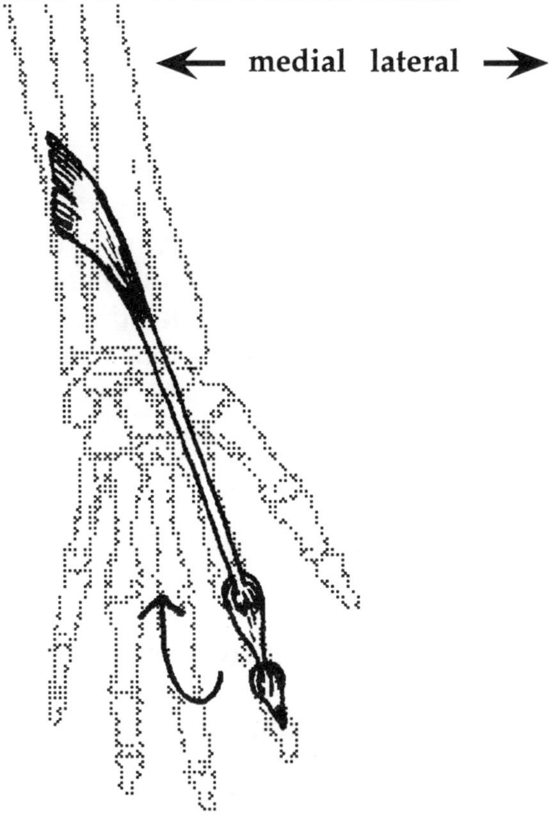

Your extensor digiti minimi is superficial, lying between your extensor carpi ulnaris and extensor digitorum. Its belly is difficult to differentiate from the extensor digitorum. Leaving all your other fingers curled, straighten out your little one. Watch or feel for its contracting belly and taut tendon just medial to your extensor digitorum, lateral to your ulna (Fig. 12-49b and 12-50).

At your wrist, its tendon is immediately lateral to that of your extensor carpi ulnaris and the round head of your ulna. Extend your fingers to make the tendons across the hand tense and easy to palpate. Feel the two tendons passing over your fifth metacarpal and proximal phalanx of your little finger. The medial one is the tendon of your extensor digiti minimi. The lateral one is the tendon of your extensor digitorum.

- This discussion completes the superficial layer of muscles in the posterior-lateral compartment of our forearm. The deep layer has five muscles. We will discuss them from medial and distal to lateral and proximal.

Extensor Indicis

Our **extensor indicis** extends our index finger. It originates from our posterior distal ulna and interosseus membrane (Fig. 12-49 and 12-55), and traverses obliquely across the back of our hand to insert into the extensor expansion of our index finger. As it contracts, it will extend the MP and IP joints of our index finger.

The belly of your extensor indicis lies deep to the tendons of your extensor digitorum and digiti minimi, just lateral to your distal ulna (Fig. 12-49). Press into this area and point up with your index finger to feel the muscle contract.

Across your second metacarpal and MP joint, the tendon of your extensor indicis lies just medial to the tendon of your extensor digitorum. If it looks like just one tendon, lightly extend and abduct your index finger. Pull the superficial part of the tendon, which is your extensor digitorum, toward your thumb to separate them.

Extensor Pollicis Longus

'Pollicis' is thumb, so our **extensor pollicis longus** is a long muscle which extends our thumb. It originates from our mid-posterior ulna and interosseus membrane (Fig. 12-49 and 12-56), and crosses obliquely to insert into the dorsal base of the distal phalanx of our thumb. Since it is a long muscle, it goes all the way to the last bone of the thumb. As our extensor pollicis longus contracts, it pulls our thumb back into extension. Since this muscle also falls lateral to the midline of our wrist, it can abduct our wrist.

The belly of our extensor pollicis longus lies deep to our extensor carpi ulnaris and extensor digiti minimi and is difficult to feel.

At our wrist, the tendon of our extensor pollicis longus emerges from underneath our extensor digitorum tendons, just medial to the dorsal tubercle of our radius, It crosses over the very end of the tendons of our extensor carpi radialis brevis and longus (Fig. 12-49b).

Strongly abduct and extend your thumb to create an indentation at the posterior base of your thumb (Fig. 12-65), just distal to the styloid process of your radius. This is called the **anatomical snuff box.** (We found the scaphoid bone in the bottom of this box.) The tendon of your extensor pollicis longus forms the medial wall of your snuff box. Follow it over the posterior surface of your first metacarpal and proximal phalanx of your thumb to the base of your thumb's distal phalanx.

Figure 12 - 56

A posterior view of the extensor pollicis longus. Arrows indicate the direction of movement of the wrist and thumb.

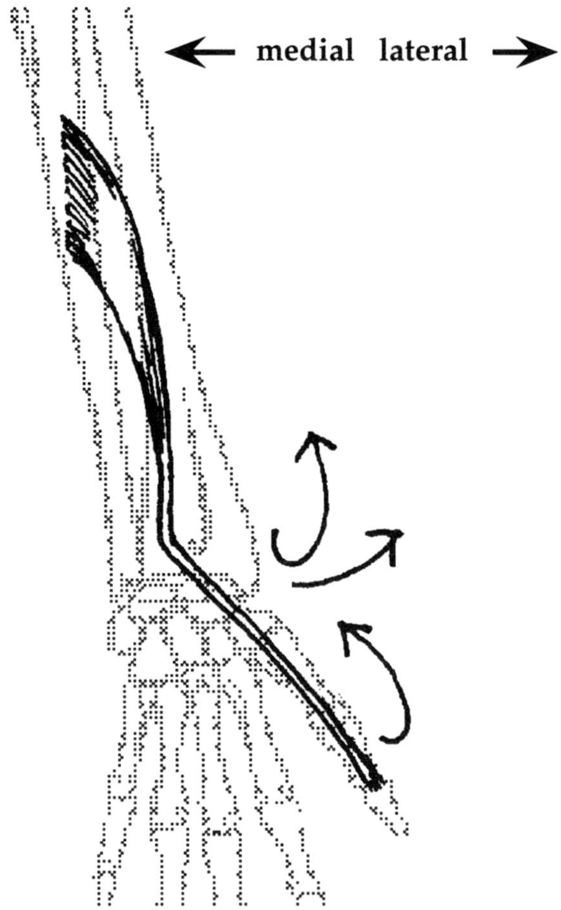

Extensor Pollicis Brevis

The name **extensor pollicis brevis** tells us that our thumb is extended by a short muscle. Our extensor pollicis brevis originates from our distal posterior radius and interosseus membrane (Fig. 12-49 and 12-57). Its tendon passes lateral to the styloid process of our radius to insert into the dorsal base of the proximal phalanx of our thumb. This is the shorter of the two thumb extensors. It originates more distally in the forearm and inserts more proximally in the thumb.

As our extensor pollicis brevis contracts, it extends the MP joint of our thumb. Passing lateral to our thumb and wrist, it will also abduct our thumb and wrist.

To palpate the belly of your extensor pollicis brevis, relocate the tendons of your extensor digitorum just before they cross your

Figure 12 - 57

A posterior view of the extensor pollicis brevis.

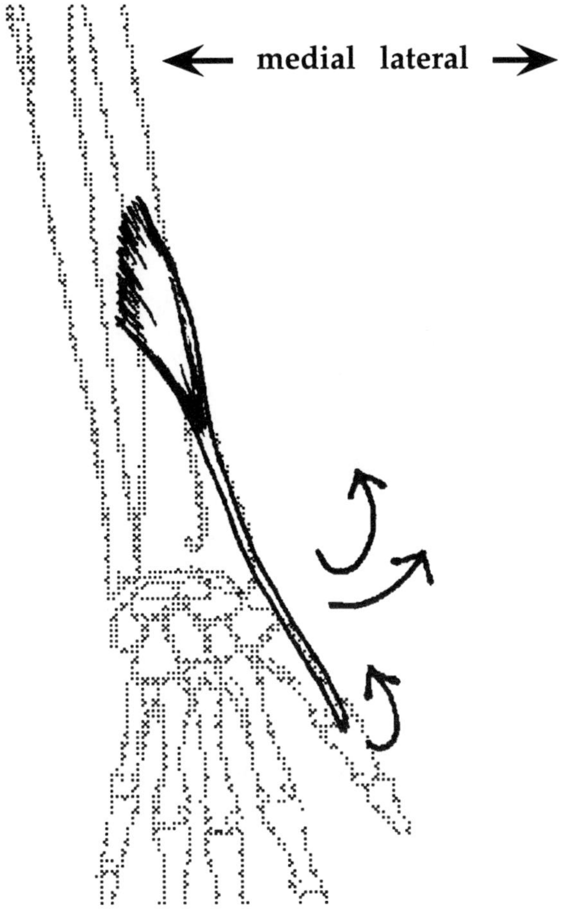

Figure 12 - 58

A posterior view of the abductor pollicis longus.

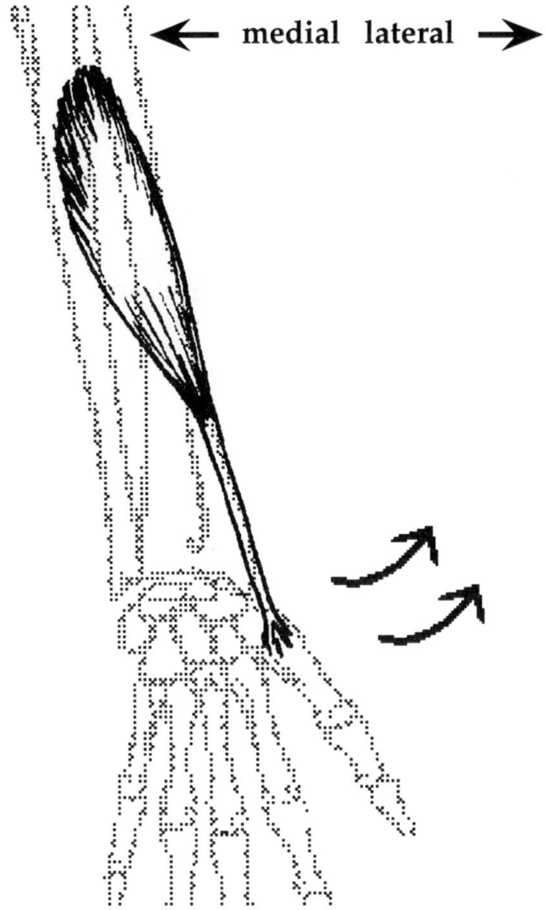

wrist. The belly of your extensor pollicis brevis is underneath and immediately lateral to these tendons (Fig. 12-50). Lay your fingers over this area while extending your thumb to feel the muscle work. Watch while you extend your thumb. A hollowness develops as the muscle withdraws, shortening toward its origin. Follow the muscle over the tendons of your extensor carpi radialis brevis and longus. Feel its tendon anchored to the lateral side of the styloid process of your radius.

Open your anatomical snuff box. The tendon of your extensor pollicis brevis forms the lateral wall of the box. Follow it across the lateral surface your first metacarpal to the base of the proximal phalanx of your thumb.

Abductor Pollicis Longus

The name **abductor pollicis longus** tells us that we have a longer muscle which abducts our thumb. Our abductor pollicis longus originates from our mid-posterior ulna, radius, and interosseus membrane (Fig. 12-49 and 12-58), just proximal to our thumb extensors. It inserts into the lateral base of our first metacarpal. This muscle contracts to abduct our thumb pulling our first metacarpal laterally over our trapezium. Our abductor pollicis longus also can abduct our wrist.

The belly of your abductor pollicis longus is the mound which lies lateral to your extensor digitorum, distal to your extensor carpi radialis brevis, and proximal to your extensor pollicis brevis (Fig. 12-49b and 12-65). Press into that area as you abduct your thumb to feel the muscle contracting. Follow its belly over the tendons of your extensor carpi radialis brevis and longus.

Feel the tendon of your abductor pollicis longus just anterior to the tendon of your extensor pollicis brevis, over the lateral surface of the styloid process of your radius. Follow it the short distance across the carpal bones before it inserts into the lateral base of your first metacarpal. Extend your thumb to pull the tendon of the extensor pollicis brevis posteriorly and distinguish it from the tendon of your abductor pollicis longus.

Supinator

Our **supinator** is the most proximal and lateral muscle in the deep layer of the posterior compartment of our forearm. It originates from the lateral side of the olecranon process of our ulna (Fig. 12-49a and 12-59a). It wraps around the lateral side of our radius to insert into our anterior-medial radius just distal to the radial tuberosity (Fig. 12-59b).

The name of this muscle tells us its action, it supinates the forearm, spinning our radius laterally to lie parallel to our ulna in anatomical position.

Your supinator is difficult to palpate since it is mostly covered by the superficial muscles of the posterior-lateral compartment (Fig. 12-49c). It is even difficult to get a sense of its bulging contraction because we tend to abduct and extent our wrist while we supinate so the muscles lying over the supinator are contracting at the same time it is.

To isolate your supinator, flex your elbow as far as it goes, so your forearm rests vertically against your chest and upper arm. Let your wrist droop in flexion. Let your whole arm, hand, and fingers stay as relaxed as possible as you supinate your forearm. Curl your other fingers around the posterior-lateral aspect of your proximal radius as you lightly supinate and pronate your forearm. You may feel the supinator contract deep in this position.

HAND MUSCLES

Much of the power for hand and finger action comes from muscles in our forearm. A lot of bulging muscles within the hand would interfere with its activity. Our hand has only a few small muscles for delicate finger and thumb manipulation. These muscles often are called the **intrinsic muscles of the hand** which says that they are located in the hand itself in contrast to the forearm muscles which move the hand from a distance.

Figure 12 - 59

(a) A posterior view and **(b)** an anterior view of the supinator. Arrows show the movement of the radius.

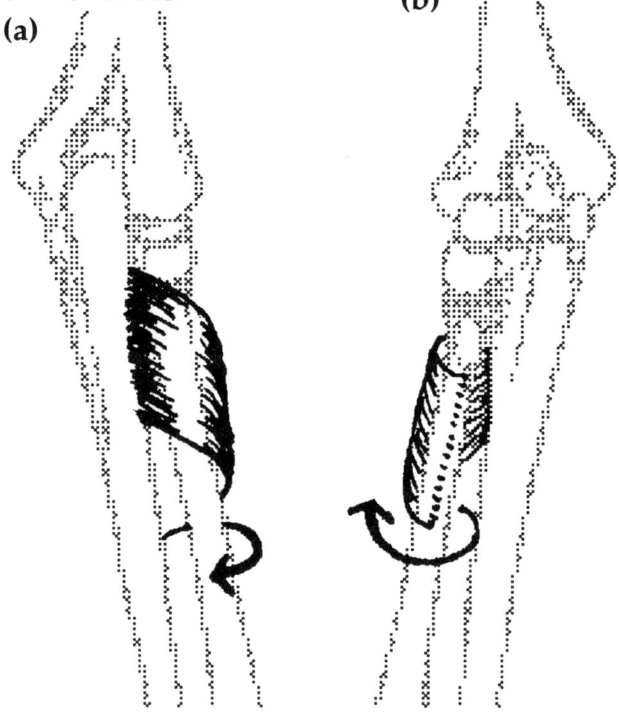

The intrinsic muscles of our hand can be divided into three groups: our **thenar eminence,** our **hypothenar eminence,** and our **midpalmar muscles.**

THENAR EMINENCE

Our **thenar eminence** is the little wad of muscle anterior and distal to our first metacarpal. It contains four muscles which are oriented transversely across our hand and move our thumb. We will discuss them from proximal to distal.

Opponens Pollicis

Our **opponens pollicis** originates from our flexor retinaculum and trapezium, the carpal bone at the base of our thumb. It curls around our first metacarpal to insert into its lateral side (Fig. 12-60).

The name of this muscle tells us its action. An 'opponens' is one who opposes. Opponents in a match stand to face each other. 'Pollicis' says that it is our thumb that is made to oppose. As our opponens pollicis contracts, it turns our thumb around so that its ventral surface faces or opposes the ventral surface of the palm of our hand. Our thumb nail is turned outward. This is called opposition of our thumb.

Most of the belly of your opponens pollicis is covered by the other muscles of your thenar eminence. It is most palpable at its insertion. Feel the lateral edge of your first metacarpal, just anterior to the tendon of your extensor pollicis brevis. The little bulge of muscle anterior to the bone is the insertion of your opponens pollicis (Fig. 12-64 and 12-65). It is about the width of a fingertip. Oppose your thumb to feel it contract.

Abductor Pollicis Brevis

In our forearm we have a long muscle which abducts our thumb. Our **abductor pollicis brevis** is a shorter helper in our hand. It originates from our flexor retinaculum, scaphoid, and trapezium, and inserts into the lateral side of the base of the proximal phalanx of our thumb (Fig. 12-61).

As our abductor pollicis brevis contracts, it pulls our thumb laterally, perpendicular to the palm of our hand.

Find your abductor pollicis brevis just distal or medial to your opponens pollicis, a fingertip width from your metacarpal (Fig. 12-64 and 12-72). Pull your thumb out to the side to feel it contract. Follow it past the lateral side of the MP joint of the thumb. The width of your abductor pollicis brevis again is about one fingertip width.

Flexor Pollicis Brevis

In our forearm we have a long muscle which flexes our thumb. Our **flexor pollicis brevis** is a shorter helper in our hand. It originates from our flexor retinaculum, capitate, and trapezium (Fig. 12-62). It inserts into the medial side of the base of the proximal phalanx of our thumb, between our abductor pollicis brevis and the tendon of our flexor pollicis longus. As the flexor pollicis brevis contracts, it flexes the carpal-metacarpal and MP joints of out thumb.

Figure 12 - 60

An anterior view of the opponens pollicis. The arrow shows how it moves the thumb.

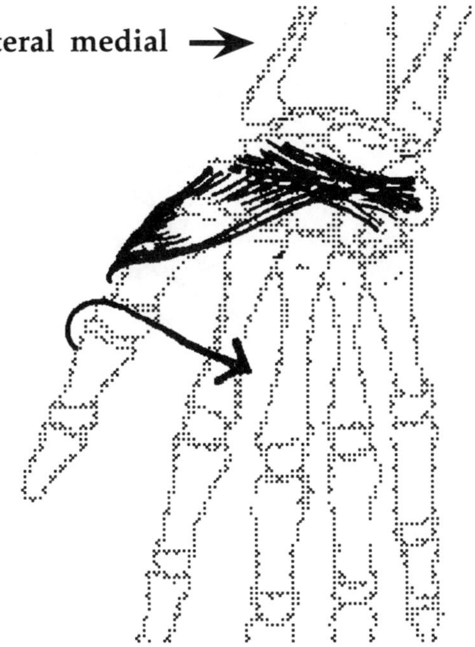

Figure 12 - 61

An anterior view of the abductor pollicis brevis. The arrow shows how it moves the thumb.

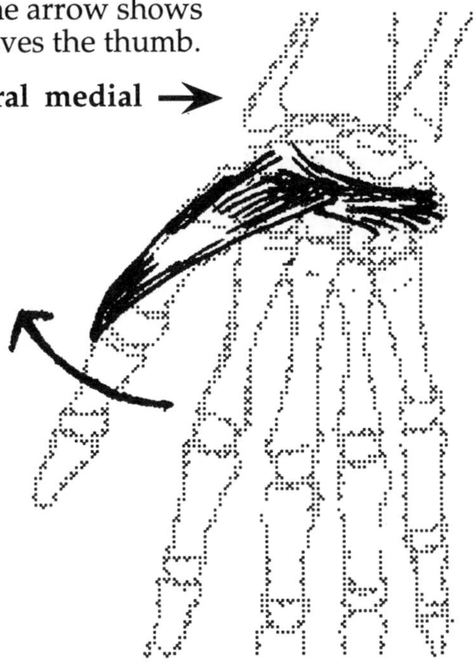

Your flexor pollicis brevis is distal to your abductor pollicis brevis. It again is about one fingertip in width, two fingertips over from your metacarpal (Fig. 12-64 and 12-72). Press lightly into its belly and flex the MP joint of your thumb to feel it contract. You may see its horizontal bulge heading toward the center of the palm of your hand.

Figure 12 - 62
An anterior view of the flexor pollicis brevis.

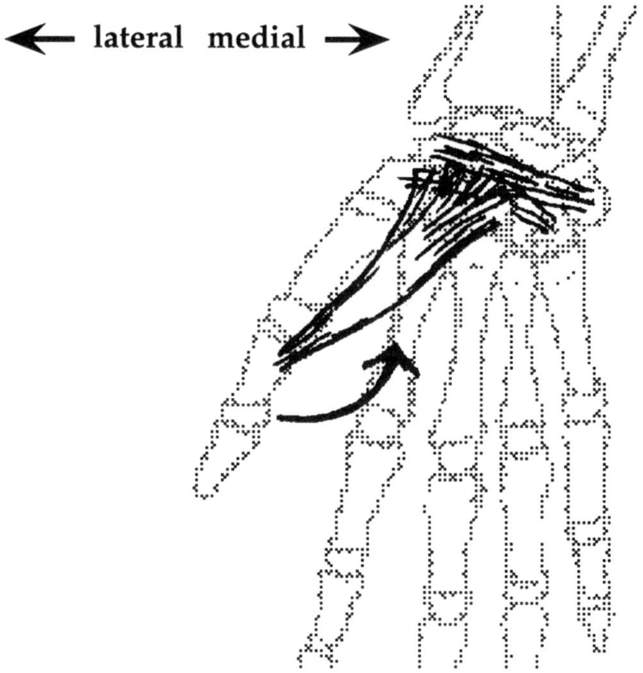

Figure 12 - 63
An anterior view of the adductor pollicis.

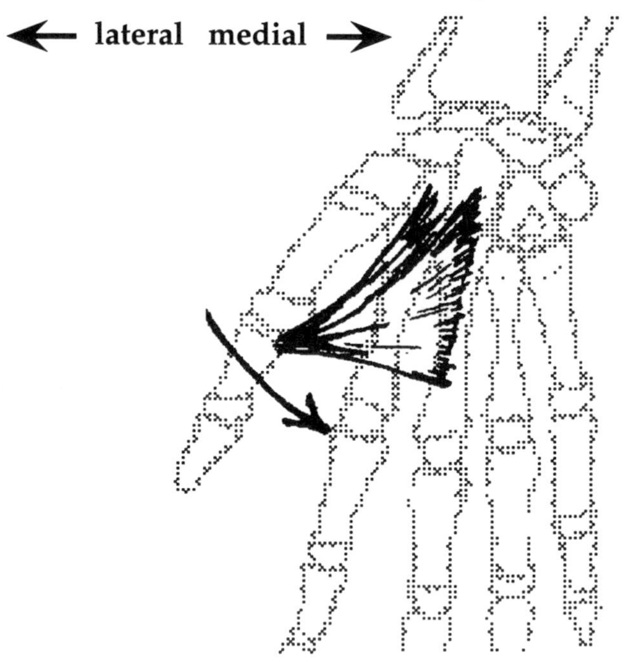

Adductor Pollicis

Our **adductor pollicis** is the deepest and most distal muscle of our thenar eminence.

It originates from the anterior surface of our third metacarpal, the base of our third and second metacarpal, our capitate, and trapezoid (Fig. 12-63). It inserts into the medial side of the base of the proximal phalanx of our thumb, medial and posterior to the tendon of our flexor pollicis longus. As the adductor pollicis contracts, it performs the action of its name. It adducts our thumb, pulling it against the side of our first finger.

Most of the belly of your adductor pollicis is hidden under your flexor and abductor pollicis brevis. Palpate its distal edge just distal to your flexor pollicis brevis (Fig. 12-64). Squeeze the web of your thumb between your thumb and index finger, right next to the head of your first metacarpal. The edge of the web is only skin, but take a deeper bite to include muscle. The first muscle is the adductor pollicis close to its insertion (Fig. 12-65).

- Place your four fingertips longitudinally, from your wrist toward your fingers, across your thenar eminence (Fig. 12-64). Start with your index finger next to the lateral side of the first metacarpal. It will cover your opponens pollicis. Your middle finger will cover your abductor pollicis brevis. Your ring finger will cover your flexor pollicis brevis. Your little finger will cover the distal edge of your adductor pollicis.

Figure 12 - 64
An anterior view of the hand, with fingertips placed from proximal to distal on the opponens pollicis, the abductor pollicis brevis, the flexor pollicis brevis, and the adductor pollicis.

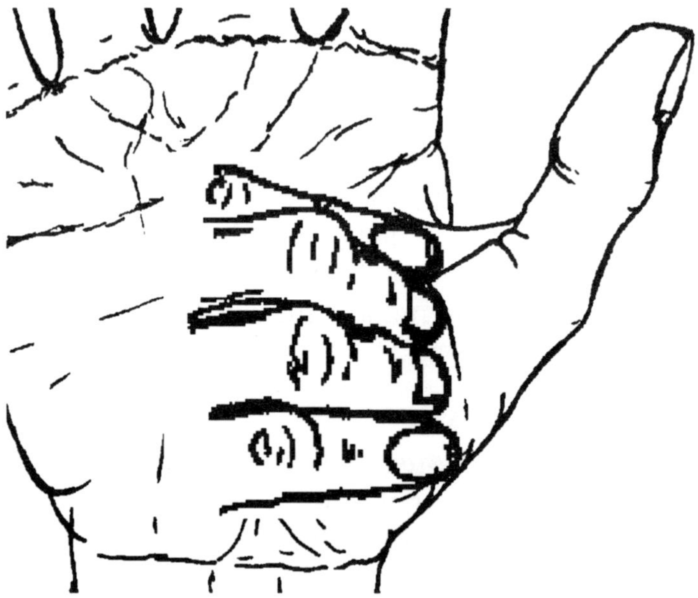

Figure 12 - 65
A lateral view of the hand, labeling appropriate muscles and tendons.

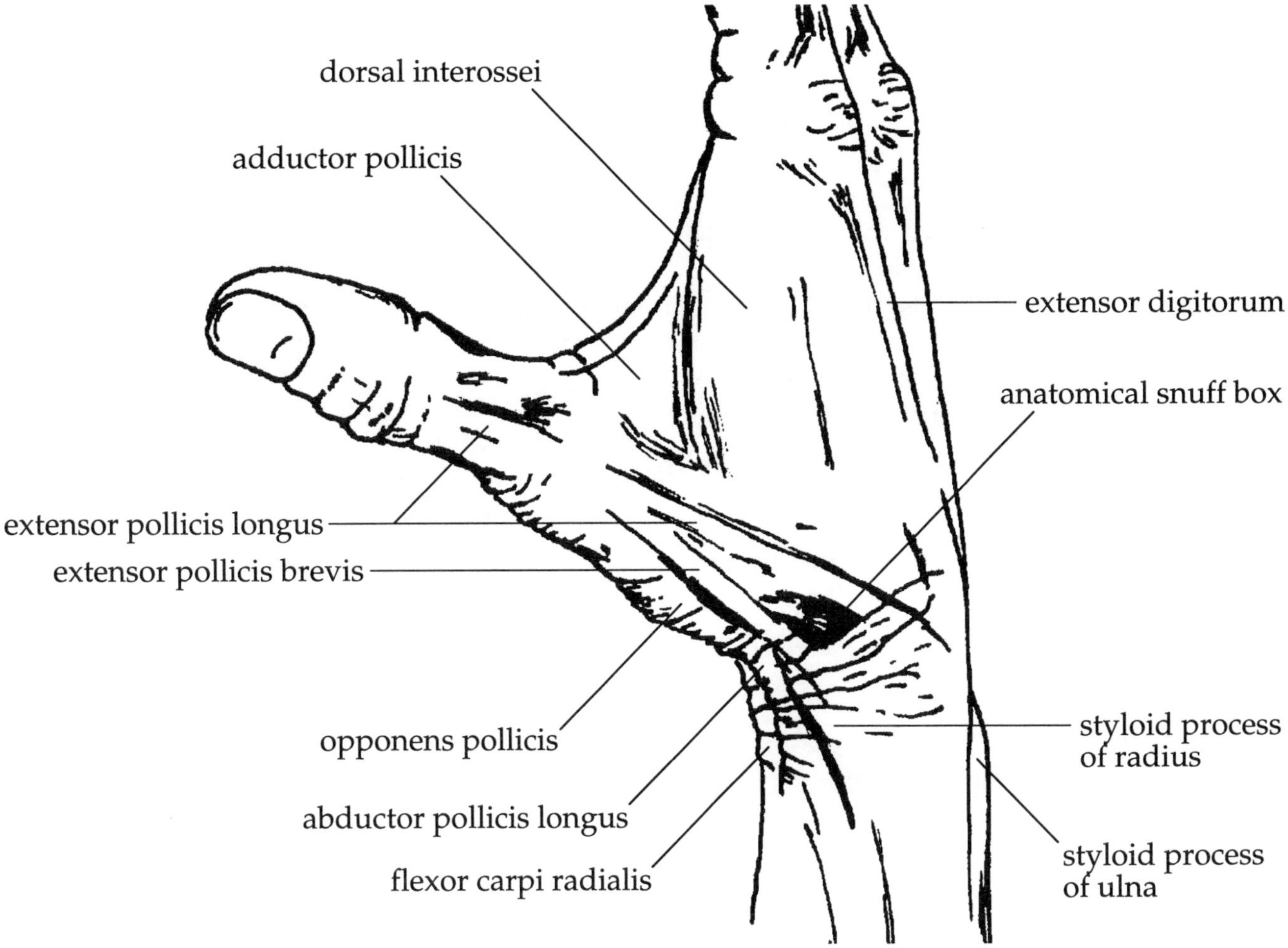

HYPOTHENAR EMINENCE

Our **hypothenar** eminence is the little wad of muscle medial and anterior to our fifth metacarpal. It contains three muscles which are oriented longitudinally in our hand and move our little finger. Their actions on our little finger are the same as the first three actions of our thenar eminence on our thumb, opposing, abducting, and flexing it. We will discuss them from medial to lateral.

Abductor Digiti Minimi

Our **abductor digiti minimi** originates from our pisiform bone in the proximal medial corner of our hand (Fig. 12-66). It inserts into the medial side of the base of the proximal phalanx of our little finger. The muscle's name tells us its action. It abducts our little finger, pulling it medially away from our other fingers.

Figure 12 - 66
An anterior view of the abductor digiti minimi.

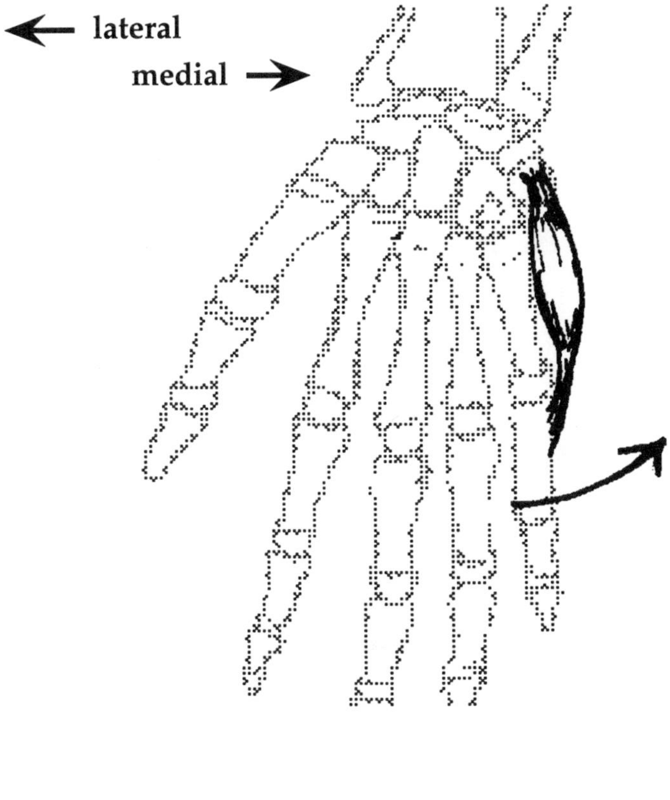

Feel the posterior aspect of your fifth metacarpal. Find the muscle belly just medial to it (Fig. 12-72). This is the belly of your abductor digiti minimi. Follow it from your pisiform bone to your fifth proximal phalanx. Abduct your little finger to feel it contract.

Flexor Digiti Minimi

Our **flexor digiti minimi** originates from our flexor retinaculum and the hook of our hamate (Fig. 12-67). It inserts into the anterior side of the base of the proximal phalanx of our little finger, just lateral to the insertion of our abductor digiti minimi. The muscle's name tells us its action. It flexes the MP joint of our little finger, curling it forward toward the palm of our hand.

Feel your flexor digiti minimi directly anterior to your fifth metacarpal (Fig. 12-72). Press your little finger toward your palm to feel the muscle contract.

Opponens Digiti Minimi

Our **opponens digiti minimi** originates from our flexor retinaculum and the hook of our hamate (Fig. 12-68). It inserts into the medial edge of our fifth metacarpal, deep to our abductor digiti minimi. The muscle's name tells us its action. It opposes our little finger, pulling it around to face our thumb.

Your opponens digiti minimi is difficult to palpate, lying deep to your abductor and flexor digiti minimi. Pull your abductor digiti minimi forward, and then, from the medial side, press into the anterior surface of your fifth metacarpal. You are over the insertion of your opponens digiti minimi. Oppose your little finger to get a sense of it contracting.

> We often say that humans have opposable thumbs but really opposition involves our thumb and little finger, both turning simultaneously to face each other, forming a pocket in the palm of our hand.

MIDPALMAR MUSCLES

Our **midpalmar muscles** lie between adjacent metacarpals in the palm of our hand. There are three muscles in this group. Each muscle is divided into three or four bundles, one bundle lying in each of our intermetacarpal spaces. We will discuss them from anterior to posterior.

Figure 12 - 67

An anterior view of the flexor digiti minimi.

← **lateral** **medial** →

Figure 12 - 68

An anterior view of the opponens digiti minimi.

← **lateral** **medial** →

Lumbricals

'Lumbrical' means like an earth worm. How these muscles reminded someone of an earthworm I can only guess. They are relatively long and thin like the shape of a worm, or their subtle movement may have seemed like a worm creeping under the skin.

Our **lumbricals** are skeletal muscles which do not attach to bones. They originate in the palm of our hand from the tendons of our flexor digitorum profundus (Fig. 12-69). They go dorsally past the lateral side of our four fingers to insert into the tendons of our extensor digitorum and extensor expansion over our proximal phalanx.

As our lumbricals contract, they take force from our flexing muscles and transfer it to our extending muscles across the IP joints, so they juggle between the amount of flexion and extension in all the joints of our fingers. This is essential for the tiny, delicate moves that make our hand the amazing and diverse tool that it is. When our lumbricals are fully contracted, they would flex our MP joints and extend both IP joints of our four fingers.

Figure 12 - 69

(a) A lateral and **(b)** an anterior view of the lumbricals showing their orgin on the flexor digitorum profundus and their insertion on the extensor digitorum.

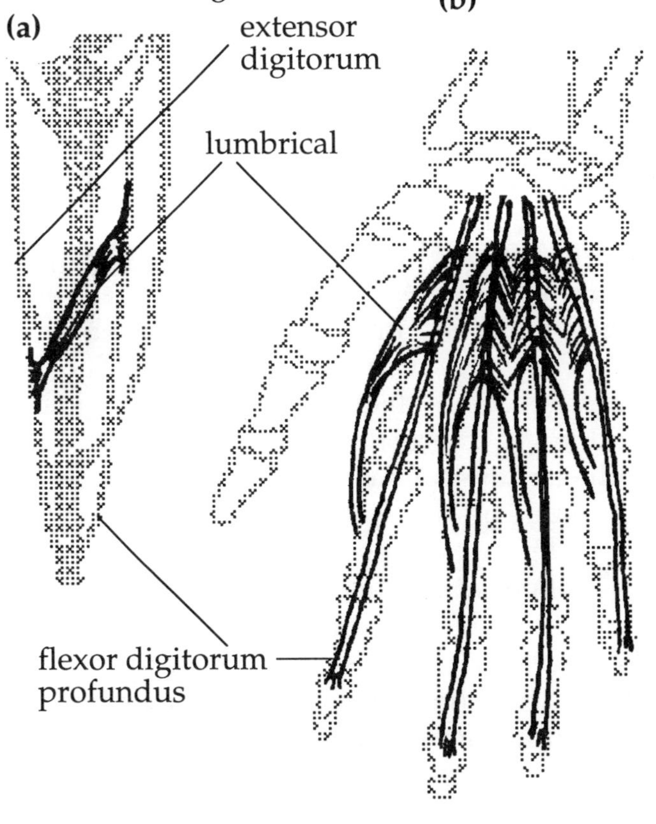

To palpate your lumbricals, press your thumb into the pocket of the palm of your other hand. Let your fingers flex lightly around your thumb. Then keeping the MP joints of your fingers flexed, extend both IP joints. Feel the muscles contracting under your thumb. The tendons of your lumbricals may be felt where they cross the lateral side of the proximal phalanx of each finger. Support the sides of one proximal phalanx between your other thumb and index finger. Put your awareness into your thumb on the lateral side of the phalanx. Leaving your MP joint flexed, extend your IP joints. Just as you initiate extension, the tendon of the lumbrical flickers under your thumb. It is easiest to feel on the third and fourth fingers. (If you feel something flickering under your index finger, it is one of the interossei muscles.)

Palmar Interossei

'Inter' means between. 'Ossei' refers to bones. This muscle is in the palm side of our hand between the bones, namely our metacarpals.

Our **palmar interossei** originate in three bundles from the medial side of our second metacarpal and the lateral side of our fourth and fifth metacarpals (Fig. 12-70). Each bundle inserts into the same side of the proximal phalanx and the dorsal expansion of its digit.

As it contracts, it pulls its finger toward the middle finger, adducting it at its MP joint. Notice that our middle or third finger has no palmar interossei. It is the midline toward which our other fingers adduct. Since our palmar interossei insert into our extensor expansion, they also will assist in flexion of the MP joints and extension of both IP joints.

To palpate your palmar interossei, press into the proximal end of the intermetacarpal space on either side of your third metacarpal. Adduct your fingers as hard as you can. You may feel the bulge of your palmar interossei as they contract. They are not easy to feel since they are deep to your lumbricals and the dense

Figure 12 - 70

An anterior view of the palmar interossei.

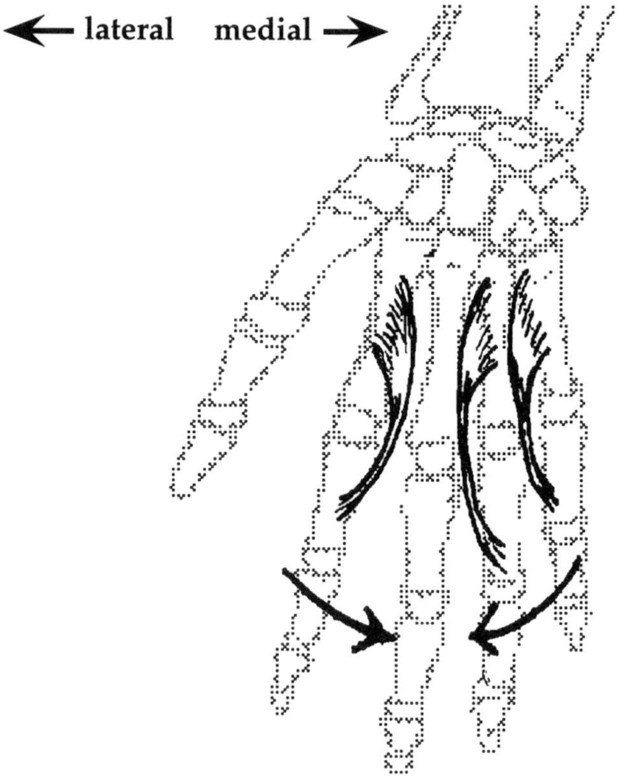

Figure 12 - 71

An anterior view of the dorsal interossei.

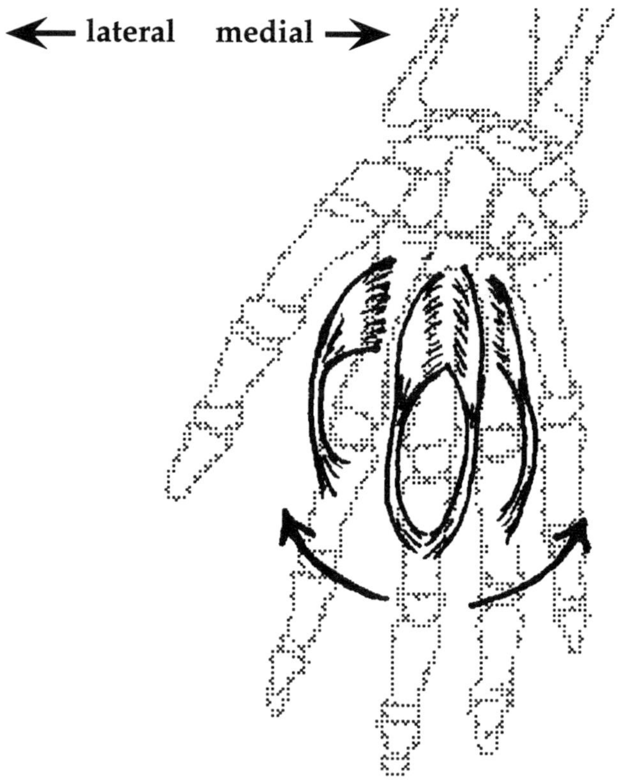

connective tissue covering your palm.

Dorsal Interossei

Our **dorsal interossei** are four bundles of muscles between our metacarpals in the dorsal side of our hand. They originate from the facing surfaces of adjacent metacarpals (Fig. 12-71).

They insert into the lateral side of the base of the proximal phalanx of our second and third fingers, and the medial side of the base of the proximal phalanx of our third and fourth fingers.

As our dorsal interossei contract, they abduct our second and fourth fingers. Our third finger has one interossei on each side so it can be pulled to either side depending on the balance of the two muscles. Our third finger is the midline from which our other fingers move away. Since our dorsal interossei insert into the dorsal expansion, they also assist in extension of the IP joints of the three fingers.

Palpate your dorsal interossei by pressing into the proximal end of your dorsal intermetacarpal space. The interossei between your first and second metacarpal, which abducts your index finger, is the easiest to feel just lateral to the whole length of your second metacarpal. Feel its tendon crossing lateral to your second MP joint. Watch and feel the back of your hand. Abduct your fingers and wiggle your middle finger from side to side to see and feel your dorsal interossei working. Spring your fingers energetically between adduction and abduction. Notice when the MP joints tend to be more flexed or extended. Our adductors are palmar so they tend to flex, while our abductors are dorsal and tend to extend.

• From our discussion, we see that all three of our midpalmar muscles juggle the tiny movements between flexion and extension of the fingers. Hold a pen and pretend to write or draw. These tiny movements of the fingers are the midpalmar muscles at work.

Figure 12 - 72

A surface view of the **(a)** anterior and **(b)** medial hand, labeling the superficial muscles.

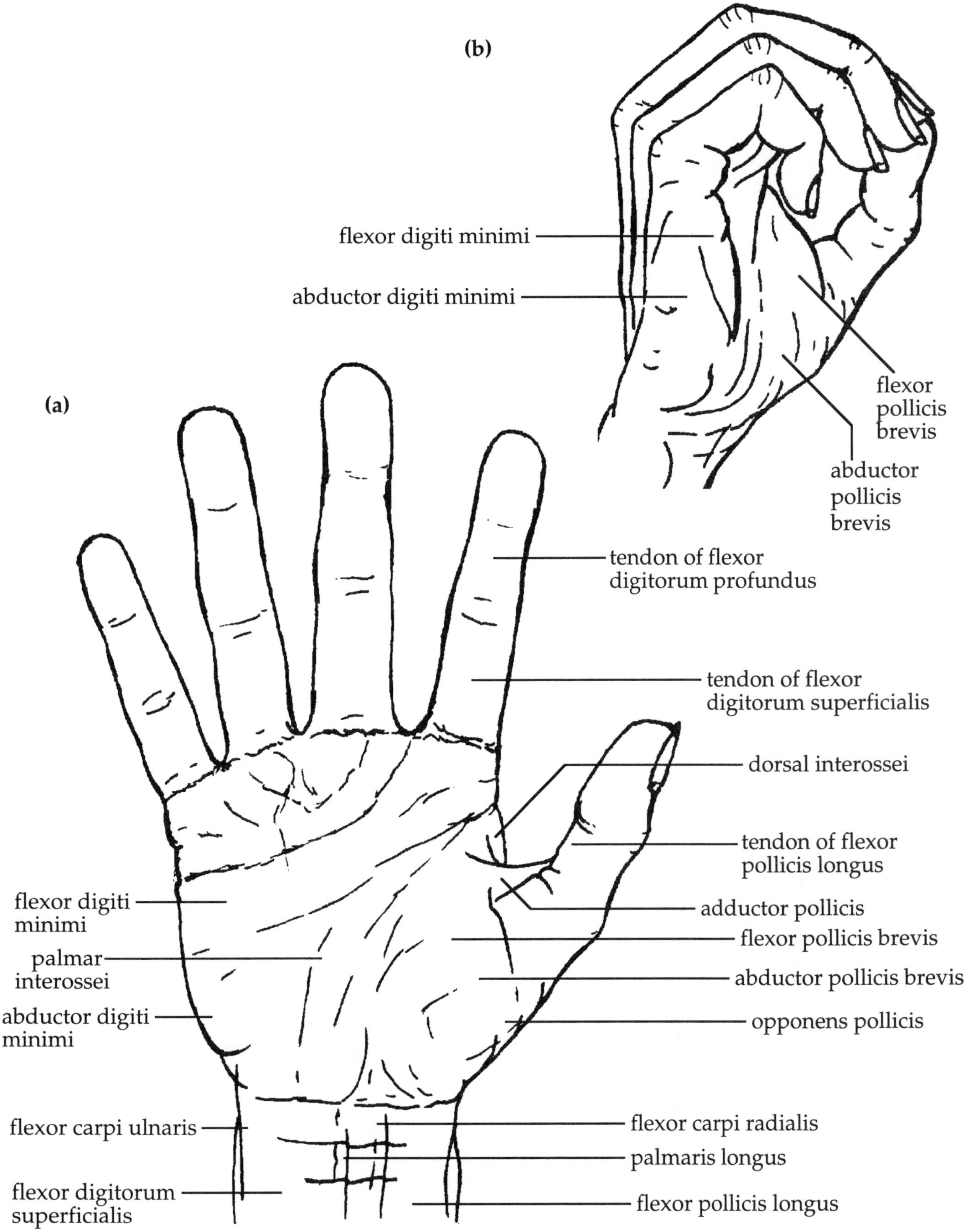

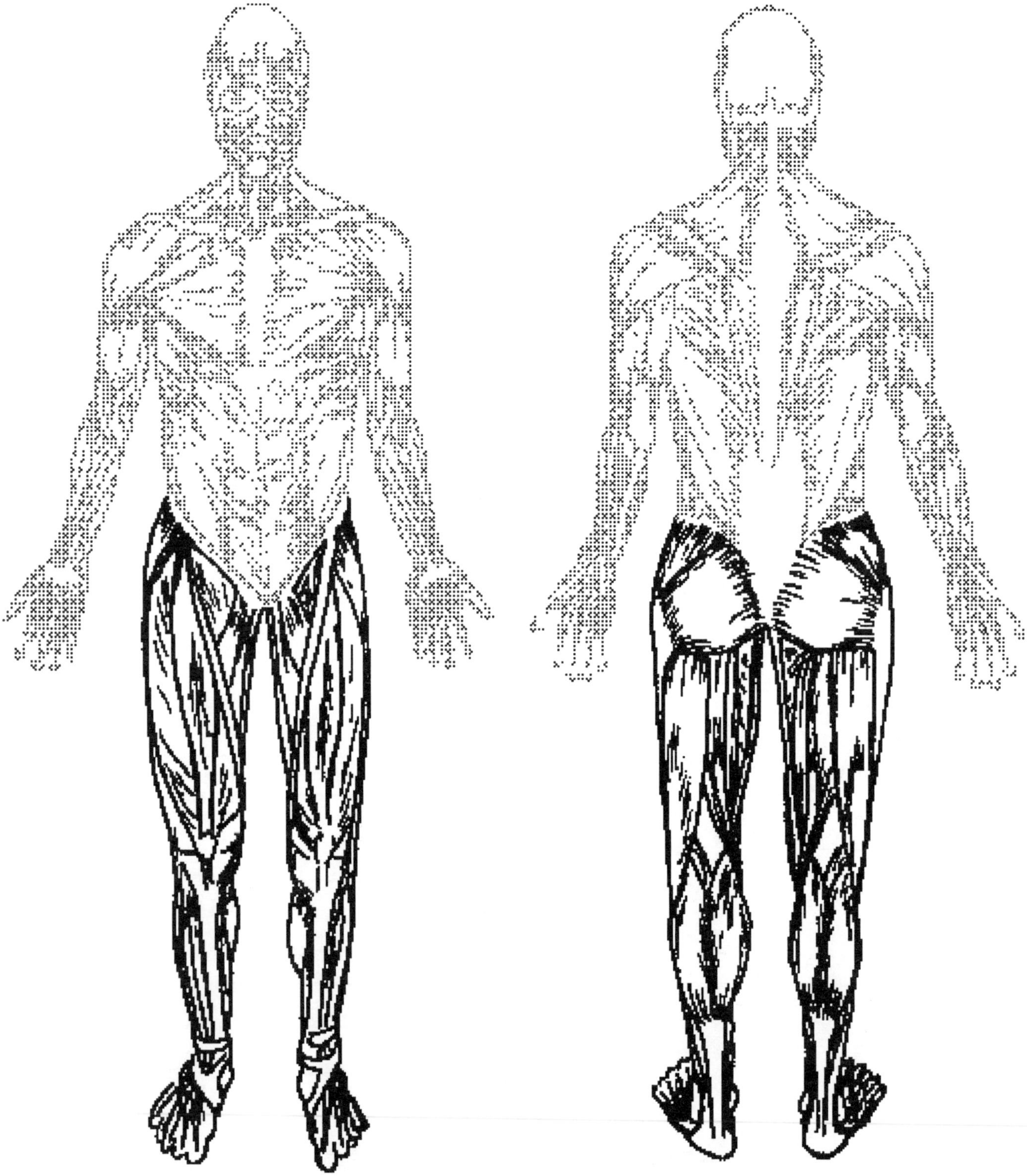

chapter 13: **Lower Appendage Muscles**

As we raise ourselves upward to balance on our two lower appendages, we depend on their stability. In contrast to our upper appendage, the design of our lower appendage sacrifices mobility in order to increase stability. Every joint of our lower appendage exhibits this characteristic, beginning with our sacroiliac joints and continuing into our feet. For example, the irregular interlocked surfaces of our sacrum and ilium offer little potential for movement compared to the pivoting of our sternoclavicular joint. Our sacrum and pelvis together form a complete bony ring (Fig. 9-1) with limited intrinsic movement compared to the extensive range of motion of our pectoral girdle. Our hip has a deeper socket than our glenohumeral joint, limiting the distance traveled in any direction. Our knee rotates only a few degrees compared to the 180 degrees spinning in our radius. Flexion of our ankle is very limited and our big and little toes do not oppose.

When our skeleton has less potential for movement, fewer muscles are needed to produce movement.

When our skeleton itself is already stable, fewer muscles are needed for support. For example, the discussion of our upper appendage muscles began with five major muscles that stabilized our pectoral girdle onto our axial skeleton. In our lower appendage there are no such muscles.

Our psoas major, psoas minor, quadratus lumborum, and a small part of our gluteus maximus cross our sacroiliac joint, but affect it only indirectly. Their main focus is on our intervertebral joints or hip.

• In contrast, the overall muscle mass of our lower appendage exceeds that of our upper appendage, even though the number of muscles is less. Since our lower appendage supports the whole weight of our body, the muscles it does have tend to be large.

> Joints with little or no muscle support are very dependent on their ligament support. Ligament injury to these joints may lead to long-term instability since there is no alternate support. There are no muscles to compensate for the loss.

POSTERIOR PELVIC WALL MUSCLES

Our **psoas major, psoas minor,** and **quadratus lumborum** are discussed in chapter 11 as muscles of our posterior abdominal wall. Their main effect is across our lumbar vertebra, only indirectly influencing the position and support of our pelvis. Only our psoas major affects our hip. We will not review these muscles in this chapter. One muscle, the **iliacus,** crosses from our internal pelvis to our femur accompanying our psoas major.

Iliacus

Our **iliacus** forms the posterior-lateral wall of our pelvic cavity, originating from the internal surface of our ilium (Fig. 13-1).

The muscle fibers pass inferiorly, joining with our psoas major to cross over our pubic bone, under our inquinal ligament and insert into the lesser trochanter of our femur (Fig. 13-19a and 13-20b). As our iliacus contracts, it pulls our lesser trochanter around to the front and upward, producing lateral rotation and flexion of our hip.

Within your pelvic cavity, the belly of your iliacus is difficult to palpate, since it is covered by abdominal viscera and your anterior-lateral abdominal wall. After your iliacus passes under your inquinal ligament, it is palpable in your proximal thigh between your anterior and medial thigh compartments. To palpate it, locate your sartorius near its origin. Feel into the hollowness immediately medial to your sartorius, in the crease between your abdomen and thigh (Fig. 13-14 and 13-20c). Strongly flex your hip and feel a short band of muscle tense into the hollowness. This is a glimpse of your iliacus passing between your sartorius and pectineus.

• Since our psoas major and iliacus have a common pelvic position and insertion, they may be grouped together as our **iliopsoas.** The tendon of our psoas major lies immediately medial to the belly of our iliacus and is not distinguishable from it on palpation.

Figure 13 - 1

An anterior view of the iliacus with arrows showing the direction of movement of the hip.

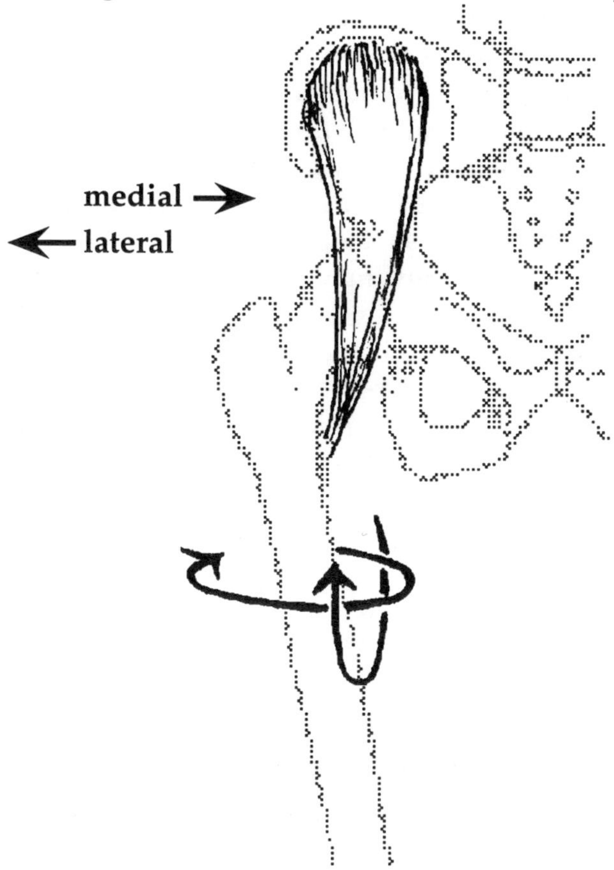

GLUTEAL AREA MUSLCES

Our **gluteal area** is the area of our buttock, a circle circumferenced by our coccyx, sacrum, iliac crest, the greater trochanter of our femur, and the fold between our buttock and thigh. Four muscles fill this area, our **gluteus maximus, gluteus medius, gluteus minimus,** and **tensor fascia lata.**

• Lying over the surface of our lateral gluteal and thigh muscles, we have a dense connective tissue called the **fascia lata** or **iliotibial band** (Fig. 13-2). 'Fascia' is a connective tissue assosciated with muscle. 'Lata' refers to lateral. 'Iliotibial' tell us about its bony attachments. The fascia lata begins on the center of our liliac crest, crosses over our gluteus medius, lateral to our hip, and attaches onto the greater trochanter of our femur. It continues down our thigh, over our vastus lateralis and across our knee, ending on our lateral, proximal tibia. It supports our hip and knee, more like a ligament than a fascia.

The most palpable aspect of our fascia lata is its distal end (Fig. 13-18). Stand with your knees slightly bent. Slide your fingers from back to front over the tissue just above the lateral side of your knee. Right at the back there is a vertical, tense cord, then a hollow stripe, then another vertical, tense cord, and then the softness around the front of your knee. The second cord, in the middle of your lateral distal thigh is the end of your fascia lata.

Figure 13 - 2

A lateral view of the fascia lata.

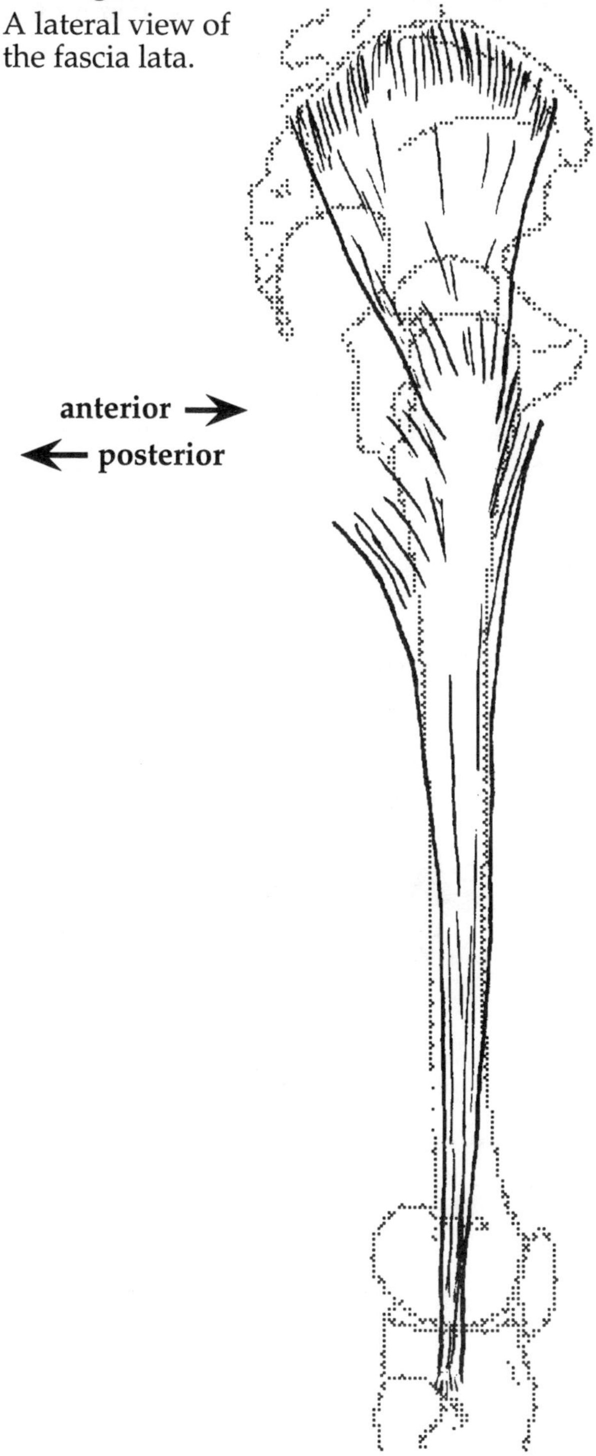

Gluteus Maximus

Our **gluteus maximus** is the largest of our gluteal muscles. It originates from our lateral sacrum and posterior iliac crest (Fig. 13-3 and 13-17). Its muscle fibers pass laterally and inferiorly to insert into our fascia lata and femur, inferior and posterior to our greater trochanter (Fig. 13-19).

As our gluteus maximus contracts, it pulls our greater trochanter around and back toward our posterior iliac spines, combining lateral rotation and extension of our hip.

A large muscle implies lots of use, but kicking backward is not something we do routinely. So why do we have such a giant muscle? Remember that extension of our hip is not just swinging backward from anatomical position. When our thigh is close to our chest, the whole movement away from our chest toward anatomical position is extension of our hip. Standing up out of a chair, climbing up stairs, propelling forward during walking or running all include extension of our hip and put our gluteus maximus to work.

In four-legged posture the torso and pelvis lie horizontal. To stand upright we pivot our torso and pelvis to balance vertically over our legs. This movement involves extension of our hips. Whether our legs move under our pelvis or our pelvis moves over our legs, the movement at our hip joint is the same. So it becomes obvious that many daily activities work our gluteus maximus.

Our gluteus maximus is a large superficial muscle and easy to palpate. Stand up to make it easiest. Place your fifth metacarpal and little finger over your sacrum and coccyx. Pick up the muscle under your thumb and fingers. All this muscle is your gluteus maximus (Fig. 13-18, 13-19c, and 13-19d). Slide your hand down until your fingers curl into the fold between your buttock and thigh. Your thumb is in a shallow indentation near the center of your gluteal area. All this muscle still is your gluteus maximus. Follow it laterally toward your femur and feel how it ends in the thickness just below your greater trochanter.

Figure 13 - 3

(a) A posterior and **(b)** a lateral view of the gluteus maximus with arrows showing the movement at the hip.

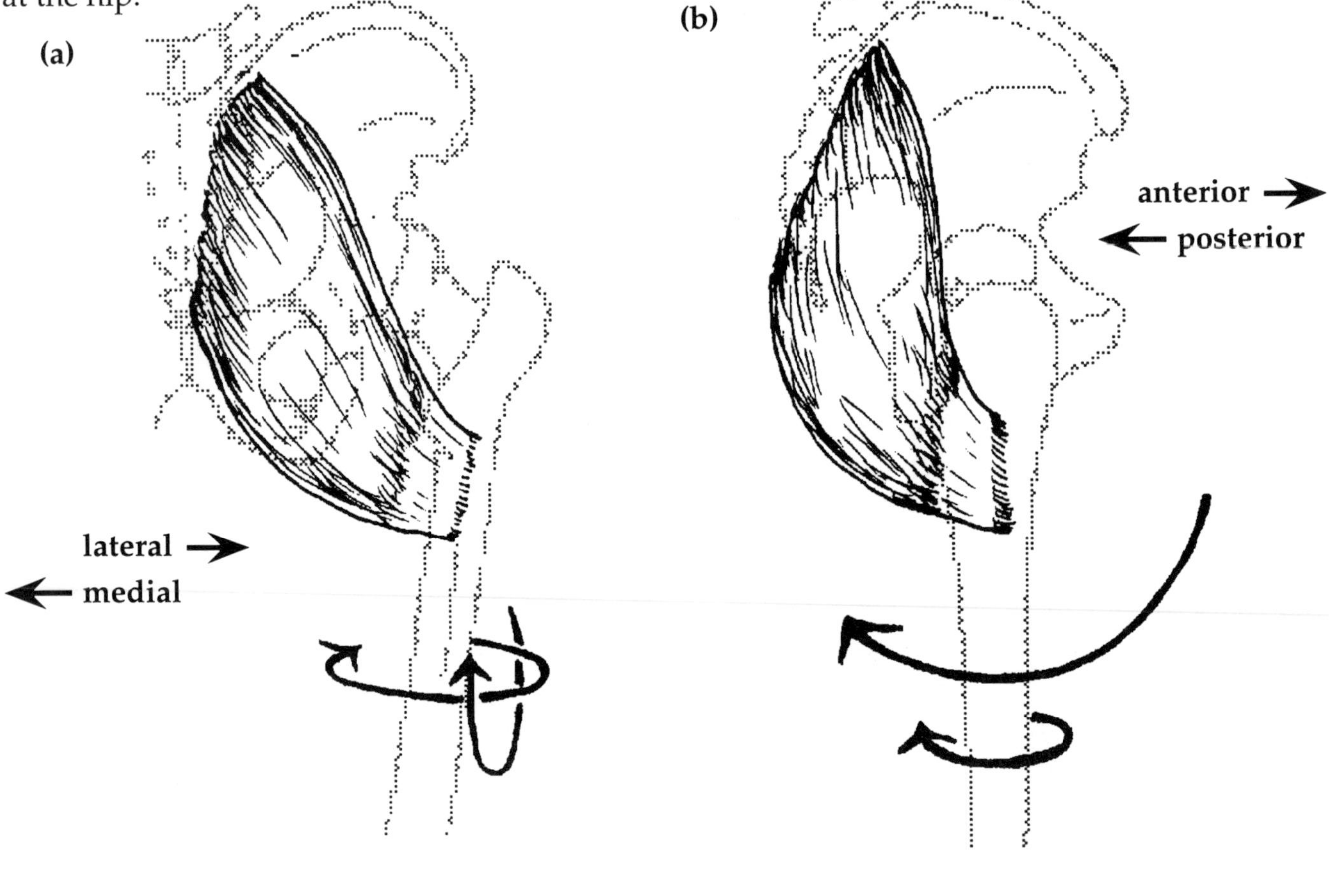

Gluteus Medius

Our **gluteus medius** is the middle-sized muscle of our gluteal area. It originates from the external surface of our ilium, just inferior to the center of our iliac crest (Fig. 13-4a, 13-4b, and 13-17). Its muscle fibers pass downward, burying the head and neck of our femur to insert into the top of its greater trochanter (Fig. 13-19a).

Since the predominant orientation of our gluteus medius is lateral to our hip joint, its main action is to abduct our hip, swinging our leg out to the side.

• Again, we have a large muscle filling the deep hollow in this area of our skeleton. But who has kicked out to the side to abduct their hip even once today? Remember that abduction of our hip simply means that our lateral pelvis and lateral thigh are being brought closer together. This happens when we swing our leg out to the side (Fig. 9-29b). It also happens if we lean our pelvis out to the side over a supporting leg. Our muscle can be used in either direction.

Imagine your pelvis as a horizontal table. The table supports the weight of your body through your vertebral column, resting on its center. The table is supported by your two legs at its lateral edges (Fig. 13-5a). But what happens when you lift one leg? One-half of the support of your table is removed. Most tables would topple over (Fig. 13-5b). What prevents our pelvis from falling downward on the side where its support is removed?

If we lift our right leg and our pelvis falls down on the right side, what movement happens across our left hip? Adduction. The medial aspect of our pelvis and thigh are approaching each other. What kind of force could prevent this movement? A force in the direction of abduction. When we lift our right leg and support ourself on our left leg, our left gluteus medius produces enough abducting force to overcome the adducting force of gravity taking advantage of our one-legged stance (Fig. 13-5c). When we stand on one leg, the abducting force of our gluteus medius on the supporting side prevents the unwanted dropping or adduction of our hip.

Figure 13 - 4

(a) A posterior and **(b)** a lateral view of the gluteus medius, and **(c)** a lateral view of the gluteus minimus, with arrows showing their movements at the hip.

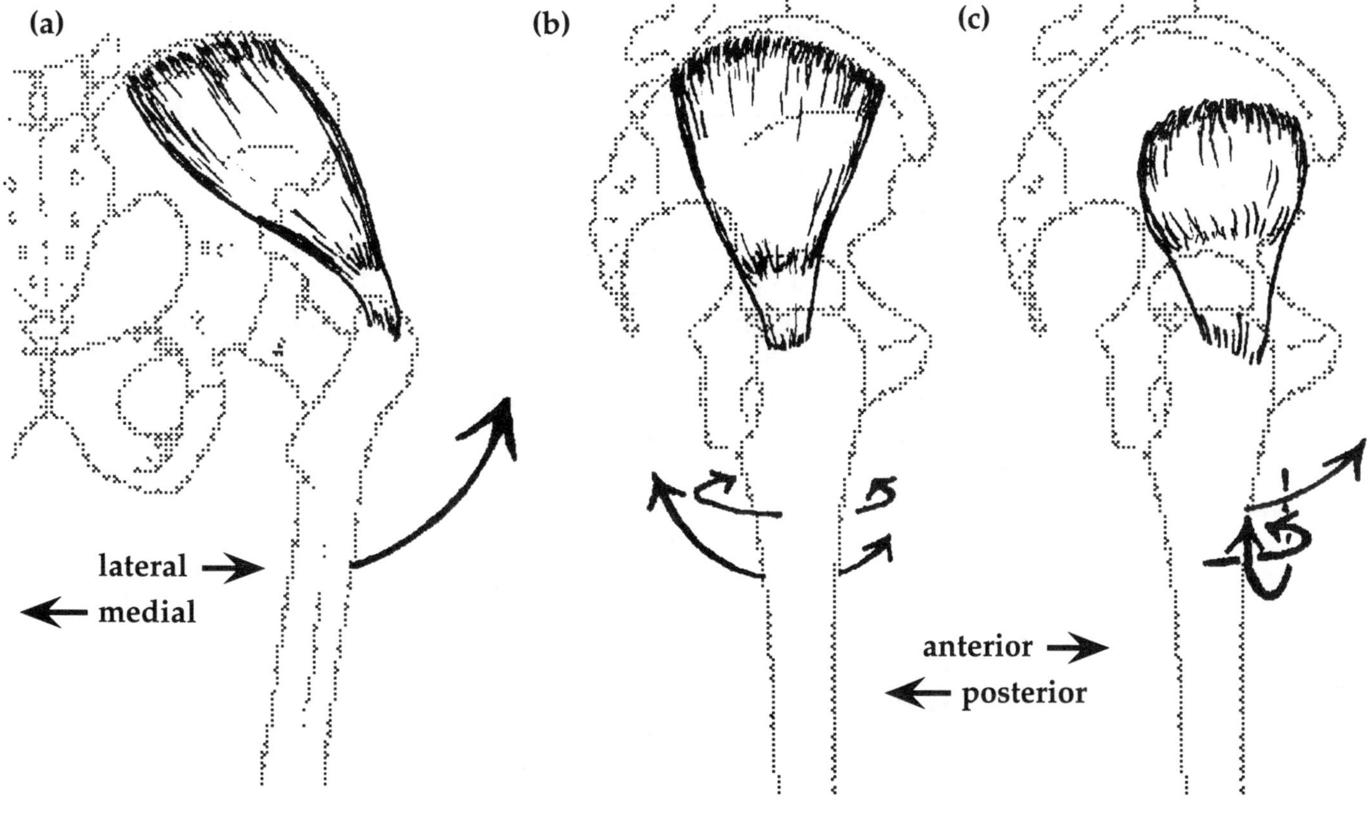

Figure 13 - 5

(a) When standing on two legs, the lateral support of the pelvis is stable. **(b)** When one leg is lifted, that side of the pelvis will drop, adducting the opposite hip. **(c)** The gluteus medius and minimus of the adducted hip can contract to prevent the adduction, thus stabilizing the pelvis horizontally.

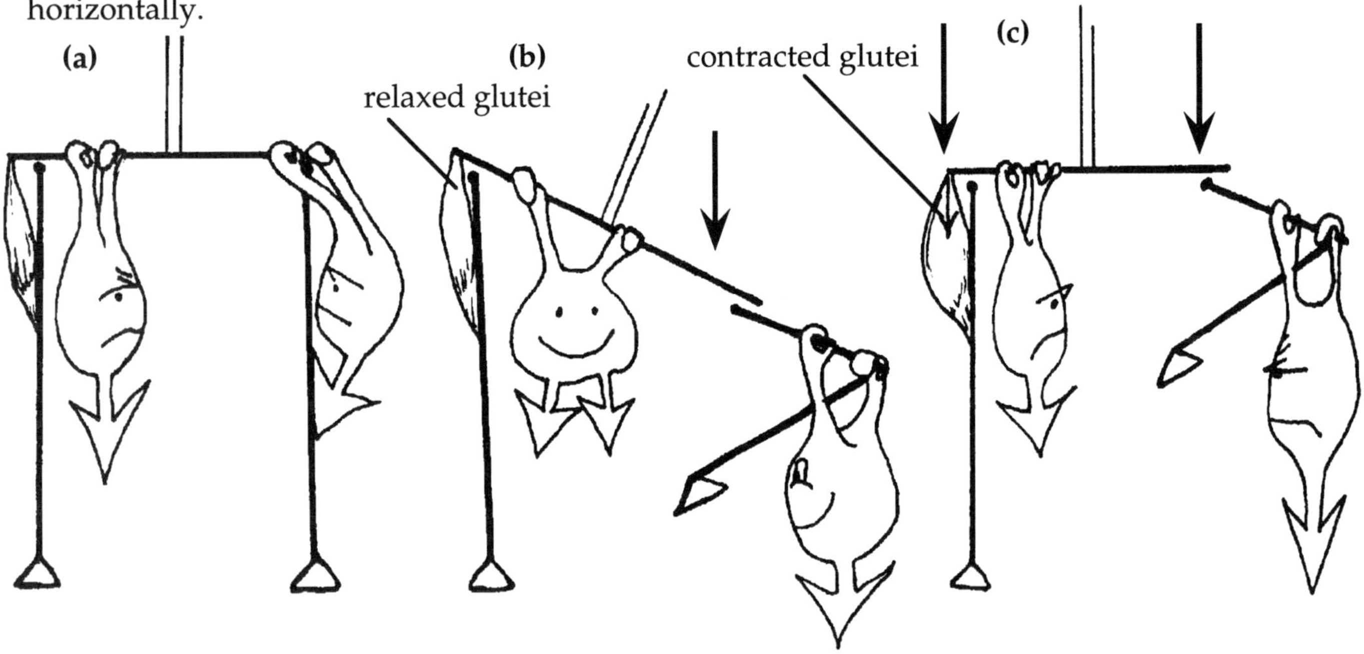

• Our gluteus medius is oriented around our hip joint much like our deltoid around our glenohumeral joint (Fig. 13-4 and 12-27). This gives our gluteus medius a diversity of action similar to our deltoid. While the whole muscle abducts, its **posterior** part may extend and laterally rotate our hip while its **anterior** part may flex and medially rotate our hip.

• Our whole gluteus medius is palpable as it lies immediately under our skin, not covered by any other muscle. Put the heel of your hand on the center of your iliac crest, at your side. Let your fingers point straight down to touch the top of the greater trochanter of your femur. In this position the belly of your gluteus medius is under your hand (Fig. 13-18, 13-19c, 13-19d, and 13-20c). Stand up. Press into the belly of one gluteus medius. Lift your opposite leg to feel the tension coming into the gluteus medius as it holds your pelvis level.

Gluteus Minimus

Our **gluteus minimus** is the smallest of our gluteal muscles. It originates from the external surface of our ilium, inferior to the origin of our gluteus medius (Fig. 13-4c and 13-17). Its fibers pass inferiorly to insert into the top of the greater trochanter of our femur in front of the insertion of our gluteus medius.

Our gluteus minimus assists our gluteus medius in most of its functions, abducting, flexing, and medially rotating our hip, and stabilizing our pelvis when standing on one leg. Since it inserts anterior to our gluteus medius, it loses most of its potential in the posterior functions of extension and lateral rotation of our hip.

Put the heel of your hand on the center of your iliac crest and your fingertips on your greater trochanter. The muscle right under your hand is your gluteus medius and deep to that, your gluteus minimus (Fig. 13-20). Your gluteus minimus is part of what fills the concavity in this area of the skeleton. Your gluteus minimus is mainly palpable through your gluteus medius, and difficult to separate from it. Since any action of the gluteus minimus also is performed by the gluteus medius, performing a movement will not help to separate them. The insertion tendons of the two muscles also are difficult to separate, but the gluteus minimus is the most anterior part.

Tensor Fascia Lata

Our **tensor fascia lata** (tensor fasciae latae in plural) is a muscle which can tense or pull on our fascia lata. It originates from our anterior iliac crest, immediately posterior to our anterior superior iliac spine (Fig. 13-6 and 13-17). It passes obliquely downward and backward to insert into our fascia lata in the area of the greater trochanter of our femur. In a sense, our fascia lata becomes the ongoing insertion tendon of our tensor fascia lata, allowing the muscle to influence our knee.

As our tensor fascia lata contracts, it tenses our fascia lata to stabilize our knee during walking. It also pulls forward and upward on our greater trochanter to medially rotate, abduct and flex our hip.

To palpate your tensor fascia lata, find your anterior superior iliac spine. Press inward just below that spine and slide in a posterior direction for a distance of about three finger-widths. The muscle that rolled under your fingers is the origin of your tensor fascia lata (Fig. 13-18, 13-20b, and 13-20c). Follow it inferiorly and posteriorly to your greater trochanter. To feel it contract, press into its belly while you medially rotate your hip.

Sit with your hip flexed and laterally rotated. Find your anterior superior iliac spine and greater trochanter. Imagine a line joining these two points. Rub your fingers back and forth perpendicular to it. The vertical fibers under your fingers are your tensor fascia lata.

Figure 13 - 6

(a) An anterior and **(b)** a lateral view of the tensor fascia lata muscle and the fascia lata into which it inserts.

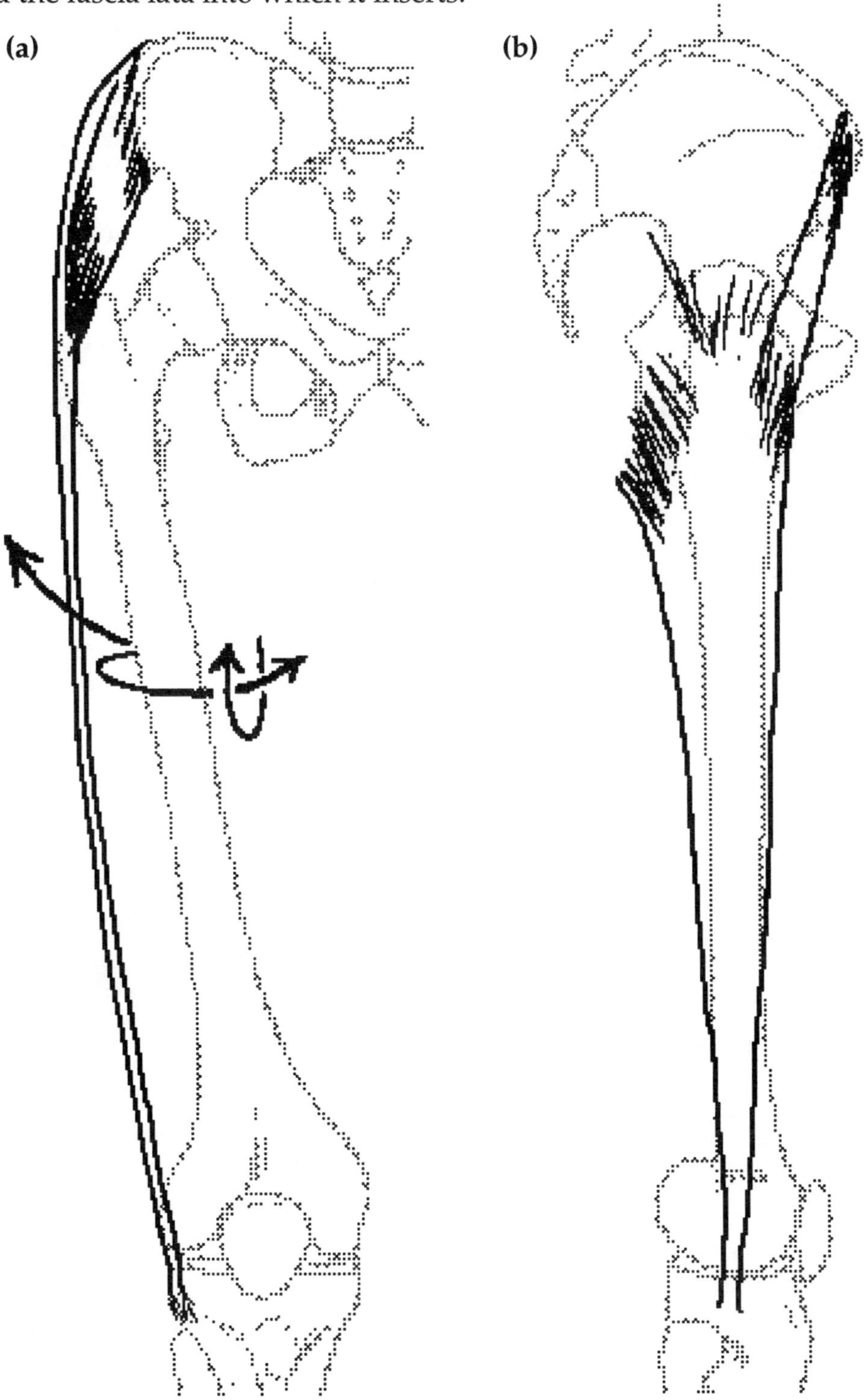

In humans, the gluteal muscles are all quite large. In four-legged animals whose basic stance has a horizontal pelvis and a flexed hip, even the largest gluteal muscle is a little wisp (Fig. 13-7).

Figure 13 - 7
The flexed hip posture and the four-legged stability of a horse reduces the work requirement of the gluteal muscles, leaving them relatively smaller than in humans.

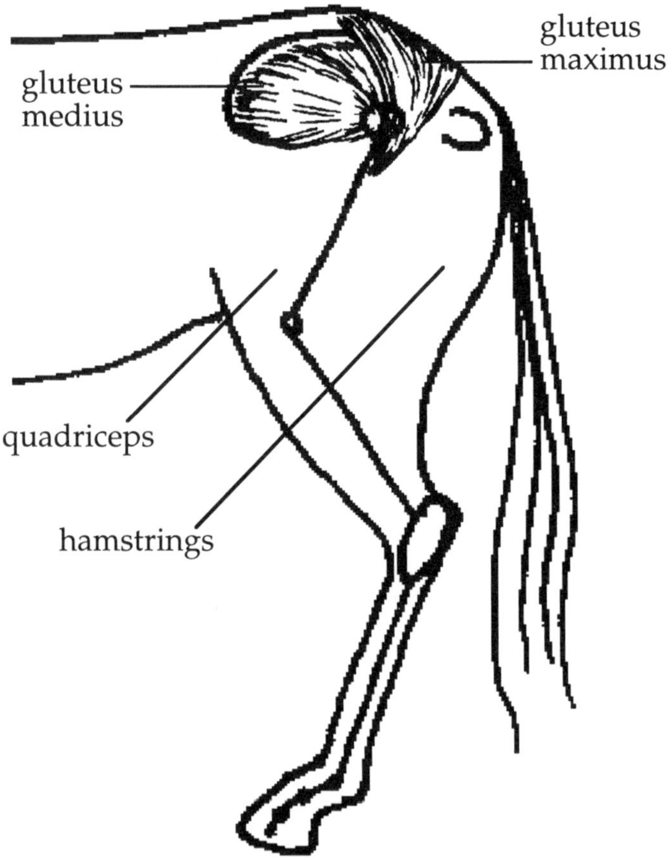

DEEP HIP LATERAL ROTATORS

Six muscles are grouped together because of a similar location and function. These muscles are deep to our gluteus maximus and laterally rotate our hip and so are called the **deep lateral rotators of the hip.** We will treat five of these muscles as a single entity, and the sixth muscle, our **piriformis,** we will discuss separately.

• Within the collection of anatomical knowledge, the names of the five muscles are **gemellus superior, gemellus inferior, obturator internus, obturator externus,** and **quadratus femoris.** These muscles originate from our inferior ischium, the bone around our obturator foramen and the membrane covering our obturator foramen (Fig. 13-8a). These muscles pass horizontally across our hip to insert into the posterior greater trochanter of our femur. Horizontal muscles have their greatest mechanical advantage in horizontal movements. The horizontal movement of our hip is rotation. The specific orientation of these muscles allows them to laterally rotate our hip.

To palpate these muscles, locate your ischial tuberosity and the greater trochanter of your femur. Curl under the inferior edge of your gluteus maximus and press upward between the two bones (Fig. 13-19b and 13-19d). Laterally rotate your hip to feel the five deep lateral rotators of your hip spring out against your fingers.

Figure 13 - 8
A posterior view of **(a)** the five deep lateral rotators of the hip and **(b)** the piriformis.

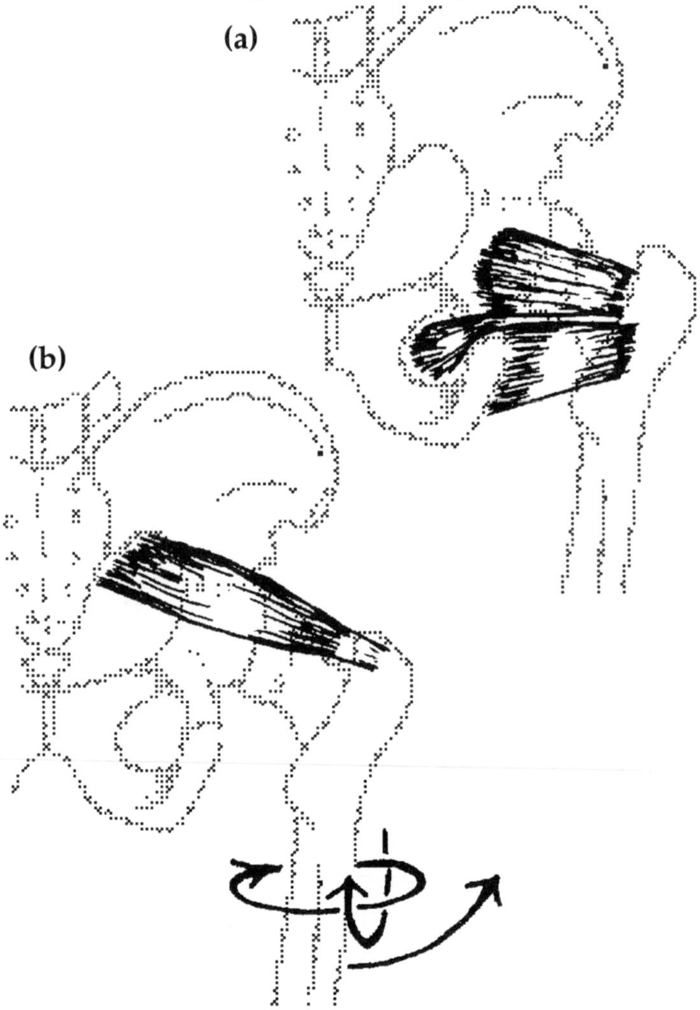

Piriformis

The one muscle which deserves separate investigation is our **piriformis.** A 'pirum' is a pear and 'forma' refers to form or shape, so the name of this muscle implies that it is pear-shaped. Its bulging belly, narrowing into its insertion tendon, is pear-shaped, similar to many other muscles. I am not sure why this muscle particularly deserves the name.

Our piriformis is the most superior of the six deep lateral rotators of our hip. It originates from our anterior sacrum and passes laterally through our greater sciatic notch to lie posterior to our pelvis (Fig. 13-8b). Our sciatic nerve passes deep to our piriformis, but superficial to the other five deep lateral rotators of our hip. Our piriformis continues horizontally to insert into the top of the greater trochanter of our femur. Its belly lies just below the posterior edge of our gluteus medius and minimus (Fig. 13-19b).

Our piriformis' action is similar to our posterior gluteus medius, laterally rotating, extending, and abducting our hip (Fig. 13-8b).

Our piriformis is difficult to distinguish from the gluteus muscles lying over and adjacent to it. Find the indentation just inferior to your posterior inferior iliac spine and lateral to your sacrum. This is your sciatic notch where your piriformis first emerges. A straight line from this notch across to the top of the greater trochanter of your femur would mark your piriformis (Fig. 13-19d). If you press your fingers into this line and laterally rotate your hip, you may feel your piriformis through your gluteus maximus.

THIGH MUSCLES

All the muscles of our thigh can be divided into three compartments: ante**rior, posterior,** and **medial** (Fig. 13-9a). Many of these muscles cross both our hip and knee, some cross only one of these joints.

To understand the actions of our thigh muscles, it is important to remember that our ventral and dorsal surfaces rotate 180 degrees in our thigh. Ventral is anterior at our hip but posterior at our knee, while dorsal is posterior at our hip and anterior at our knee (Fig. 8).

• Before we discuss any specific muscles in our thigh, let us look at their basic action potentials.

Figure 13 - 9

A cross section through the thigh showing **(a)** the three compartments and **(b)** the individual muscles in each compartment.

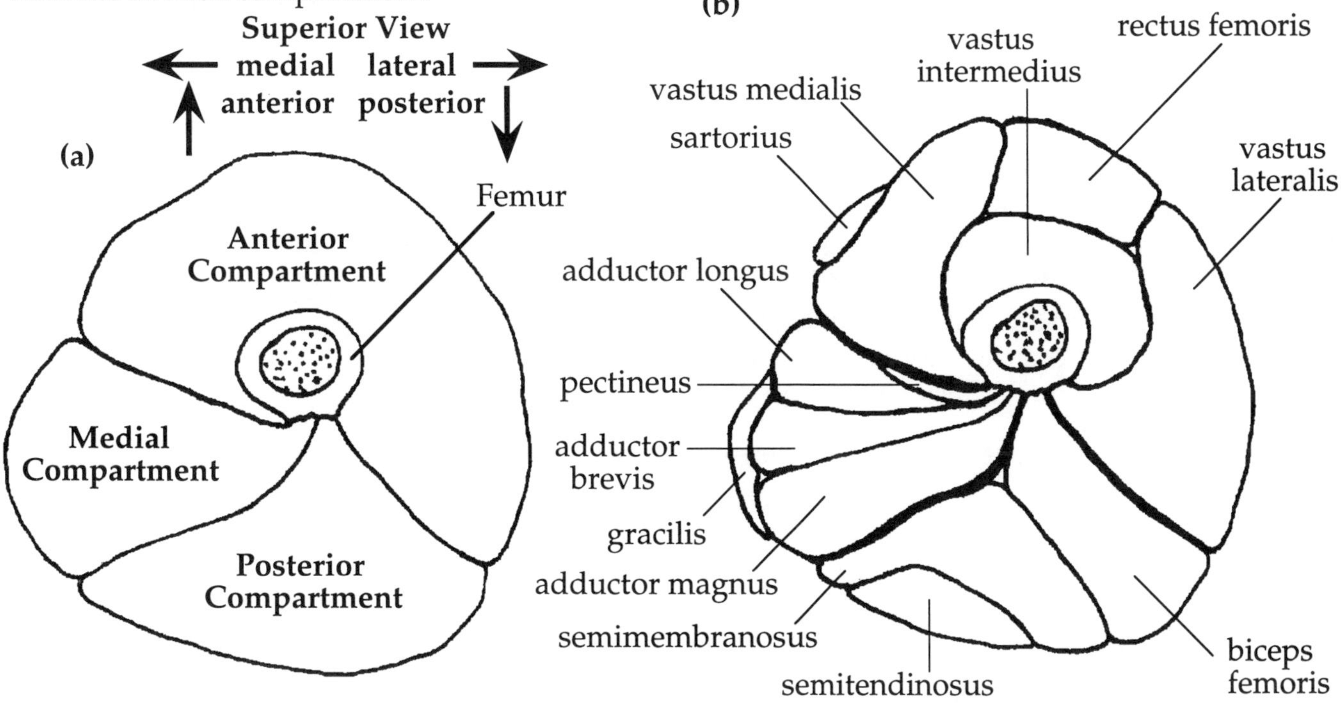

We have just said that anterior is defined as vental at our hip and dorsal at our knee. If **anterior compartment muscles** cross our hip, their contraction would bring ventral surfaces together, flexing our hip. If they cross our knee, they would approximate dorsal surfaces, extending our knee.

Since posterior is dorsal at our hip and ventral at our knee, **posterior compartment muscles** will extend our hip and flex our knee.

Medial compartment muscles lie medially in our thigh. Across our hip, their contraction would approximate our medial pelvis and thigh, adducting our hip. If our medial compartment crossed our knee, it also would try to adduct our knee. This is not a normal knee movement and our medial compartment, in fact, does not cross our knee.

These are the basic muscle actions which can be deduced from their compartment location. Specific muscles have other specific functions depending on their specific attachments.

ANTERIOR COMPARTMENT

Our anterior compartment is the largest of the three, and includes the anterior and lateral aspects of our thigh (Fig. 13-9a). It begins with five muscles (Fig. 13-9b). One of these muscles mutinies, however, and joins our posterior compartment at its insertion.

Sartorius

A 'sartor' is a tailor who, in the good old days before chairs and tables, would sit on the ground with his legs crossed in front of him as he sewed. Taking this position uses all four actions of our **sartorius,** and so it was named after the tailor. Our sartorius is the longest and thinnest muscle of our anterior compartment. It originates from our anterior superior iliac spine (Fig. 13-10 and 13-17). It spirals medially and posteriorly as it approaches our knee, leaving the anterior compartment, crossing over the medial compartment, joining the posterior compartment behind our knee. It then comes forward to insert anterior and medial on our proximal tibia (Fig. 13-10, 13-20b, 13-20c, and 13-32).

Figure 13 - 10

An anterior view of the sartorius. The upper arrows show movements of the hip, while the lower arrows show movements of the knee.

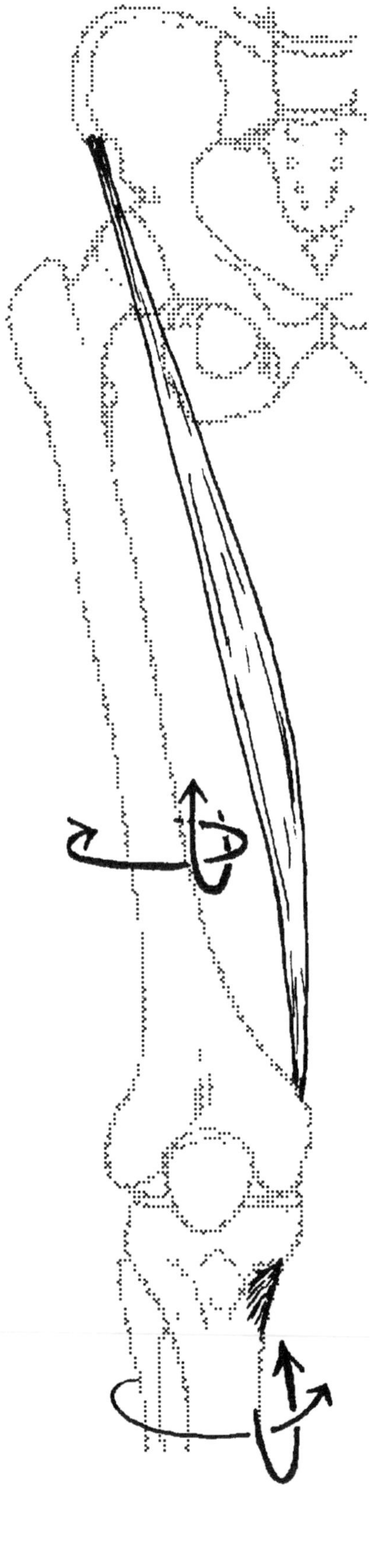

As our sartorius contracts, it performs anterior compartment actions across our hip, flexing it, and posterior compartment actions across our knee, flexing it (Fig. 13-10).

As our sartorius wraps around our thigh, it gains mechanical advantage for horizontal movement, laterally rotating our femur at our hip while medially rotating our tibia at our knee. These four moves combined point our knee up and out to the side and bring our heel toward our groin in the classic yogi position.

Our sartorius is superficial and palpable through most of its length. To feel its origin, find your anterior superior iliac spine. Move just inferior to the spine and flex your hip. Two muscles pop out on your proximal thigh. The medial one is your sartorius (Fig. 13-14, 13-18, 13-20b, and 13-20c).

To palpate your sartorius along its oblique path around your thigh, flex and laterally rotate your hip, but leave your knee straight. In this position, you may be able to see or feel the tense band of the contracting muscle passing from your anterior superior iliac spine to behind your medial knee. If not, imagine a line along this path. Slide back and forth perpendicular to this line. The band that you bump over is your contracting sartorius.

QUADRICEPS FEMORIS

The four remaining muscles of our anterior compartment often are considered as one muscle, and are called the **quadriceps femoris.** 'Quadri' means four and 'ceps' refers to heads which, for muscles, means their origin. Each of our quadriceps muscles has its own origin, but they all share one common insertion. All their insertion tendons join together to cross anterior to our knee and insert into our tibial tuberosity (Fig. 13-11).

- Our patella is a bony knob within and surrounding the quadriceps insertion tendons. It functions to maintain the anterior alignment of our quadriceps insertion and is further discussed in chapter 9, Lower Appendage Skeleton.

Since our four quadriceps muscles merge into one tendon across our knee, they all perform one common knee action, the predicted anterior compartment action of extending our knee. For three of these muscles this is their only action.

To feel your quadriceps insertion, press your fingertips into the tissue just above your patella. Extend your knee and feel the tendon tense under your fingers. Feel how the tendon penetrates into your patella. Relax your leg. Press into the tissue between your patella and tibial tuberosity. Extend your knee again to feel the tendon as it leaves the patella and heads for your tibia. This last part of the tendon sometimes is called the **patellar ligament** (Fig. 13-11, 13-18, and 13-20).

- The four muscles of our quadriceps are our **rectus femoris, vastus lateralis, vastus intermedius,** and **vastus medialis.** We will examine the individual characteristics of each.

Rectus Femoris

As in rectus abdominis or erector spinae, 'rectus' refers to the erect, vertical orienation of this muscle. 'Femoris' tells us which bone it is adjacent to.

Our **rectus femoris** originates from our anterior inferior iliac spine (Fig. 13-11c and 13-17). It runs down in front of our femur, through our patella, to insert into our tibial tuberosity (Fig. 13-32).

Our rectus femoris typifies the anterior compartment of our thigh, lying in the center front of our thigh and crossing both our hip and our knee. It performs both the predicted anterior compartment actions, flexing our hip and extending our knee (Fig. 13-11c).

Your rectus femoris is superficial and palpable throughout its length. To feel its origin, find your anterior superior iliac spine. Slide just inferior to the spine and flex your hip. Two muscles pop out on your proximal thigh. The lateral one is your rectus femoris (Fig. 13-14, 13-20b, and 13-20c), just lateral to your sartorius. Follow its belly down the center front of your thigh to its common quadriceps insertion.

Vastus Lateralis

'Vastus' tell us that this is a vast or large muscle. 'Lateralis' tells us that it is lateral in position. Our **vastus lateralis** orginates from our posterior femur along the linea aspera, from just below our greater trochanter to our lateral epicondyle (Fig. 13-11b and 13-19). It wraps around the lateral side of our femur to join the common quadriceps insertion.

The only joint crossed by our vastus lateralis is our knee. Our vastus lateralis performs the one common quadriceps function, extending our knee (Fig. 13-11b).

Figure 13 - 11

An anterior view of the quadriceps from deep to superficial, **(a)** being the vastus intermedius, **(b)** the vastus lateralis and vastus medialis, and **(c)** the rectus femoris.

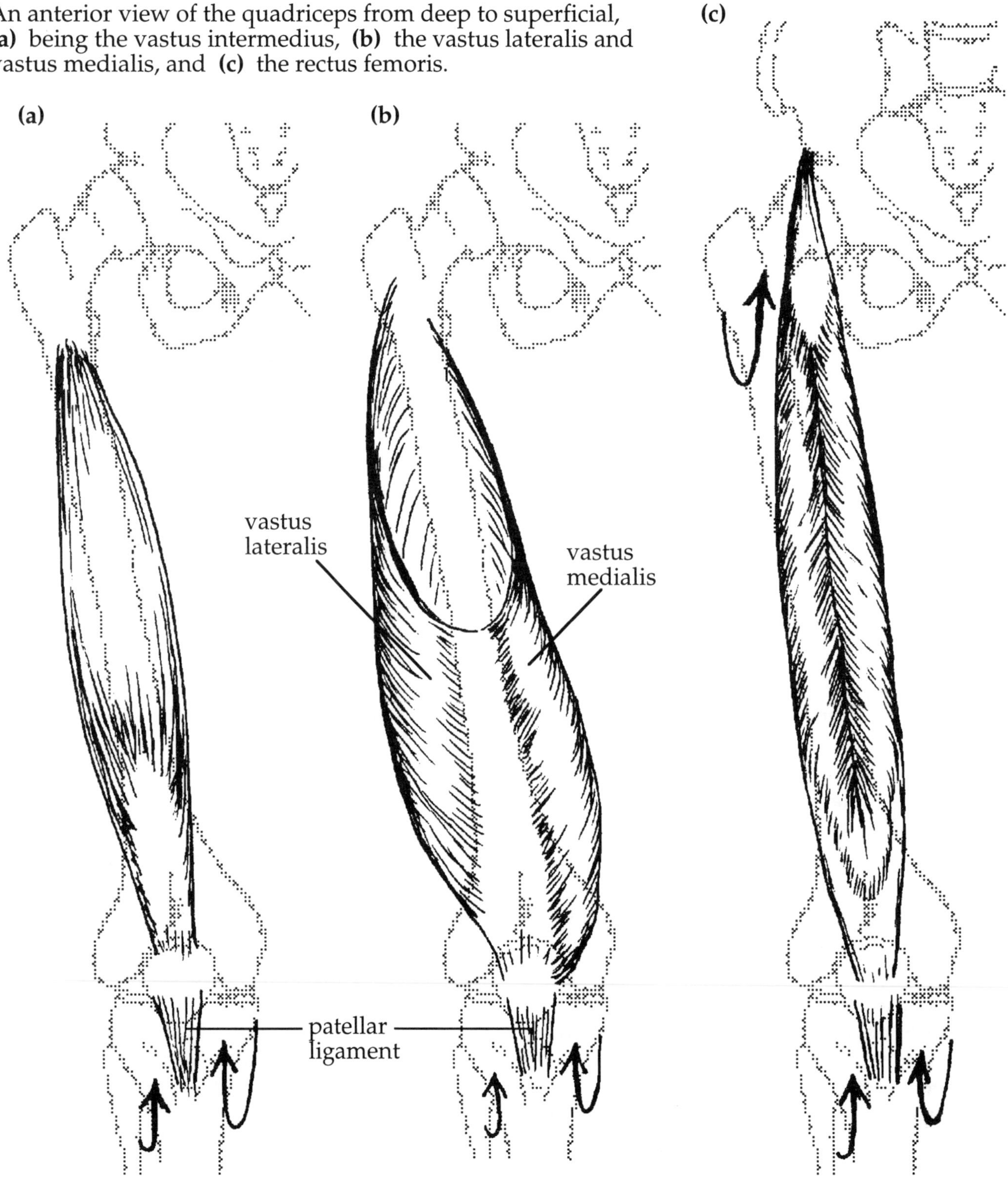

The origin of your vastus lateralis is covered by your thigh's posterior compartment. The belly of your vastus lateralis is palpable in your lateral thigh (Fig. 13-18 and 13-20). Find your greater trochanter. Slide down to your lateral epicondyle. All this muscle, for the width of your hand is your vastus lateralis. Extend your knee to feel it contract.

The belly of your vastus lateralis is covered by your fascia lata (Fig. 13-18 and 13-20). As your vastus lateralis passes anterior toward its insertion, your fascia lata continues straight down across your lateral knee. Do not confuse this as the insertion of your vastus lateralis.

Vastus Intermedius

'Inter' means between. 'Medius' means middle. Our **vastus intermedius** is another large muscle which is placed in the middle, between two other large muscles.

Our vastus intermedius originates from our anterior and lateral femur, from just below our greater trochanter to the patellar surface (Fig. 13-11a). It lies deep to our rectus femoris. It inserts with the common insertion of our quadriceps and extends our knee. In palpation, our vastus intermedius cannot be separated from the rectus femoris lying over it.

Vastus Medialis

Our **vastus medialis** is the most medial of our three large muscles. It originates from our posterior femur along the linea aspera, from just below the neck to the medial epicondyle (Fig. 13-11b, 13-19a, and 13-19b). It wraps around the medial side of our femur to join the common quadriceps insertion. The only joint it crosses is our knee, so it performs the one common quadriceps function, extending our knee.

The origin of your vastus medialis is covered by the posterior and medial muscle compartments of your thigh. To palpate the belly of your vastus medialis, locate your sartorius and rectus femoris. Feel them wedge apart in your lower thigh. All the muscle between them is your vastus medialis (Fig. 13-20b and 13-20c). Press into it as you extend your knee to feel the muscle contract.

MEDIAL COMPARTMENT

Our medial compartment begins with five muscles (Fig. 13-9b). One of them mutinies and, like our sartorius, joins our posterior compartment at its insertion. Proximally, our medial compartment is quite large, but as the muscles end, it dwindles to a point ending on the adductor tubercle of our femur.

Adductor Longus

Our **adductor longus** originates immediately below our pubic tubercle (Fig. 13-12c and 13-17). It lies obliquely in our upper thigh, crossing inferiorly and laterally to insert into our posterior femur, in the middle of the linea aspera (Fig. 13-19a). As our adductor longus contracts, it pulls our femur forward and medially to flex and adduct our hip. Since the insertion of our adductor longus approaches our femur from the medial side, it also pulls our posterior femur medially. Our anterior femur then will face laterally. This action is defined as lateral rotation of our hip.

Your adductor longus is most palpable at its origin. Find your pubic tubercle. Feel the dense cord passing from this tubercle down your thigh. This cord may include two tendons. To separate the two, flex your hip upward from a sitting position and squeeze your knees together. The most anterior of the two tendons is your adductor longus. Follow it down your medial thigh until it dissappears under your sartorius in the middle of your thigh (Fig. 13-20c and 13-20d).

Pectineus

Our **pectineus** is the smallest muscle in our medial compartment. In Latin, a 'pecten' is a comb. To be 'pectinate' means to be toothed like a comb. Our pectineus is one of the teeth of our comb (Fig. 13-14).

Our pectineus originates from the superior ramus of our pubic bone just lateral to our pubic tubercle and adductor longus (Fig. 13-12a and 13-17). It lies obliquely in our upper thigh, passing inferiorly and laterally to insert into our posterior femur, just distal to our lesser trochanter (Fig. 13-19a).

As our pectineus contracts, it pulls our lesser trochanter toward our pubis, flexing, adducting, and laterally rotating our hip.

To palpate your pectineus, sit with your hips flexed. Find your pubic tubercle and the dense cord heading from the tubercle down your thigh. Press your fingers into the hollow immediately lateral to the cord, in the crease between your abdomen and thigh (Fig. 13-20b and 13-20c). Squeeze your knees together, creating an adducting force. The short band of muscle that springs up into your fingers is the only glimpse you get of your pectineus.

Adductor Brevis

Our **adductor brevis** originates from the inferior ramus of our pubic bone, just posterior to our adductor longus (Fig. 13-12b and 13-17). It then lies obliquely in our upper thigh, passing inferiorly and laterally to insert into our posterior femur along the proximal end of the linea aspera (Fig. 13-19).

As our adductor brevis contracts, it adducts and laterally rotates our hip. The origin of the muscle is too far posterior to give it an advantage in flexion of our hip.

Figure 13 -12

An anterior view of **(a)** the pectineus, **(b)** the adductor brevis, and **(c)** the adductor longus. The insertion of each muscle is dotted to indicate its posterior attachment.

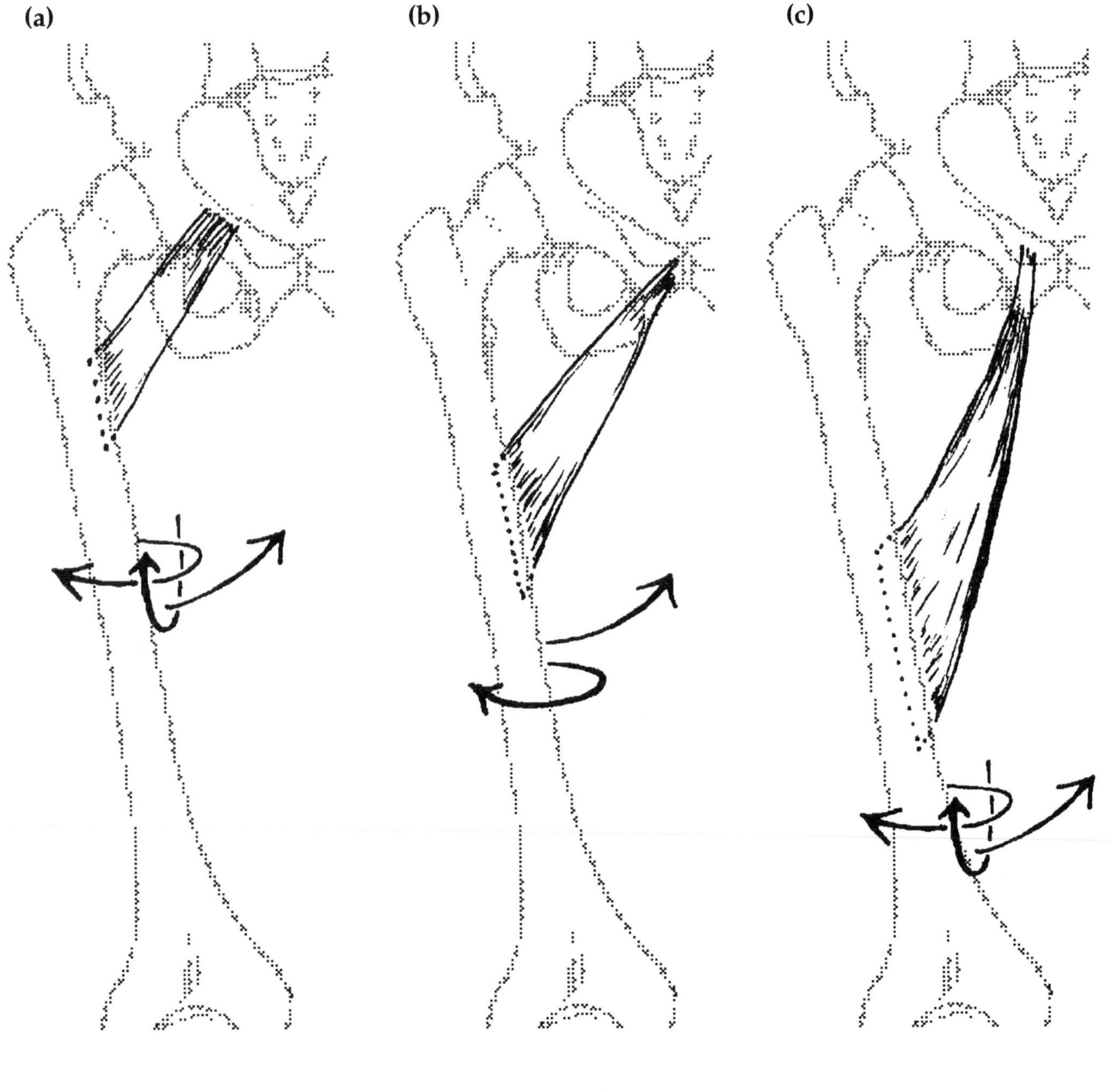

Find the origin of your adductor brevis immediately posterior to the origin of your adductor longus (Fig. 13-20a and 13-20b). Stand up. With your thumb on your adductor longus, slide your fingers backward about an inch to another dense cord heading downward from the inferior pubic ramus. This is your adductor brevis. Squeeze your knees together to make it easier to feel. It quickly plunges laterally, disappearing into the whole mass of muscle as it heads for your femur.

Adductor Magnus

'Magnus' and 'magnitude' have the same root, meaning large or great in size. Our **adductor magnus** is the largest and most posterior of our medial compartment muscles.

Our adductor magnus originates from the inferior ramus of our pubis and ischium, and our ischial tuberosity (Fig. 13-13b, 13-13c, and 13-17). It fans open to insert along the entire length of the linea aspera on our posterior

Figure 13 - 13

An anterior view of **(a)** the gracilis and **(b)** the adductor magnus, and **(c)** a posteroir view of the adductor magnus.

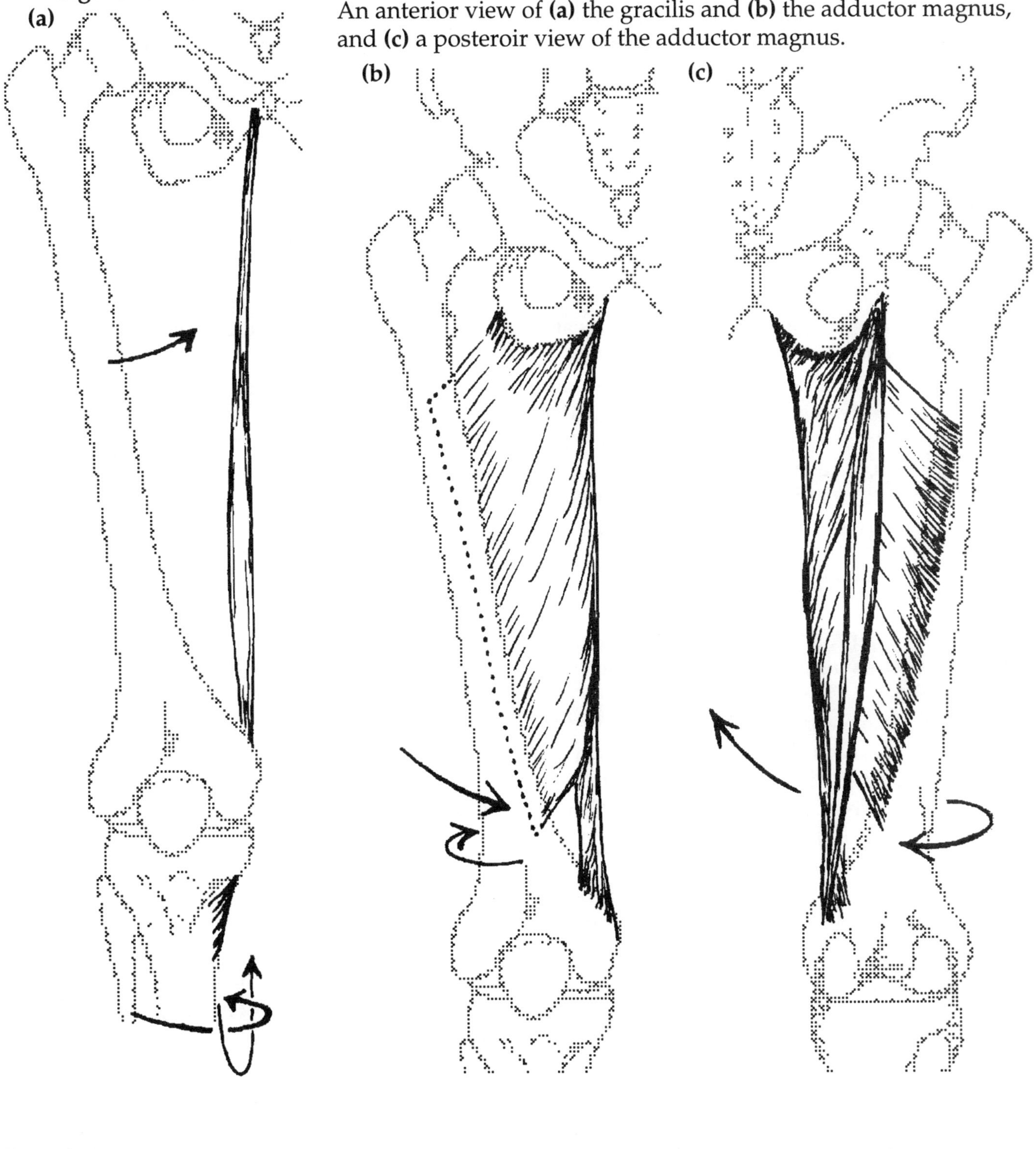

femur, from the level of our lesser trochanter to our adductor tubercle at the top of the medial epicondyle of our femur (Fig. 13-19a). Along this insertion, one small spot is skipped to make a canal for our femoral artery to pass into our posterior thigh (Fig. 13-13b and 13-13c).

As our adductor magnus contracts, it works with all our medial compartment muscles to adduct and laterally rotate our hip. Because of its more posterior origin, it also assists in extension of our hip.

To palpate the origin of your adductor magnus, begin at your adductor brevis and move backward along the inferior pubic and ischial rami. All the muscle that you cross until you come to the ischial tuberosity is your adductor magnus (Fig. 13-19b, 13-19c, and 13-19d). Follow the muscle down your thigh toward the medial epicondyle of your femur. In the middle of your thigh it is crossed by your gracilis and sartorius and becomes difficult to follow.

The distal end of your adductor magnus again is palpable. Flex your knee to create more space between the distal end of your sartorius and vastus medialis. Feel across the hollow space between these muscles, just proximal to the medial epicondyle of your femur (Fig. 13-20a, 13-20b, and 13-31). The longitudinal cord that you bump over is the distal end of the insertion of your adductor magnus.

Gracilis

'Gracilis' means slender. Our **gracilis** is a long, slender muscle. It is the most superficial and medial muscle in our medial compartment.

Our gracilis originates below our pubic tubercle, just lateral to our pubic symphysis (Fig. 13-13a and 13-17). It passes inferiorly and posteriorly across the medial edge of our adductor brevis and magnus to join our posterior compartment. It passes behind our knee to insert into our proximal anterior-medial tibia, just medial to the insertion of our sartorius (Fig. 13-32).

As our gracilis contracts, it adducts our hip and assists in medial rotation and flexion of our knee.

To palpate the orign of your gracilis, go back to the origin of your adductor longus and brevis. Separate the two tendons at your pubic tubercle by flexing your hips upward from a sitting position and squeezing your knees together. The anterior tendon was your adductor longus. The posterior and medial tendon is your gracilis. The wispy fibers that you cross on your way from your adductor longus to your adductor brevis is the continued origin of your gracilis. Follow your gracilis down your thigh and posterior to your knee (Fig. 13-14, 13-19c, 13-19d, 13-20b, and 13-20c). Here its tendon is just medial to that of your semitendinosus and superficial to that of your semimembranosus.

• The muscles of our medial compartment can be arranged into three layers from front to back (Fig. 13-9b). The most anterior layer contains our adductor longus and pectineus. The middle layer contains only our adductor brevis. The posterior layer contains our giant adductor magnus. Our gracilis descends along the medial edge of all three layers. In cross section, they look like the capital letter E.

In another sense, our medial compartment muscles all fit into two layers. Our adductor magnus forms the whole posterior layer going all the way from top to bottom, Our pectineus, adductor brevis, and adductor longus form the anterior layer, in that order from top to bottom. Our gracillis again lies over the medial edge of both layers.

POSTERIOR COMPARTMENT

The muscles in the posterior compartment of our thigh often are called our **hamstrings.** Proximally, there are three muscles (Fig. 13-9b). They are joined by our sartorius and gracilis to end as a group of five muscles.

Biceps Femoris

'Bi' means two and 'ceps' refers to heads, or in the case of muscles, origins. 'Femoris' tells us that this muscle is adjacent to our femur. We have a two-headed muscle in our leg as well as our arm. The **long head** of our **biceps femoris** originates from our ischial

'Groin' means a branching or splitting apart, like the branching of a tree or the prongs on a fork or comb. At our groin, our body splits into two legs. Each groin area also can be visualized as a comb. Our pubic bone and inguinal ligament are the transverse base or handle of the comb (Fig. 13-14). All the muscles, veins, arteries, and nerves passing longitudinally down our thigh are seen as the teeth of the comb. The order of the teeth from medial to lateral is our gracilis muscle, adductor longus muscle, adductor brevis muscle, pectineus muscle, femoral vein, femoral artery, femoral nerve, the tendon of our psoas major, our iliacus muscle, the lateral cutaneous nerve of our thigh, our sartorius muscle, rectus femoris muscle, and tensor fascia lata muscle.

Figure 13 - 14
All the groin structures emerging beneath the inguinal ligament can be seen as teeth on a comb.

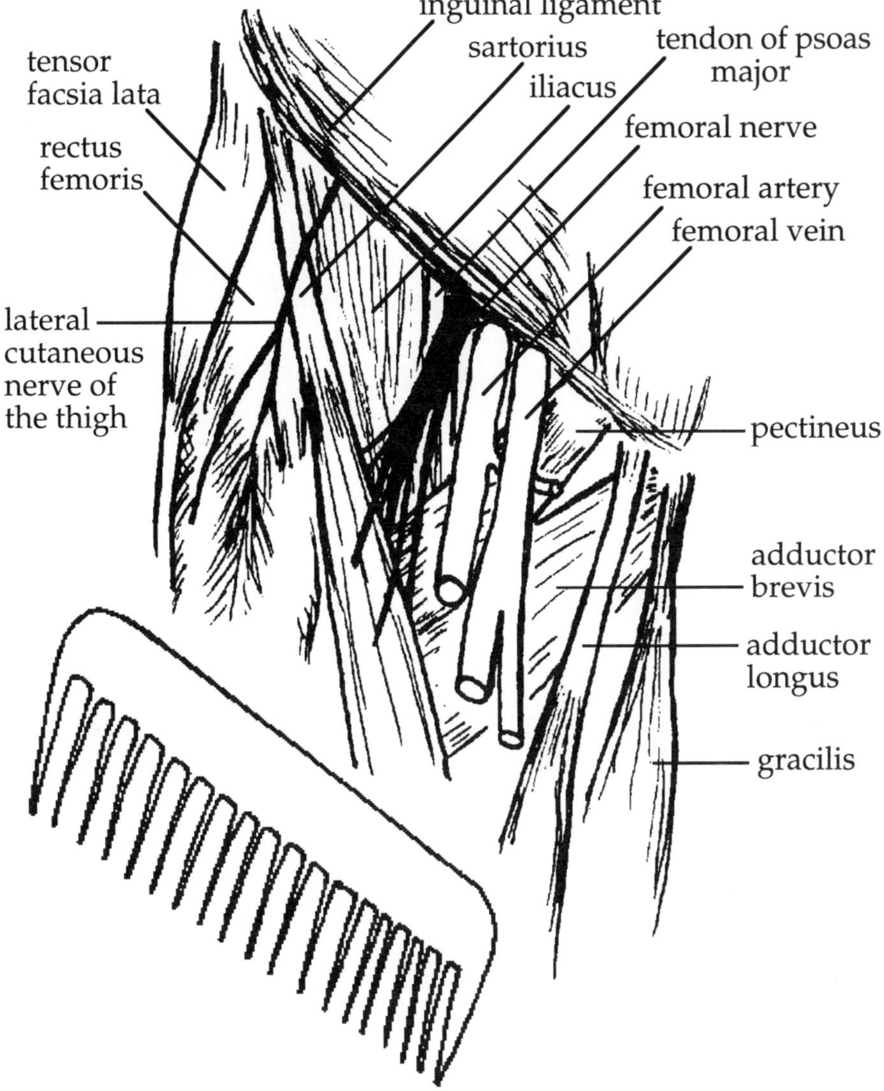

tuberosity (Fig. 13-15a and 13-17). It passes downward, dorsal to our hip and is joined by the **short head** originating from our central posterior femur, between the origin of our vastus lateralis and the insertion of our medial muscle compartment (Fig. 13-19a and 13-19b). Both heads of our biceps femoris cross ventral and lateral at our knee to insert into the head of our fibula and our lateral tibia (Fig. 13-29).

As this muscle contracts, it flexes our knee (Fig. 13-15a). When our knee is flexed, it laterally rotates our tibia over our femur. The long head of our biceps femoris also assists in extension of our hip.

Your biceps femoris is easiest to palpate at its insertion. Flex your knee to 90 degrees. Posterior to your knee on the lateral side, feel the bold tendon attached to the head of your fibula (Fig. 13-18, 13-19b, 13-19c, and 13-19d). This is the insertion tendon of your biceps femoris. From this tendon, slide medially behind your knee to dip into your popliteal fossa. From this tendon, slide anteriorly, on the lateral side of your knee, through a narrow indentation to your fascia lata. Follow this tendon proximally to its origin on your ischial tuberosity. Stand up. Bend forward as you explore. Your biceps femoris will contract to balance the force of gravity. It feels like a thick, tense cord through the back of your thigh.

• The two remaining muscles of the posterior compartment of our thigh are called the **semitendinosus** and the **semimembranosus,** a muscle collection which is half tendon and half membrane. Our semitendinosus is thin and cord-like relative to our semimembranosus which is a bulky mass all the way down to its point of insertion.

Semitendinosus

Our **semitendinosus** is the most superficial of the two muscles. It originates from our ischial tuberosity with the long head of our biceps femoris (Fig. 13-15b and 13-17). It passes down our posterior-medial thigh, with its insertion tendon lying posterior to the medial

Figure 13 - 15

A posterior view of **(a)** the biceps femoris, **(b)** the semitendinosus, and **(c)** the semimembranosus.

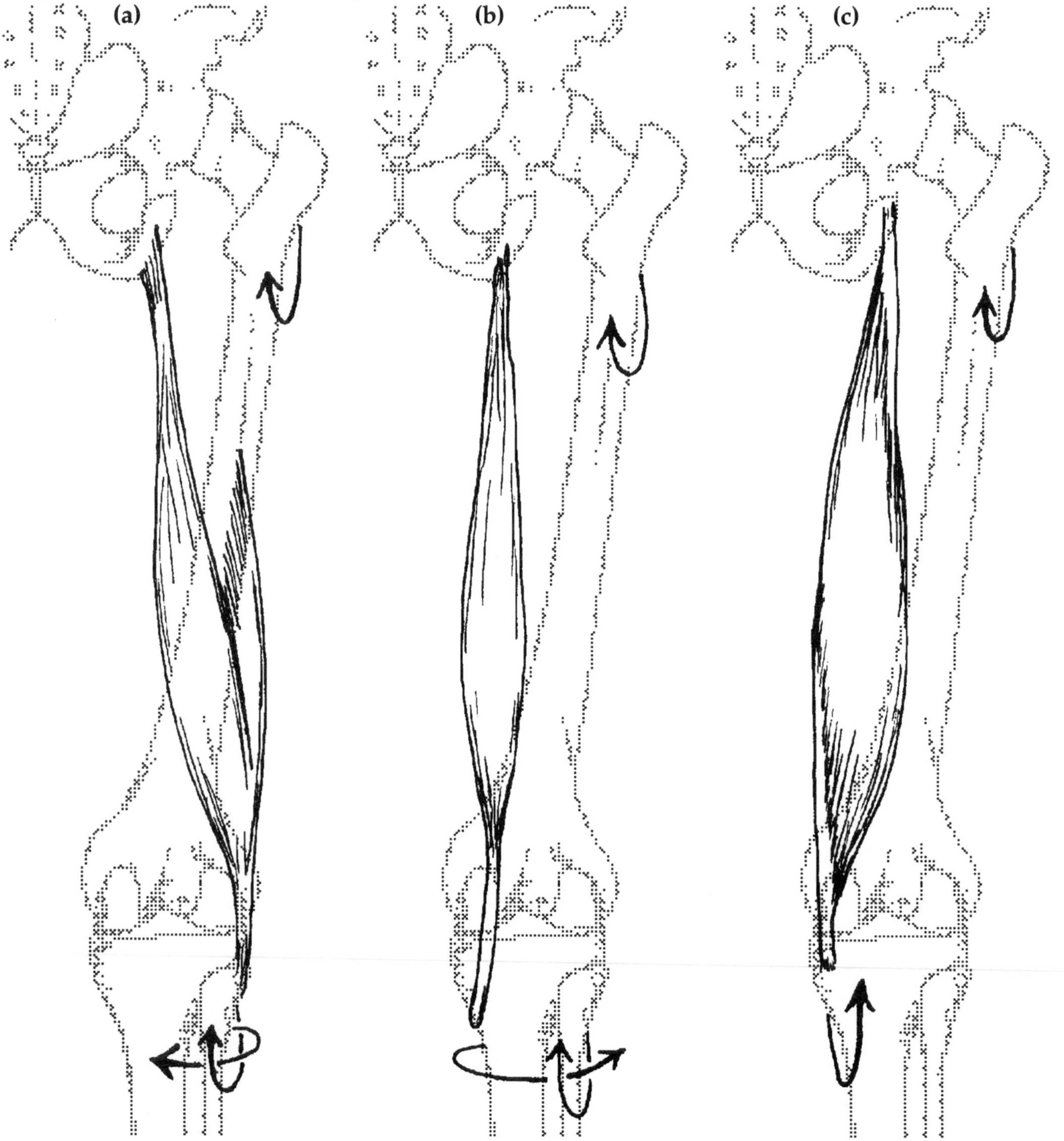

condyle of our femur and tibia. Its insertion tendon continues downward, wrapping around our medial tibia to insert into our proximal, anterior-medial tibia, just medial to the insertion of our gracilis (Fig. 13-32a, 13-19c, 13-19d, and 13-29c).

As our semitendinosus contracts, it pulls our whole leg back toward our ischial tuberosity, extending our hip and flexing our knee. Since our semitendinosus wraps around our tibia, it also medially rotates our tibia over our femur when our knee is flexed.

To palpate your semitendinosus, place your fingers in the popliteal fossa behind your knee. Flex your knee and slide your fingers medially. The first tendon you bump into is the semitendinosus (Fig. 13-19c and 13-19d). You may be able to follow the tendon a short distance toward its tibial insertion, but it tends to disappear as it wraps medial to your tibia. The skin over the proximal, anterior-medial tibia often is thickened as it protects the insertion of your semitendinosus, gracilis, and sartorius.

Go back to the tendon on the medial edge of your popliteal fossa. Follow it proximally. From a standing position, lean forward to tense the muscle and make it easier to feel. Close to your ischial tuberosity, your semitendinosus blends with your biceps femoris so that their origins on the ischial tuberosity are not distinguishable.

Semimembranosus

Our **semimembranosus** is the more bulky and deep of the two muscles. It originates from our ischial tuberosity, just lateral to the origin of our biceps femoris and semitendinosus (Fig 13-15c and 13-17). Our semimembranosus passes deep to our biceps femoris and semitendinosus (Fig. 13-19c) and continues down our thigh behind our semitendinosus, crossing posterior to the medial condyle of our femur and tibia. It inserts on our proximal, posterior-medial tibia, immediately after crossing our knee joint (Fig. 13-29a).

As our semimembranosus contracts, it assists the other hamstrings in extending our hip and flexing our knee.

Because our semimembranosus is deep and inserts so close to the joint, it does not pop out predominantly as it first contracts. Go

Figure 13 - 16

A medial view of the pelvis and thigh showing three muscles. Each muscle originates from a different pelvic bone, but all insert into the same place on the tibia, forming an inverted tripod.

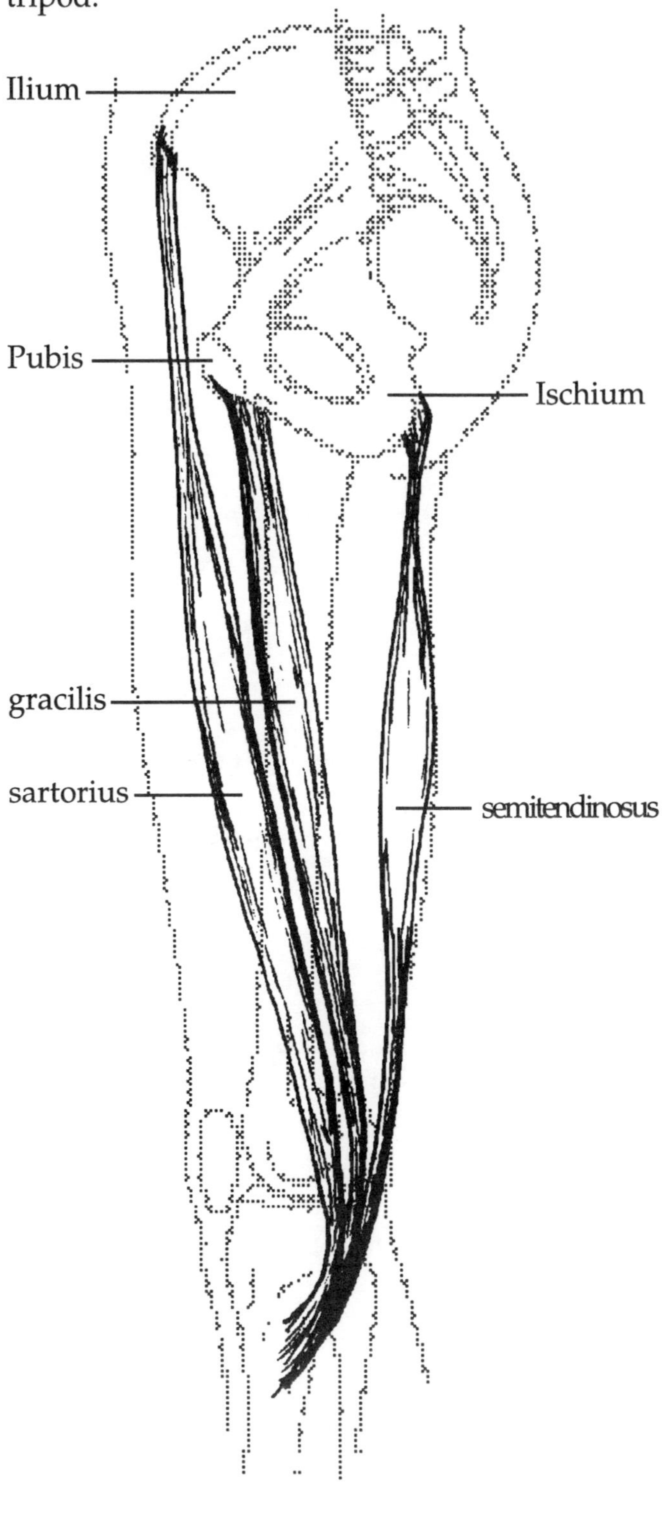

back to your semitendinosus and gracilis tendons medial to your popliteal fossa. Press your fingers deeper into the popliteal fossa. Support the medial side of your leg with your thumb. Tense these muscles. The deep thick rope is your semimembranosus. From a standing position, lean far forward at your hips so your hamstrings must work hard against gravity. The increased tension brings out the semimembranosus. It seems to drown out the gracilis. You may feel it as the predominant tendon just lateral to the tendons of your sartorius and gracilis. You may be able to follow its belly some distance proximally through your thigh, but its origin blends with that of your other hamstrings.

- You may have noticed an interesting phenomena. From each of our pelvic bones and its associated muscle compartment in our thigh, one long, thin muscle inserts into that broad flat area on our proximal anterior-medial tibia (Fig. 13-16). From our ilium and anterior thigh, we have our sartorius. From our pubis and medial thigh, we have our gracilis. From our ischium and posterior thigh, we have our semitendinosus. These three muscles form a team, shaped as an **inverted tripod.** Their job is to balance our pelvis over our tibia. Each muscle can pull as needed in its direction to maintain the balance of our pelvis, which in turn receives all the weight of our upper body through our vertebrae. Our femur becomes a floating strut which keeps the distance between our pelvis and tibia.

Figure 13 - 17

A lateral view of the pelvis indicating muscle origins with diagonally-lined patches and muscle insertions with dotted patches. Notice that the pelvis is predominantly a place for muscle origins.

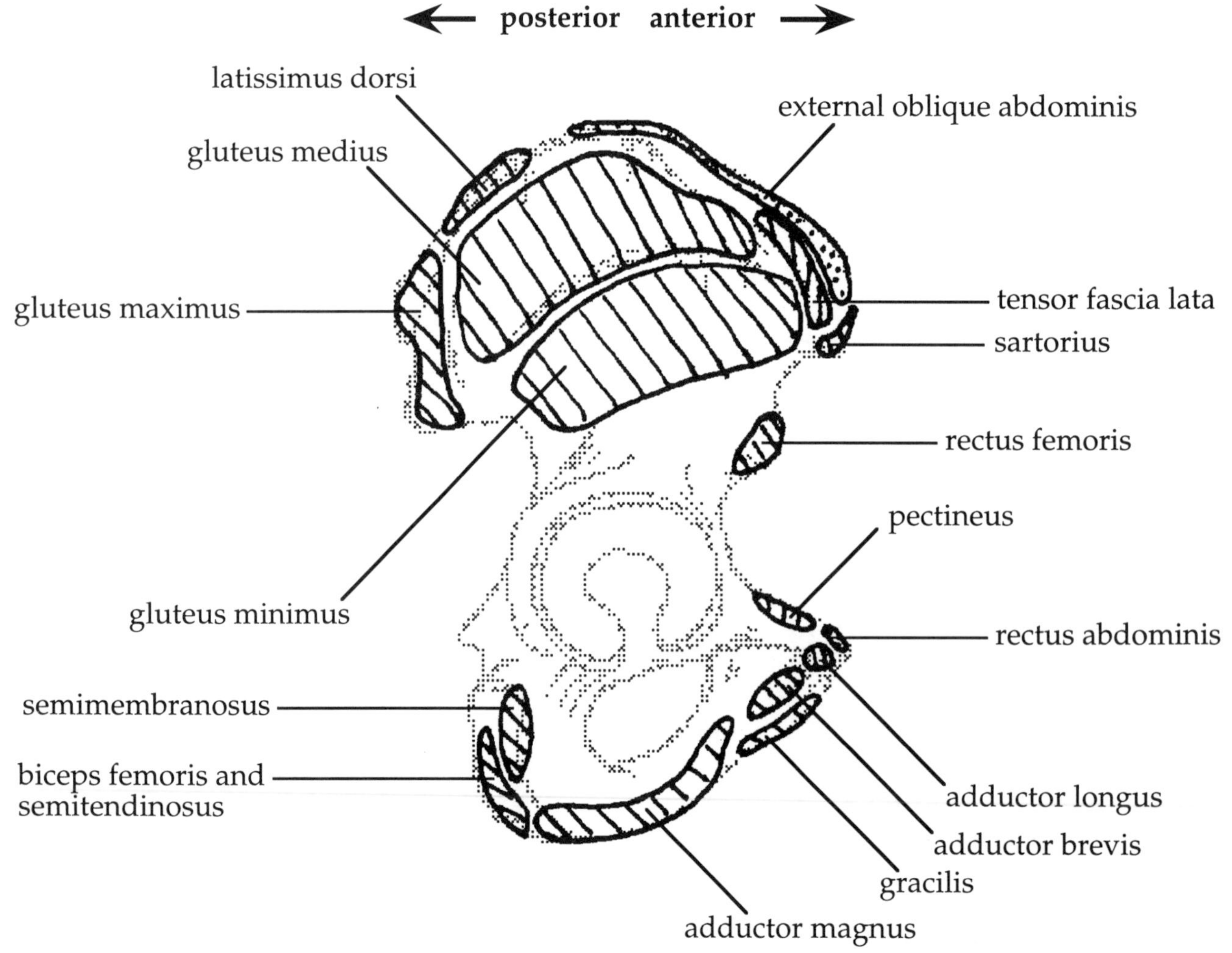

Figure 13 - 18

A lateral view of the pelvis and thigh showing **(a)** the superficial muscles and **(b)** the skin surface.

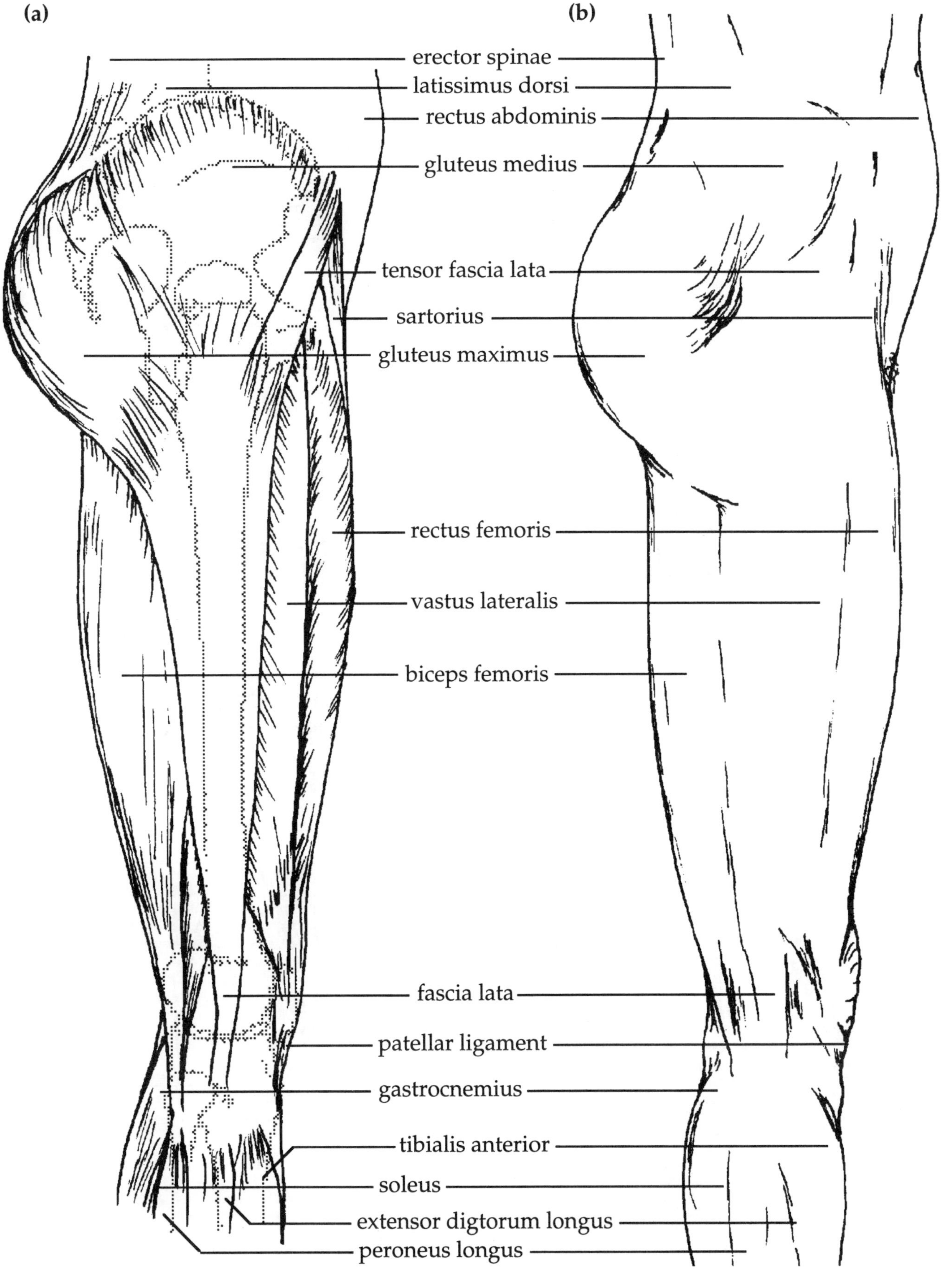

Figure 13 - 19

(a) A posterior view of the femur indicating muscle origins with diagonally-lined patches and muscle insertions with dotted patches.

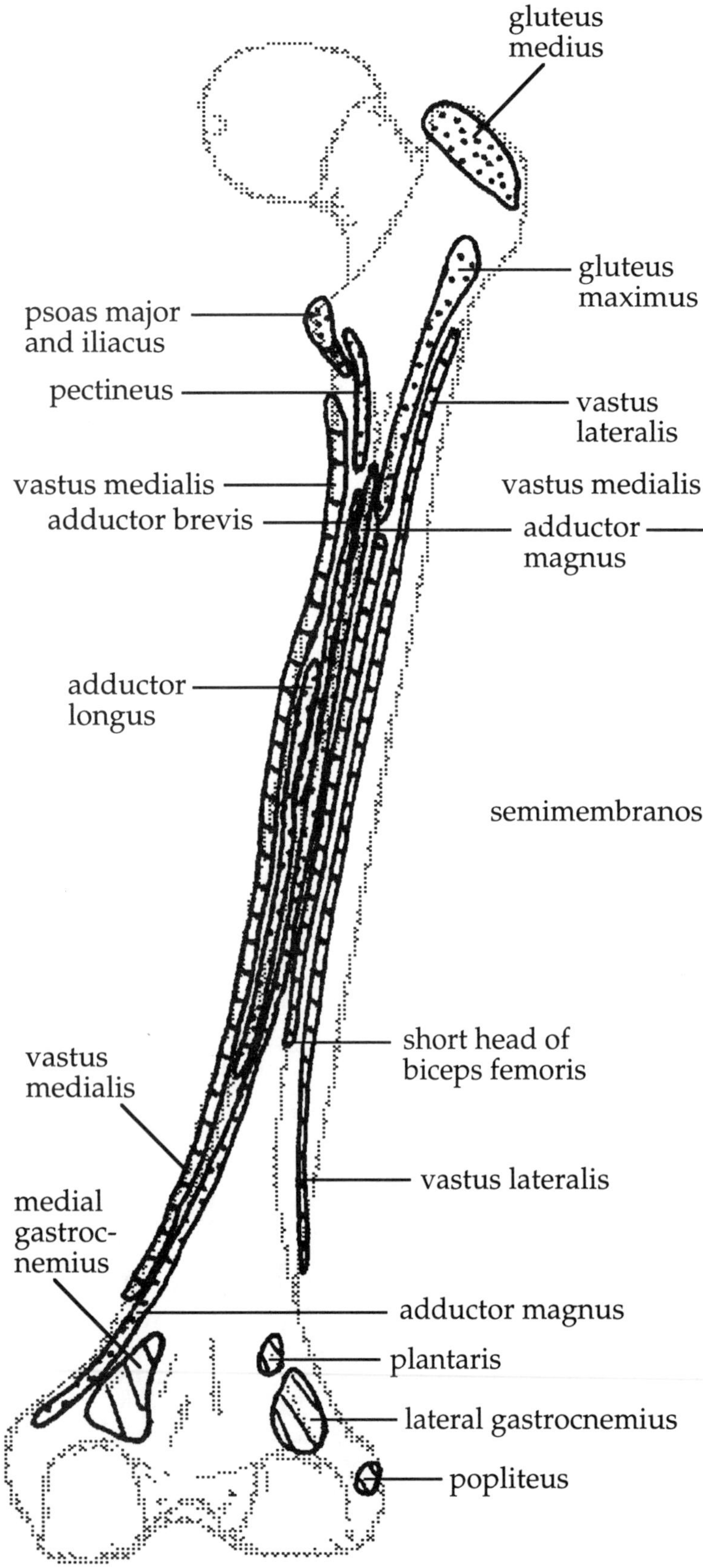

(b) A posterior view of the deep muscles of the pelvis and thigh.

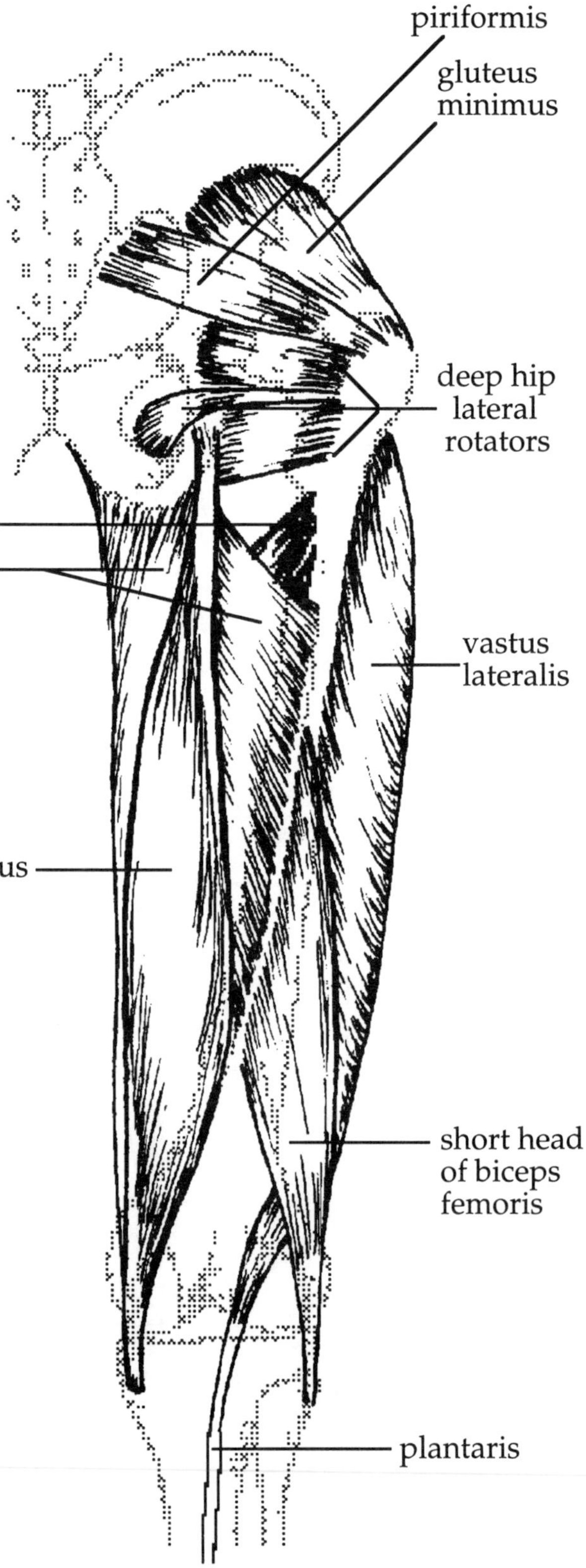

Figure 13 - 19

(c) A posterior view of the superficial muscles of the pelvis and thigh.

(d) A posterior surface view of the pelvis and thigh.

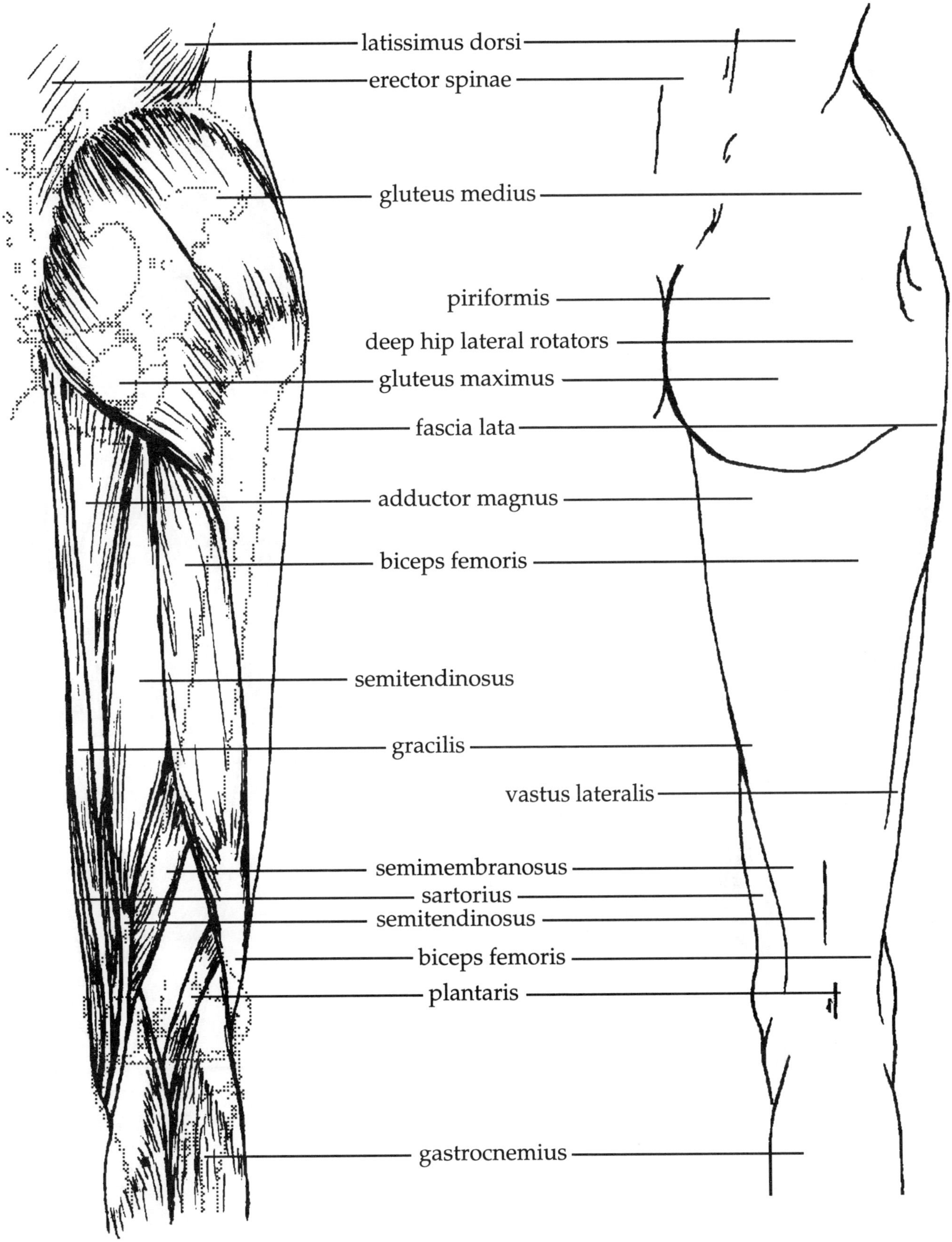

Figure 13 - 20

(a) An anterior view of the deep muscles of the pelvis and thigh.

(b) An anterior view of the superficial muscles of the pelvis and thigh.

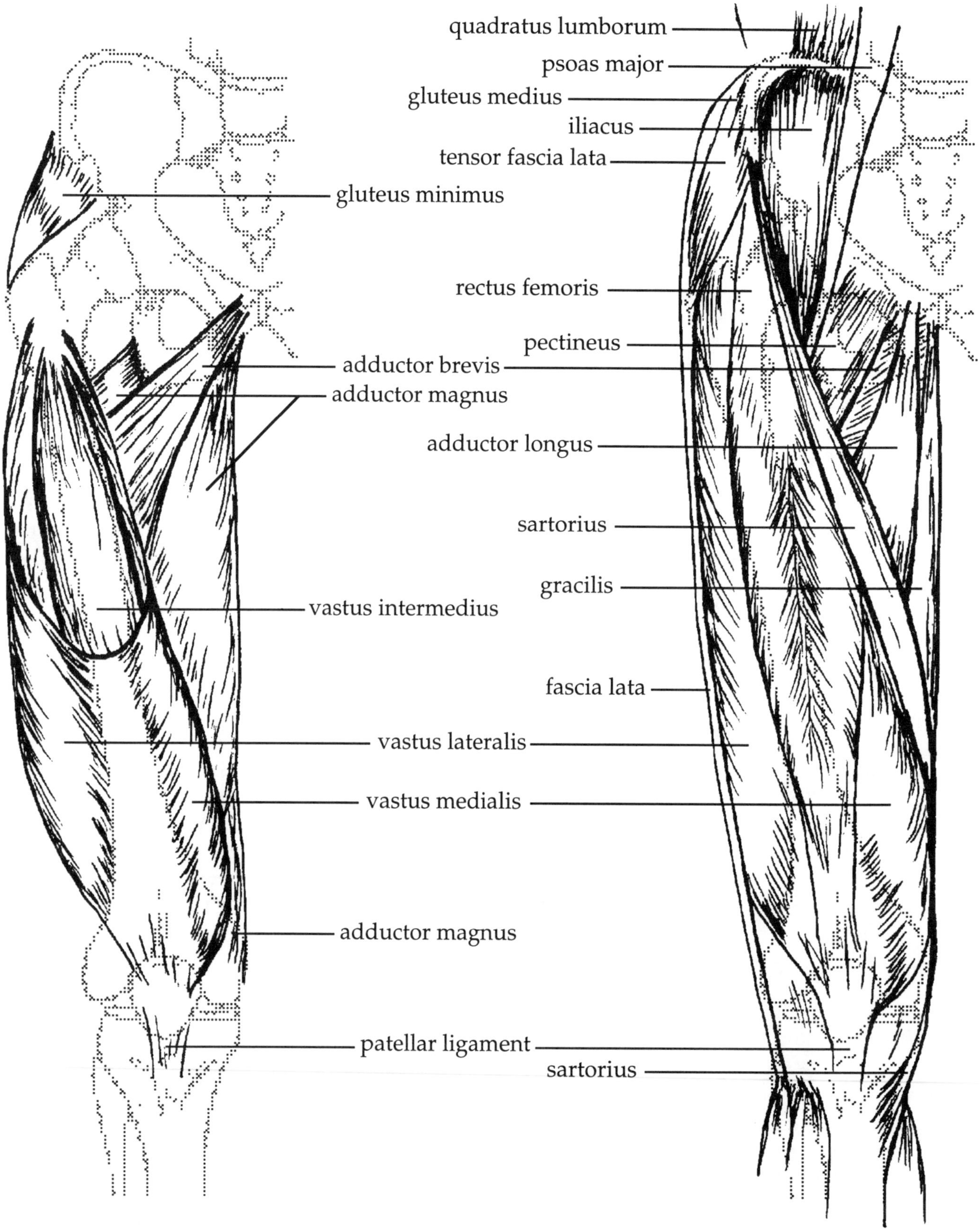

Figure 13 - 20
(c) An anterior surface view of the pelvis and thigh.

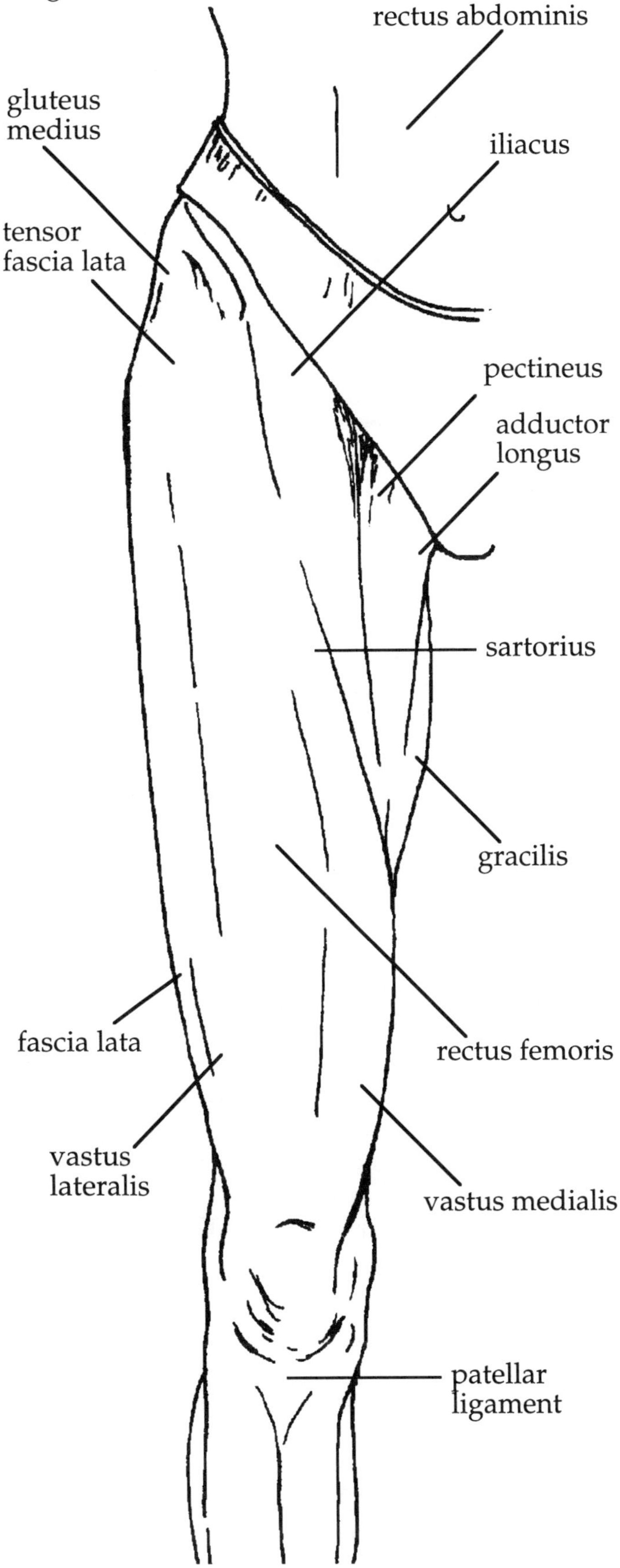

LOWER LEG MUSCLES

The muscles of our lower leg cross one or several joints all the way from our knee to the tips of our toes. These muscles all can be divided into three compartments: **posteroir, lateral,** and **anterior** (Fig. 13-21a). Our exposed tibia takes up the medial space.

Before we discuss any specific lower leg muscles, we will examine their basic action potentials.

Remember that the ventral and dorsal surfaces are twisted around our leg so that ventral is posterior and dorsal is anterior (Fig. 9). Muscles of our posterior compartment cross ventral to their joints and pull ventral surfaces to face each other, flexing our joints. Muscles of our anterior compartment cross dorsal to their joints and approximate dorsal surfaces, extending their joints. Muscles in our lateral compartment cross lateral to their joints. This generally would produce abduction, but most of the joints involved do not abduct.

- A unique motion of our foot is **eversion** and **inversion** (Fig. 9-40). Any muscle from any compartment which goes to or around the lateral side of our foot will evert it, including the muscles of the lateral compartment. Any muscle from any compartment which goes to or around the medial side of our foot will invert it. Inversion and eversion of our foot are important actions when moving over sloped or uneven surfaces. They give us an adjustment to bumps or slopes across our foot.

These are the basic muscle actions which can be deduced from the general muscle location. Specific muscles have other specific functions depending on their specific attachments.

POSTERIOR COMPARTMENT

There are seven muscles in the posterior compartment of our lower leg (Fig. 13-21b). Three muscles, our **gastrocnemius, plantaris,** and **soleus,** are superficial and long, inserting into our heel. One muscle, our **popliteus.** is deep and short, crossing only our knee. The popliteus is not shown on Figure 13-21b since it ends superior to the cut of this illustration.

Figure 13 - 21

A cross section through the lower leg showing **(a)** the three compartments and **(b)** the individual muscles in each compartment.

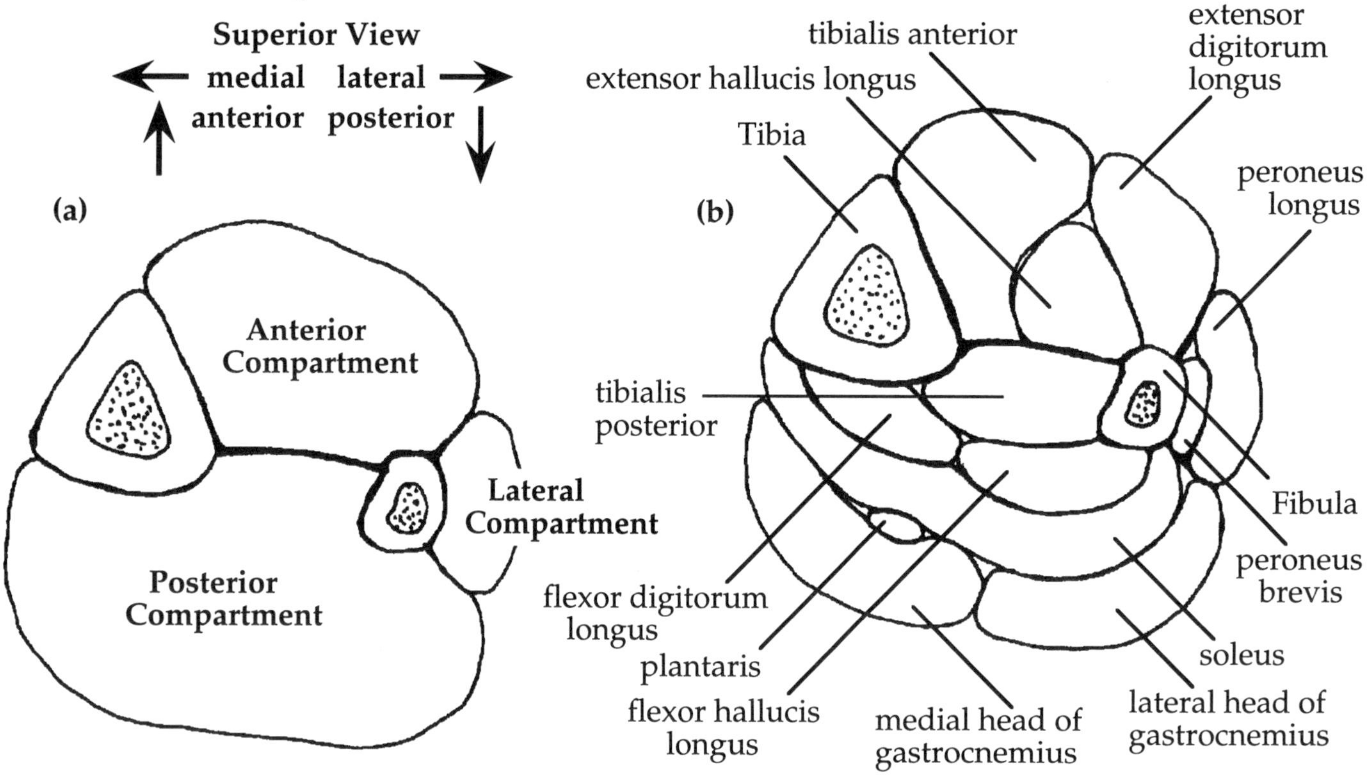

Three muscles, our **tibialis posterior, flexor digitorum longus,** and **flexor hallucis longus,** are deep and long. They cross posterior to our medial malleolus and medial to our calcaneus, into the sole of our foot.

Gastrocnemius

'Gastro' means belly. From ancient origins to present manifestations, belly is akin to bulge. 'Nemius' comes from the word for leg. So our **gastrocnemius** forms a prominent bulge on the back of our lower leg.

Our gastrocnemius has two heads. The **medial head** originates from our posterior femur, just above the medial condyle (Fig. 13-22a and 13-19a), while the **lateral head** originates from our posterior femur, above and lateral to the lateral condyle. The two bellies of our gastrocnemius bulge about half way down the back of our lower leg. At that point they join together into one common tendon which inserts onto our posterior calcaneus. This tendon is called our **achilles tendon.**

As our gastrocnemius contracts, it pulls our heel up toward our posterior thigh, producing flexion of our knee and ankle (Fig. 13-22a). Flexion of our ankle also is called plantar flexion.

The immediate origins of our gastrocnemius are covered by our posterior thigh muscles, but its bellies are superficial and palpable in our upper posterior calf (Fig. 13-29c, 13-29d, 13-30, 13-31, 13-32b, and 13-32c). Tuck the heel of your hand up behind your knee. Let your thumb and little finger wrap around opposite sides of your calf while your other fingers point down your leg. The muscle you are encompassing is the two bellies of the gastrocnemius.

Still holding the muscle, stand up on your toes. Feel this delineate the inferior edge of the bellies in the middle of your calf.

Find the dense cord standing about an inch behind your ankle bones. This is your achilles tendon. Follow it down to its attachment into your posterior superior calcaneus.

Figure 13 - 22
A posterior view of **(a)** the gastrocnemius, **(b)** the plantaris, and **(c)** the soleus along with its namesake, the sole of a shoe.

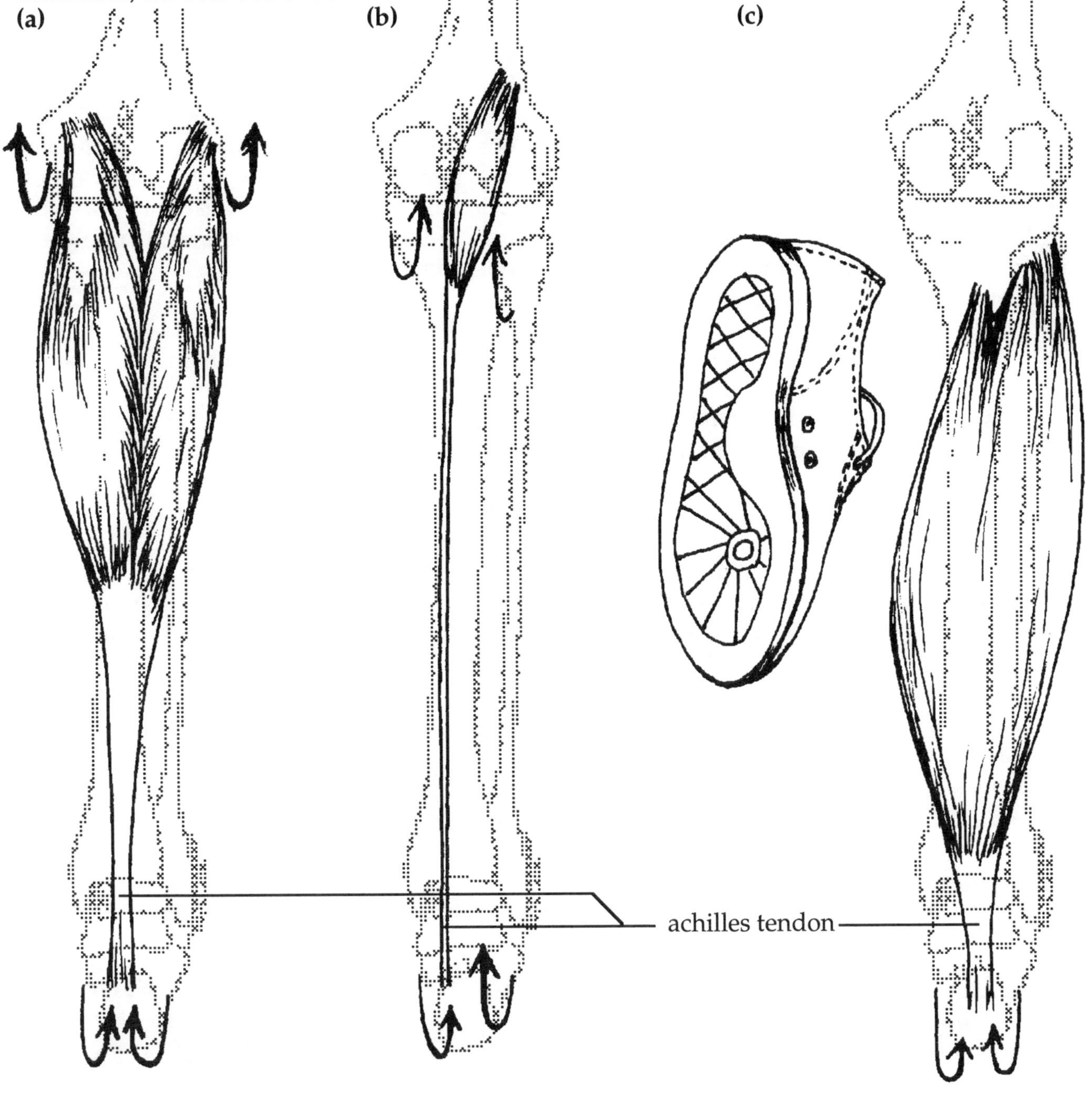

Plantaris

Our **plantaris** is our lower appendage counterpart of our palmaris longus. In many animals, it is larger than the gastrocnemius. In humans, it is vestigial.

It originates from our posterior femur, just above the lateral condyle just superior to the origin of the lateral head of our gastrocnemius (Fig. 13-22b and 13-19a). Our plantaris lies deep to our gastrocnemius and inserts into our posterior calcaneus as the medial part of our achilles tendon.

The orientation of our plantaris is similar to that of our gastrocnemius, so its action also is similar, assisting in flexion of our knee and ankle (Fig. 13-22b). The large plantaris of the four-legged population is used to lock the joints so the animals will not collapse when they fall asleep in a standing position.

Through most of its length, our

plantaris is covered by our gastrocnemius and is not palpable. Its proximal end may be palpable as it forms the lateral-inferior border of our popliteal fossa (Fig. 13-23). Flex your knee to 90 degrees. Block your lower leg so your knee cannot bend further, but continue to exert a flexing force. Feel a liitle cord and fleshy mass in the center of your popliteal fossa between the two heads of your gastrocnemius. This is the proximal part of your plantaris.

Watch from behind as someone runs away from you. Their plantaris often will bounce out prominently behind their knee.

Soleus

The root of 'soleus' means flat and solid. Our **soleus** muscle typifies these qualities in its shape and action. It is wide but flat, like the sole of a shoe or the sole fish (Fig. 13-22c). It works as a major supporter of our ankle, solidifying our stance.

Our soleus originates from our proximal posterior tibia and fibula (Fig. 13-22c and 13-29a). It inserts into our posterior calcaneus as the third contributor to our achilles tendon.

Our soleus is the strongest flexor or plantar flexor of our ankle. It is a slower acting muscle and a stalwark in balancing our body over our feet. If we start to tip forward, our soleus is stretched. When our brain receives this message, it orders our soleus to contract, flexing our ankle, pulling us back upright.

The proximal end of our soleus is covered by our gastrocnemius, but after our gastrocnemius ends in mid-calf, our soleus predominates. To palpate your soleus, locate both bellies of your gastrocnemius. Move around both sides of the distal aspect of these bellies to just in front of them, but still behind your tibia and fibula (Fig. 13-29c, 13-29d, 13-30, 13-31, 13-32b, and 13-32c). Grab the floor with your toes and lift your heel slightly to feel your soleus contract. Find the inferior edge of your gastrocnemius and move into the wad of muscle below it. This is the distal half of your soleus. Follow it down as it narrows into your achilles tendon.

> Our **popliteal fossa** is a diamond-shaped concavity formed by muscles posterior to our knee. At this point we have discussed all the muscles which outline its shape. The muscle forming its proximal lateral border is our biceps femoris (Fig. 13-19c and 13-23). The muscles forming its proximal medial border are our semitendinosus and semimembranosus. The muscles forming its distal lateral border are the lateral head of our gastrocnemius and plantaris. The muscle forming the distal medial border is the medial head of our gastrocnemius.

Popliteus

Popliteus means the ham of the knee and refers to the shape of this muscle. Our popliteus is a deep muscle behind our knee, just below our popliteal fossa.

It originates just lateral to the lateral condyle of our femur (Fig. 13-24 and 13-19a). It passes obliquely across our knee to insert into our proximal, posterior-medial tibia (Fig. 13-29a and 13-29b). The oblique orientation of our popliteus gives it a capacity for vertical and horizontal movements. Vertically, it flexes our knee. Horizontally, it rotates our tibia medially over our femur, unlocking our extended

Figure 13 - 23
A posteroir view of the knee outlining the popliteal fossa.

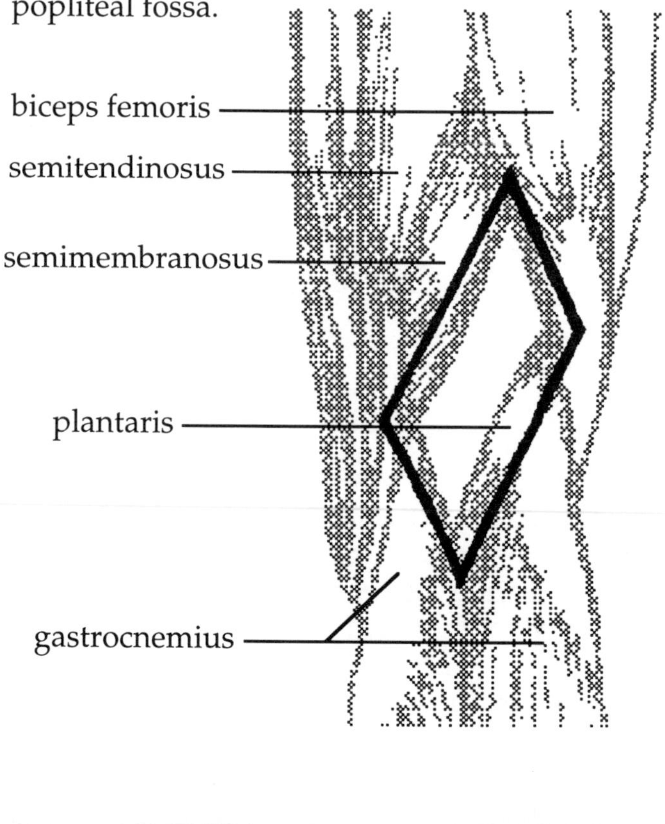

knee to initiate knee flexion.

Since our popliteus is deep to larger muscles, it is difficult to palpate clearly. From the medial side of your lower leg, press just posterior to your tibia, below the insertions of your semitendinosus. Block the medial side of your foot so it cannot move, but attempt to medially rotate the tibia. The tension which develops against your fingers is the effort of your popliteus toward medial rotation.

Figure 13 - 24
A posterior view of the popliteus muscle.

Flexor Digitorum Longus

Our **flexor digitorum longus** originates from our posterior tibia, lateral and distal to the origin of our soleus (Fig. 13-25a and 13-29a). Its belly runs down behind our tibia. Its tendon begins behind our medial malleolus and continues around our medial calcaneus to the sole of our foot. Here it divides into four tendons and insert into the distal phalanx of toes two through five.

This muscle is the counterpart of the flexor digitorum profundus in our forearm. It lies deep in the sole of our foot and inserts into the distal phalanx of our toes.

The name of this muscle tell us its action. It flexes our digits. It crosses our ankle, MP, PIP, and DIP joints, and can flex all these joints. Since our flexor digitorum longus crosses medial to our foot, it also has leverage to pull up on the medial side of our foot, inverting our foot. Since our flexor digitorum longus travels the length of our foot, it supports both longitudinal arches of our foot. Exercises which incorporate flexion of our toes strengthen and build our longitudinal arches. Examples of such exercises are picking up objects with our toes or walking barefoot, focusing on our toes for forward propulsion.

To palpate your flexor digitorum longus, go to the medial side of your lower leg. Press into the muscle immediately posterior to your tibia, about half way down (Fig. 13-29b, 31, 32b, and 32c). Curl your toes to feel the muscle contracting. Watch the space between your medial malleolus and achilles tendon (Fig. 13-40b). Extend your toes. The space bulges as the muscle belly is stretched into it. Flex your toes. The space hollows as the muscle belly contracts, pulling up toward its origin. Lay your thumb across the space. Alternately flex and extend your toes to feel the tendon move. Your flexor hallucis longus may be part of what you feel in this space. It is posterior and lateral to your flexor digitorum longus. Unless you have learned to move your big toe separate from your other toes, it is hard to distinguish the two toe flexing muscles.

Tibialis Posterior

As its name blatantly states, our **tibialis posterior** lies posterior to our tibia. It orginates from our lateral tibia, interosseous membrane, and medial fibula (Fig. 13-25b and 13-29a). Its belly lies deep in the center of the posterior compartment of our lower leg (Fig. 13-21b).

Its tendon crosses under our flexor digitorum longus, behind our medial malleolus, medial to our calcaneus, into the plantar surface of our foot. Here it inserts into our navicular bone, our three cuneiform bones, our cuboid, and the base of metatarsals two through four.

Since our medial malleolus holds our tibialis posterior behind the pivot point of our ankle, this muscle pulls the sole of our foot backward. Remember that back is ventral in our leg, so ventral surfaces are approaching each other in this movement which is called flexion or plantar flexion of our ankle. Since our tibialis posterior crosses medial to our foot, it also lifts the medial side of our foot, thus producing inversion.

The proximal end of our tibialis posterior is deep and not palpable, but its distal end emerges between our tibia and our flexor digitorum longus (Fig. 13-29b). Press into the muscle immediately behind your tibia, above your medial malleolus. Invert and flex your ankle to feel a tendon pop out against your fingers. This is the tendon of your tibialis posterior. Follow it over the posterior edge of your medial malleolus. The tendon almost slips across the malleolus as the muscle contracts. It is held back by your flexor retin-

Figure 13 - 25

A posterior view of the leg showing the **(a)** flexor digitorum longus, **(b)** tibialis posterior, and **(c)** flexor hallucis longus. **(a)** and **(c)** include an inferior view of the foot, showing the insertion tendon.

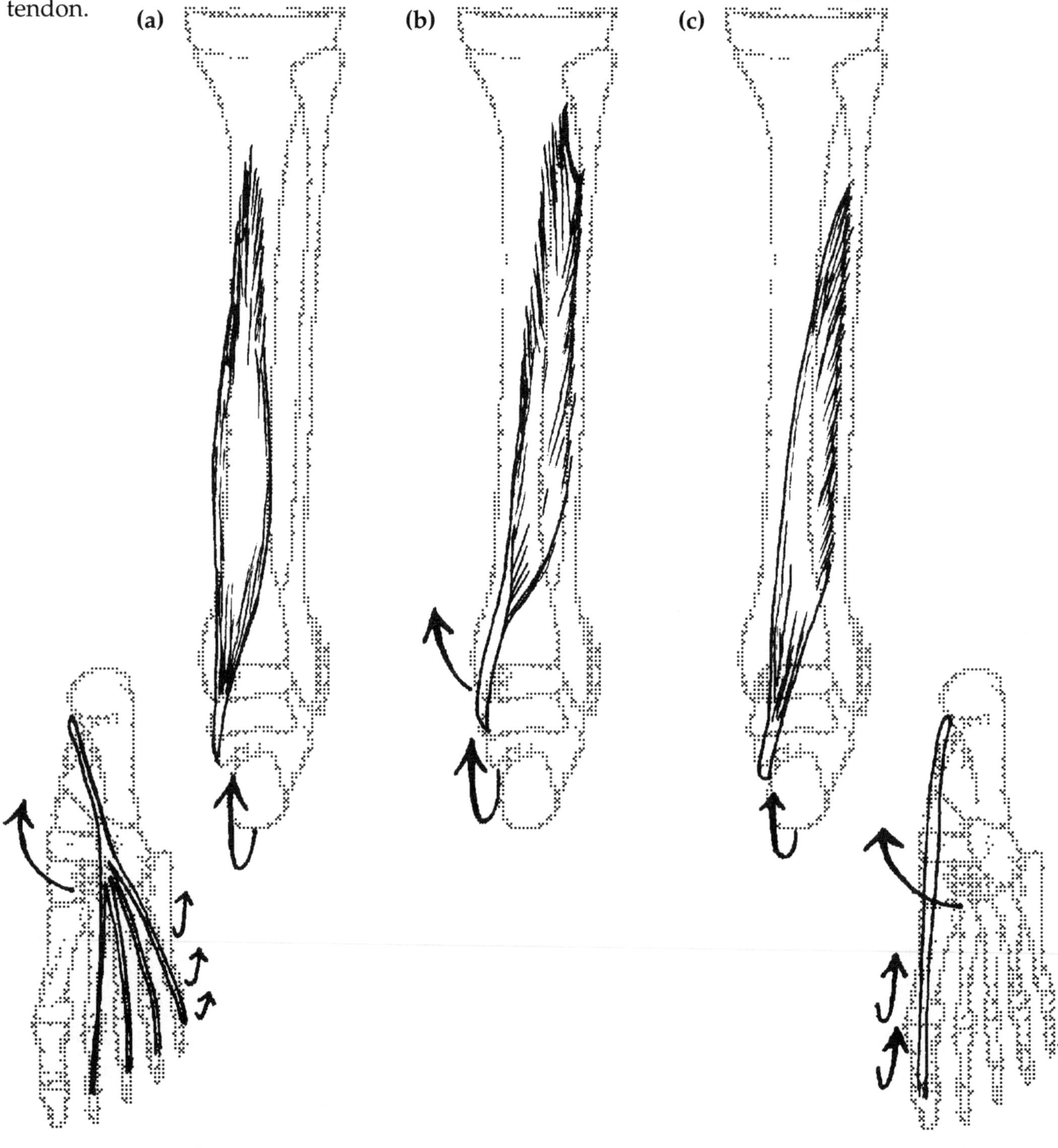

aculum, which is discussed later. To find the tendon as it continues on to your navicular bone, lengthen the space between your medial malleolus and navicular by extending and e-verting your foot. See the tendon stand out between the two bones (Fig. 13-40a and 13-40b).

Flexor Hallucis Longus

'Hallucis' is the word for big toe, so we know by its name that this muscle flexes our big toe. Our **flexor hallucis longus** originates from the lower two-thirds of our posterior fibula (Fig. 13-25c and 13-29a). Its tendon passes medially behind our talus to the medial side of our calcaneus. It runs forward through the sole of our foot to insert into the distal phalanx of our big toe (Fig. 13-39b and 13-40b).

Lying on the ventral side of our ankle and toe joints, this muscle flexes our ankle and big toe when it contracts. Crossing into the sole of our foot from the medial side, it inverts our foot and traveling the length of our foot on the medial side, it supports our medial longitudinal arch.

Most of the belly of our flexor hallucis longus is hidden by our soleus (Fig. 13-29b and 13-29c). Grasp the lower part of your soleus between your thumb and fingers. Focus lateral to your soleus as you flex and extend your toes. The movement you feel is the contraction of your flexor hallucis longus. Watch and press between your achilles tendon and lateral malleolus. Alternately flex and extend your toes to see and feel the movement of the tendon of your flexor hallucis longus.

> From medial to lateral in the lower leg, the order of the deep, posterior muscle origins is flexor digitorum longus, tibialis posterior, and flexor hallucis longus (Fig. 13-29a), while the order of their tendons behind the ankle is tibialis posterior, flexor digitorum longus, and flexor hallucis longus (Fig. 13-29b).
>
> From superior to inferior, the order of muscle palpation along the medial-posterior tibia is semimembranosus, popliteus, soleus, flexor digitorum longus, and tibialis posterior (Fig. 13-29a and 13-31).

LATERAL COMPARTMENT

There are two muscles in the lateral muscle compartment of our lower leg. The name of both muscles begins with the word peroneus. The root of 'peroneus' means a thin, piercing object like a pin or awl. This is the shape of our fibula, the bone of origin of both these muscles. This also is the shape of the two muscles. They are long and thin and their tendons seem to pierce into our foot as they approach their insertions (Fig. 13-26).

Peroneus Longus

Our **peroneus longus** is the longer of the two muscles. It originates from our lateral proximal fibula (Fig. 13-26a and 13-29a). Its tendon lies over our peroneus brevis, hooks behind our lateral malleolus (Fig. 13-29b and 13-30), wraps lateral to our calcaneus and cuboid and continues medially across the sole of our foot to insert into the ventral surface of our medial cuneiform and the base of our first metatarsal.

Our peroneus longus is held ventral to the pivot point of ankle movement by our lateral malleolus, so it flexes our ankle. It approaches the sole of our foot from the lateral side, and so everts our foot. It passes transversely across our foot, and so supports the transverse arch of our foot.

To palpate your peroneus longus, find the head of your fibula. Aim straight down the lateral side of your fibula, toward your lateral malleolus. Slide half way down your lower leg. The muscle you have just crossed is the belly of your peroneus longus (Fig. 13-29b, 13-30, and 13-32). Press your fingertips into this muscle. Evert your foot to feel the muscle contract under your fingers. Flex your ankle to delineate your soleus muscle just behind your peroneus longus. Your extensor digitorum longus is just anterior to it.

Palpate the tendon of your peroneus longus directly above the lateral malleolus of your fibula. Follow it downward over the posterior edge of your malleolus. Just below your lateral malleolus, find two tendons. Evert your foot to make them stand out. The tendon of your peroneus longus is the more posterior and inferior of the two (Fig. 13-30 and 13-33b). It disappears into the sole of your foot.

Peroneus Brevis

Our **peroneus brevis** is the shorter of the two muscles in the lateral compartment of our lower leg. Its origin comes later and its insertion comes sooner than that of our peroneus longus.

Our peroneus brevis originates from our distal lateral fibula (Fig. 13-26b and 13-29a), deep to the tendon of our peroneus longus. Its tendon hooks behind our lateral malleolus and then runs forward along the lateral side of our foot to insert into the base of our fifth metatarsal (Fig. 13- 33b and 13-40a).

The actions of our peroneus brevis are similar to those of our peroneus longus, flexing our ankle and everting our foot. With its brief longitudinal orientation on the side of our foot, our peroneus brevis lends support to the lateral longitudinal arch of our foot.

To palpate your peroneus brevis, squeeze on either side of your peroneus longus tendon over the distal half of your fibula (Fig. 13-30). Evert your foot to feel your peroneus brevis working.

Figure 13 - 26
A lateral view of the lower leg showing **(a)** the peroneus longus and **(c)** the peroneus brevis, each with their namesake awl. **(b)** An inferior view of the foot showing the distal end of the peroneus longus tendon.

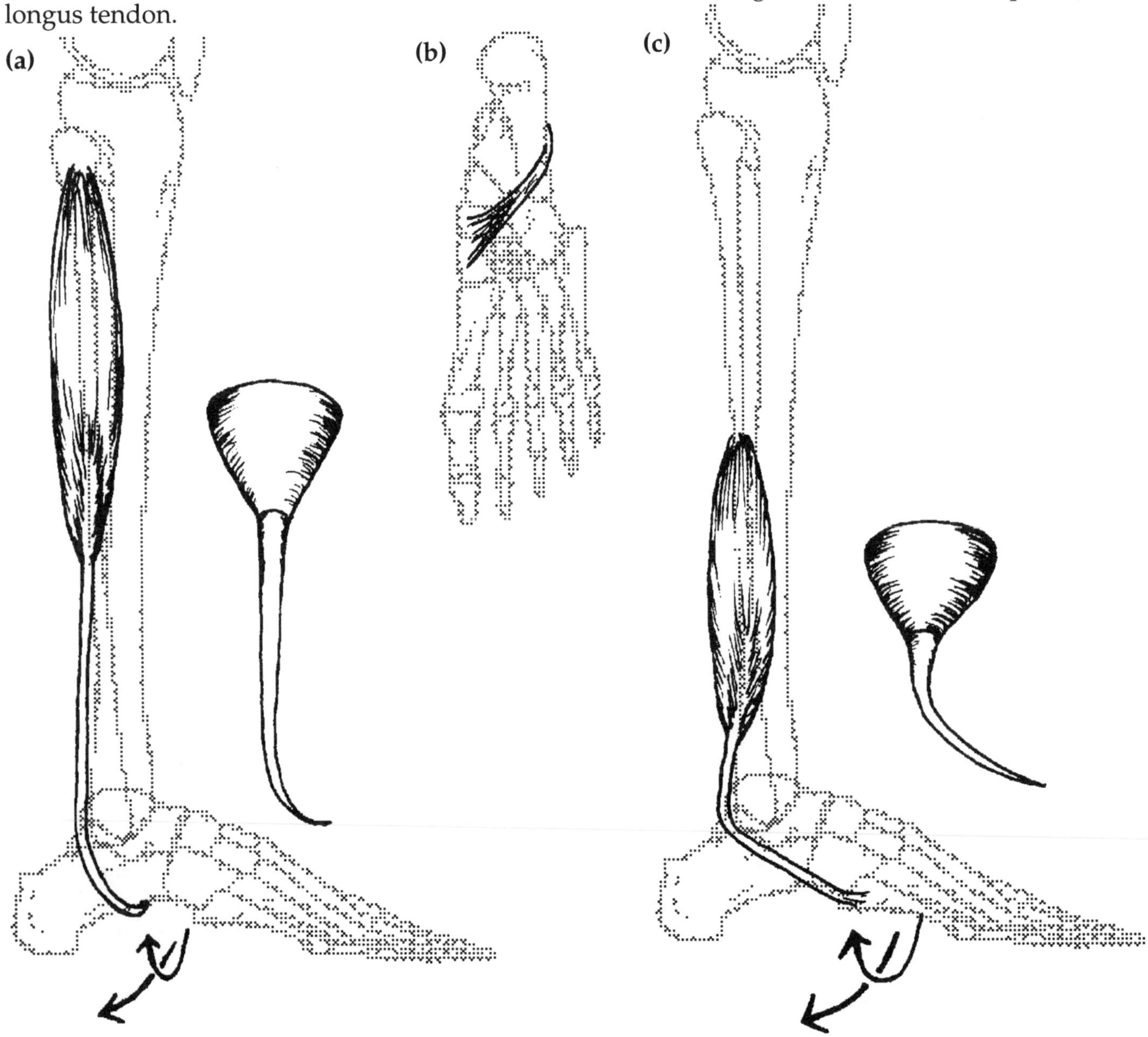

The most obvious tendon over your lateral malleolus is the tendon of your peroneus longus. By flexing your ankle and pushing backward on the tendon of your peroneus longus, you may be able to expose a deeper, slightly anterior tendon. This is the tendon of your peroneus brevis (Fig. 13-30).

Below your lateral malleolus, the tendon of your peroneus brevis is the most obvious of the two palpable tendons. Find it anterior and superior to the tendon of your peroneus longus and follow it along the side of your foot to the protruding base of your fifth metatarsal (Fig. 13-33b).

ANTERIOR COMPARTMENT

There are four muscles in the anterior compartment of our lower leg. Three of them, our **tibialis anterior, extensor digitorum longus,** and **extensor hallucis longus,** are the antagonistic mates to the three deep, long muscles in the posterior compartment of our lower leg. One of these muscles, our **peroneus tertius,** is the antagonistic mate to our other peroneal muscles.

Tibialis Anterior

The name **tibialis anterior** tells us about the location of this muscle. Our tibialis anterior originates from the lateral side of our tibia and our anterior interosseus membrane (Fig. 13-27a and 13-32a). Its belly lies just lateral to the shaft of our tibia and its tendon crosses anterior to the distal end of our tibia to insert into the dorsal aspect of our medial cuneiform and the base of our first metatarsal just opposite the insertion of our tibialis posterior and peroneus longus (Fig. 13-40.

As our tibialis anterior contracts, it pulls up on the dorsum of our foot, extending or dorsiflexing our ankle (Fig. 13-27a). Its pull is focused on the medial side of the foot so it may also invert our foot. Its upward pull on our medial tarsal and metatarsal bones also supports our medial longitudinal arch.

To palpate your tibialis anterior, follow the anterior ridge of your tibial shaft all the way from its lateral epicondyle to half way down your shin. The muscle immediately lateral to this ridge is your tibialis anterior (Fig. 13-30, 13-31, 13-32b, and 13-32c). Extend your ankle to feel it contract. Follow the tendon downward. About one handwidth above your ankle, feel how the tendon crosses in front of your tibia, heading boldly across your ankle to its medial insertion. Extend your ankle to feel the whole tendon clearly.

> Three muscles converge on the medial tarsal and metatarsal bones of our foot: our tibialis posterior from our posterior compartment, our peroneus longus from our lateral compartment, and our tibialis anterior from our anterior compartment (Fig. 13-31). They form a stirrup which supports all the arches of our foot. They also provide the qualities of an inverted tripod, balancing our lower leg over our foot.

Extensor Digitorum Longus

The name **extensor digitorum longus** tells us about the action of this muscle and from that we can make an educated guess about its location. As an extensor of our toes, it must lie dorsal to our toes. Since it is long, it probably begins in our lower leg rather than our foot.

Our extensor digitorum longus originates from the lateral epicondyle of our tibia and our proximal, anterior fibula (Fig. 13-27b, 13-30, and 13-32). It passes dorsal to our ankle and runs the length of our foot to insert into the dorsal surface of the middle and distal phalanges of toes two through five.

As its name implies, our extensor digitorum longus extends toes two through five (Fig. 13-27b). Crossing dorsal to our ankle, it assists in extension of our ankle, and since it is predominantly lateral in our foot, it assists in eversion of our foot.

Find the belly of your extensor digitorum longus in your proximal lower leg, between your tibialis anterior and peroneus longus (Fig. 13-30 and 13-32). Alternately extend and evert your foot to distinguish these two boundary muscles. Extend your ankle and toes to feel your extensor digitorum longus work.

While working the muscle, feel its tendon stand out in front of your ankle, just lateral to the tendon of your tibialis anterior (Fig. 13-40a). See how the tendon divides into four branches in the dorsum of your foot. Follow one branch into each of your four lateral toes.

Extensor Hallucis Longus

The name **extensor hallucis longus** tells us that this muscle is long and extends our big toe. With this much information, where should we go looking for the muscle? To be long it must start in our lower leg. To extend it must be dorsal to our big toe.

Our extensor hallucis longus originates from our anterior fibula and adjacent interosseous membrane, inferior and medial to our extensor digitorum longus (Fig.13-27c and 13-32). Its belly lies deep to our tibialis anterior and extensor digitorm longus and its tendon emerges between their two tendons just in front of our ankle (Fig. 13-40a). It runs the length of our foot to insert into the distal phalanx of our big toe.

As our extensor hallucis longus contracts, it extends the MP and IP joints of our big toe (Fig. 13-27c) and assists in extension of our ankle. Its medial location allows it to also invert our foot.

The belly of your extensor hallucis lon-

Figure 13 - 27
An anterior view of **(a)** the tibialis anterior, **(b)** the extensor digitorum longus, **(c)** the extensor hallucis longus, and **(d)** the peroneus tertius.

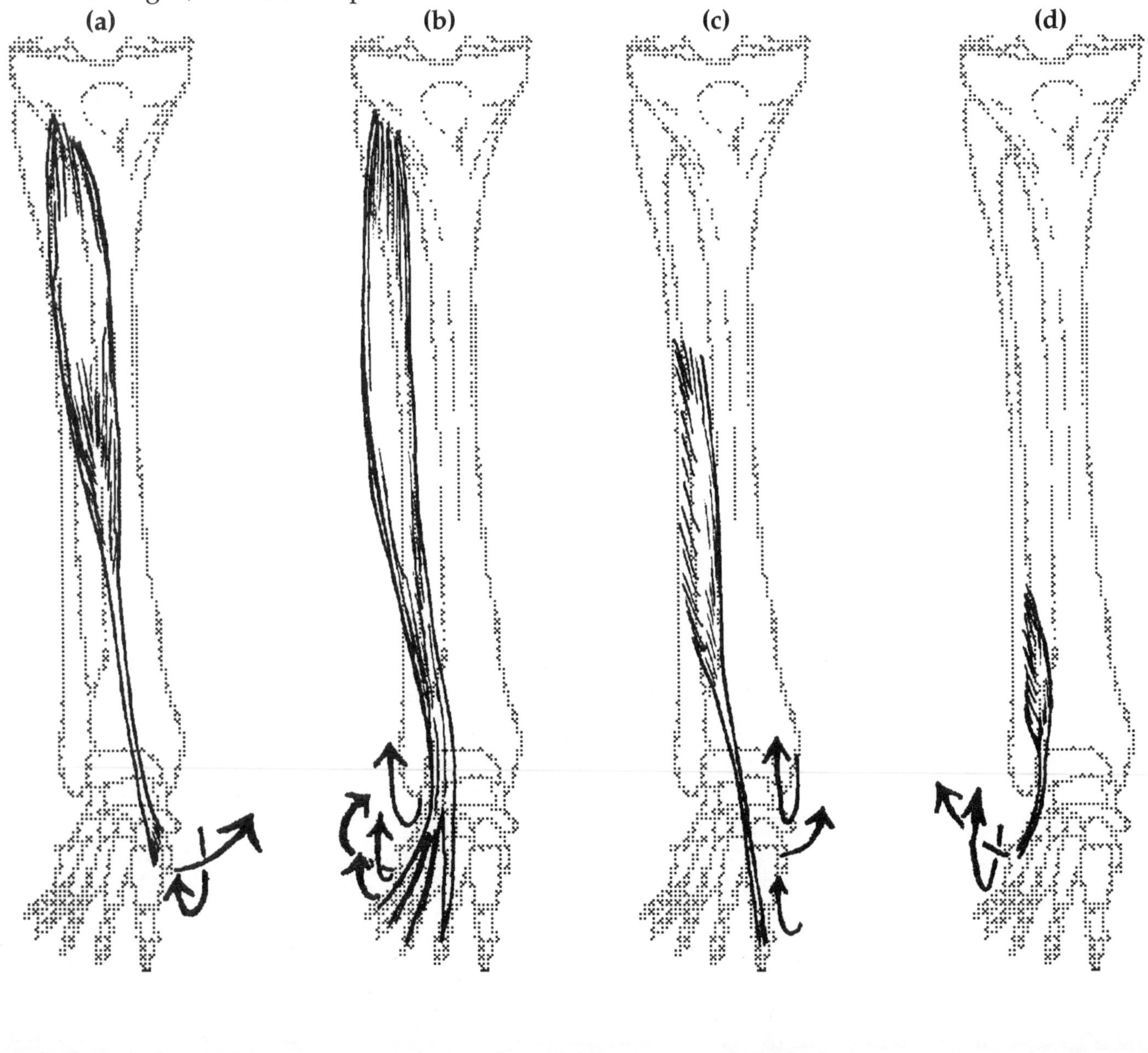

gus is deep and hard to palpate but you can begin to feel its tendon in front of your ankle. Press there, between the tendons of your tibialis anterior and extensor digitorum longus (Fig. 13-32b and 13-32c). Extend your big toe. Feel the tension in the tendon of your extensor hallucis longus and see it stand out across your ankle, the dorsum of your foot, and the proximal phalanx of your big toe.

Relax your toe and strum across the tendon on top of the foot. It is slack and mobile. Lightly extend your big toe again and feel the tendon snap to attention with the first sign of movement.

Peroneus Tertius

The root of 'tertius' is three. We have already discussed two muscles which are named by their origins from our stick-pin fibula, this muscle is the third one. Our **peroneus tertius** orginates from our distal anterior fibula and the adjacent interosseous membrane (Fig. 13-27d and 13-32a). It crosses dorsal to our ankle to insert into the dorsal surface of the base of our fifth metatarsal. With this position, our peroneus tertius extends our ankle and everts our foot as it contracts.

The belly and tendon of your peroneus tertius is continuous with the inferior and lateral aspect of the extensor digitorum longus and difficult to differentiate from it. Find the lateral malleolus of your fibula. Follow your fibula upward several inches until it is buried in muscle. Just as your fibula disappears, slip into the muscle in front of it. This is the belly of your peroneus tertius (Fig. 13-30 and 13-32). Extend your ankle and evert your foot to feel it contract.

Across your ankle, the tendon of your peroneus tertius is not distinguishable from that of your extensor digitorum longus, but as the latter divides to go to the toes, the peroneus tertius goes its own way to the base of your fifth metatarsal. Feel the transverse orientation of your peroneus tertius tendon as it crosses your foot, between the tendon to your little toe and the base of your fifth metatarsal (Fig. 13-30, 13-33b, and 13-40a). Evert your foot to tense the tendon and make it more palpable.

As we step forward, we automatically extend our ankles and toes to clear our foot over the approaching surface. Weakness in the anterior muscle compartment of our lower leg leaves our foot dragging and leads to stumbling and stubbing of our toes. This is called **footdrop.**

RETINACULA

Similar to our hand, our foot has many muscles which begin in our lower leg, cross our ankle, and insert some distance into our foot. As our ankle bends, these muscles would tend to take the shortest distance between their origin and insertion, webbing out away from our ankle bones (Fig. 13-28). This condition would stretch out our socks, create a nightmare for shoemakers, and reduce the effectiveness of our muscles' actions. To avoid this mess, our ankle has dense bands of tough,

Figure 13 - 28

Tendons crossing dorsal to the ankle will tend to web out from the ankle as it extends.

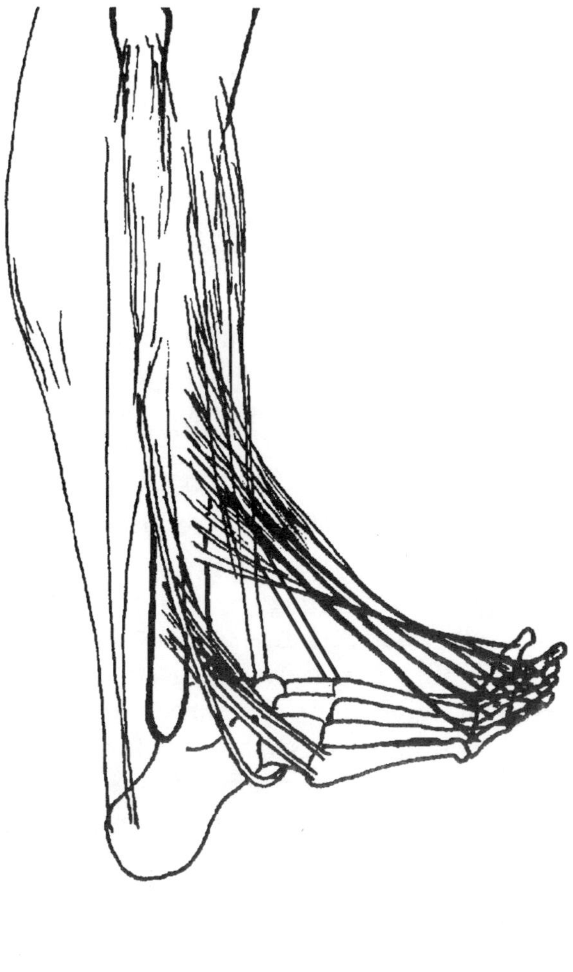

nonstretchy connective tissue which wrap around it like a bandage, holding all our tendons against our bones (Fig. 13-30, 13-31, and 13-32b). These bands are called **retinacula,** as in our hand, although they are tougher and more extensive in our foot because of increased weights and forces. Each compartment of our lower leg is secured by its own retinaculum.

In our posterior compartment, the three superficial muscles inserting through our achilles tendon have no retinaculum. They are allowed to stand away from the bone (Fig. 13-30), increasing their mechanical advantage for ankle flexion. Relative to muscled beings, skeletons seem to have long heels since they are missing the musculo-tendonous web behind their ankles.

The three deep muscles are held behind our medial malleolus by a retinaculum. It bridges across the tendons, attaching to the medial malleolus of our tibia and our medial calcaneus (Fig. 13-31). It is called the **flexor retinaculum** since the muscles it constrains flex their joints.

The two peroneal muscles of the lateral compartment of our lower leg are held behind the lateral malleolus of our fibula by our **peroneal retinaculum** which crosses from our lateral malleolus to the lateral side of our calcaneus (Fig. 13-30).

Our extensor muscles pose the greatest threat because the range of extension is greater than the movement in any other direction. Two retinacula hold these muscles down against their bones. Our **superior extensor retinaculum** is two or three inches wide and crosses over our extensor tendons just above our ankle, attaching firmly into our anterior tibia and fibula (Fig. 13-30, 13-31, and 13-32b). Our **inferior extensor retinaculum** is a Y-shape which droops over our proximal foot. The stem of the Y begins on the lateral side of our foot imbedded into our lateral calcaneous (Fig. 13-30). It continues up and over all the extensor tendons. The upper arm of the Y attaches into the medial malleolus of our tibia. The lower arm falls around the medial side of our foot to blend into the dense connective tissue of the sole of our foot (Fig. 13-31).

Figure 13 - 29

(a) A posterior view of the tibia and fibula showing muscle origins in diagonally-lined patches and muscle insertions in dotted patches. ← **medial lateral** →

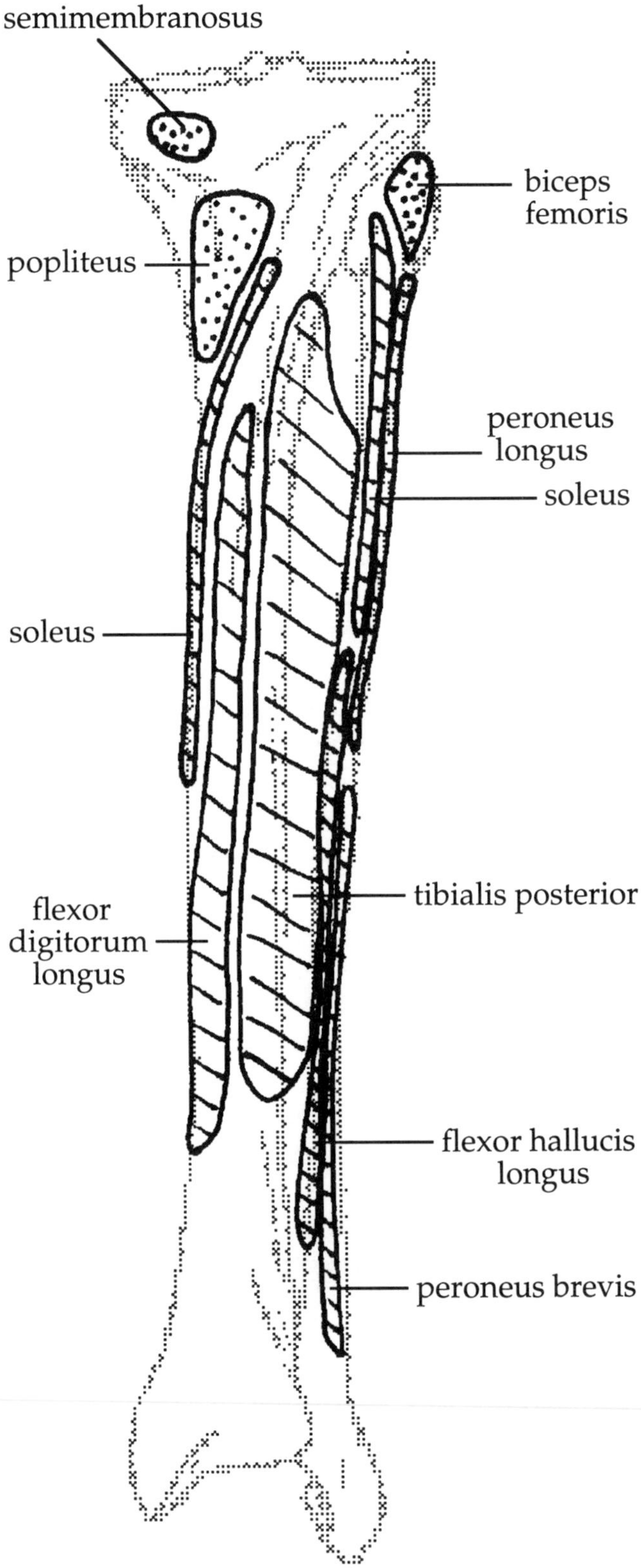

Figure 13 - 29

(b) A posterior view of the lower leg showing the deeper muscles.

(c) A posterior view of the lower leg showing the superficial muscles.

(d) A surface view of the posterior lower leg.

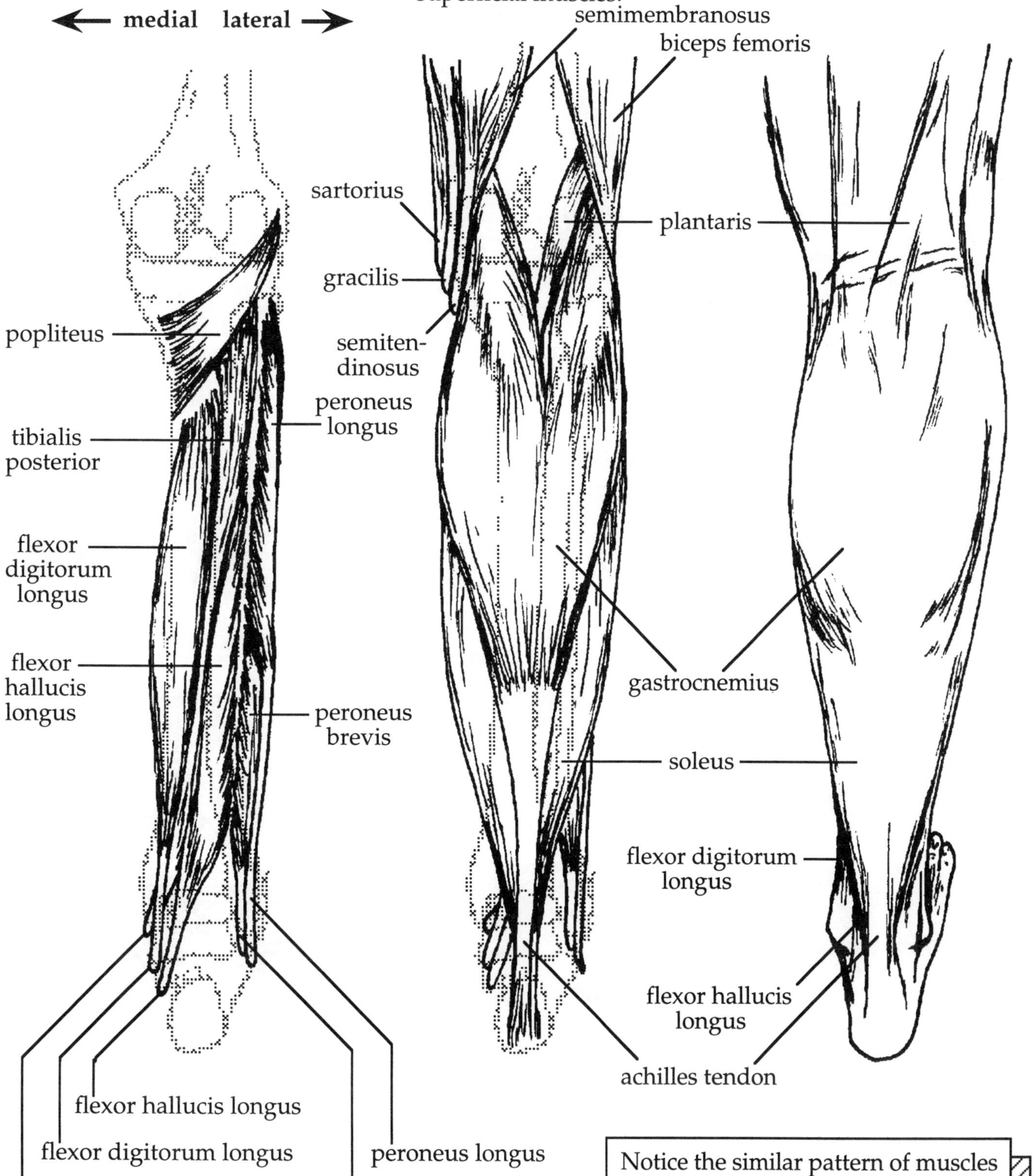

Notice the similar pattern of muscles around the medial knee (Fig. 13-29c) and the medial ankle (Fig. 13-29b).

Figure 13 - 30

A lateral view of the muscles of the lower leg.

Figure 13 - 31

A medial view of the muscles of the lower leg.

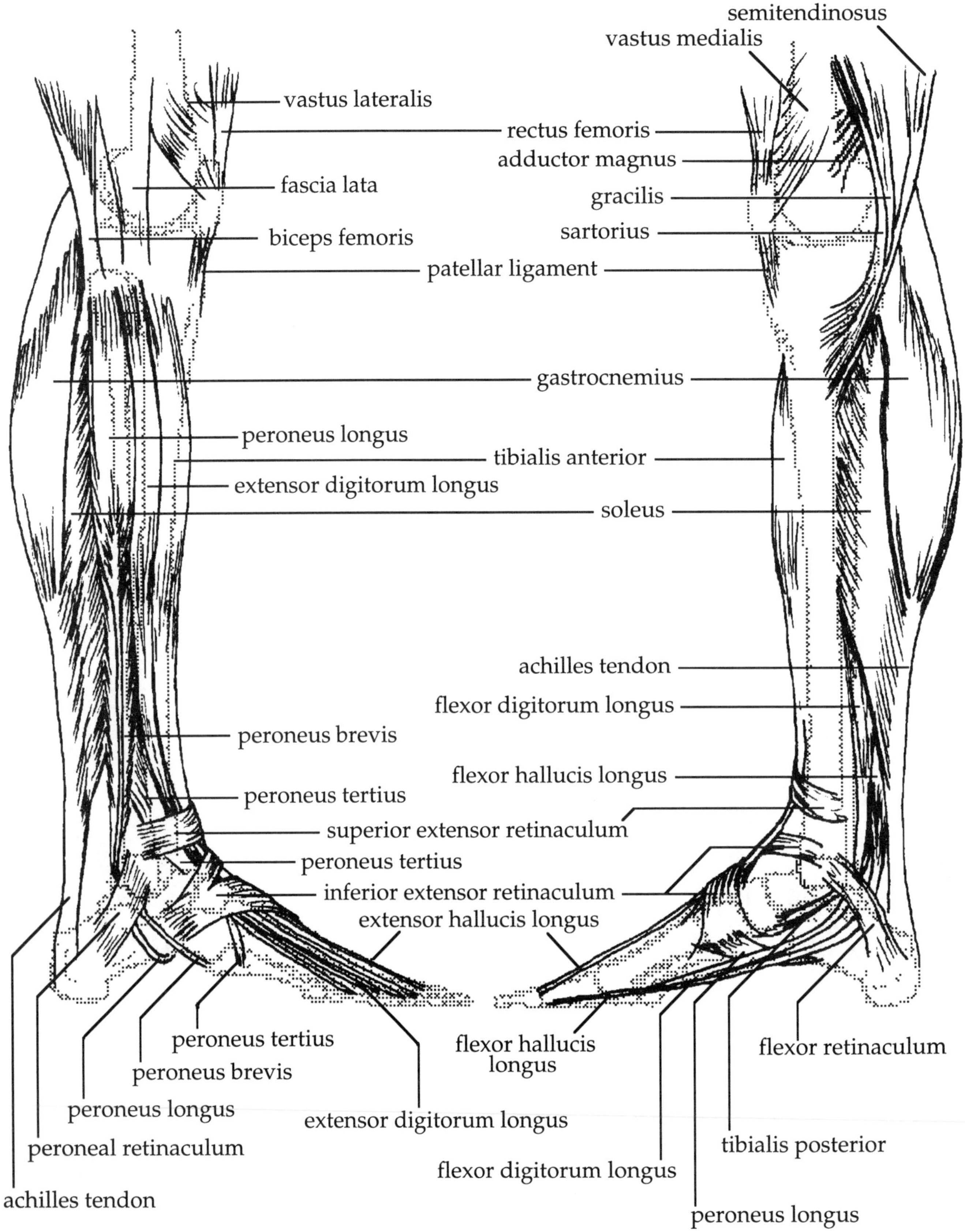

Figure 13 - 32

(a) An anterior view of the tibia and fibula showing muscle origins in diagonally-lined patches and muscle insertions in dotted patches.

(b) An anterior view of the muscles of the lower leg.

(c) A surface view of the anterior lower leg.

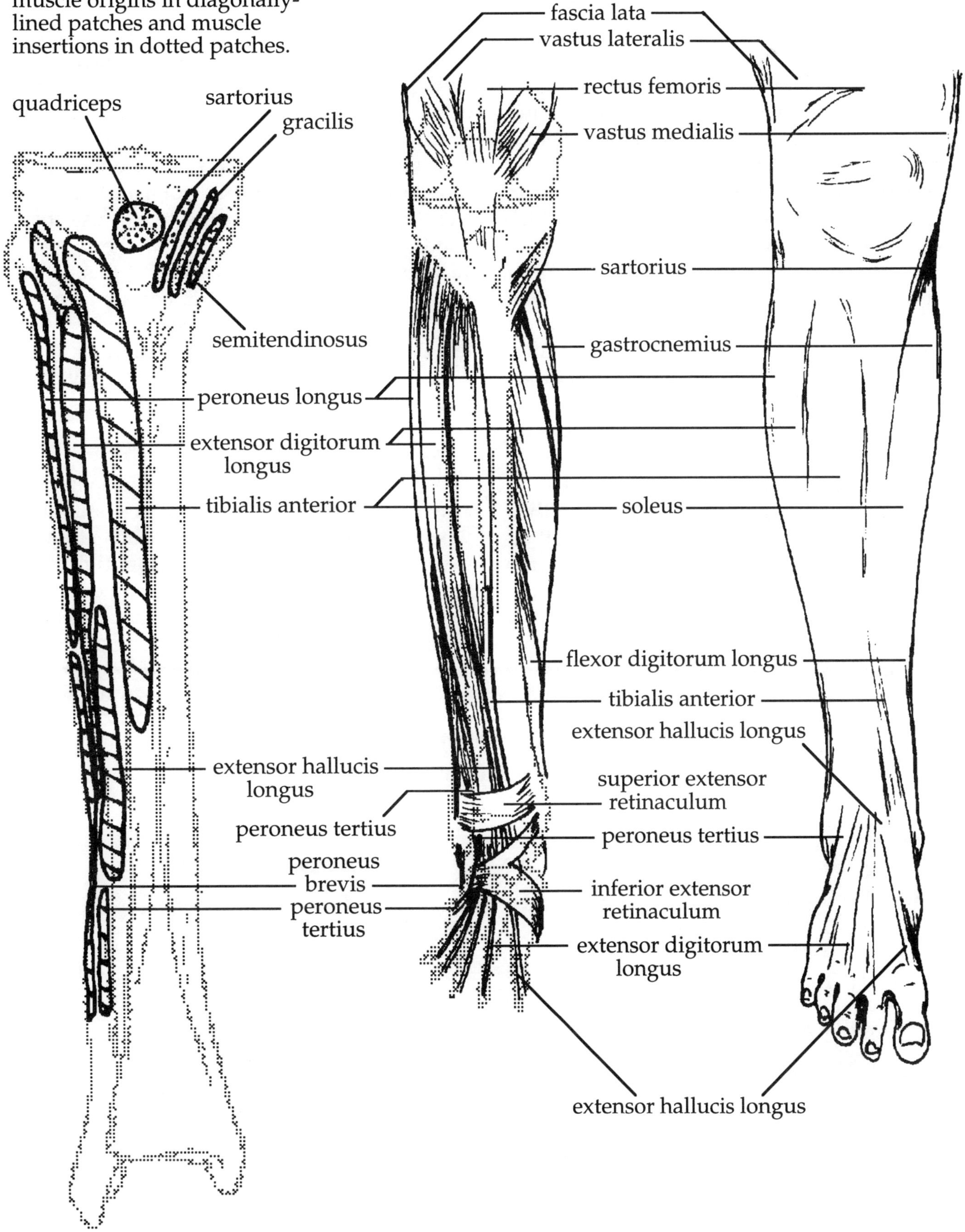

FOOT MUSCLES

Since we already have discussed our hand, the muscles of our foot are most easily summarized by saying that they essentially are the same as the muscles of our hand. Our foot has much the same movement and dexterity potential as our hand. Our foot and hand continue the ongoing saga of stability versus mobility and exemplify the rule that tissues respond to stress and use. It is the manner in which we use these two instruments which changes their actual abilities.

We continually stand and walk on our feet, while we perfect tiny, precise moves with our hands. It is no surprise then, that our feet strengthen the structures which provide support and stability, while our hands strengthen mobility and precision control.

However, there are people doing with their feet, anything that you can imagine doing with your hands, such as turning door knobs, cooking, playing guitar, or sewing with a needle and thread. These people spend their lives sitting on their butts with their feet out in front of them practicing and working, usually forced by a lack or inability of arms or hands.

For most of us, even if we spend our life sitting on our butts, our feet just hang down there somewhere like wooden clumps.

It would take two things to change the ability of our feet, if we wanted to: **1.** stop stomping around on them, and **2.** start practicing delicate movements with them. The younger we are when we change, the better our chance for success. It takes years of practice for our nervous system to develop delicate and precise coordination of any muscle group, whether in our hands or our feet. Children are not expected to use their hands with delicate precision until they are 6 to 10 years old. On the other side of this, when hands are forced to do hard, coarse, physical work, they often become thicker, stiffer, and less delicate, for their own protection.

- On the dorsum of our foot we have one muscle which has no clear hand equivalent.

Extensor Digitorum Brevis

Since extension of our toes is so important to prevent stumbling or toe stubbing when stepping forward, we have an extra toe-extending muscle. Our **extensor digitorum brevis** originates from our lateral anterior calcaneus and inserts into the dorsal connective tissue over our four medial toes (Fig. 13-33).

Figure 13 - 33

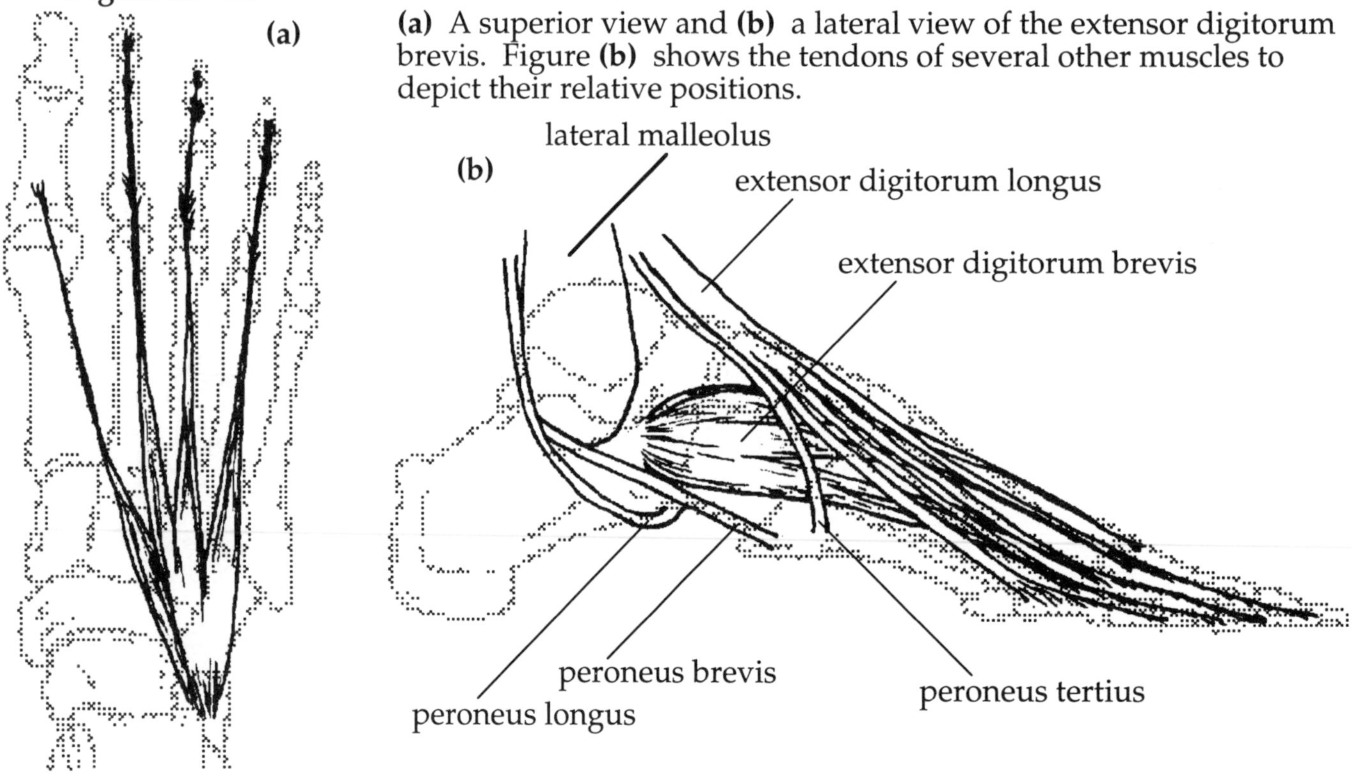

(a) A superior view and **(b)** a lateral view of the extensor digitorum brevis. Figure **(b)** shows the tendons of several other muscles to depict their relative positions.

This time our little toe is left out. Contraction of our extensor digitorum brevis performs its name, extending our four medial toes.

Palpate the belly of your extensor digitorum brevis in the proximal, lateral dorsum of your foot. Locate the inferior edge of your lateral malleolus and the tendons of your peroneus brevis, peroneus tertius, and extensor digitorum longus. Press into the place surrounded by these structures. This is the belly of your extensor digitorum brevis (Fig. 13-33b and 13-40a). Extend your toes to feel it bulge. Its belly and tendons slip under the tendons of your extensor digitorum longus on their way to your toes.

• The muscles in the sole of our foot often are divided into four layers. Alternately, we can divide these muscles into the same three groups as we did in our hand; one group of muscles for our big toe, one group of muscles for our little toe and all those in between.

Individual muscle palaption is much more difficult in our foot than in our hand, since the connective tissue over the sole of our foot is so well developed and our muscles are not practised to do their moves independently.

HALLUCIS MUSCLES

Our big toe has three special muscles.

Abductor Hallucis

Our **abductor hallucis** originates from our medial calcaneus and inserts into the medial side of the proximal phalanx of our big toe (Fig. 13-34). As it contracts, it abducts or prevents lateral displacement of our big toe and supports the medial longitudinal arch of our foot. Our abductor hallucis is the most superficial, medial, and proximal muscle in the sole of our foot. It covers the tendons of our tibialis posterior, flexor digitorum longus, and flexor hallucis longus (Fig. 13-38b and 13-38c).

To palpate your abductor hallucis, go just in front of your heel in the medial instep of your foot (Fig. 13-40b), and inferior to your navicular and medial cuneiform bones. Press into this muscle tissue and watch your big toe being pulled medially into abduction.

Flexor Hallucis Brevis

Our **flexor hallucis brevis** originates laterally in our foot, from our cuboid and lateral cuneiform, deep to our flexor digitorum brevis and longus (Fig. 13-34 and 13-38a). It runs obliquely through our foot to insert into the medial and lateral sides of the base of the proximal phalanx of our big toe.

With its oblique orientation and medial insertion, the medial part of our flexor hallucis brevis has the potential to oppose our big toe, rotating it medially to face the sole of our other toes. In this way it serves as a counterpart to our hand's opponens pollicis and sometimes is called the **opponens hallucis.**

Figure 13 - 34

An inferior view of the foot showing the three short muscles which move the big toe.

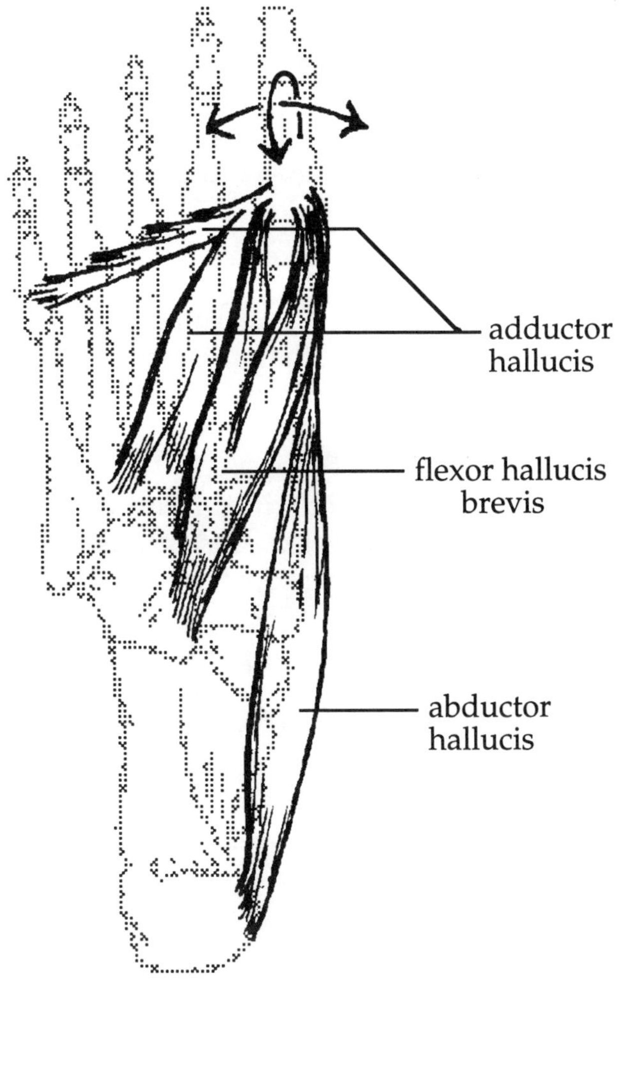

Our flexor hallucis brevis flexes the MP joint of our big toe, and is important in propelling us forward and maintaining balance.

The origin of our flexor hallucis brevis is deep and not palpable. Its belly lies distal and medial in the sole of our foot, deep to the tendon of our flexor hallucis longus (Fig. 13-38c). Press into the little wad of muscle just inferior to your first metatarsal (Fig. 13-40b). This is the belly of your flexor hallucis brevis. Squeeze into this muscle and flex your big toe to feel it working. Feel the tendon of your flexor hallucis longus stand out over it.

Adductor Hallucis

Our **adductor hallucis** has two distinct origins, one from the base of our second, third and fourth metatarsals, and one from the joint capsules of our four lateral MP joints (Fig. 13-34). All these muscle slips insert onto the lateral base of the proximal phalanx of our big toe.

As our adductor hallucis contracts, it adducts and flexes our big toe, prevents spreading of our distal metatarsals, and supports the transverse arch of our foot.

This muscle lies deep to our flexor digitorum brevis and longus and is not palpable (Fig. 13-38a).

DIGITI MINIMI MUSCLES

There are two special muscles for our little toe. They form the little wad of muscle along the lateral sole of our foot.

Abductor Digiti Minimi

Our **abductor digiti minimi** originates from our lateral calcaneus and inserts into the lateral base of the proximal phalanx of our little toe (Fig. 13-35, 13-38c, and 13-40a). As it contracts, our abductor digiti minimi abducts and flexes our little toe, and supports the lateral longitudinal arch of our foot.

Find the belly of your abductor digiti minimi in the proximal, lateral sole of your foot, in front of your heel, inferior to your lateral calcaneus and cuboid (Fig. 13-40a and 13-40b). Squeeze into the muscle belly and watch it pull your little toe out to the side.

Flexor Digiti Minimi Brevis

Our **flexor digiti minimi brevis** originates from our cuboid and the base of our fifth metatarsal (Fig. 13-35). It follows the inferior surface of our fifth metatarsal to insert into the lateral base of the proximal phalanx of our little toe. As it contracts, it flexes our little toe.

Since the insertion of both our abductor digiti minimi and flexor digiti minimi brevis wraps to the lateral side of our little toe, they will tend to rotate it medially to face the rest of our foot. In this sense, they form the couterpart to our hand's opponens digiti minimi. The lateral part of our flexor digiti minimi brevis sometimes is called the **opponens digiti minimi.**

Figure 13 - 35

An inferior view of the foot showing the two short muscles which move the little toe.

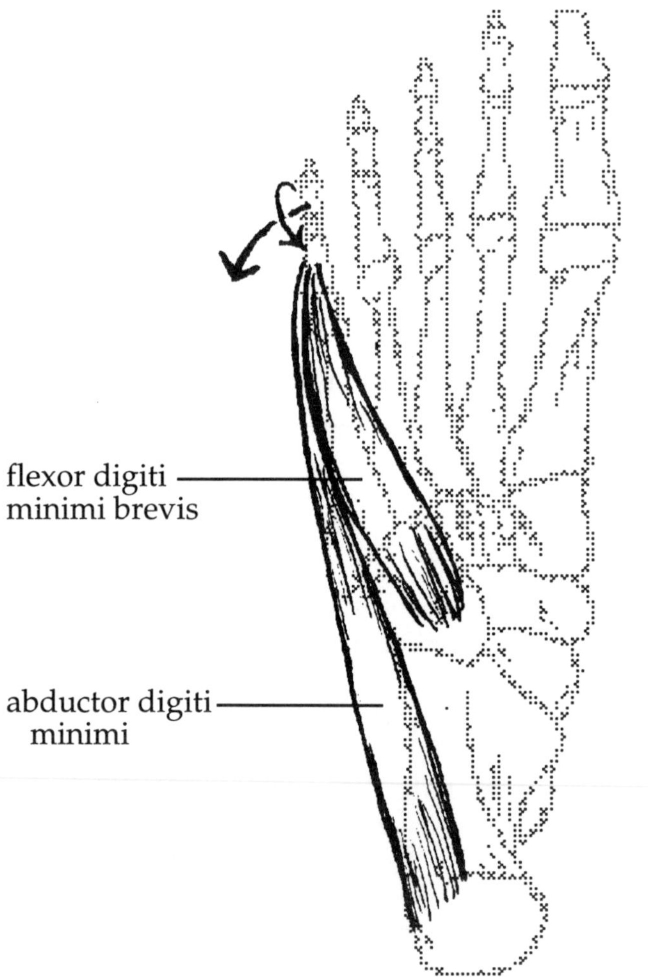

Find the belly of your flexor digiti minimi brevis in the distal, lateral sole of your foot (Fig. 13-38b and 13-40a), inferior to your fifth metatarsal. Squeeze into the muscle belly and watch it pull down your little toe.

MIDPLANTAR MUSCLES

There are five muscles in the center of the sole of our foot. We will discuss them from superficial to deep.

Flexor Digitorum Brevis

Our **flexor digitorum brevis** is the most superficial muscle in the central sole of our foot. It lies immediately deep to the dense connective tissue in our skin, between the abductors of our big and little toes (Fig. 13-36a and 13-38c). It covers over the tendons of our flexor digitorum longus. Our flexor digitorum brevis originates from our inferior calcaneus. It traverses the length of the sole of our foot to insert into the middle phalanx of toes two through five. The insertion splits to allow the tendon of our flexor digitorum longus to continue on to the distal phalanx of our toes. Because of its insertion and superficial position, our flexor digitorum brevis is considered the counterpart of the flexor digitorum superficialis in our forearm.

As our flexor digitorum brevis contracts, it flexes our MP and PIP joints of toes two through five. It also is a major supporter of both longitudinal arches of our foot.

To palpate your flexor digitorum brevis, press into the center of the sole of your foot. See how the pressure already curls your toes downward. Actively flex your toes to feel the muscle working.

Figure 13 - 36

An inferior view of the foot showing **(a)** the flexor digitorum brevis, **(b)** the quadratus plantae, and **(c)** the lumbricals.

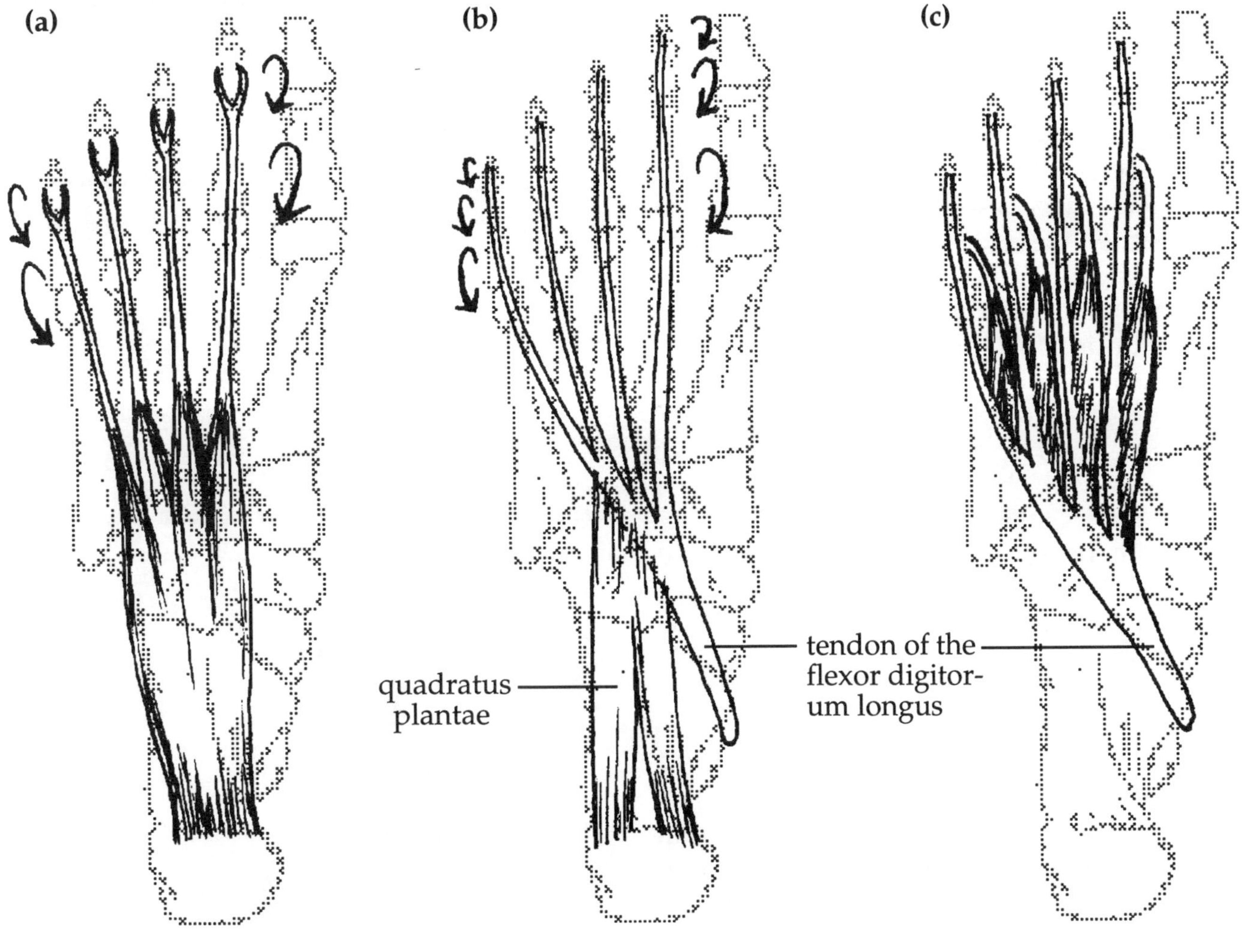

Quadratus Plantae

'Quadratus' refers to a flat, four-sided structure like our quadratus lumborum or pronator quadratus. 'Planta' refers to the sole of the foot which is repeatedly planted down onto the ground. Our **quadratus plantae** is a quadrangular muscle in the sole of our foot.

Our quadratus plantae originates from our inferior calcaneus (Fig. 13-36b and 13-38b), just deep to the origin of our flexor digitorum brevis, and inserts into the tendons of our flexor digitorum longus.

As our quadratus plantae contracts, it enhances the effect of our flexor digitorum longus, flexing all the joints of toes two through five and supporting both longitudinal arches of our foot.

Your quadratus plantae is just in front of your heel in the center of the sole of your foot. It is not palpable separately from the flexor digitorum brevis lying over it.

Lumbricals

The **lumbricals** of our foot are similar to the lumbricals of our hand. Four muscle slips originate from the tendons of our flexor digitorum longus in the sole of our foot (Fig. 13-36c and 13-38b). Each muscle slip passes medial to its respective toe and inserts into the tendons of our extensor digitorum longus and the extensor expansion of its respective toe, in the area of our proximal phalanx.

As our lumbricals contract, they transfer flexion force into extension, flexing our MP joint, but extending both IP joints of toes two through five. In this way they can juggle the delicate moves of our toes, as required.

The lumbricals in my feet are not palpable. Perhaps if I had more dexterous feet, I could find them in a similar manner as in my hands.

Plantar Interossei

The plantar surface of our foot is the sole, which we plant onto the ground. 'Inter' means between. 'Ossei' refers to bones. Our **plantar interossei** lie between our metatarsal bones, toward the sole of our foot, but deep to our flexor tendons and lumbricals.

These muscles originate in three bundles, one from the medial side of each of our third, fourth, and fifth metatarsals (Fig. 13-37 and 13-38a). They cross the MP joint to insert into the medial side of the proximal phalanx of the same toe.

As our plantar interossei contract, they pull the three lateral toes toward the second toe, adducting them. Notice that the non-moving axis for adduction in our foot is our second toe, compared to the third finger in our hand.

Due to my unskilled foot and the numerous overlying muscles, my plantar interossei are not palpable. If your foot knows a few tricks, you might find yours in a similar manner as the palmar interossei of your hand.

Dorsal Interossei

Our **dorsal interossei** lie between our metatarsal bones, toward the dorsum of our foot. They originate in four bundles from facing surfaces of adjacent metatarsals (Fig. 13-38a and 13-39) and insert into the medial side of the proximal phalanx of our second toe and the lateral side of the proximal phalanx of

Figure 13 - 37

An inferior view of the plantar interossei muscles.

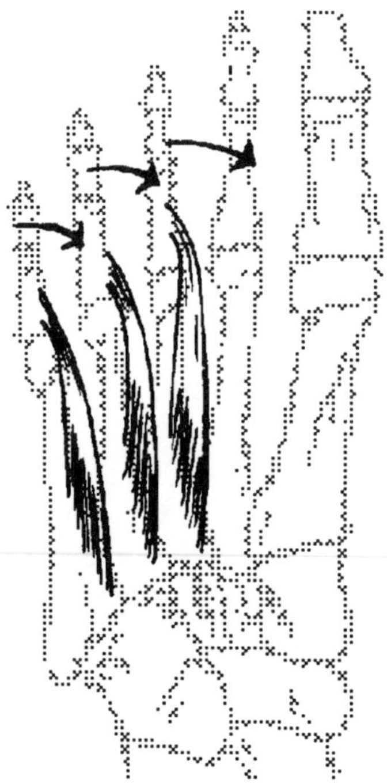

our second, third, and fourth toes.

Our second toe has a dorsal interossei on each side, so it can be pulled medially or laterally, waving to the world and forming the midline of the foot in terms of adduction and abduction. Our third and fourth toes are abducted by their dorsal interossei.

Again, in my unskilled foot it is impossible to differentiate my dorsal interossei through the extensor tendons on the dorsum of my foot. If you can wave with your second toe, you might feel yours on either side of your second metatarsal.

Figure 13 - 39

An inferior view of the dorsal interossei muscles.

Figure 13 - 38

An inferior view of the foot showing three layers of muscles, which are **(a)** deep, **(b)** middle, and **(c)** superficial.

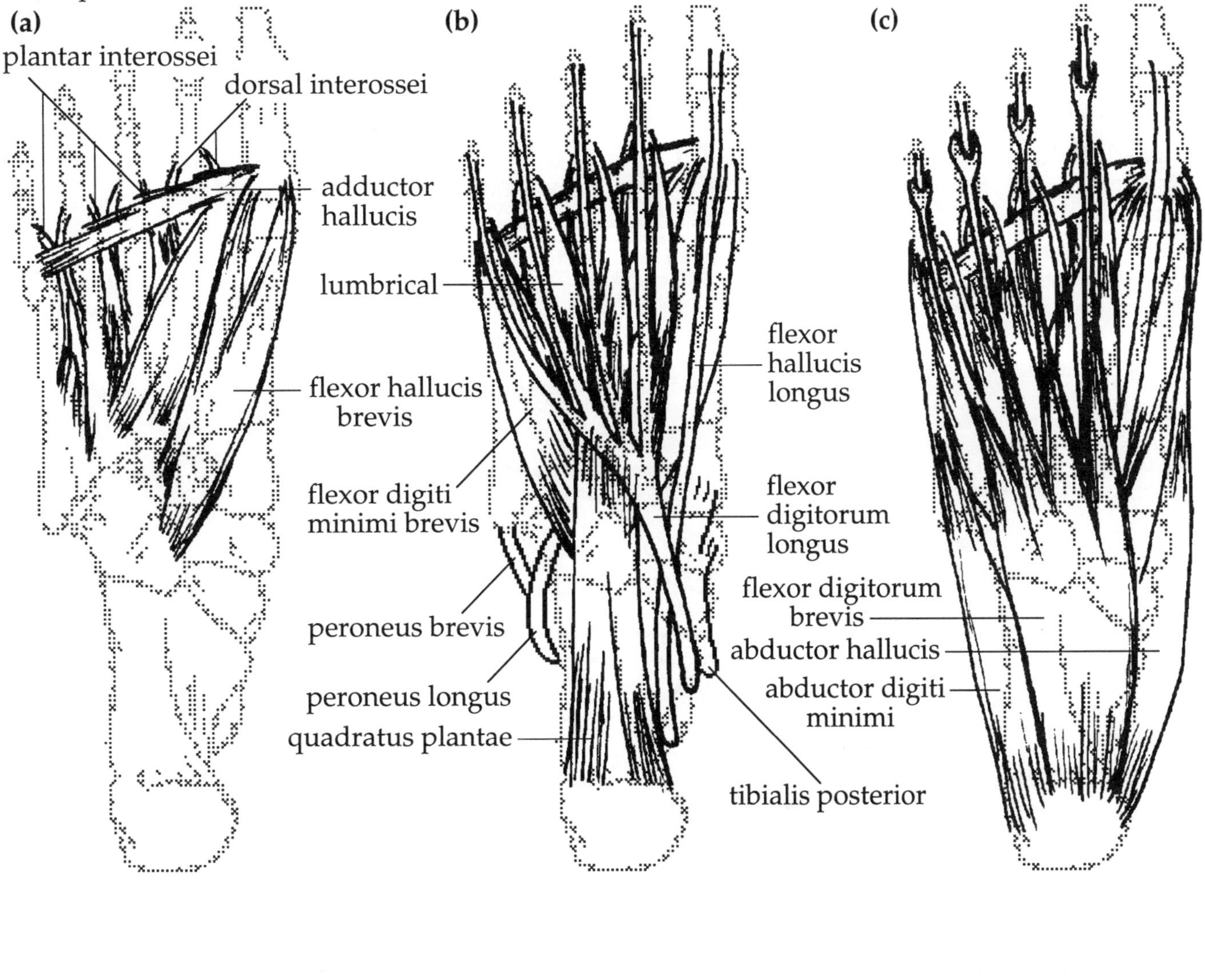

Figure 13 - 40

(a) A superior or dorsal view and **(b)** a medial view of the surface of the foot.

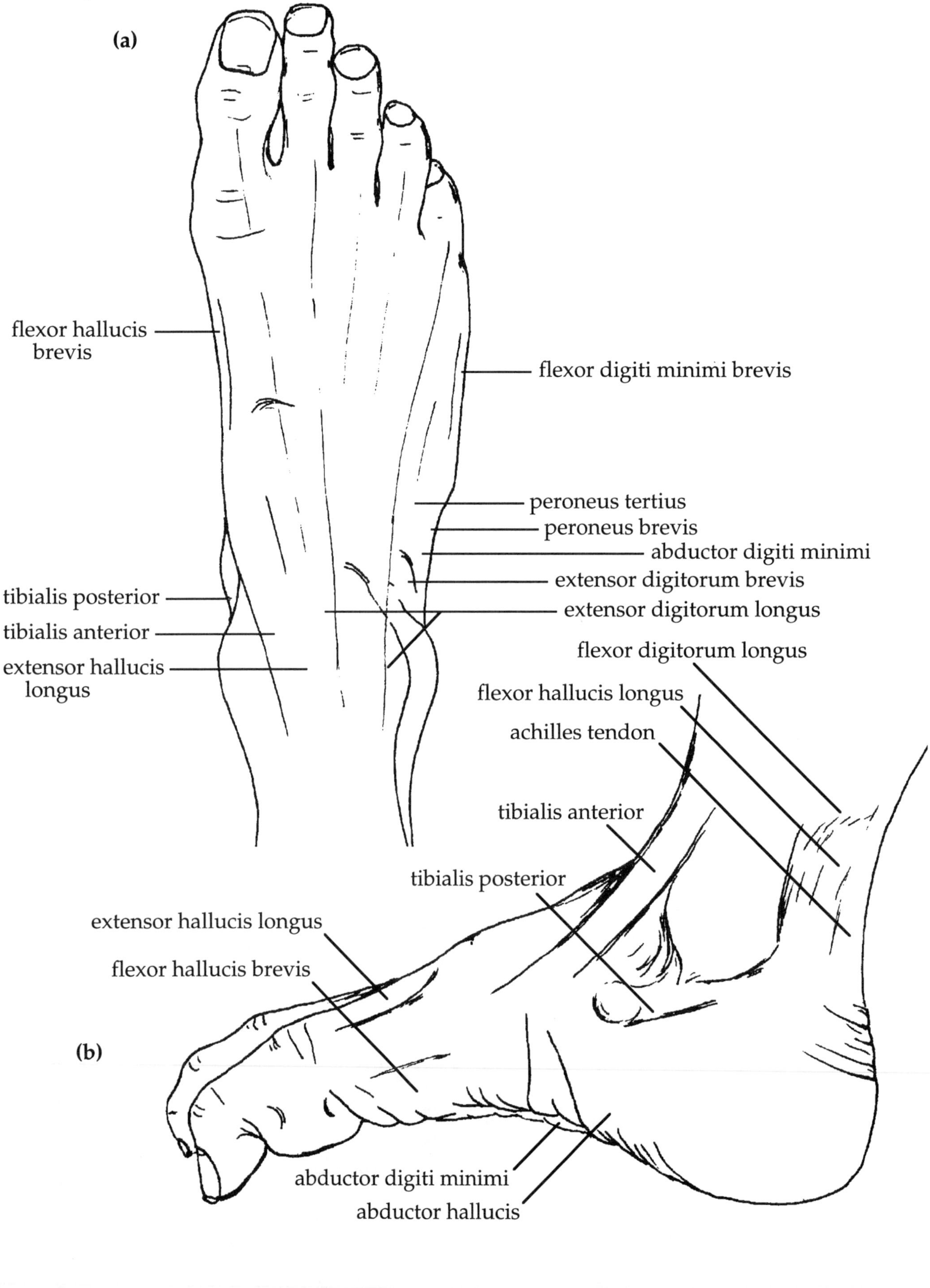

APPENDIX 1

OUR ILLNESS PROCESS

When the normal process of our physical nature, our anatomy and physiology, is disrupted, we call this an illness, disease, or pathology. Symptoms are the perceptible changes, the manifestations of our illness, from the most subtle sensations to the most severe pain or disability.

In this discussion we address two questions about illness. **1.** Is an illness imposed on us from the outside, beyond our control, or do we have some participation or responsibility in our illness? And **2.** is an illness purely physical or may it also have roots in aspects of our being such as our thoughts, emotions, judgements, spirit, energy, or imagination?

- We are not an omnipotent power in our universe nor even in our own life. Things happen to us that we cannot control, or so it seems. Sometimes it feels as though we cannot even have the smallest influence on things that create major changes in our life. We construct an explanation for these things that happen to us, a story to hold and integrate these things into our life, to give them meaning and a place of significance. Our belief systems help us with this.

Another viewpoint is that a butterfly, flapping its wings, fluttering over a tea bush in China, affects the air currents over North America. A single thought in my brain creates an energy pattern that reverberates endlessly through space and time. In subtle and dramatic ways, we can change what happens to us, to those around us, and to the planet we live on, more than we can ever know or imagine.

In a sense, the things we cannot control in our life are these reverberations from everybody and everything, reflecting back upon us and passing through us.

Following these viewpoints, we can propose two theories of the cause and process of illness. **1.** In one theory, we are victims of afflictions that are imposed on us beyond our influence or control. We handle the illness as best we can, we suffer through it, we get stronger, we get weaker, we get angry, we become more tolerant, and so forth. We deal with the effects of having the affliction. We can find our peace with it. **2.** In the second theory, we are participants in bringing our illness on, allowing it to enter our being. Our habits or patterns of eating, sleeping, thinking, judging, feeling, imagining, behaving, and so forth, play an active part in causing our illness, sometimes a big part. Our illness is a response to how we live. We still have to deal with the effects of having the illness, but we can perhaps unravel the story, or parts of the story, leading to our illness. We can understand why it is there or where it comes from, fitting it into the larger picture of our life. With these threads we can perhaps learn to weave our life into more wholeness. Our illness becomes our teacher, our tool, our opportunity to live our life in a fuller way. This may or may not reverse or heal our illness process, but we will be more at peace and more integrated into the purpose of our life.

Some people opt for the first theory of the cause of illness because the second one adds value judgements such as right and wrong, righteousness and guilt, vengeance, justice and punishment into the construct of illness. To take these attitudes is a common response to anything that happens. We can gloat or squirm about anything that happens to us. We can strut or we can besiege and batter ourself. We can bog ourself in the mire of righteousness or guilt, or both.

There is, however, another possible response. We can just do what we do and understand it. We can accept who we are and what we have with love, compassion, and honesty. We can ackowledge and define our participation in our own life and the lives of those around us. We can find the parts of our life that we do control. Then we have the option of influencing a change in what happens to us. We may influence the course of an existing illness or prevent illnesses from affecting us. We can change our behavior patterns or habits to better serve us.

I am not suggesting that we can utterly control our life. Outside forces participate in our formation and development. We may do all the right things and still get sick. We may do all the wrong things and never get sick.

The actual cause and process of illness lies somewhere between the two theories of being totally **controlled from the outside,** by God, nature, circumstance, viruses, and other people, or being totally **controlled from the inside,** by ourself. Each illness has its own place on this continuum. Finding our own place of balance in this is the journey of our lifetime. It is very personal, a process that we only can do for ourself.

• The first theory of illness traditionally has been our model. If we choose the second theory of illness, what would it look like? How can we untangle our life? How can we discover how we participate in and influence our illness, or any situation for that matter? In the second theory we take a journey with our illness.

Our body always is telling us what is happening. It is the physical print out of our whole being. We may feel synchronous, effortless, flowing, clumsy, cluttered, painful, stiff, uneasy, tired, heavy, thick, stuck, and so on. These sensations are all messages from our body, telling us about our present condition, showing us how what we are doing or thinking is affecting us, giving clues about what leads us to our present state, good or bad.

When a friend is telling us how she/he is doing, what do we do? Turn away, paying no attention? Put tape across her/his mouth? Not usually. We listen, with interest, empathy, and openness, supporting them with comments, questions, or ideas.

Do we give our own body the same care and respect? Is our body on our list of friends? Do we listen to it with interest and openness? Do we ask it questions to clarify the information?

Too often we give our body the turn away treatment, telling it "Don't burden me with your troubles, I'm busy." We pay no attention, as if our body has nothing to say to us.

If the complaints are too loud to ignore, we still shut it up, almost literally putting tape across its mouth, layers of masking tape. Symptoms are seen as problems which need to be eradicated rather than signals of a deeper process. We block aches and pains with aspirin. We smother an itchy rash in cortisone cream. We mask our anxiety with valium. We snuff out sleeplessness with sleeping pills. We arouse from fatigue and lethargy with caffeine. We drown our unworthiness and loneliness in alcohol, and so forth. Then we are surprised when our body gives out on us. No other friend would have stood by us for so long.

Our body tries harder to get our attention. Our pain gets worse. Our rash gets worse. Our fatique gets worse. Finally we cannot run away or cover it up any more. We have to stop and say, "What is happening to me?" and at least in part, "What am I doing to myself?" Often the illness has become quite progressed before we even notice it. Our body may have been whispering clues, compensating, and running on reserve for a long time.

We can better serve ourself by paying attention to our body. Anywhere along the path we can stop and ask, "How am I feeling? Why do I feel this way? What is this feeling telling me about myself? What am I doing that leads to this?" and the important follow-up question, "What can I do differently to feel different?" "What can I do to feel better?"

We do not have to wait for discomforting sensations. When we feel good, we can ask "What am I doing that is helping me feel so good?"

We need to ask, then we need to listen to what our body says, then we need to act on what we learn, to have the motivation and courage to address the issues and make changes in how we live.

The simplest level of asking, listening and acting is in our physical aspect. How is my body doing? How do my physical behaviors influence my physical well being? If I get adequate rest, my resistance is up and I fight off a cold. If I do not smoke cigarettes, my chances of getting emphysema are less. If I am attentive to my driving, my chances of being injured on the road are less. If I do not excess in low-density lipoproteins, my chances of arterial disease are less. If I take in nutritious food, I can build a healthier body.

• If we see illness as purely physical, this may be a far as we go. If we recognize that we are more than a physical body, that we include thoughts, emotions, judgements, imagination, energy, and spirit, and that all these aspects affect each other, then our illness and health can begin in any of these aspects.

Our body speaks for all our aspects. It is the physical manifestation of our totality, our vital integrity. When there is stress, fear, constriction, or blockage in any aspect, it can become physically manifest. When there is lightness, flowing, security, or purpose in any aspect, this also can become physically manifest.

The concept of body language has been discussed for some time, how the posture and movement of our body speaks of our emotions or thoughts, our need to protect or project ourself in the outside world, and so forth. Our health or illness is an extension of our body language, expressing the ongoing play of our emotions, thoughts, protection, projections and so on. Our health or illness is like constructive criticism, challenging us to take care of ourself, to clarify our intentions, to go deeper into our life, to grow to our fullest potential. Our health or illness can show us our path, from where we have been in the past, what is a good place to move to, in order to live our life more fully.

To unravel our story, we need to listen to all our aspects, and always remember that we are not omnipotent, that outside forces are always influencing us.

Many common phrases express the interconnection of our body, mind, emotion, spirit, and so on. These phrases can give us ideas about the non-physical aspects of disease. For example, when we are angry at someone, we may say that that person "gets our gall" or "galls us". Then I wonder about gall stones, how many of them are solidified, unexpressed anger? When something creates irritation or frustration, we may call it "a pain in the neck". The thing itself does not actually hurt our neck, but our irritation and frustration does. When someone has deep, intense grief and soon dies, we may say that they "died of a broken heart".

We can ask our own questions to discover our own connections. There are many places to investigate. One is to discover how our body is affected by our other aspects, to see how different states affect us. When our thoughts are negative or positive, how does our body feel differently? When we imagine a streak of good luck versus someone blocking our path, how does our body feel differently? When we condemn or evaluate an idea, how does our body feel differently? When we are carried by fear or by joy, how does our body feel differently?

Another place from which to listen is "What in us is creating this thought, emotion, judgement, imagination, and so on? Why do we choose the responses we do? Do we condemn something because we feel put down and want to get back or because it actually would not work? Do we imagine a bad scenario because we do not deserve any better? Do we think negatively because we need to cut ourself down? Why do we need to cut ourself down? What are we defending?"

Another way to look is to expand our options, to see that any story has many sides, many possibilities, to see that we choose a certain one, and to ask why we make that choice. This can free us to make other choices and so change the path of our life. We can challenge ourself to think of another explanation for what has happened. We can imagine a different end to our story. We can find one good quality in the thing we have just condemned. There are endless questions and when we just listen with acceptance and compassion, the answers begin to unravel our story, and to show us how we are choosing our life and what our choices are creating in our body.

Sometimes we have to juggle between our aspects to decide what is really good for us. Sometimes I stay out dancing way past my bedtime. My body may be a little tired the next day, but my being holds a sense of peace and magnificence that overrides any fatigue. When my dance is a pure expression from my vital integrity, not entangled with needs or hopes, my body is carried with that and often has no suffering.

The more we practice listening and responding to our body's message, the more easily we will hear its subtle whispers, the earlier in an illness we may consciously enter and influence its process.

- Returning to the symptoms of an illness process, I am not suggesting that we should never intercept a symptom, that aspirin, valium, cortisone, massage, acupuncture, and so on should never be used to palliate symptoms. Palliating the symptom can help us through a crisis, it can already begin to shift the disease process, it can relieve us enough to be able to stop and listen, but it should never be viewed as the complete answer. When we have intercepted, the absence of symptoms does not mean the end of the disease. The interception does not fix the underlying process, it only removes the uncomfortable, disturbing manifestations.

In the middle of a dripping cold, I have taken an antihistamine to avoid splattering my students with sneezes. The antihistamine stops my symptoms, I feel better, but it does not take away the cause of my cold.

Our symptoms are our body's communication to us about our underlying process. Stopping our symptoms blocks our body's ability to communicate, so that we can no longer rely on how we feel. For example, after a gruesome dental extraction, the dentist wrote a codeine pain prescription for me and warned me that this situation often became infected. I asked him how I would know if it got infected. He said that the pain would increase. To my next question, "If I am using this pain killer, how will I know that the pain is worse?" the dentist had no answer.

When we have intercepted our symptoms, then in spite of feeling better, we must remember that an illness process is still present. We should not use our improved condition to party or work long into the night. We should continue to take every care to promote the healing of our body.

There is irony in intercepting the symptoms of our illness process. While we say that it relieves us enough to ask and listen, to unravel our life story, we have blocked or muffled what we are trying to listen to. Its like asking our friend to speak through the tape. We need to be very present to actually get the message. Interception also disguises the urgency of the message, and allows it to be not actually so important in our life.

It is precisely our uncomfortable, distressing symptoms that are our greatest motivators to change our life. If we remove symptoms, we have to create our own motivating force out of our own will. When we remove symptoms before we activate our will, it is easy to slip back into our self-destructive routines, be they physical, emotional, spiritual, moral, energetic, imaginative, or whatever.

If we change our perceptions about symptoms, to really accept them as messages and tools rather than problems, it can help us to be more present and benefit from our symptoms, to not be so easily overwhelmed by them. For example, pain is a symptom we just cannot stand, but <u>pain is a great tool for healing</u>.

<u>Pain gets our immediate attention</u> like nothing else. It is sometimes amazing how far something progresses before we notice it. When it hurts, we finally pay attention. Sometimes it is too bad that things do not hurt a lot sooner, then they would not progress so

far.

Pain motivates us into action. It is sometimes amazing what we do or do not do to ourself, or what we allow to happen. When something hurts, then we finally make a change. For some of us, it has to hurt worse and worse for a long time before we get moving.

Sadly, too often the only thing we do is kill the pain rather than investigate the cause of it. When the pain is gone, we have no motivation to investigate or work with it. It is no longer there to work with. We carry on with all the things we do and never take the opportunity to learn what we do that is hurtful.

Pain helps us to focus and concentrate our attention and energy. Our pain can bring us into ourself. It helps us to listen intently and gather our energy into the purpose of healing.

Pain is sensory imput to our nervous system. We can choose our response to it. We can respond with fear, dread, anger or denial. We can respond with patience, attentiveness, and compassion for ourself. Our response determines the next expisode of our story. Either we release hormones of fear or anger, which close and contract our body, or we release hormones of compassion, which increase openness and flow in our body. Any symptom can benefit us in all these ways.

• I have continually referred to the process of an illness. There is generally a long continuum from being perfectly healthy to being obviously sick. Our body compensates for small inconveniences. We can miss one night of sleep occasionally. We have large reserve capacities. Our normal resting breath averages 500 milliliters but there are 3000 milliliters of space still available if we need more air. It can take a long time for a negative thought process to really wear a groove in our body.

At first, as our body begins to give out, there are subtle sensations that we do not even relate to as symptoms. Two days before the onset of a cold I have felt a tension in my face or an itchiness in my nose.

Gradually we move into obvious symptoms. That may be when we first recognize that something is not quite right. We may feel like we just got hit from nowhere. It makes us feel vulnerable and out of control. We hope it will just go away as mysteriously as it came. Our symptoms get worse until we finally do something to shift the course of our illness.

Healing from an illness reverses the same continuum. Gradually our symptoms diminish. We slowly feel better until we are symptom free. At this point we may say that we are all better, but our reserves still may be very depleted, so we need to continue to take care of ourself until we have fully regained our health.

• We should never use the disappearance or progress of our illness as a measuring stick for our capacity to consciously reweave our life. Sometimes the physical process initiated in our body cannot be reversed or leaves a permanent scar.

I met a tall, handsome man in his forties with Lou Gherig's disease, a disease which is progressively debilitating and usually fatal in less than a decade. If I can summarize our conversation, for this man, getting Lou Gherig's disease was the best thing that had ever happened to him. Before this disease he was a lost alcoholic with no meaning or direction in his life. Now he just glows. He knows why he is here, he is into what he is doing and he is doing what he is into. He still works on healing his body, but whether he does or not, the most meaningful part of his life is the few years since the onset of the disease rather than the decades that went before.

We have all heard someone say that a certain illness was the best thing that happened to them. It can wake us up and give us the insight, the energy, and the courage to get on with our life, to get through whatever is blocking us.

Our life is a giant tapestry that we can unravel and reweave bit by bit. May our life flow with greater ease, deeper joy, and gracious bounty.

APPENDIX 2

ROLES OF OUR HEALER

In response to our two polar theories of illness process, we have two polar concepts of the role of our healer. Any situation sorts out for itself where it will stand on the line between the two poles.

- When we see our illness as an affliction imposed on us, unrelated to how we live, then we tend to look to our healers to extricate our illness, to make it disappear as mysteriously as it came. We do not need to try to understand anything. We put our healers in charge of our life. We want them to take responsibility for our life. We give them control. We surrender our power, our self-knowledge, our perception, and our intuition.

What our healer says is 'the answer'. No further questions. No other possible answers. Maybe we get a second opinion, but we give the person whose opinion we accept an omnipotent position in our life.

We do anything our healer says, almost. Often the more outrageous a treatment is, the more we go along with it. We allow parts of our body to be cut out. We ingest potent chemicals. We have our bones cracked over each other. We let them poke us full of needles.

If we are advised to perform a subtle shift in our daily routine, we are less apt to do it, we are suspicious, we do not trust it to really have an effect. It is the least important part of our treatment, and the first thing that we do not have time for. To not drink ice water, to eat the last full meal of our day before 6:00 p.m., to take a few minutes of silence each day, most people are not paying attention to these things. Most people are doing the opposite. They all seem healthy, so what difference can it make?

We ask our healers to tell us how we are doing. We doubt our own perceptions and feelings. We trust what they say more than our own sensations. Our healers also may trust their own observations and lab tests more than what we say.

We tend to equate the words "I am doing everything I can" with "Everything possible is being done". It may be true that that person is doing everything he/she can. Whether or not this is everything possible depends on the expanse or limitation of that person.

When we are in need of healing, we are frequently called a patient. This word fits perfectly into this model of the healer/healee relationship. According to Webster, a patient is "one who is passively affected, submissive, long suffering, bearing pain or trial without murmuring."

When our treatment is complete, we expect to go back to our daily routine just as it was before our illness, or as close as possible.

Whether our healers receive our praise or condemnation often depends more on their ability to hold an omnipotent position in our eyes than on the outcome of our illness. If we believe that they have done everything possible in the best way, they receive our praise whether we get better or die. On the other hand, if we believe that our treatment could have been done better, they receive our condemnation whether we get better or die.

- When we see our illness as our teacher or our tool in the journey of our life, we are not a silent, submissive, passive patient, and the role of our healer is different.

In a certain way our healers are above us, in that they have certain specialized capacities beyond our own. They know things that we do not know. They can do things that we cannot do. But they carry their knowing and doing as a respectful participant in our journey, not as the ruler over us. Our relationship with our healer should empower us, not accentuate our helplessness. While we may depend upon our healers for certain things, our relationship with them should help us grow more confident and capable in knowing and taking care of ourself. It allows both the healer and the client to be intelligent, capable, fallible, and respected human beings.

- Our healers can be our facilitator, one who lessens the labor of. They make our labor easier in many ways, but they do not take over our labor. We need to do our own work. We hold the reins, we hold our own power.

Our healers can be our teacher. They are a resource of information. They can instruct us, imparting theoretical and practical knowledge, clarifying and filling out our understanding and capacity to work with our process. They can tell us their point of view of the cause, risk factors, process, and treatment possibilities of our illness. They can tell us how different parts of our body are built or function. They may give us homework, teaching us exercises, visualizations, or meditations to practice, self-massage techniques or little everyday procedures or skills that might help us. They may recommend reading materials such as this book.

With this approach to illness, we take all this information as their contribution, rather than the one and only true answer. We use it with an open, observant mind, testing how it works for us.

Our healers can participate as a guide through our process. They are a fellow explorer, joining our exploration. As an outside observer they may be able to put things in a broader perspective. They may encourage our questions and participation. They may ask the questions we are missing. They may point out our possible biases, the sides we never look at or the aspects we overlook, increasing the possibilities in our choices and answers. They may suggest things to do, different habits, different ways of thinking and imagining. They are someone for us to push against, to test our ideas. In this way we increase our self-awareness, understanding the behaviors, thoughts, and feelings that might be entangled in our process.

Our healer is our servant, looking for what will further our cause, while remaining subject to our command.

Our healer may be our comforter, making us more comfortable by being present with us wherever we are, empathizing with us, sharing compassion, palliating our symptoms, reducing our fears, and increasing our understanding.

Our healers may give us energy, through their attention, with their hands, or through different treatment modalities they practice.

Our healers may offer various treatment modalities to strengthen our bodies, to palliate symptoms, to influence the process of our illness. These modalities add to our tools of healing. Surgery, medications, nutritional supplements, bone adjustments, or acupuncture, may become part of our plan. Our healers are the expert in these modalies and may do them to us, but we understand what they can and cannot do. We are free to have our own experience of them and see if they work for us.

Our healer may give us new or different feeling experiences, providing alternatives to our previous experience, making different behavior patterns easier for us.

Our healers often can help us see ahead. They have walked similar roads before with other people. They can give us information on the outcome of similar, previous trials or lack thereof. They have a log of precedents. We need to judge how closely our story matches the precedent.

In a sense our healer is like another symptom of our illness process, helping us to be aware of ourself, to ask helpful questions, to focus our attention and energy, to encourage and motivate us, to give us feedback about what they see.

Any one healer may not know or do all these things, but he/she has an understanding of playing a role in this kind of picture.

Our healer can lay all these things out with us, but we must sort it out for ourself, pick and choose what fits us, what we can work with. We have to interpret and integrate it into our own life. We are the only one who can know what it feels like to take it into ourself. We are the only one who knows our deepest secrets and motivations We are the only one who can feel what rings true for us. We are the only one who can change our ways.

We should feel free to weigh or test everything we receive against our own self-knowledge, perceptions, and intuition. We should feel free to question everything and understand everything to our satisfaction. We participate in the process. We are confident in our own capacity. We can look to ourself and see how we are doing.

• I recently had a pain in one of my teeth. Having not felt good with my last dentist, I picked a new one from the yellow pages. He found a crack in my tooth and treated it, telling me very briefy what the situation was and what I should expect if it was appropriately treated. The original pain was gone, but there was another pain that came up suddenly now and then, so I went back.

The dentist examined my tooth and said that it was probably weakened by the crack so he wanted to shave off an area of healthy enamel to reduce the stress on the tooth.

My guts turned into a knot inside me. If my tooth was already weakened, shaving off healthy enamel seemed like the worst possible solution. All kinds of images crowded my imagination. I felt my teeth withering in my mouth. I saw myself as an old lady with a jagged, toothless frown. I supposed that this dentist paid his bills by shaving peoples teeth down to nothing and then being so sorry that all there was left to do was expensive crowns or caps. My hands were sweating.

My brain continued waging war. "This does not at all make sense. A horrible thing is being done to me. Take control. Do not be a helpless victim" versus "He is the dentist. He knows what he is doing. Just trust him. Surrender your need to control, you old bitch." All my personal defenses and projections started coming up.

Finally I went back to the beginning, emptied all those emotions as much as possible and said, "It doesn't make sense to me. If my tooth is already weakened, how can shaving off more enamel help?"

His first response was "I won't do it if you don't want me to."

I clarified my position. I was not arguing, I was needing more information. In less than a minute he explained the situation in more detail. His concept was congruent with my understanding of physics and mechanics. With this information his suggestion did make sense to me. I could even see how, after the original crack, this procedure may, in fact, reduce the stress on my tooth and lengthen its life.

I went home in perfect peace and that tooth has served me without pain to this very day.

I did not have to be a victim. I did not have to be a bitch. I did not have to be haunted forever by this gnaw in my guts about what had been done to me. I did not have to spend days researching in the dental college library to justify his side or mine.

What did I have to do? I had to give myself the right to question and understand. I had to act on that right. I had to be clear on what was happening, what I really needed, to not get caught in his or my own personal defenses and projections.

That sounds like a lot of work just to get rid of a pain in my tooth when I could have just trusted the guy in the first place.

If I <u>had</u> just trusted him, that may be the easiest. But once my trust is broken, <u>once the question arises, when I allow myself to acknowledge it and follow it, that is what leads me back to my trust and a fuller way of being in my world</u>.

My capacity for myself is increased, specifically about teeth, but also in a general way, how to better serve myself and those I relate to in a way that empowers us both. Now I do trust that dentist more, with my teeth, as well as with my whole person, that he can meet me wherever I am. And I am more whole as a person. I am more knowledgeable about how teeth work and how they are stressed. I understand the difference between the different pains the tooth gave me, so I can interpret my symptoms more clearly. I see my defenses and projections more clearly so I am better able to untangle my life and meet my real needs.

• How much of a controlling versus participatory role our healer might best take or how much we should give over, is different in each case. Where a case sits on the line between the two poles depends on many factors.

What is the illness? How acute is it? How long have we had this particular illness? How experienced are we in working with illness process in general as well as this illness in particular? How experienced is the healer in working with illness process in general as well as this illness in particular? How long have we been working with this healer, how well do we know and trust each other?

How much are we generally involved in our own life? If we never ask "Why?" for ourself, if we just trust without having to understand, then we are lost if we are expected to participate. Our healer needs to take a more active leading role, at least in the beginning, guiding us into a self-discovery process.

If we are already effective in our own self-discovery process, in unravelling our life, then we just need input here and there We need the dentist to shave off the enamel. We need the acupuncturist to really push a stagnant energy. We need antibiotics for our pneumonia. Maybe we could make it without these services. We could just always watch what and how we were chewing. We could work more with our own energy flow. But our process through an illness, or anything else in life, really can be enhanced by using the expertise of others. Others can promote our health and life, reducing the chronicity, the scar, the extent, the diversity, and the recurrence of our illness. When I talk about participating and trusting ourself, I am not saying we should do everything on our own, but whoever we do it with, our own understanding, capacity, and power should be increased. Our life should be fuller after an illness. If we feel emptier, like the illness took something out of us, we should question who we gave ourself away to.

• While this role of our healer does not hold the omnipotent place wherein we stake our unerring faith, it still is a place of great honor and great humility. It is a great honor to be allowed so deeply into another's process and personal life.

For me as a healer I must look deeply into the weaving of my own life, to be continually untangling, clarifying, and reweaving, so I can meet my clients as openly, honestly, and clearly as possible. Being a healer challenges me to separate my issues from theirs so I can be present with my clients, not feeling threatened or put off by their defenses and projections, but to see through to what they really need, rather than responding with my own defenses and projections.

Being a healer challenges me to communicate at my clients' level, so they really understand. It challenges me to communicate honestly, to rid my vocabulary of devious words that are ploys to make them think I know what I do not, to make them think all is well when I am not sure, to create a certain image that is maybe only half the truth or covers over a haunting truth. It challenges me to unload my statements, to mean what I say and to say what I mean.

In a sense I am always using my clients in my own game, they are integral to my own fulfilment. For me to continue my work, I have to get something in exchange. My clients provide that something, at all the different levels, to make my work meaningful for me. They give me honor. They give me humility. They pay my bills. My goal is to be deeply aware of when and how I am using my clients for my own game, to do what I do as consciously as possible.

• It is interesting that we often call our healers physicians, which technically means one who serves the physical well-being. It is no wonder that our conventional modalities of practicing medicine focus so strongly on our physical aspect.

In our minds, and even in some dictionaries, however, we define physicians as persons skilled in the art of healing. We think of them as serving our total well-being. No wonder we get tangled unaware into the concept that our disease is exclusively a physical problem and needs only a physical remedy. No wonder there are so many pathologies that go untended and have no cure. We might do better to expand our concept of who practices the art of healing and which of our aspects needs to be addressed in order to be healed.

We have divided our servants into those who serve our physical needs versus those who serve our spiritual needs versus those who serve our moral or emotional needs, and so forth. To create these specialties is good, each discipline is such a large area. The problem is when we see these people as serving totally unrelated problems, thus fragmenting our vital integrity. In ourself, we can tie our work with our physician, our clergy, our police, our day dreams, our night dreams, and our dance class into the same process.

Some say that <u>love is the greatest healer there is</u>. This supports the idea that all our aspects are involved in our illness.

The healers of some cultures do focus more on how a disease is rooted in aspects of our being other than the physical. We often laugh at their modalities of healing and call them ignorant, primitive, unsubstantiated, superstitious, inhumane, barbaric, savage, or demonic. We laugh, and sometimes, I think, we are afraid, because they touch a place inside of us that we do not know.

TIDBIT BOX CONTENTS

INDEX OF TITLE PAGE ILLUSTRATIONS

listed by page numbers

GLOSSARY

As a general rule, to pluralize a word that ends in 'a', add an 'e' after the 'a', for example, 'vertebra' is singular while 'vertebrae' is plural. To pluralize a word that ends in 'us', drop the 'us' and add 'i', for example, 'bronchus' is singular while 'bronchi' is plural.

A- means without, lack of or deficient. In front of any word, it negates that word.
Abduction has the same root as abduct, to take away. It is movement in a side-to-side direction, moving body parts away from each other or away from the median of our torso. (116)
Active transport across a cell membrane is a process of transporting which consumes cell energy but gives the cell control over what is being transported. (46)
Adduction has the same root as add, to put together or unite. It is movement in a side-to-side direction, bringing body parts together, toward each other or toward our torso. (116)
Adeno- refers to a gland.
Alg-, -algia means pain.
Anatomy studies form and structure. (14)
Angi- refers to vessels, either blood or lymph.
Antagonistic comes from two words, 'anti' meaning against and 'ergon' meaning work. Antagonistic muscles are not against the principle of work, but work against each other, pulling in opposite directions, stabilizing a joint. (212)
Ante- means before.
Anterior is the relative position which is in front of, the prefix 'ante-' meaning before. (27)
Anti- means against.
Antigen comes from two words, 'anti' meaning against and 'gennan' meaning to produce or generate. An antigen is part of the protein in a cell membrane. When it is foreign, we produce antibodies against it, leading to the destruction of the foreign cell. (37)
Arthr- refers to a joint, as in arthritis.
Atoms are the particles which differentiate the basic elements found on this planet. The atoms within any specific element are all the same, but each element has its own kind of atom. (25)
In **atrophy**, the 'a' negates the trophy. 'Trophe' means nourishment. In atrophy there is wasting or decrease in cell size or function for any reason, not necessarily related to its nutritional status. (222)
Auto- means self.
Avascular means that there are no blood vessels in the area. (58)
Basal metabolic rate is the rate at which basic cell processes run to just keep the cell alive when we are completely at rest.
Basal metabolic temperature is our basic body temperature which comes from the amount of heat released in our body by the energy it takes to do the minimal cell processes which just keep our cells alive when we are completely at rest. (53)
Basal metabolism is the minimal, basic cell processes needed to keep our cells alive even when we are completely at rest, physically, emotionally and digestively, having not eaten in the last 12 hours. (53)
-bi- refers to two, double, or both.
Brachi- refers to the upper part of our arm.
Bronchi- refers to the lower part of our airway, from our throat down.
In Greek, a **bursa** is a leather sac. In our body it is the name of the connective tissue sac which lies between layers of soft tissues around a synovial joint. It contains a viscous fluid which lubricates the movement of the layers over each other. (114)
Carbohydrates are a group of chemicals which are made in plants by using the sun's energy to bind together carbondioxide and water, forming the structure of the plant. Sugars, saccharides, glycogen, starches, destrins and cellulose are all carbohydrates. Most carbohydrates in our body are converted to glucose and burned for the energy for most of our cells metabolic processes. (22)
Carcin- refers to cancer.
Cardi- refers to our heart.

Catabolic enzymes are chemicals which assist the breakdown of complex substances into simpler substances, 'cata-' meaning downward or against. (38)
-cep-, -cephal- refers to a head, which for a muscle, is its origin.
-cerebro- refers to our brain.
Chole- refers to our bile or gall.
Chondro- refers to cartilage.
Circumduction is the swinging of a body part such that it describes a cone-shaped figure. The apex of the cone is at the proximal joint of the swinging part, while the distal end of the swinging part draws a circle. (162)
Collagen is made of two words, 'kolla' meaning glue and 'gennan', to produce or generate. It is found in many connective tissues, generating the glue that holds together the structure of our body. It is a strong, non-elastic, insoluble protein. (78)
In **compact bone** the calcified fibers of our bone are all pressed together with no space between them visible to the naked eye. There is usually a single, central marrow space. (104)
A **condyle** is the end of a bone within a synovial joint, covered with articular cartilage. (118)
Connective tissue is one of the four tissues which make up our body. Its general function is to connect all our bits and pieces together, forming a continuous weaving of protein fibers throughout our body. (66)
A **cross section** is the surface exposed by cutting any part of our body perpendicular to its length. It may be abbreviated as x-section.
-cut- refers to our skin.
-cyt- is a word part meaning the cell.
Depression is the act of pressing down and can refer to our affect on any structure being lowered relative to the rest of our body. (128)
-derm- refers to our skin or an outer layer.
Diffusion is the spreading or broad dispersion of molecules as a result of their kinetic energy creating incessant motion. This is one form of passive transport across membranes. (48)
Distal has the same root as distant, meaning far away. It is the relative position which is farthest away from either our torso or the inflow end of a tubular organ. (27)
Dorsal is our body surface which is least sensitive and most protected in itself. It is the one we most instincitively put forward to protect ourself. It may have coarser, longer hair, increased pigment or more superficial bones.(27)
-ec-, -ecto- means out of or outside of.
Electrolytes are atoms or molecules which become more stable in solution by taking on or giving off electrons, becoming electrically charged, either positive or negative. (52)
Elevation is the act of lifting upward and refers to any structure being lifted relative to the rest of our body. (128)
Endo- means inside.
Epi- means among, upon, on, beside, above or over.
An **epicondyle** is a bony prominence close beside or above a condyle or synovial joint. (118)
An **epiphysis** is a center of ossification at each end of a long bone. 'Physis' means growth, so an epiphysis is beside the growth area of the bone. 'Epiphyses' is the plural form and 'epiphyseal' is pertaining to an epiphysis. (105)
Epithelial tissue is one of the four tissues that make up our body. It is made of closely packed cells, forming selectively permeable barriers on the inner or outer surfaces of organs. It also forms the main structure of any organ which absorbs or secretes anything. (58)
Eryth-, erythro- means red.
Eversion is lifting the lateral or outside of our foot, twisting our foot so its sole faces laterally or outwards. The 'e' is for exit, going out. (193)
Exo- means without or outside of.
Exocytosis comes from three word parts, 'exo' meaning outside of, 'cyt' refering to cell and 'osis' is a condition or process of. Literally translated, exocytosis is the condition of being outside of the cell and more accurately refers to the process of the cell putting something out of itself, like spitting out. (48)
Extension is movement in the forward/backward direction in which ventral surfaces separate, moving away from each other and dorsal surfaces approach each other. (116)
Extra- means outside of, beyond, or in addition to.
Extracellular fluid is the fluid outside of our cells, including the fluid directly between our cells (interstitial fluid) as well as the fluid in-

side our blood and lymph.

Facilitated diffusion is diffusion which needs facilitation, some kind of assistance. The facilitator in each case is different. An example of facilitated diffusion is glucose, which is too big to pass through the spaces in our cell membrane. Insulin is the facilitator in this case, enlarging the spaces to allow diffusion. (51)

Fascia is connective tissue which wraps, supports, and separates muscle fibers. It also binds our skin to its underlying tissues. (208)

Fats, also called lipids, are a group of chemicals which are not soluble in water. They are made of an alcohol and fatty acids, forming different greases or oils in our body. They are essential for building cell membranes and all our cell's internal structures. Fats form about 15 percent of our body weight. (21)

Filtration is a process of separating certain particles out of a liquid by passing the liquid through a porous membrane which allows the liquid to pass through but retains the particles. The success of filtration depends on the size of the holes in the membrane and the force which pulls or pushes the liquid through. (52)

A **flaccid** muscle is relaxed beyond a healthy state of relaxation. It has no tone, no nerve message, and not one motor unit stimulated. It is completely limp. This is always pathological. (220)

Flexion is movement in the forward/backward direction in which ventral surfaces approach each other and dorsal surfaces separate, moving away from each other. (116)

A **foramen** is a hole through a bone. (118)

A **fossa** is a shallow depression or concave shape. (118)

Glucose is the simple sugar which is the most frequent product of carbohydrate digestion in our digestive canal. It is absorbed into our blood and used by our body for fuel and the building of more complex carbohydrates. (22)

Glyc- refers to anything which tastes sweet and may be used specifically in reference to glucose.

Glycerol comes from a Greek word meaning sweet. It is a form of alcohol used in making fats. (22)

Glycogen comes from the word for sweet and 'gennan', to produce or generate. It is made out of sugars, storing them in our body until we need them for fuel. (43)

Gyn- refers to the female or a woman.

Hem-, hemat- refers to blood.

Hemi- means half.

Hepat- refers to our liver.

Hetero- means other or different.

Hist-. histio- refers in a general way to tissues.

Homeo-, homo- means unchanging, like, the same, or steady.

Homeostasis comes from 'homeo' and 'stasis', literally meaning unchanging in a stance or standing, and refers to our body's continual function to move towards a specific internal environment. It is the process of our dynamic equilibrium. (22)

Horizontal abduction refers to the movement of our arms being pulled away from each other while in a horizontal position. (162)

Horizontal adduction refers to the movement of our arms being pulled toward each other while in a horizontal position. (162)

Hyper- means beyond or excessive.

Hyperextension is the movement of any joint in the direction of extension, passing beyond anatomical position for that joint. (116)

Hypertrophy comes from two words, 'hyper' meaning excessive or beyond and 'trophe' meaning nourishment. It refers to any increased stimulation of a cell, leading to an excessive size or function of the cell, not necessarily related to its nutritional status. (222)

Hypo- means under, below or deficient.

Hypoglycemia can be literally translated from its three word parts. 'Hypo' means the state of being deficient, 'glyc' is glucose and 'emia' is the blood, i.e. low blood glucose. (52)

Ileo- relates to our ileum, the last part of our small intestine.

Ilio- relates to ilium, our upper pelvic bone.

Inferior is the relative position which is below. (27)

Infra- means below.

The **insertion of a muscle** is one of its two bony attachments. Generally the insertion is the end which is less stable, is more likely to move, and has the narrower, more concentrated, and longer tendon. (210)

Inter- means between, among, or mutually.
Intercellular substance is any substance which is between cells, 'inter' meaning between. (21)
Interstitial fluid is the fluid in the space between adjacent cells. 'Inter' means between and 'stitial' refers to things standing close together like people in a crowded stadium. In this case, the crowd is the cells of some tissue. (36)
Intra- means within or inside.
Intracellular fluid is the fluid inside our cells. (36)
Inversion is lifting the medial or in side of our foot, twisting our foot so its sole faces medially or inward. (193)
Ions are atoms or molecules which become more stable in solution by taking on or giving off electrons, becoming electrically charged, either positive or negative. (52)
Iso- means equal, like, or no change.
-itis means inflammation.
Kinetic energy is the energy of motion and refers to the energy of the incessant motion of all the microscopic particles of matter on this planet. (48)
Lacri- refers to tears.
Lateral is the relative position which is to the side, farthest away from the median or midplane of our body. (27)
Lateral rotation is a rotational movement around the central axis of a bone in a direction which brings the ventral surface of that body part forward and lateral. It is also called **external rotation**. In lateral rotation of our scapula, the inferior angle of our scapula swings laterally, away from our vertebrae. This is also called **upward rotation** of our scapula since its acromion process swings upward. (161)
Leuko- means white.
Ligaments are the dense, collagenous bands of connective tissue which bind bone to bone, stabilizing and maintaining alignment across a synovial joint. 'Liga' means binding as in a ligation. (113)
Lipids are a group of chemicals which are not soluble in water. They form greases or oils made of an alcohol and fatty acids. Lipid can be used interchangeably with fat. (21)
Lipo- refers to fat.
A **longitudinal section** is the surface exposed by cutting a part of our body along its length.
-lysis, lyso- means the breakdown, loosening, or dissolving of.
Macro-, mega- means large or great.
Mal- means bad or abnormal.
-malacia means softening.
Medial is the relative position which is closest to the median or midplane of our body. (27)
Medial rotation is a rotational movement around the central axis of a bone in a direction which brings the ventral surface of that body part forward and medial. This is also called **internal rotation**. In medial rotation of our scapula, the inferior angle of our scapula swings medially, toward our vertebrae. This is also named **downward rotation** of our scapula since its acromion process swings downward. (161)
Our **median** is the midplane of our body in the side-to-side direction. (27)
Meta- means after, beyond, or change.
Metabolism is the state of changing and refers to the sum of all the physical and chemical processes of energy and material transformation that take place in a living cell.
Micro- means small.
A **midline** is any line in the plane that divides our body into two equal halves, right and left. (27)
A **molecule** is formed by combining atoms into a more complex chemical with unique characteristics not held by any of the separate atoms. (52)
Muscle tissue is one of the four tissues which make up our body. Its unique characteristic is its ability to create a force of contraction and thereby shorten its length, physically moving whatever it is attached to. (82)
Myel- refers to bone marrow or the spinal cord.
Myo- refers to muscle.
A **neck** is a narrow length of bone supporting a rounded end, or a constricted part of an organ. (118)
Nephro- refers to kidney.
Nerve tissue is one of the four tissues which make up our body. Its unique characteristics are its ability to: **1.** respond to many different

kinds of stimuli, both internal and external, **2.** synthesize, evaluate and sometimes bring to consciousness the sense of these stimuli, and **3.** create our response to the stimuli. (90)

Neuro- refers to some aspect of nerve tissue.

Normal tone is the small but continuous tone or tension of contraction from which a healthy muscle never rests, not even in sleep. It is not enough tension to create movement but it helps to hold our postures and keeps our muscles in readiness to contract. (220)

-oma means a tumor.

Opposition is the movement in the hand whereby our thumb and fourth and fifth digits rotate forward to face each other like opponents. When wrapping our hand around an object, they press in opposite directions, providing the pressure and friction to hold the object. (164)

An **organ** is that part of a living structure by which some particular function or work is done. It is in a localized, contained package. (19)

Organelles are the internal cell structures which perform all the cell functions. Each organelle plays a specific role. (164)

The **origin of a muscle** is one of its two bony attachments. Generally the origin is the end which is more stable, is less likely to move, and has the broader, shorter tendon. (211)

-osis is a process or a particular mode, condition, or state of being.

Osmosis is basically the diffusion of water, with water moving passively from an area where there is relatively more water (less concentrated solution) to an area of less water (more concentrated solution). The force which moves the water across the membrane is the momentum of the perpetual motion created by its kinetic energy. (50)

-oss-, -osteo- pertains to bone.

Ossification is the process of converting another tissue into bone. (110)

Palpation is to find or locate by feeling. (13)

Para- means beside, near, or just beyond.

Passive transport across a cell membrane is a process of transporting which consumes no cell energy and gives the cell no control over what is being transported. The cell is passive in the process and accepts what happens, whether it likes it or not. (46)

Patho- refers to disease. (14)

peri- means around.

-phag- means to eat.

Phagocytosis literally means the process of the cell eating. 'Osis' is the process of, 'cyt' the cell, and 'phagos' means to eat. (47)

A **phospholipid** is a molecule made of phosphorus, oxygen, and lipid. (35)

Physiology studies normal function. (14)

Pinocytosis literally means the process of the cell drinking. 'Osis' is the process of, 'cyt' the cell, and 'pinos' means to drink. (47)

Poly- means much or many.

Post- means after or behind.

Posterior is the relative position which is behind. (27)

Pre- means before or in front of.

The **prime mover** for a specific movement is the muscle which has the best mechanical advantage to do that movement with the least effort. (211)

A **process** is a long, finger-like bony protrusion. (118)

Pronation comes from prone, lying face or belly down. It is the medial rotation of our radius crossing over our ulna, turning the palm or face of our hand downward. (163)

Proteins are a group of organic chemicals containing nitrogen. They are made out of amino acids and make up about 20 percent of the weight of our body. They form the strength of our connective tissue and the contractile element of our muscle tissue. (21)

Protraction is the forward movement of our scapula or pectoral girdle, so our scapula moves away from our vertebrae. When we are 'for' something happening, we are 'pro' that thing. (128)

Proximal has the same root as proximity, meaning nearest to. It is the relative position which is nearest to either our torso or the inflow end of a tubular organ. (27)

Retinaculum (plural is retinacula) means halter. It is a band of dense, collagenous connective tissue which functions like a halter to hold tendons in place against their underlying bones while their muscles contract, to prevent the tendons from webbing outward as their joint bends. (294)

Retr- means backward or located behind.

Retraction is the backward movement of our scapula or pectoral girdle, so our scapula approaches our vertebrae. To 'retract' means to take back. (128)

Rotation is the movement of spinning around the central axis of bone. (116)

-rrhage, -rrhea means a flow or discharge.

Scar tissue is healed connective tissue, replacing damaged connective tissue or other tissue which cannot reproduce itself. (76)

-sclero- means hard.

Semi- means half.

The **shaft of a bone** is its longest, cylindrical aspect spanning the space between its two knobby ends.

A **spastic** muscle has some degree of excessive tone, from mild to severe. It is working overtime, creating too much force of contraction with too many nerve messages, too many motor units stimulated, consuming valuable nutrients, and creating stress in our body. (220)

Spongy bone is spongy in appearance only, having many small holes dispersed among the calcified fibers, but maintaining the same strength as any part of that bone. The many holes are filled with red bone marrow where blood cells are reproducing. (104)

A **sprain** occurs when a force across a joint exceeds the strength of the ligament support causing excessive or abnormal movement of the bones, damaging the ligaments to some degree, ranging from mild to severe. (115)

-stasis means to stop or stand still.

Sub- means under, beneath, or below.

Super-, supra- means above.

Superior is the relative position which is above. (27)

Supination comes from supine, lying face or belly up. It is the lateral rotation of our radius to lie parallel to our ulna, turning the palm or face of our hand upward. (163)

Suture means a seam and is used to name the closely bound fibrous joint between two bones of our skull. (122)

Symptom originates from 'syn' meaning together and 'pipto' to fall. It is what comes together when we fall ill, out of connection with ourself. Symptoms are the manifestations or perceptible changes in bodily structure or function, indicating the presence of an illness, the evidence from which we infer the nature of the disease.

Synergistic comes from two words, 'syn' meaning together and 'ergon' meaning work, so synergistic muscles are muscles which work together to perform the same movement. (211)

A **system** is a group of structures or organs in our body, related to each other by participating in the same general functions. (19)

Systemic pertains to our body as a whole or something that is spread extensively throughout our body.

A **tendon** is a dense, collagenous band of connective tissue which binds our skeletal muscle fibers to our bones, conveying the force of muscle contraction into a bone to move it. (208)

Tissue comes from the Latin word meaning to weave and refers to a group of similar cells and the intercellular substances which weave them into a cohesive structure. Tissues group together to form organs, each tissue performing a particular aspect of the organ's job. Four tissues make up our entire body: epithelial, connective, muscular and nervous. (21)

Toxic means to resemble or be caused by poison. In our body, anything in excess acts like a poison and can be said to be toxic, including water, oxygen, vitamin A, and so on. (23)

Transverse is the direction going horizontally across our body from side to side. Anything lying in that direction may be described as being transverse.

Tri- means three.

-troph- means to nourish.

-tropic refers to changing or influencing, usually in a stimulating, tonifying way.

A **tubercle** is a rounded bony prominence, often tending to be smaller. (118)

A **tuberosity** is a rounded bony prominence, often tending to be larger. (118)

Ventral is our body surface which is most sensitive and vulnerable. It is the one we most instinctively protect. It may have finer, shorter hair, decreased pigment or softer, less bony flesh. (27)

INDEX

REFERENCES

Anthony, RN, BA, MS and Thibodeau, PhD: Textbook of Anatomy and Physiology, 11th edition: C. V. Mosby Company, 1983.

Arey, Leslie Brainerd, PhD, ScD, LlD: Developmental Anatomy, 7th edition: W. B. Saunders Company: 1965.

Boyd, J. D. MA, MSc, MD: Textbook of Human Anatomy: MacMillan: 1966.

Copenhaver, MD, Bunge, MD and Bunge, PhD: Bailey's Textbook of Histology, 16th edition: The Williams & Wilkins Company: 1971.

Dorland's Pocket Medical Dictionary, 21st edition: W. B. Saunders Company: 1968.

Fong, BS, MS, Ferris, BS, MS and Skelley, RN, MS: Body Structures and Functions, 6th edition: Delmar Publishers Inc: 1984.

Fowler, Ira: Human Anatomy: Wadworth Publishing Company: 1984.

Gordon, Malcolm S: Animal Physiology: Principles and Adaptations, 2nd edition: The Macmillan Company: 1972.

Guyton, Arthur C., M.D: Textbook of Medical Physiology, 5th edition: W. B. Saunders Company: 1976.

Harrison's Principles of Internal Medicine, 12th edition: McGraw-Hill Inc: 1991.

Hoppenfeld, Stanley, MD: Physical Examination of the Spine and Extremities: Appleton-Century-Crofts: 1976.

Jawetz, PhD, MD, Melnick, PhD and Adelberg, PhD: Review of Medical Microbiology, 10th edition: Langs Medical Publications: 1972.

Juhan, Deane: Job's Body: Station Hill Press: 1987.

McNaught, MB, ChB, PhD and Callander, FFPh, MMAA, AIMBI: Illustrated Physiology, 3rd edtition: Churchill Livingstone: 1975.

Memmler, MD and Wood, RN, BS, PhN: The Human Body in Health and Disease, 5th edition: J. B. Lippincott Company: 1983.

Romanes, G. J. PhD, MB, ChB, FRCSEd, FRSE: Cunningham's Manual of Practical Anatomy, 13th edition: Oxford University Press: 1969.

Seeley, PhD/Stephens, PhD/Tate, DA: Anatomy and Physiology, 2nd edition: Times Mirror/Mosby College Publishing: 1992.

Selye, Hans, M.D: The Stress of Life: McGraw-Hill Book Company: 1956.

Sieg, MEd, OTR and Adams, MOT, OTR: Illustrated Essentials of Musculoskeletal Anatomy, 2nd edition: Megabooks: 1985.

Scientific American: Vertebrate Structures and Functions: W. H. Freeman and Company: 1974.

Spence, Alexander P, PhD: Basic Human Anatomy: The Benjamin/Cummings Publishing Company: 1982.

Taber's Cyclopedic Medical Dictionary, 16th edition: F. A. Davis Company: 1985.

Tortora and Grabowski: Principles of Anatomy and Physiology, 7th edition: Harper Collons College Publishers: 1993.

Webster Encyclopedic Dictionary: Portland House: 1989.

Webster's New World Dictionary: William Collins + World Publishing Co: 1977

ORDER FORM

Books may be ordered from:

Rose Bud Publishing
17126 Mt. Woodson Rd.
Ramona CA 92065

Price list:

1-2 books	$34.95 US per book
3-9 books	$31.95 US per book
10-24 books	$27.95 US per book
25 or more books	$20.95 US per book

Orders shipped within the state of California, please add 7.75% sales tax.
If you have a resale number, please indicate it on the form and omit tax.

For shipping add $2.00 for the first book and $1.00 for each additional book.

Make cheques payable to Charlene Penner.

For more information call (619) 789-2053.

Book Title	Quantity	Individual Price	Total Price
Dry Bones or Wet Noodles?			
		Subtotal	
Resale Number		Tax (0.0775 x Subtotal)	
		Shipping	
		Total	

Please Print Clearly

Name ____

Address ____

City ______________________ State ___________ Zip ___________

Phone __(______)________-__________ Best time ________________

Copy this page and mail with your cheque.